Fundamentals of Stochastic Signals, Systems and Estimation Theory with Worked Examples

Second Edition

Branko Kovačević, Ph. D.
Željko Đurović, Ph. D.

Fundamentals of
Stochastic Signals, Systems
and Estimation Theory
with Worked Examples

Second Edition

Springer

ACADEMIC MIND

Branko Kovačević, Ph. D.
Željko Đurović, Ph. D.
Faculty of electrical engineering, University of Belgrade

Fundamentals of Stochastic Signals, Systems and Estimation Theory with Worked Examples

Second Edition

Reviewers
Srđan Stanković, Ph. D.
Miroslav Mataušek, Ph. D.

(c) 2008
AKADEMIC MIND, Belgrade, Serbia
SPRINGER-VERLAG, Berlin Heidelberg, Germany

Design of cover page
Zorica Marković, Academic Painter

Printed in Serbia by
Planeta print, Belgrade

Circulation
700 copies

ISBN 978-86-7466-323-3
ISBN 978-3-540-70990-9

Library of Congress Control Number: 2008931771

Preface

This book is intended for undergraduate students, graduate students and practical engineers, specializing in the areas of electrical communications, signal processing and automatic control. It has emerged from various lecture notes, in Serbian and in English, that we have compiled and used in different courses at Faculty of Electrical Engineering, University of Belgrade, since 1980. The specializing in the areas of electrical communications, signal processing and automatic control required background is a course in a probability theory and basic linear dynamic systems and signals theory (transfer functions and state space). Appendix A provides a short review of Laplace and Z-transform for those who need it. Moreover, one advantage of modern control theory is that it employs matrix algebra, which results in a simplification in notation and mathematical manipulations when dealing with multivariable systems. Therefore, a brief review of the matrix algebra is presented in Appendix B. Generally speaking, the goal of this textbook is to clarify the terms digital filtering, Kalman filtering and other types of estimation techniques appearing quite often in the present-day professional literature. In the electrical engineering curriculum, a course in Stochastic signals, systems and estimation is introduced to accompany underlying courses in Control systems, Communication systems and Signal processing. In addition, many problems in wireless communications, networking, electronics, photonics, power systems and robotics are now studied from the stochastic signals and systems point of view. This textbook can be also used by other engineering students interested in these topics, especially biomedical, aerospace, civil, traffic, mechanical and industrial engineering students.

The main theme of the text concerns fundamental concepts underlying stochastic signal, or linear stochastic system, modeling and analysis, as well as model-based signal processing. Two popular stochastic models, the polynomial (or transfer function) model and state space model, are employed in schemes that lead to estimation of unknown signal/system model parameters or states. Equations of requiring usefulness are displayed in tabular form for easy access. Many examples are used to illustrate the concepts and to emphasize intuition, and the reader is shown how to write software implementations of estimators for use on a computer. The experiments are performed using the MATLAB package. MATLAB represents good learning tool that helps readers to get a better understanding of main theoretical concepts. It is especially useful for studing hihg-order dynamic systems with random inputs.

The structure of the book is as follows.

Chapter 1 begins by giving a short review of probability theory and random variables. A review of the most commonly used probability density functions is presented in appendix C. Various model descriptions of stochastic signals are illustrated in Chapter 2. The important concepts a Gauss-Markov process and a state vector in a stochastic stetting are discussed. Properties of covariance functions and spectra are also presented.

Chapter 3 covers the analysis of linear discrete-time stochastic systems, with an emphasis on second order function and spectral density in a system. Spectral factorization is also treated. It is concerned with how to proceed from a specified spectrum to a filter description of a signal. Two popular stochastic models, the auto-regressive moving-average model with exogenous inputs (ARMAX) and the state space model, are also introduced. It is shown how these models can be used to simulate stochastic processes.

Chapter 4 is devoted to analysis of linear dynamical continuous-time systems whose inputs are stochastic processes. The transfer function (polynomial) and Gauss-Markov (state-space) *models are examined. Fundamental spectral factorization theorem is discussed in terms of its practical applications. This chapter, coupled with chapter 3, provide the minimum background for further developments.*

Chapter 5 discusses estimation theory and means of assessing estimator performance. The general techniques of minimum variance, least squares, maximum a posterior and maximum likelihood estimation are developed. The theory of linear estimation, when the estimates are constrained to be linear functions of the observations, is also given, showing that the optimal estimator depends only upon the first and second-order statistics of the random variables involved. These important features bring great simplifications to the always thorny questions of determination of suitable mathematical models of physical problems.

In Chapter 6 the classical Wiener estimator design is discussed in terms of the innovations approach, and it is shown how some of the digital signal processing techniques (e.g. finite and infinite impulse response filters) can be considered as Wiener filter design techniques. Systems theory techniques for both discrete and continuous Wiener filter design are also developed, and simulated examples are discussed.

Kalman filtering (state space) design is developed theoretically in Chapter 7 using the classical minimum mean-square error technique. The derivation of the processor is coupled with more sophisticated innovations approach. The practical aspects of Kalman filter design are discussed, and simulated examples are also given. Finally, the classical Wiener filter is linked back to the Kalman estimator.

Chapter 8 discusses the extensions of the Kalman filtering technique to solve problems it was not directly designed to solve. In section 8.1. is shown how to use the linear Kalman filtering approach with minor modifications to solve the coloured-noise-source problems. Nonlinear estimatiors, using the linearized and extended Kalman filters, are developed in section 8.2, as well. Finally, in section 8.3. linear state estimators are improved by simultaneously estimation of the uncertain system parameters and/or noise statistics together with the system states. This results in parameter and noise adaptive filtering. Multiple-model estimation, representing an approach which allows for many possible values of model parameters and noise levels has been also presented in this section. Numerical approach to implementing the Kalman type estimators on a digital computer is presented through the simulated examples.

The colleagues from the Signals and Systems Department, have contributed directly or indirectly to these lecture notes. Moreover, the feedback we have received from the students over many years has been very valuable in compiling the manuscript.

Belgrade, 2007.   *Authors*

CONTENT

REVIEW OF THE THEORY OF PROBABILITY AND RANDOM VARIABLES

In any communication or control system it is necessary that some features of the signals that the system must process should be known a priori. A review of those mathematical tools of probability theory which are most applicable to the study of stochastic signals, in communication and control science, is presented in this chapter. The presentation will be intentionally brief, assuming that the reader already has some familiarity with these topics, and is intended mainly as a refresh chapter.

1.1 Some probability theory

Definition of probability: We begin with a space Ω of a physical experiment σ, whose outcomes are named ω and depend on chance. Over the space Ω, and its subsets , we define a probability function P, which assigns a positive number between 0 and 1 to each countable combination of subsets Ω_i in Ω to which an outcome ω may belong (outcome ω is also called elementary event, while a subset Ω_i is named event) [20, 39, 42, 44, 47,53].

The function P has the following properties:

- the probability of some outcome that is certain is equal to 1, that is: $P(\Omega) = 1$; Ω-certain event
- the probability of an outcome which may result from events $\Omega_i \cap \Omega_j = \Omega_i \Omega_j$ which have no common points (intersection or product of Ω_i and is empty) is the sum of the probabilities of Ω_i and Ω_j: $P(\Omega_i + \Omega_j) = P(\Omega_i \cup \Omega_j) = P(\Omega_i) + P(\Omega_j)$ where $\cup$ or $+$ is the symbol for the sum (or union) of subsets, while $\cap$ is the symbol for their product or intersection.

Definition of σ-algebra: σ-algebra $\mathfrak{I}$ is a set of subsets Ω_i of for which probabilities P are defined (only those events Ω_i from Ω for which the probability P is defined form $\mathfrak{I}$).

Thus, a physical experiment σ is specified by three concepts: $\Omega, \mathfrak{I}, P$, or equivalently $\sigma : (\Omega, \mathfrak{I}, P)$.

Definition of conditional probabilities: Given an event B with nonzero probability, $P(B) > 0$, we define *a conditional probability of A assuming B* by

$$P(A/B) = \frac{P(AB)}{P(B)}$$

Example 1.1 (probability space) : Our experiment is the single tossing of a coin generating the outcomes h (head) and t (opposite side) and forming the space $\Omega = \{\omega_1, \omega_2\}$; $\omega_1 = h$; $\omega_2 = t$. With these elements we can from $2^2 = 4$ sets Ω_i, which are known as events: $0, \{h\}, \{t\}, \Omega$; 0-empty set or impossible event; Ω-certain event or probability space. We now assign to each event the probability:

$$P(t) = P(h) = \frac{\text{favourable number of outcomes}}{\text{total number of outcomes}} = \frac{1}{2}$$

Moreover, $\{h\}$ and $\{t\}$ are mutually exclusive (they have not common points), so that

$$\Omega = h \cup t \Rightarrow P\{\Omega\} = P\{h\} + P\{t\} = \frac{1}{2} + \frac{1}{2} = 1$$

what is consistent with the definition of probability. Finally, Ω and 0 are mutually exclusive, yielding $P(0 \cup \Omega) = P(0) + P(\Omega)$, and since $0 \cup \Omega = \Omega$ with $P(\Omega) = 1$, we conclude $P(0) = 0$. Thus, the σ-algebra is the set $\Im = \{\Omega, 0, \{h\}, \{t\}\}$.

Example 1.2 (probability space): Our space consists of the six faces of a die $\Omega = \{f_1, f_2, ..., f_6\}$. Now Ω has $2^6 = 64$ subsets Ω_i: $0, \{f_1\}, ..., \{f_6\}, \{f_1, f_2\}, ..., \{f_5, f_6\}, \{f_1, f_2, f_3\}, ..., \Omega$. If the space consists of a finite number of elements n, then the total number of its subspaces (or subsets) is 2^n. We assume a fair die, i.e.

$$P\{f_i\} = \frac{1}{6} = \frac{\text{favourable outcomes}}{\text{total outcomes}} \; ; \; i = 1, ..., 6$$

Let us define $B = \{\text{even}\}$ and $A = \{f_2\}$, then the corresponding probabilities are

$$P\{B\} = P\{f_2, f_4, f_6\} = \frac{\text{number of favourable outcomes}}{\text{total number of outcomes}} = \frac{3}{6} \; ; \; P(A) = P(f_2) = \frac{1}{6}$$

Furthermore, the event

$$AB = \{f_2\} \cap \{f_2, f_4, f_6\} = \{f_2\} \Rightarrow P\{AB\} = P\{A\} = \frac{1}{6}$$

while the conditional probability for the event A assuming the event B is

$$P(A/B) = P(f_2/\text{even}) = \frac{P(f_2 \cap \text{even})}{P(\text{even})} = \frac{P(f_2 \cap B)}{P(B)} = \frac{P(f_2)}{P(B)} = \frac{\frac{1}{6}}{\frac{3}{6}} = \frac{1}{3}$$

Total probability: We are given n mutually exclusive events $A_1, ..., A_n$ ($A_i \cap A_j = 0$ or $A_i A_j = 0$, $i, j = 1, ..., n; \; i \neq j$) whose sum (union) $\sum_{i=1}^{n} A_i = \bigcup_{i=1}^{n} A_i = \Omega$. With B an arbitrary event in Ω, $B \subset \Omega$, we have [42, 44]

$$P(B) = P(B/A_1)P(A_1) + \cdots + P(B/A_n)P(A_n) = \sum_{i=1}^{n} P(B/A_i)P(A_i)$$

Proof: From Fig. 1.1, we have

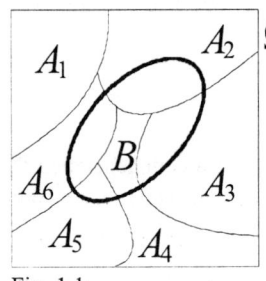

Fig. 1.1

$$B = B \cap \Omega = B\Omega = B\sum_{i=1}^{n} A_i = \sum_{i=1}^{n} BA_i$$

From the fact that A_i has no common points with A_j, it follows that BA_i has no common points with BA_j, yielding

$$P(B) = P\left(\sum_{i=1}^{n} BA_i\right) = \sum_{i=1}^{n} P(BA_i)$$

From the definition of conditional probability, we also have

$$P(B/A_i) = \frac{P(BA_i)}{P(A_i)} \Rightarrow P(BA_i) = P(B/A_i)P(A_i)$$

so that $P(B) = \sum_{i=1}^{n} P(BA_i) = \sum_{i=1}^{n} P(B/A_i)P(A_i)$ which completes the proof.

Example 1.3 (total probability): A box contains 2000 components of which 5% are defective (denoted as d). A second box contains 500 components of which 40% are defective. Two other boxes contain 1000 components each, with 10% defective components. We select at random one of the boxes and remove from it at random a single component. What is a probability that it is defective?

$$\text{Box 1: } 1900g \; ; \; 100d \; ; \quad \text{Box 2: } 300g \, ; \, 200d$$
$$\text{Box 3: } 900g \, ; \quad 100d \; ; \quad \text{Box 4: } 900g \; ; \; 100d$$

where g is a good component and d is a defective one. Thus the probability space Ω has 4500 elements. Let us define the following events:

event A_i: select the i-th box $(i=1,...,4)$

event B: select a defective (d) element (the number of such elements is 500)

Then, clearly: $\sum_{i=1}^{4} A_i = \Omega \; ; \; A_i A_j = 0; \; i \neq j = 1,...,4$. By random selection of a box one means that

$P(A_i) = \dfrac{1}{4} \, ; \, i = 1,...,4$. Once a box is selected, the probability that the removed element is defective

(d) equals the ratio of the number of defective elements to the total number of elements in this box, i.e.

$$P(B/A_i) = \frac{\text{number of defective elements}}{\text{total number of elements}} \; ; \; i = 1,...,4$$

or equivalently

$$P(B/A_1) = \frac{100}{2000} = 0.05 \; ; \; P(B/A_2) = \frac{200}{500} = 0.4 \; ; \; P(B/A_3) = P(B/A_4) = \frac{100}{1000} = 0.1$$

so that

$$P(B) = P(B/A_1)P(A_1) + \cdots + P(B/A_4)P(A_4) = 0.1625$$

Bayes Theorem (Formula): This theorem permits us to evaluate the probabilities $P(A_i/B)$ (a posterior probabilities) of the events A_i in terms of the probabilities $P(A_i)$ (a priori probabilities) and the conditional probabilities $P(B/A_i)$. It is given by the formula [42]

$$P(A_i/B) = \frac{P(B/A_i)P(A_i)}{P(B)} \; ; \; P(B) = \sum_{i=1}^{n} P(B/A_i)P(A_i)$$

Proof: Using the definition of conditional probability, we have

$$\left. \begin{aligned} P(A_i/B) = \frac{P(A_iB)}{P(B)} &\Rightarrow P(A_iB) = P(A_i/B)P(B) \\ P(B/A_i) = \frac{P(A_iB)}{P(A_i)} &\Rightarrow P(A_iB) = P(B/A_i)P(A_i) \end{aligned} \right\} \Rightarrow$$

$$P(A_i/B)P(B) = P(B/A_i)P(A_i) \Rightarrow P(A_i/B) = \frac{P(B/A_i)P(A_i)}{P(B)}$$

where $P(B)$ is given by the formula for total probability.

Example 1.4 (Bayes formula): We examine the selected component in the preceding example, and we find it defective (d). What is the probability that it was taken from the box 2 ?

From the preceding example 1.3, we have

$$P(B) = 0.1625; P(B/A_2) = 0.4 \; ; \; P(A_2) = 0.25$$

$$P(A_2/B) = \frac{P(B/A_2)P(A_2)}{P(B)} = \frac{0.4 \cdot 0.25}{0.1625} = 0.615$$

Independent events: Two events A and B are called independent if

$$P(AB) = P(A)P(B).$$

The events A_1, A_2, A_3 are called independent if

$$P(A_1A_2) = P(A_1)P(A_2) \; ; \; P(A_1A_3) = P(A_1)P(A_3) \; ;$$
$$P(A_2A_3) = P(A_2)P(A_3) \; ; \; P(A_1A_2A_3) = P(A_1)P(A_2)P(A_3)$$

The total number of equations that are required to establish the independence of n events is equal to: $2^n - (n+1)$.

Remark: If the events A_i, A_j are independent in pairs, i.e. $P(A_iA_j) = P(A_i)P(A_j) \; ; \; i \neq j$, it does not follow that they are independent.

1.2 Random variables and distributions

Definition of random variable and its distribution: One-dimensional (scalar) real random variable $X(\cdot)$ is a real-valued function which maps the probability space Ω into the real line, such that to each outcome ω in Ω we associate a real value $X(\omega)$ [8, 23, 42, 44, 53].

The probability that a chance experiment $\sigma : (\Omega, \Im, P)$ maps into a value $X(\omega)$ which is less than or equal to the constant a is given by

$$P\{\omega : X(\omega) \leq a\} = F_X(a) \tag{1.1}$$

and the function $F_X(\cdot)$ is named *the distribution function* of the random variable X.

Properties of the distribution function are the following:
a) $F_X(-\infty) = 0$, $F_X(+\infty) = 1$; since $\{\omega : X(\omega) \leq -\infty\}$ is empty set and $\{\omega : X(\omega) \leq \infty\} = \Omega$ is the certain event.
b) it is a nondecreasing function of a: $F_X(a_1) \leq F_X(a_2)$ for $a_1 \leq a_2$; since $A_1 = \{\omega : X(\omega) \leq a_1\} \subset A_2 = \{\omega : X(\omega) \leq a_2\}$ for $a_1 \leq a_2$, so that $P(A_1) \leq P(A_2)$.
c) it is continuos from the right

Definitions of probability density function (pdf): If $F_X(\cdot)$ is a smooth function, we define its derivative $f_X(\cdot)$ as *the probability density function (pdf)*, which has the property [36, 42, 52]

$$f_X(\xi) = \frac{dF_X(\xi)}{d\xi} \; ; \; F_X(a) = \int_{-\infty}^{a} f_X(\xi) d\xi \tag{1.2}$$

so that

$$P\{a \leq X(\omega) \leq b\} = \int_{a}^{b} f_X(\xi) d\xi = F_X(b) - F_X(a)$$

From the monotonous of $F_X(a)$ follows that $f_X(a) \geq 0$, and since the whole space Ω maps into the real line somewhere, we have

$$\int_{-\infty}^{\infty} f_X(\xi) d\xi = 1 \tag{1.3}$$

which is known as normality condition for pdf.

Two common pdf functions which are commonly used are the uniform pdf and the normal (Gaussian) pdf. Additionally, the Appendix C describes some of the most common pdf's.

Uniform pdf: a random variable having a uniform pdf has zero probability of having any value outside of finite range between lower limit l and upper limit u $(l \leq x \leq u)$, and $f_X(\cdot)$ is constant inside the range. Because of the normality condition (1.3) this constant is $1/(u-l)$. A sketch of the uniform pdf is given in Fig. 1.2

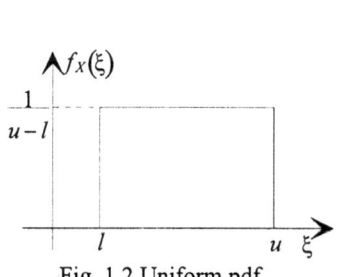

Fig. 1.2 Uniform pdf   Fig. 1.3 Normal (Gaussian) pdf

Normal pdf: The probability density function of one-dimensional normal random variable is given by the equation

$$f_X(\xi) = \frac{1}{\sqrt{2\pi}\sigma_X} \exp\left\{-\frac{1}{2}\frac{(\xi - \mu_X)^2}{\sigma_X^2}\right\} \tag{1.4}$$

and it is shown in Fig. 1.3. The importance of the normal pdf derives mainly from the following facts:

1) The distribution of a random variable based on events which themselves consists of a sum of a large number of independent random events are accurately approximated by the normal law (this statement is known as the central limit theorem of mathematical statistics). For example, such distribution describes electrical noise caused by thermal motions of a large number of particles, as in a resistor.

2) If two random variables have (jointly) normal distributions, then their sum (union) also has a normal distribution. As an extension of this point, if the random input to a linear system is normal distributed, then the distribution of its output signal is also normal.

Example 1.5 (discrete type random variable): Our experiment is the tossing of a coin. As in the example 1.1, we have:

$$\Omega = \{h,t\} \;;\; P\{h\} = p = \frac{1}{2} \;;\; P\{t\} = q = \frac{1}{2} \;;\; p+q = 1$$

We define the random variable $X(\cdot)$ by: $X(h)=1$ and $X(t)=0$, and we shall determine its distribution function $F_X(a)$.

a) If $a \geq 1$ then $\{X(\omega) \leq a\} = \Omega$, since $X(h) = 1 \leq a$, $X(t) = 0 \leq a$ and $P\{X(\omega) \leq a\} = P\{\Omega\} = 1$.

b) If $0 \leq a < 1$ then $\{X(\omega) \leq a\} = \{t\}$, since $X(t) = 0$ and $P\{X(\omega) \leq a\} = P\{t\} = q$

c) If $a < 0$ then $\{X(\omega) \leq a\} = 0$ and $P\{X(\omega) \leq a\} = P\{0\} = 0$.

Thus, we have

$$F_X(a) = \begin{cases} 1 & ; \ a \geq 1 \\ q & ; \ 0 \leq a < 1 \\ 0 & ; \ a < 0 \end{cases}$$

If the distribution function of random variable X is of a staircase form with discontinuities at the points a_i we then say that X is of *discrete type*. Denoting by p_i the jump of $F_X(a)$ at the point a_i, we have

$$F_X(a) = \sum_i p_i h(a - a_i)$$

where $h(\cdot)$ is the unit step function ($h(a) = 0$ for $a < 0$ and $h(a) = 1$ for $a \geq 0$). So, the probability density function of discrete type random variable $X(\cdot)$ takes the form:

$$f_X(a) = \frac{dF_X(a)}{da} = \sum_i p_i \delta(a - a_i)$$

where $\delta(\cdot)$ represents the unit impulse (delta) function ($\delta(a) = 0$ for $a \neq 0$, $\delta(0) = \infty$ and $\int_{-\infty}^{\infty} \delta(a) da = 1$). The discrete type random variable $X(\cdot)$ is completely described by the discrete values a_i that it can takes, as well as the corresponding probabilities $p_i = P\{X(\omega) = a_i\}$.

Example 1.6 (continuous type random variable): A telephone call occurs at random in the interval $(0, T)$, and in the probability space Ω of this example we define the random variable X by $X(t) = t$. Thus, t has a double meaning: it is the outcome (time of call) of the experiment, and the value of the random variable at this outcome, i.e. $\omega = t$ and $X(t) = t$ (the random variable X has as domain and range the interval $(0, T)$). We shall find its distribution function $F_X(a)$, using the following consideration:

a) If $a > T$ then $\{X(\omega) \leq a\} = \{0 \leq t \leq T\} = \Omega \Rightarrow F_X(a) = P\{\Omega\} = 1$.

b) If $0 \leq a < T$ then

$$\{X(\omega) \leq a\} = \{0 \leq t \leq a\} \Rightarrow F_X(a) = P\{X(\omega) \leq a\} = \frac{a}{T} = \frac{\text{favourable number of trials}}{\text{total number of trials}}.$$

c) If $a < 0$ then $\{X(\omega) \leq a\} = 0 \Rightarrow F_X(a) = P\{0\} = 0$.

We say that the random variable X is of *continuous type* if the corresponding distribution function $F_X(a)$ is a continuos function of a (it does not follow that it has derivative for every a, but the number of points at which it is not differentiable are countable).

Frequency interpretation of $F_X(a)$ and $f_X(a)$: Suppose that our experiment performed n-times. At a given outcome ω of the experiment the random variable X takes a value $X(\omega)$. Given a number a, we denote by $n(a)$ the total number of trials such that $X(\omega) \leq a$; we then have

$$F_X(a) = P\{X(\omega) \leq a\} = \frac{n(a)}{n} = \frac{\text{favourable number of trials}}{\text{total number of trials}}$$

To determine $f_X(a)$, we count the number of trials such that $a < X(\omega) \leq a + \Delta a$; with $\Delta n(a)$ the number of these trials, we conclude:

$$P\{a < X(\omega) \leq a + \Delta a\} = \int_a^{a+\Delta a} f_X(\alpha) d\alpha \approx f_X(a)\Delta a = \frac{\Delta n(a)}{n}; \ \Delta a \text{ sufficient small}$$

1.3 Expectations

By the vary nature of a variable $X(\omega)$ whose values are dependent on chance, we cannot discuss a formula for the calculation of its values x. To describe a random variable, we instead discuss average, or expected, values. Such concept is contained in the idea of expectation [20, 39, 47].

Definition: The expected value of a function $g(\cdot)$ of a random variable X whose pdf is $f_X(\cdot)$ is defined as

$$E\{g(X)\} = \int_{-\infty}^{\infty} g(\xi) f_X(\xi) d\xi = \overline{g(X)} \tag{1.5}$$

Important special cases are the following:

1) if $g(X) = X$ then we have the mean value

$$E\{X\} = \int_{-\infty}^{\infty} \xi f_X(\xi) d\xi = \overline{X} = \mu_X \tag{1.6}$$

For example, for the uniform pdf, the mean is given by

$$\overline{X} = \int_{-\infty}^{\infty} \xi f_X(\xi) d\xi = \int_l^u \frac{\xi}{u-l} d\xi = \frac{1}{u-l} \frac{\xi^2}{2} \bigg|_l^u$$

$$= \frac{1}{u-l} \left[\frac{u^2}{2} - \frac{l^2}{2} \right] = \frac{u+l}{2} \tag{1.7}$$

Since the pdf $f_X(\cdot)$ has the intuitive properties of a histogram of relatively frequency of occurrence of a particular values of the random variable X, the mean is a weighted average of the random variable values. In other words, this is the point around which the most of the random variable realizations is grouped.

2) if $g(X) = (X - \overline{X})^2$, then the expected value is the average of the square of the variation of the variable X from its mean, and this number is called the *variance* of X, denoted as $\text{var}\{X\}$ or σ_X^2. Thus, for the uniformly distributed random variable $X(\cdot)$, we have

$$\sigma_X^2 = E\left\{\left(X-\overline{X}\right)^2\right\} = \int_{-\infty}^{\infty}\left(\xi-\overline{X}\right)^2 f_X\left(\xi\right)d\xi = \int_l^u\left(\xi-\overline{X}\right)^2\frac{1}{u-l}d\xi$$

$$= \frac{1}{u-l}\frac{\left(\xi-\overline{X}\right)^3}{3}\Bigg|_l^u = \frac{1}{u-l}\frac{\left(\xi-\dfrac{l+u}{2}\right)^3}{3}\Bigg|_l^u = \frac{\left(u-l\right)^2}{12} \tag{1.9}$$

The square root of the variance is called the *standard deviation* and is assigned by the symbol σ_X, or sometimes $\text{std}(X)$.

3) if $g(X) = X^2$, then we compute by (1.5) the *mean-square value* $\overline{X^2}$

$$\overline{X^2} = E\left\{X^2\right\} = \int_{-\infty}^{\infty}\xi^2 f_X\left(\xi\right)d\xi \tag{1.10}$$

The relation between the mean-square value and variance is given by

$$\sigma_X^2 = \int_{-\infty}^{\infty}\left(\xi-\overline{X}\right)^2 f_X\left(\xi\right)d\xi = \int_{-\infty}^{\infty}\left(\xi^2-2\xi\overline{X}+\overline{X}^2\right)f_X\left(\xi\right)d\xi$$

$$= \int_{-\infty}^{\infty}\xi^2 f_X\left(\xi\right)d\xi - 2\overline{X}\int_{-\infty}^{\infty}\xi f_X\left(\xi\right)d\xi + \overline{X}^2\int_{-\infty}^{\infty}f_X\left(\xi\right)d\xi \tag{1.11}$$

$$= \overline{X^2} - 2\overline{X}\,\overline{X} + \overline{X}^2 = \overline{X^2} - \overline{X}^2$$

4) $g(X) = X^n$; $n \geq 0$; then we compute by (1.5) the *non-centered moment of the order n*

$$m_n = E\left\{X^n\right\} = \int_{-\infty}^{\infty}\xi^n f_X\left(\xi\right)d\xi \tag{1.12}$$

5) $g(X) = \left(X-\overline{X}\right)^n$; then we compute by (1.5) the *centered moment of the order n*

$$\mu_n = E\left\{\left(X-\overline{X}\right)^n\right\} = \int_{-\infty}^{\infty}\left(\xi-\overline{X}\right)^n f_X\left(\xi\right)d\xi \tag{1.13}$$

The higher central moments reflect important attributes of the pdf of a random variable: the third moment $(n=3)$ indicates skewness of the pdf (Fig. 1.4), while the fourth moment $(n=4)$ measures the steepness of the pdf's peak (Fig. 1.5).

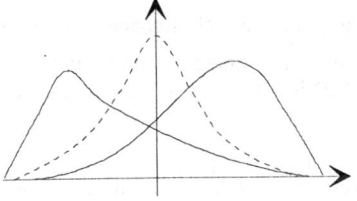
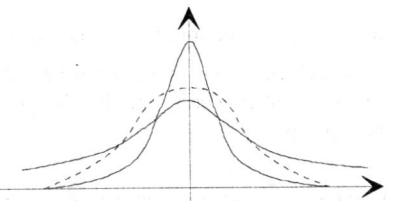

Fig. 1.4 ⎯⎯ Skewed pdf ($\mu_3 \neq 0$);   Fig. 1.5 pdf's with non-Gaussian kurtosis;

- - - Gaussian pdf ($\mu_3 = 0$)   that is μ_4 moment

Link between centered and non-centered moments: We shall now express μ_n in terms of m_n. From

$$(x \mp a)^n = \sum_{r=0}^{n} \binom{n}{r} (\mp 1)^r a^r x^{n-r}$$

one obtains

$$\mu_n = E\left\{(X - \bar{X})^n\right\} = E\left\{\sum_{r=0}^{n} \binom{n}{r} (-1)^r \bar{X}^r X^{n-r}\right\}$$

$$= \sum_{r=0}^{n} \binom{n}{r} (-1)^r \bar{X}^r E\left\{X^{n-r}\right\} = \sum_{r=0}^{n} \binom{n}{r} (-1)^r \bar{X}^r m_{n-r}$$

One can similarly determine m_n from μ_k:

$$m_n = E\left\{X^n\right\} = E\left\{\left((X - \bar{X}) + \bar{X}\right)^n\right\} = E\left\{\sum_{r=0}^{n} \binom{n}{r} \bar{X}^r (X - \bar{X})^{n-r}\right\}$$

$$= \sum_{r=0}^{n} \binom{n}{r} \bar{X}^r E\left\{(X - \bar{X})^{n-r}\right\} = \sum_{r=0}^{n} \binom{n}{r} \bar{X}^r \mu_{n-r}$$

Example 1.7 (normal density): The pdf of a normal random variable X

$$f_X(x) = \frac{1}{\sqrt{2\pi} b} \exp\left(-\frac{(x-a)^2}{2b^2}\right)$$

is symmetric about $x = a$, so that $\bar{X} = E\{X\} = a$. To find its variance we differentiate

$$\int_{-\infty}^{\infty} \exp\left(-\frac{(x-a)^2}{2b^2}\right) dx = b\sqrt{2\pi}$$

with respect to b

$$\int_{-\infty}^{\infty} \frac{(x-a)^2}{b^3} \exp\left(-\frac{(x-a)^2}{2b^2}\right) dx = \sqrt{2\pi}$$

Hence

$$\sigma_X^2 = E\left\{(X-a)^2\right\} = \frac{1}{b\sqrt{2\pi}} \int_{-\infty}^{\infty} (x-a)^2 \exp\left(-\frac{(x-a)^2}{2b^2}\right) dx = \frac{1}{b\sqrt{2\pi}} \sqrt{2\pi} b^3 = b^2$$

We shall now show that if X is normal with zero-mean, then

$$m_n = E\{X^n\} = \begin{cases} 1 \cdot 3 \cdots (n-1)\sigma^n & ; \ n - \text{even} \\ 0 & ; \ n - \text{odd} \end{cases}$$

Differentiating

$$\int_{-\infty}^{\infty} \exp\left(-\alpha x^2\right) dx = \sqrt{\frac{\pi}{\alpha}}$$

k times with respect to α, we find

$$\int_{-\infty}^{\infty} x^{2k} \exp\left(-\alpha x^2\right) dx = \frac{1 \cdot 3 \cdots (2k-1)}{2^k} \sqrt{\frac{\pi}{\alpha^{2k+1}}}$$

and with $\alpha = \dfrac{1}{2\sigma^2}$ we obtain

$$\int_{-\infty}^{\infty} x^{2k} \exp\left(-\frac{x^2}{2\sigma^2}\right) dx = \frac{1 \cdot 3 \cdots (2k-1)}{2^k} \sqrt{\pi \left(2\sigma^2\right)^{2k+1}} = 1 \cdot 3 \cdots (2k-1)\sqrt{2\pi}\,\sigma^{2k+1}$$

so that

$$E\left\{X^{2k}\right\} = \frac{1}{\sqrt{2\pi}\sigma} \int_{-\infty}^{\infty} x^{2k} \exp\left(-\frac{x^2}{2\sigma^2}\right) dx = 1 \cdot 3 \cdots (2k-1)\sigma^{2k} = m_{2k}$$

Moreover,

$$E\left\{X^{2k+1}\right\} = \frac{1}{\sqrt{2\pi}\sigma} \int_{-\infty}^{\infty} x^{2k+1} \exp\left(-\frac{x^2}{2\sigma^2}\right) dx = 0$$

since the function under the integral is odd.

6) *Tchebycheff inequality:* The variance σ_X^2 of a random variable X gives an estimate of the concentration of $f_X(x)$ near its center of gravity (mean value) $\bar{X}$ [2, 42]. Suppose that X is arbitrary random variable with the pdf $f_X(x)$ and finite variance σ_X^2. Then

$$P\left\{\left|X - \bar{X}\right| \ge k\sigma_X\right\} \le \frac{1}{k^2}$$

where $\bar{X} = E\{X\}$, or if we take $k\sigma_X = \varepsilon$,

$$P\left\{\left|X - \bar{X}\right| \ge \varepsilon\right\} \le \frac{\sigma_X^2}{\varepsilon^2}$$

and

$$P\left\{\left|X - \bar{X}\right| < \varepsilon\right\} = 1 - P\left\{\left|X - \bar{X}\right| \ge \varepsilon\right\} \ge 1 - \frac{\sigma_X^2}{\varepsilon^2}$$

Thus, the probability that a random variable X takes values in the interval $\left(\bar{X} - \varepsilon, \bar{X} + \varepsilon\right)$ centered at the point $\bar{X}$ is close to 1 provided $\sigma_X \ll \varepsilon$. Let us give the proof of the Tchebycheff inequality.

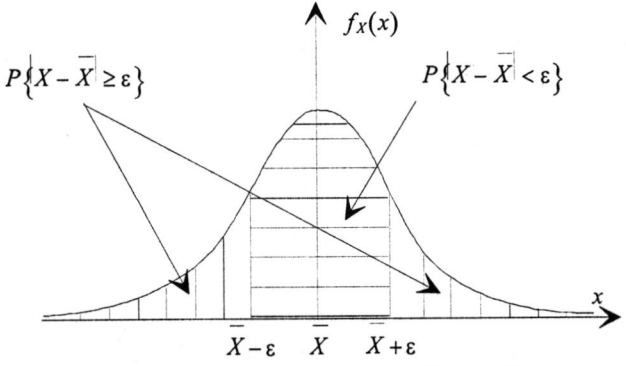

Fig. 1.6 Illustration of Tchebycheff inequality

Proof: Since

$$\sigma_X^2 = \int_{-\infty}^{\infty}(x-\bar{X})^2 f_X(x)\,dx \geq \int_{|x-\bar{X}|\geq k\sigma_X}(x-\bar{X})^2 f_X(x)\,dx \geq \int_{|x-\bar{X}|\geq k\sigma_X}(k\sigma_X)^2 f_X(x)\,dx$$

$$= k^2\sigma_X^2 \int_{|x-\bar{X}|\geq k\sigma_X} f_X(x)\,dx = k\sigma_X^2 P\{|X-\bar{X}|\geq k\sigma_X\}$$

we conclude

$$P\{|X-\bar{X}|\geq k\sigma_X\}\leq \frac{1}{k^2}$$

which completes the proof.

7) Characteristic function: If one takes $g(X)=\exp(j\omega X)$ in (1.5), where $j^2=-1$, it will be obtained the characteristic function of a random variable X, denoted as $\Phi_X(\omega)$, i.e. [2, 36, 42]

$$\Phi_X(\omega)=E\{\exp(j\omega X)\}=\int_{-\infty}^{\infty}\exp(j\omega x)f_X(x)\,dx$$

We note that

$$\Phi_X(0)=\int_{-\infty}^{\infty}f_X(x)\,dx=1$$

and since $f_X(x)\geq 0$, we have

$$|\Phi_X(\omega)|=\left|\int_{-\infty}^{\infty}f_X(x)\exp(j\omega x)\,dx\right|\leq \int_{-\infty}^{\infty}f_X(x)\,dx=1,$$

or equivalently

$$|\Phi_X(\omega)|\leq 1.$$

If $f_X(x)$ is even function, i.e. $f_X(-x)=f_X(x)$, then $\Phi_X(\omega)$ is real and even; that is

$$\Phi_X(\omega)=\int_{-\infty}^{\infty}f_X(x)\cos(\omega x)\,dx$$

The pdf $f_X(x)$ can be expressed in terms of $\Phi_X(\omega)$ as

$$f_X(x)=\frac{1}{2\pi}\int_{-\infty}^{\infty}\Phi_X(\omega)\exp(-j\omega x)\,d\omega$$

The proof is based on the properties of the delta function $\delta(x)$:

$$\frac{1}{2\pi}\int_{-\infty}^{\infty}\exp(-j\omega x)\,dx=\delta(x)\ ;\ \int_{-\infty}^{\infty}f(y)\delta(x-y)\,dy=f(x)$$

Hence

$$\frac{1}{2\pi}\int_{-\infty}^{\infty}\Phi_X(\omega)\exp(-j\omega x)\,d\omega=\frac{1}{2\pi}\int_{-\infty}^{\infty}\left[\int_{-\infty}^{\infty}f_X(y)\exp(j\omega y)\,dy\right]\exp(-j\omega x)\,d\omega$$

$$=\frac{1}{2\pi}\int_{-\infty}^{\infty}f_X(y)\int_{-\infty}^{\infty}\exp(j\omega(y-x))\,d\omega\,dy$$

$$=\int_{-\infty}^{\infty}f_X(y)\frac{1}{2\pi}\int_{-\infty}^{\infty}\exp(-j\omega(x-y))\,d\omega\,dy$$

$$=\int_{-\infty}^{\infty}f_X(y)\delta(x-y)\,dy=f_X(x)$$

If $f_X(x)$ is even, then $\Phi_X(\omega)$ is real and even, so that

$$f_X(x) = \frac{1}{2\pi} \int_{-\infty}^{\infty} \Phi_X(\omega) \cos(\omega x) d\omega$$

Convolution: Given two functions $f_1(x)$ and $f_2(x)$, integral

$$\int_{-\infty}^{\infty} f_1(\tau) f_2(x-\tau) d\tau = \int_{-\infty}^{\infty} f_1(x-\tau) f_2(\tau) d\tau$$

is known as *the convolution* of $f_1(x)$ and $f_2(x)$. This integral defines a function $f(x)$, and the above operation is often written in the form

$$f(x) = f_1(x) \otimes f_2(x).$$

We shall show later (see section 1.8) that if $f_1(x)$ and $f_2(x)$ are the pdf's of two independent random variables X_1 and X_2, then $f(x)$ is the pdf of their sum $X_1 + X_2$. Furthermore, if $f_1(x)$ and $f_2(x)$ are the pdf's functions, and $\Phi_1(\omega)$ and $\Phi_2(\omega)$ are the characteristic functions of $f_1(x)$ and $f_2(x)$, then

$$\Phi_X(\omega) = \Phi_1(\omega) \Phi_2(\omega)$$

is the characteristic function of the convolution $f(x) = f_1(x) \otimes f_2(x)$.

Proof: By definition

$$\Phi_X(\omega) = \int_{-\infty}^{\infty} \exp(j\omega x) f(x) dx = \int_{-\infty}^{\infty} \exp(j\omega x) \left[\int_{-\infty}^{\infty} f_1(y) f_2(x-y) dy \right] dx$$

$$= \int_{-\infty}^{\infty} f_1(y) \int_{-\infty}^{\infty} \exp(j\omega x) f_2(x-y) dx dy$$

With $x - y = \tau$, the last integral equals

$$\int_{-\infty}^{\infty} f_1(y) \exp(j\omega y) dy \int_{-\infty}^{\infty} \exp(j\omega\tau) f_2(\tau) d\tau = \Phi_1(\omega) \Phi_2(\omega)$$

Moment theorem: The derivatives of the characteristic function of a random variable X is related to its moments $m_n = E\{X^n\}$ by

$$\frac{d^n \Phi_X(0)}{d\omega^n} = j^n m_n \ ; \ m_n = E\{X^n\}$$

Proof: By expanding the exponential

$$\exp(j\omega x) = 1 + j\omega x + \cdots + \frac{(j\omega x)^n}{n!} + \cdots$$

we have

$$\Phi_X(\omega) = E\{\exp(j\omega X)\} = \int_{-\infty}^{\infty} f_X(x) \left[1 + j\omega x + \cdots + \frac{(j\omega x)^n}{n!} + \cdots \right] dx$$

Assuming that the above integration is valid, we obtain

$$\Phi_X(\omega) = 1 + j\omega m_1 + \cdots + \frac{(j\omega)^n}{n!} m_n + \cdots$$

On the other hand, by expanding $\Phi_X(\omega)$ directly at the point $\omega = 0$, we have

$$\Phi_X(\omega) = \Phi_X(0) + \Phi_X'(0)\omega + \cdots + \frac{1}{n!}\frac{d^n\Phi_X(0)}{d\omega^n}\omega^n + \cdots \quad ; \quad \Phi_X(0) = 1$$

so that

$$\frac{1}{n!}\frac{d^n\Phi_X(0)}{d\omega^n} = \frac{j^n m_n}{n!} \Rightarrow \Phi_X^{(n)}(0) = j^n m_n$$

which completes the proof.

Example 1.8 (moment theorem) : The characteristic function of a normal random variable X with mean $\overline{X}$ and variance σ_X^2 is given by

$$\Phi_X(\omega) = \exp\left(j\overline{X}\omega - \frac{1}{2}\sigma_X^2\omega^2\right).$$

Proof: We assume first that $\overline{X} = 0$. In this case, the moments of X are given by (see example 1.7)

$$m_{2k} = 1 \cdot 3 \cdots (2k-1)\sigma_X^{2k} \quad ; \quad m_{2k+1} = 0$$

Using the moment theorem, we have

$$\Phi_X(\omega) = 1 + j\omega m_1 + \cdots + \frac{(j\omega)^n}{n!} m_n + \cdots$$

$$= 1 - \frac{1}{2}\sigma_X^2\omega^2 + \cdots + (-1)^k \frac{1 \cdot 3 \cdots (2k-1)}{(2k)!}\sigma_X^{2k}\omega^{2k} + \cdots$$

However,

$$\frac{1 \cdot 3 \cdots (2k-1)}{(2k)!} = \frac{1}{2^k k!}$$

so that

$$\Phi_X(\omega) = \sum_{k=0}^{\infty}(-1)^k \frac{a^k}{k!} = \exp(-a) \quad ; \quad a = \frac{1}{2}\sigma_X^2\omega^2$$

Furthermore, we observe that the random variable $X + \overline{X}$ is normal with mean $\overline{X}$ and variance σ_X^2, so that

$$E\left\{\exp\left(j\omega(X + \overline{X})\right)\right\} = \exp(j\omega\overline{X})E\left\{\exp(j\omega X)\right\}$$

where

$$E\left\{\exp(j\omega X)\right\} = \exp\left(-\frac{1}{2}\sigma_X^2\omega^2\right)$$

which completes the proof.

1.4 Function of one random variable

The function $Y = g(X)$, where X is a random variable, is also a random variable, defined not directly for each experimental outcome, but indirectly via the random variable X and the deterministic function $g(\cdot)$. Thus, the domain of $g(X)$, where X is a random variable, is the set Ω of all experimental outcomes ω, whereas the domain of the deterministic function $g(X)$ is a set of real numbers [17, 30, 36, 42].

Determination of the distribution: Given a real number b, we denote by I_b the set of all real numbers x such that $g(x) \leq b$. We maintain that $\{Y(\omega) \leq b\} = \{X(\omega) \in I_b\}$, and from the above follows

$$F_Y(b) = P\{Y(\omega) \leq b\} = P\{X(\omega) \in I_b\}$$

Thus, to determine the distribution function of the random variable $Y = g(X)$, $F_Y(b)$ for a given b we must find the set I_b and the probability that the random variable X is in I_b.

Example 1.10 (distribution of function of random variable): Let $Y = g(X) = \dfrac{1}{X^2}$

If $b > 0$, then the above equation $Y = g(X)$ has two solutions

$$a_1 = -\frac{1}{\sqrt{b}} \; ; \; a_2 = \frac{1}{\sqrt{b}}$$

and $g(X) \leq Y$ if $X \leq a_1$ or $X \geq a_2$. Hence

$$F_Y(b) = P\{Y(\omega) \leq b\} = P\{\{X(\omega) \leq a_1\} \cup \{X(\omega)\} \geq a_2\}$$

$$= P\left\{X(\omega) \leq -\frac{1}{\sqrt{b}}\right\} + P\left\{X(\omega) \geq \frac{1}{\sqrt{b}}\right\} = F_X\left(-\frac{1}{\sqrt{b}}\right) + 1 - F_X\left(\frac{1}{\sqrt{b}}\right)$$

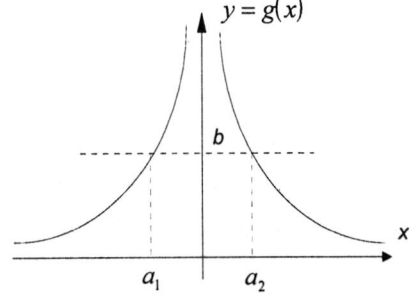

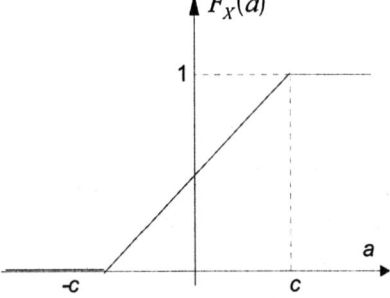

Fig. 1.7 Plot of the function $y = g(x)$ Fig. 1.8 Distribution function of X

Here is used the fact that the events $\{X(\omega) \leq a_1\}$ and $\{X(\omega) \geq a_2\}$ are mutually exclusive. If random variable X is uniformly distributed between $-c$ and c, as in Fig. 1.8, then

$$F_X(a) = \frac{a}{2c} + \frac{1}{2} \text{ for } |a| \leq c$$

so that

$$F_Y(b) = 1 - \frac{1}{c\sqrt{b}} \text{ for } b \geq \frac{1}{c^2} \; ; \; F_Y(b) = 0 \text{ for } b < \frac{1}{c^2}.$$

Determination of the pdf: We shall now determine the pdf f_y of the random variable $Y = g(X)$ in terms of the pdf f_X of the random variable X. Additionally, we shall assume that the random variable X is a continuous type, and $g(X)$ is a continuous function which is not equal a constant over any interval (this means that for a given b the equation $b = g(X)$ has a countable number of roots $a_1, \ldots, a_n$).

To find $f_Y(b)$ for a given b, we solve the equation $b = g(X)$ for X in terms of b. If $a_1, \ldots, a_n$ are all its real roots, then

$$f_Y(b) = \frac{f_X(a_1)}{|g'(a_1)|} + \cdots + \frac{f_X(a_n)}{|g'(a_n)|} \; ; \; g'(x) = \frac{dg(x)}{dx}$$

Clearly, the number $a_1, \ldots, a_n$ depend on b. If for a certain b the equation $b = g(x)$ has no real roots, then $f_Y(b) = 0$.

Proof: To prove the above formula, let us consider Fig. 1.9.

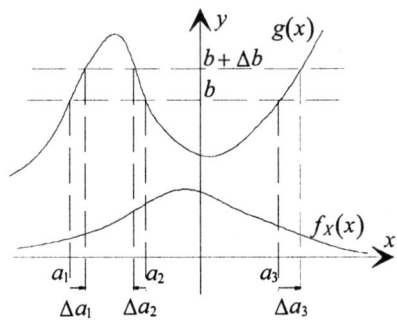

Fig. 1.9 Plot of $y = g(x)$ and plot of $f_X(x)$

To avoid generalities, we shall assume that, for a given $Y = b$, the equation $Y = g(X)$ has three roots a_1, a_2 and a_3, as in Fig. 1.9. Then we have

$$P\{b < Y(\omega) < b + \Delta b\} = f_Y(y)\Delta b \text{ for small } \Delta b.$$

Therefore, to find $f_Y(b)$ it suffices to find all values X such that $b < g(X) < b + \Delta b$. As we see from Fig. 1.9 , the above is true for

$$a_1 < X < a_1 + \Delta a_1 \; ; \; a_2 + \Delta a_2 < X < a_2 \text{ and } a_3 < X < a_3 + \Delta a_3$$

where $\Delta a_1 > 0$, $\Delta a_2 < 0$ and $\Delta a_3 > 0$. Hence

$$P\{b < Y(\omega) < b + \Delta b\} = P\{a_1 < X(\omega) < a_1 + \Delta a_1\} + P\{a_2 + \Delta a_2 < X(\omega) < a_2\}$$
$$+ P\{a_3 < X(\omega) < a_3 + \Delta a_3\}$$

where

$$P\{a_1 < X(\omega) < a_1 + \Delta a_1\} = f_X(a_1)\Delta a_1 \; ; \; P\{a_2 + \Delta a_2 < X(\omega) < a_2\} = f_X(a_2)|\Delta a_2|$$
$$P\{a_3 < X(\omega) < a_3 + \Delta a_3\} = f_X(a_3)\Delta a_3$$

Furthermore,

$$g'(a)\Delta a_i = \Delta b \; ; \; i = 1,2,3$$

so that

$$f_Y(b) = \frac{P\{b < Y(\omega) < b + \Delta b\}}{\Delta b} = \frac{f_X(a_1)}{|g'(a_1)|} + \frac{f_X(a_2)}{|g'(a_2)|} + \frac{f_X(a_3)}{|g'(a_3)|}$$

and the proof is thus completed.

Example 1.11 (pdf of a function of random variable): Consider the random variable $Y = g(X) = dX^2 \; ; \; d > 0$. If $Y = b < 0$, then the equation $b = g(X)$ has no real solutions; hence $f_Y(b) = 0$. If $b > 0$, then this equation has two solutions

$$a_1 = \sqrt{\frac{b}{d}} \; ; \; a_2 = -\sqrt{\frac{b}{d}}$$

Since

$$g'(a_1) = 2da_1 = 2\sqrt{bd} \; ; \; g'(a_2) = 2da_2 = -2\sqrt{bd}$$

we conclude that

$$f_Y(b) = \frac{f_X(a_1)}{|g'(a_1)|} + \frac{f_X(a_2)}{|g'(a_2)|} = \frac{1}{2\sqrt{bd}}\left[f_X\left(\sqrt{\frac{b}{d}}\right) + f_X\left(-\sqrt{\frac{b}{d}}\right) \right]$$

Particularly, if $f_X(\cdot)$ is even then $f_X(-a) = f_X(a)$, so that

$$f_Y(b) = \frac{1}{\sqrt{bd}} f_X\left(\sqrt{\frac{b}{d}}\right)$$

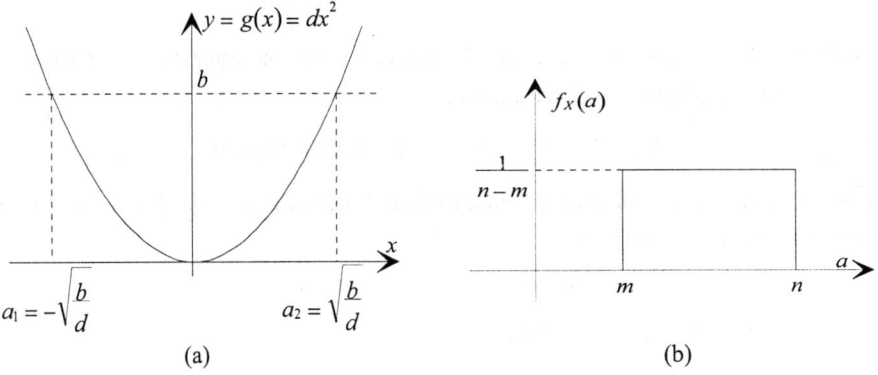

(a)   (b)

Fig. 1.10 (a) Plot of the function $y = g(x)$;   (b) Plot of the pdf of random variable X

However, if random variable X is uniformly distributed in the interval (m,n), where $m > 0$ and $n > m$, then

$$f_X\left(-\sqrt{\frac{b}{d}}\right) = 0 \; ; \; f_X\left(\sqrt{\frac{b}{d}}\right) = \frac{1}{n-m} \; ; \; \text{for } m \le \sqrt{\frac{b}{d}} \le n$$

so that

$$f_Y(b) = \frac{1}{2\sqrt{bd}(n-m)} \text{ for } dm^2 \le b \le dn^2 \text{ , and zero elsewhere.}$$

Example 1.12 (electrical circuit) : In the circuit of Fig. 1.11 $R = 1000\Omega$ is a constant resistance and the voltage source e is a random variable, uniformly distributed between 5 and $10V$. Find the pdf of the random power $w = \frac{1}{R}e^2$.

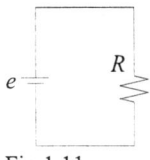

Fig.1.11

If we take $m = 5$, $n = 10$, $d = \frac{1}{R} = 10^{-3}$, in the preceding example, we have

$$f_W(b) = \begin{cases} \frac{1}{10}\sqrt{\frac{1000}{b}} \; ; \; \frac{25}{1000} \le b \le \frac{100}{1000} \\ 0 \; ; \; \text{otherwise} \end{cases}$$

1.5 More than one random variable: random vectors and multivariate distributions

Multivariate distribution and density functions: Frequently the random experiment has an outcome ω which is mapped into several random variables, say $X_1(\omega), ..., X_n(\omega)$, or organized as a column vector

$$X^T = \{X_1, ..., X_n\}$$

If we introduce some deterministic n-dimensional vector, say

$$a^T = \{a_1, ..., a_n\},$$

for this collection we can define the probability of the event that

$$X_1(\omega) \le a_1, X_2(\omega) \le a_2, ..., X_n(\omega) \le a_n$$

as

$P\{X_1(\omega) \le a_1, ..., X_n(\omega) \le a_n\} = P\{X(\omega) \le a\}$,and this is called the distribution function of the random vector X and is denoted as $F_X(a) = F_X(a_1, ..., a_n)$; that is

$$F_X(a) = F_X(a_1, ..., a_n) = P\{X(\omega) \le a\} = P\{X_1(\omega) \le a_1, ..., X_n(\omega) \le a_n\}$$

The corresponding multivariate density function $f_X(a)$ is a function of a vector argument a, which is defined by

$$f_X(a) = f_X(a_1,...,a_n) = \frac{\partial^n F_X(a)}{\partial a} = \frac{\partial^n F_X(a_1,...,a_n)}{\partial a_1 \cdots \partial a_n}.$$

Conversely, we have

$$F_X(a) = \int_{-\infty}^{a_1} \cdots \int_{-\infty}^{a_n} f_X(\xi_1,...\xi_n) d\xi_1 \cdots d\xi_n = \int_{-\infty}^{a} f_X(\xi) d\xi$$

so that the probability that a random vector $X(\omega)$ is in a box with sides a_i and b_i; $i = 1,...,n$, is given by

$$P\{a < X(\omega) \le b\} = \int_{a_1}^{b_1} \cdots \int_{a_n}^{b_n} f_X(\xi_1,\cdots,\xi_n) d\xi_1 \cdots d\xi_n = \int_a^b f_X(\xi) d\xi$$

Similarly, as in the scalar case $(n = 1)$, the pdf $f_X(a)$ satisfies the properties:

$$f_X(a) \ge 0; \quad \int_{-\infty}^{\infty} \cdots \int_{-\infty}^{\infty} f_X(a_1,...,a_n) da_1 \cdots da_n = 1.$$

Now let X be a n-dimensional random vector, a an arbitrary n-dimensional vector, Y a m-dimensional random vector, and b an arbitrary m-dimensional vector. Then, by a trivial extension of the preceding definition, we have

$$F_{X,Y}(a,b) = P\{X_1(\omega) \le a_1,..., X_n(\omega) \le a_n, Y_1(\omega) \le b_1,..., Y_n(\omega) \le b_n\}$$
$$= P\{X(\omega) \le a, Y(\omega) \le b\}$$

which is called *the joint probability distribution function* of X and Y. The joint pdf is seen to be

$$f_{X,Y}(a,b) = \frac{\partial^{n+m} F_{X,Y}(a,b)}{\partial a \partial b} = \frac{\partial^{n+m} F_{X,Y}(a_1,...,a_n,b_1,...,b_m)}{\partial a_1 \cdots \partial a_n \partial b_1 \cdots \partial b_m}.$$

The following properties are easily established:

1) $F_{X,Y}(a,b) = \int_{-\infty}^{a_1} \cdots \int_{-\infty}^{a_n} \int_{-\infty}^{b_1} \cdots \int_{-\infty}^{b_m} f_{X,Y}(\xi_1,...,\xi_n,\eta_1,...,\eta_m) d\xi_1 \cdots d\xi_n d\eta_1 \cdots d\eta_m$

       $= \int_{-\infty}^{a} \int_{-\infty}^{b} f_{X,Y}(\xi,\eta) d\xi d\eta$

2) $\int_{-\infty}^{\infty} \cdots \int_{-\infty}^{\infty} f_{X,Y}(\xi_1,\cdots,\xi_n,\eta_1,\cdots,\eta_m) d\xi_1 \cdots d\xi_n d\eta_1 \cdots d\eta_m = 1$

3) $P\{X(\omega) \le a, b \le Y(\omega) \le c\} = \int_{-\infty}^{a_1} \cdots \int_{-\infty}^{a_n} \int_{b_1}^{c_1} \cdots \int_{b_m}^{c_m} f_{X,Y}(\xi_1,...,\xi_n,\eta_1,...,\eta_m) d\xi_1 \cdots d\xi_n d\eta_1 \cdots d\eta_m$

where $\xi = [\xi_1 \; \xi_2 \; \cdots \; \xi_n]^T$ is n-vector, $\eta = [\eta_1 \; \eta_2 \; \cdots \; \eta_m]^T$ is m-vector, a is arbitrary n-vector, while b and c are arbitrary m vectors.

Marginal distribution and *density functions* follow in a simple way. Returning to the two random vectors X and Y, we consider the event $\{X(\omega) \le a, Y(\omega) \le b\}$. We call the function

$$F_{X,Y}(a,\infty) = P\{X(\omega) \le a, Y(\omega) \le \infty\}$$

the marginal probability distribution function of X. Since $\{Y(\omega) \le \infty\}$ is the certain event, we see that

$$\{X(\omega) \le a, Y(\omega) \le \infty\} = \{X(\omega) \le a\}.$$

Hence,

$$F_{X,Y}(a,\infty) = P\{X(\omega) \le a\} = F_X(a);$$

that is, it is the probability distribution function of X alone. Similarly,

$$F_{X,Y}(\infty, b) = F_Y(b).$$

If $f_{X,Y}(a,b)$ exists, we see that

$$F_{X,Y}(a,\infty) = \int_{-\infty}^{a_1} \cdots \int_{-\infty}^{a_n} \int_{-\infty}^{\infty} \cdots \int_{-\infty}^{\infty} f_{X,Y}(\xi_1,...,\xi_n,\eta_1,...,\eta_m) d\eta_1 \cdots d\eta_m \, d\xi_1 \cdots d\xi_n d\eta_1 \cdots d\eta_m$$

and since

$$F_X(a) = \int_{-\infty}^{a_1} \cdots \int_{-\infty}^{a_n} f_X(\xi_1,...,\xi_n) d\xi_1 \cdots d\xi_n$$

we see from $F_{X,Y}(a,\infty) = F_X(a)$ that

$$f_X(\xi) = f_X(\xi_1,...,\xi_n) = \int_{-\infty}^{\infty} \cdots \int_{-\infty}^{\infty} f_{X,Y}(\xi_1,...,\xi_n,\eta_1,...,\eta_m) d\eta_1 \cdots \eta_m$$

which is termed the *marginal pdf* of X (it is the pdf for X alone. In identically the same way, we obtain

$$f_Y(\eta) = f_Y(\eta_1,...,\eta_m) = \int_{-\infty}^{\infty} \cdots \int_{-\infty}^{\infty} f_{X,Y}(\xi_1,...,\xi_n,\eta_1,...,\eta_m) d\xi_1 \cdots \xi_n$$

Finally, marginal distribution of the component X_i of n-vector X is given from the joint distribution $F_X(a_1,...,a_i,....,a_n)$ as

$$F_{X_i}(a_i) = F_X(\infty, \cdots, \infty, a_i, \infty, \cdots, \infty)$$

while the corresponding marginal density is related to the joint density $f_X(a_1,...,a_i,...,a_n)$ as

$$f_{X_i}(\xi_i) = \int_{-\infty}^{\infty} \cdots \int_{-\infty}^{\infty} f_X(\xi_1,...,\xi_{i-1},\xi_i,\xi_{i+1},...,\xi_n) d\xi_1 \cdots d\xi_{i-1} d\xi_{i+1} \cdots d\xi_n$$

Expectations: The mean of the random vector $X(\omega)$ is a vector of the means of its components $X_i(\omega)$; that is

$$E\{X\} = \bar{X} = \{E(X_1),...,E(X_n)\}^T = \{\bar{X}_1,...,\bar{X}_n\}^T$$

and, likewise the deviation of the random variables from the mean-value, random vector is measured by a matrix called the covariance matrix, , defined as

$$\mathrm{cov}\{X\} = E\{(X-\bar{X})(X-\bar{X})^T\} = R_X$$

Thus, the covariance between the random variables X_i and X_j is in the i-th row and j-th column of the covariance matrix R_X, i.e.

$$\mathrm{cov}\{X_i,X_j\} = R_X(i,j) = E\{(X_i-\bar{X}_i)(X_j-\bar{X}_j)\} ; \ i,j = 1,...,n$$

The matrix

$$\Psi_X = E\{XX^T\}$$

is called the correlation matrix of X.

In an obvious way, we can extend the above definitions to include the case of different random vectors X and Y with mean vectors $\bar{X}$ and $\bar{Y}$, respectively. We define the cross-covariance matrix of X and Y as

$$R_{XY} = \mathrm{cov}\{X,Y\} = E\{(X-\bar{X})(Y-\bar{Y})^T\}$$

while the cross-correlation matrix of X and Y is defined as

$$\Psi_{XY} = E\left\{XY^T\right\}.$$

Example 1.12 (mean and covariance of random vector): Let us consider the situation where X is a 2-dimensional ($n=2$) vector, whose pdf is the two dimensional Rayleigh function

$$f_X(\xi) = f_X(\xi_1,\xi_2) = \begin{cases} 4\xi_1\xi_2 \exp\left(-\left(\xi_1^2 + \xi_2^2\right)\right) & ; \text{ for } \xi_1,\xi_2 \geq 0 \\ 0 & ; \text{ elsewhere} \end{cases}$$

Find the expectation $\bar{X} = E\{X\}$ and the covariance matrix R_X

By definition, we have

$$E\{X\} = \int_{-\infty}^{\infty} \xi f_X(\xi) d\xi_1 d\xi_2$$

$$= 4 \int_0^{\infty} \int_0^{\infty} \begin{bmatrix} \xi_1 \\ \xi_2 \end{bmatrix} \xi_1\xi_2 \exp\left(-\left(\xi_1^2 + \xi_2^2\right)\right) d\xi_1 d\xi_2 = \begin{bmatrix} E\{X_1\} \\ E\{X_2\} \end{bmatrix}$$

where $\xi = \begin{bmatrix} \xi_1 & \xi_2 \end{bmatrix}^T$. Therefore, we see that

$$E\{X_1\} = 4 \int_0^{\infty} \int_0^{\infty} \zeta_1^2\zeta_2 \exp\left(-\left(\zeta_1^2 + \zeta_2^2\right)\right) d\zeta_1 d\zeta_2$$

$$= 4 \int_0^{\infty} \zeta_2 \exp\left(-\zeta_2^2\right) \int_0^{\infty} \zeta_1^2 \exp\left(-\zeta_1^2\right) d\zeta_1 d\zeta_2 = 4 \cdot \left(\frac{1}{2}\right) \cdot \frac{\sqrt{\pi}}{4} = \frac{\sqrt{\pi}}{2}$$

Similarly, we obtain $E\{X_2\} = \dfrac{\sqrt{\pi}}{2}$, so that

$$\bar{X} = E\{X\} = \frac{\sqrt{\pi}}{2}\begin{bmatrix} 1 \\ 1 \end{bmatrix}$$

The 2×2 dimensional covariance matrix R_X is defined as $R_X = \Psi_X - \bar{X}\bar{X}^T$, where

$$\Psi_X = E\left\{XX^T\right\} = \int_{-\infty}^{\infty}\int_{-\infty}^{\infty} \begin{bmatrix} \xi_1 \\ \xi_2 \end{bmatrix} \begin{bmatrix} \xi_1 & \xi_2 \end{bmatrix} f_{X,Y}(\xi_1,\xi_2) d\xi_1 d\xi_2$$

so that

$$R_X = 4\int_0^{\infty}\int_0^{\infty} \begin{bmatrix} \xi_1^2 & \xi_1\xi_2 \\ \xi_1\xi_2 & \xi_2^2 \end{bmatrix} \xi_1\xi_2 \exp\left(-\left(\xi_1^2 + \xi_2^2\right)\right) d\xi_1 d\xi_2 - \frac{\pi}{4}\begin{bmatrix} 1 & 1 \\ 1 & 1 \end{bmatrix} = \begin{bmatrix} R_X(1,1) & R_X(1,2) \\ R_X(2,1) & R_X(2,2) \end{bmatrix}$$

Thus, we have

$$R_X(1,1) = 4\int_0^{\infty}\int_0^{\infty} \xi_1^3\xi_2 \exp\left(-\left(\xi_1^2 + \xi_2^2\right)\right) d\xi_1 d\xi_2 - \frac{\pi}{4}$$

$$= 4\int_0^{\infty} \xi_1^3 \exp\left(-\xi_1^2\right) d\xi_1 \int_0^{\infty} \xi_2 \exp\left(-\xi_2^2\right) d\xi_2 - \frac{\pi}{4} = 4 \cdot \frac{1}{2}\cdot\frac{1}{2} - \frac{\pi}{4} = 1 - \frac{\pi}{4}$$

We note that $R_X(2,2) = R_X(1,1)$. Finally,

$$R_X(1,2) = R_X(2,1) = 4 \int_0^\infty \int_0^\infty \xi_1^2 \xi_2^2 \exp\left(-\left(\xi_1^2 + \xi_2^2\right)\right) d\xi_1 d\xi_2 \, -\frac{\pi}{4}$$

$$= 4 \frac{\sqrt{\pi}}{4} \frac{\sqrt{\pi}}{4} - \frac{\pi}{4} = 0$$

and we have

$$R_X = \begin{bmatrix} 1 - \dfrac{\pi}{4} & 0 \\[2mm] 0 & 1 - \dfrac{\pi}{4} \end{bmatrix}$$

Thus, the random variables X_1 and X_2 are uncorrelated $\left(R_X(1,2) = R_X(2,1) = 0\right)$, and R_X is a diagonal matrix. The elements $R_X(1,1)$ and $R_X(2,2)$ on the main diagonal of matrix R_X are the variances of the random variables X_1 and X_2, i.e. $\sigma_{X_1}^2 = R_X(1,1)$ and $\sigma_{X_2}^2 = R_X(2,2)$. In general, R_X is a symmetric matrix, i.e. $R_X = R_X^T$.

Independence and correlation: Two random variables X_i and Y_j are said to be uncorrelated if

$$E\{X_i Y_j\} = E\{X_i\} E\{Y_j\}$$

or equivalently

$$\mathrm{cov}\left(X_i, X_j\right) = R_X(i,j) = 0$$

Thus, if X is a random vector whose components X_i $(i = 1,...,n)$ are uncorrelated, then $R_X(i,j) = 0$ for all $i \neq j$, and for this it follows that $R_X = \mathrm{cov}\{X\}$ is a diagonal matrix.

If X and Y are random vectors with components $X_i, i = 1,...,n$ and Y_j, $j = 1,...,m$ and if the equation $E\{X_i Y_j\} = \bar{X}_i \bar{Y}_j$ is satisfied for all i and j, then X and Y are termed uncorrelated random vectors. In this case $E\{XY^T\} = \bar{X}\bar{Y}^T = \Psi_{X,Y}$.

Furthermore, for uncorrelated random vectors

$$R_{X,Y} = E\left\{\left(X - \bar{X}\right)\left(Y - \bar{Y}\right)^T\right\} = E\{XY^T\} - \bar{X}\bar{Y}^T = \Psi_{X,Y} - \bar{X}\bar{Y}^T = 0$$

Additionally, if

$$\Psi_{X,Y} = E\{XY^T\} = 0$$

the random vectors X and Y are termed *orthogonal*.

For two random variables X_i and Y_j, the quantity

$$\rho_{ij} = \frac{E\left\{\left(X_i - \bar{X}_i\right)\left(Y_j - \bar{Y}_j\right)\right\}}{\sqrt{E\left\{\left(X_i - \bar{X}_i\right)^2\right\} E\left\{\left(Y_j - \bar{Y}_j\right)^2\right\}}}$$

is called the *correlation coefficient* of X_i and Y_j. It is obviously zero if X_i and Y_j are uncorrelated. We assert that $|\rho_{ij}| \leq 1$. To see this, it is sufficient to consider the case $\bar{X}_i = \bar{Y}_j = 0$. Then for α being a scalar constant, we note that

$$E\left\{\left(\alpha X_i - Y_j\right)^2\right\} = \alpha^2 E\left\{X_i^2\right\} - 2\alpha E\left\{X_i Y_j\right\} + E\left\{Y_j^2\right\} \geq 0$$

However, since the above expression is quadratic in α, necessary and sufficient condition for it to be non-negative are

$$E\left\{X_i^2\right\} \geq 0 \; ; \; 4\left[E\left\{X_i Y_j\right\} - E\left\{X_i^2\right\} E\left\{Y_j^2\right\}\right] \leq 0$$

The first condition is obviously satisfied, while the second one gives $1 - \rho_{ij}^2 \geq 0$, which completes the proof.

Analogous to the definition of independent events, we say that two random vectors X and Y are *independent* if [36, 42]

$$F_{X,Y}(a,b) = F_X(a) F_Y(b)$$

where $F_X(a) = P\{X(\omega) \leq a\}$ and $F_Y(b) = P\{Y(\omega) \leq b\}$ are the so-called marginal distributions of X and Y, respectively, and $F_{X,Y}(a,b) = P\{X(\omega) \leq a, Y(\omega) \leq b\}$ is the joint distribution. It then follows

$$f_{X,Y}(a,b) = f_X(a) f_Y(b)$$

where

$$f_X(a) = \frac{\partial^n F_X(a)}{\partial a_1 \cdots \partial a_n} \; ; \; f_Y(b) = \frac{\partial^m F_Y(b)}{\partial b_1 \cdots \partial b_m}$$

are marginal densities of X and Y, respectively, and

$$f_{X,Y}(a,b) = \frac{\partial^{n+m} F_{X,Y}(a,b)}{\partial a_1 \cdots \partial a_n \partial b_1 \cdots \partial b_m}$$

is the joint probability density function of these two random vectors.

The elements $X_i \; ; i = 1, \ldots, n$ of a random vector X are independent if the joint distribution $F_X(a)$ can be expressed in the form

$$F_X(a) = F_X(a_1, \ldots, a_{i-1}, a_i, a_{i+1}, \ldots, a_n) = \prod_{i=1}^n F_{X_i}(a_i)$$

where $F_{X_i}(a_i) = F_X(\infty, \ldots, \infty, a_i, \infty, \ldots, \infty)$ is the marginal pdf of X_i. The corresponding joint density

$$f_X(a) = f_X(a_1, \ldots, a_{i-1}, a_i, a_{i+1}, \ldots, a_n) = \prod_{i=1}^n f_{X_i}(a_i) \; ; \; f_{X_i}(a_i) = \frac{dF_{X_i}(a_i)}{da_i}$$

where

$$f_{X_i}(\xi_i) = \int_{-\infty}^{\infty} \cdots \int_{-\infty}^{\infty} f_X(\xi_1, \ldots, \xi_{i-1}, \xi_i, \xi_{i+1}, \ldots, \xi_n) d\xi_1 \cdots d\xi_{i-1} d\xi_{i+1} \cdots d\xi_n$$

is the marginal pdf of X_i.

Finally, we take note of the fact that two independent random vectors are uncorrelated, but that the opposite is not true in general. If X and Y are independent, then obviously

$$E\left\{XY^T\right\} = \int_{-\infty}^{\infty} \cdots \int_{-\infty}^{\infty} \xi \eta^T f_{X,Y}(\xi_1, \ldots, \xi_n, \eta_1, \ldots, \eta_m) d\xi_1 \cdots d\xi_n d\eta_1 \cdots d\eta_m$$

$$= \int_{-\infty}^{\infty} \cdots \int_{-\infty}^{\infty} \xi f_X(\xi_1, \ldots, \xi_n) d\xi_1 \cdots d\xi_n \int_{-\infty}^{\infty} \cdots \int_{-\infty}^{\infty} \eta^T f_Y(\eta_1, \ldots, \eta_m) d\eta_1 \cdots d\eta_m = \overline{X}\,\overline{Y}^T$$

To verify that the converse is not generally true, it is sufficient to take $n=m=1$; that is X and Y are scalars. Assume further that $f_X(\xi)$ is symmetric with $\bar{X} = 0$ and take $Y = g(X) = X^2$. Then

$$E\{XY\} = E\{X^3\} = \int_{-\infty}^{\infty} \xi^3 f_X(\xi) d\xi = 0$$

since the integrand is an odd function of the argument ξ. Since $\bar{X} = 0$ we have

$$E\{XY\} = \bar{X}\bar{Y} = 0$$

so that the random variables X and Y are uncorrelated, but by the relation $Y = X^2$ not independent.

1.6 Conditional distributions, density functions and expectations

Conditional probability distribution and density function: Of considerable importance in probability theory, and also in estimation and control, is the notation of conditional probability, wherein the probability function for an event is dependent upon the prior occurrence of some other event. Let $f_{X,Y}(a,b)$ denote the joint pdf of the random vectors X and Y, and let $f_Y(b)$ denote the marginal pdf of Y. In addition, let A denote the event $\{X(\omega) \le a\}$ and B the event $\{b \le Y(\omega) \le b + \Delta b\}$, where X and a are n-dimensional vectors, while Y, b and Δb are m dimensional vectors with $\Delta b_j > 0$; $j = 1, ..., m$. Then from the definition of conditional probability [36, 42]

$$P\{X(\omega) \le a | b \le Y(\omega) \le b + \Delta b\} = \frac{P\{X(\omega) \le a, b \le Y(\omega) \le b + \Delta b\}}{P\{b \le Y(\omega) \le b + \Delta b\}}$$

However, since for Δb_j small enough

$$P\{b \le Y(\omega) \le b + \Delta b\} = \int_{b_1}^{b_1 + \Delta b_1} \cdots \int_{b_m}^{b_m + \Delta b_m} f_Y(\eta_1, ..., \eta_m) d\eta_1 \cdots d\eta_m \approx f_Y(b_1, ..., b_m) \prod_{i=1}^{m} \Delta b_i$$

and

$$P\{X(\omega) \le a, b \le Y(\omega) \le b + \Delta b\}$$

$$= \int_{-\infty}^{a_1} \cdots \int_{-\infty}^{a_n} \int_{b_1}^{b_1 + \Delta b_1} \cdots \int_{b_m}^{b_m + \Delta b_m} f_{X,Y}(\xi_1, ..., \xi_n, \eta_1, ..., \eta_m) d\xi_1 \cdots d\xi_n d\eta_1 \cdots d\eta_m$$

$$\approx \int_{-\infty}^{a_1} \cdots \int_{-\infty}^{a_n} f_{X,Y}(\xi_1, ..., \xi_n, b_1, ..., b_m) d\xi_1 \cdots d\xi_n \prod_{i=1}^{m} \Delta b_i$$

one concludes

$$P\{X(\omega) \le a | b \le Y(\omega) \le b + \Delta b\} = \frac{\int_{-\infty}^{a_1} \cdots \int_{-\infty}^{a_n} f_{X,Y}(\xi_1, ..., \xi_n, b_1, ..., b_m) d\xi_1 \cdots d\xi_n}{f_Y(b_1, ..., b_m)}$$

The function

$$F_{X/Y}(a | Y(\omega) = b) = P\{X(\omega) \le a | Y(\omega) = b\}$$

is called the *conditional probability distribution function* of X given Y. Letting $\Delta b_i \to 0$, $i = 1, ..., m$, we obtain from the last two expressions

$$F_{X/Y}(a/b) = F_{X/Y}(a_1,...,a_n/b_1,...,b_m) = \frac{\int_{-\infty}^{a_1} \cdots \int_{-\infty}^{a_n} f_{X,Y}(\xi_1,...,\xi_n,b_1,...,b_m)d\xi_1 \cdots d\xi_n}{f_Y(b_1,...,b_m)}.$$

We then define the function

$$f_{X/Y}(a|b) = \frac{\partial^n F_{X/Y}(a|b)}{\partial a_1 \cdots \partial a_n} = \frac{\partial^n F_{X/Y}(a_1,...,a_n|b_1,...,b_m)}{\partial a_1 \cdots \partial a_n}$$

to be the *conditional pdf* of X given Y. From this definition and the preceding relation, it is clear that

$$f_{X/Y}(a/b) = \frac{\partial^n}{\partial a_1 \cdots \partial a_n} \frac{\int_{-\infty}^{a_1} \cdots \int_{-\infty}^{a_n} f_{X,Y}(\xi_1,...,\xi_n,b_1,...,b_m)d\xi_1 \cdots d\xi_n}{f_Y(b_1,...,b_m)}$$

$$= \frac{f_{X,Y}(a_1,...,a_n,b_1,...,b_m)}{f_Y(b_1,...,b_m)} = \frac{f_{X,Y}(a,b)}{f_Y(b)}$$

Similarly,

$$f_{Y/X}(b/a) = \frac{f_{X,Y}(a,b)}{f_X(a)}$$

The obtained result is termed *Bayes rule*. We note that if $f_{X,Y}(a,b)$ is given, the marginal pdf's $f_X(a)$ and $f_Y(b)$ can be computed from it, and then $f_{X/Y}(a/b)$ and $f_{Y/X}(b/a)$ follow immediately from the Bayes rule.

Conditional expectations: The conditional expected value of a scalar or vector-valued function $g(\cdot)$ of a random vector X, given another random vector Y, is defined as

$$E_X\{g(X)/Y(\omega) = b\} = \int_{-\infty}^{\infty} \cdots \int_{-\infty}^{\infty} g(\xi)f_{X/Y}(\xi/b)d\xi_1 \cdots d\xi_n$$

The subscript X on E denotes that the expected value operation, i.e. the integration, is over X. The subscript could obviously be omitted in this case without causing confusion.

The *conditional mean* of $g(X) = X$ given $Y(\omega) = b$ is defined as

$$E\{X/b\} = \int_{-\infty}^{\infty} \cdots \int_{-\infty}^{\infty} \xi f_{X/Y}(\xi/b)d\xi_1 \cdots d\xi_n$$

and the corresponding *conditional covariance matrix* can be obtained if we substitute $g(X) = (X - E\{X|b\})(X - E\{X|b\})^T$:

$$R_{X/Y} = E_X\left\{[X - E\{X/b\}][X - E\{X/b\}]^T\right\}$$

$$= \int_{-\infty}^{\infty} \cdots \int_{-\infty}^{\infty} [\xi - E\{X/b\}][\xi - E\{X/b\}]^T f_{X/Y}(\xi/b)d\xi_1 \cdots d\xi_n$$

The conditional expectation has the following properties:

1) $E\{g(X)/X(\omega) = x\} = g(x)$
2) $E\{AX/Y(\omega) = y\} = AE\{X/Y(\omega) = y\}$
3) $E_{XY}\{X + Y/Z = z\} = E_X\{X/Z = z\} + E_Y\{Y/Z = z\}$
4) $E_Y\{E_X\{X/Y(\omega) = y\}\} = E_X\{X\}$

The verification of the first three properties is relatively straightforward by using the definition of the conditional expectation. Therefore, let us prove the last property. Since by definition

$$E_X\{X/Y(\omega)=\eta\}=\int_{-\infty}^{\infty}\cdots\int_{-\infty}^{\infty}\xi\,f_{X/Y}(\xi/\eta)d\xi_1\cdots d\xi_n$$

one concludes

$$E_Y\{E_X\{X/Y(\omega)=\eta\}\}=\int_{-\infty}^{\infty}\cdots\int_{-\infty}^{\infty}E_X\{X/Y(\omega)=\eta\}f_Y(\eta)d\eta_1\cdots d\eta_m$$

$$=\int_{-\infty}^{\infty}\cdots\int_{-\infty}^{\infty}\xi\,f_{X/Y}(\xi/\eta)f_Y(\eta)d\xi_1\cdots d\xi_n d\eta_1\cdots d\eta_m$$

$$=\int_{-\infty}^{\infty}\cdots\int_{-\infty}^{\infty}\xi\,f_{X,Y}(\xi,\eta)d\xi_1\cdots d\xi_n d\eta_1\cdots d\eta_m$$

$$=\int_{-\infty}^{\infty}\cdots\int_{-\infty}^{\infty}\xi\left[\int_{-\infty}^{\infty}\cdots\int_{-\infty}^{\infty}f_{X,Y}(\xi,\eta)d\eta_1\cdots d\eta_m\right]d\xi_1\cdots d\xi_n$$

$$=\int_{-\infty}^{\infty}\cdots\int_{-\infty}^{\infty}\xi\,f_X(\xi)d\xi_1\cdots d\xi_n=E_X\{X\}$$

Here is used the Bayes rule and the fact that

$$\int_{-\infty}^{\infty}\cdots\int_{-\infty}^{\infty}f_{X,Y}(\xi,\eta)d\eta_1\cdots d\eta_m=f_X(\xi)=f_X(\xi_1,\ldots,\xi_n)$$

We remark also that it can be shown that $E\{X/Y(\omega)=b\}$ is unique.

Example 1.13 (conditional density and expectations): Suppose that X and Y are scalars whose joint pdf is cylindrical

$$f_{X,Y}(\xi,\eta)=\begin{cases}\dfrac{1}{\pi}&\text{for }\xi^2+\eta^2\leq1\\0&\text{elsewhere}\end{cases}$$

Find the marginal pdf of Y, and determine $E\{X/Y(\omega)=b\}$ and $R_{X/Y}$.

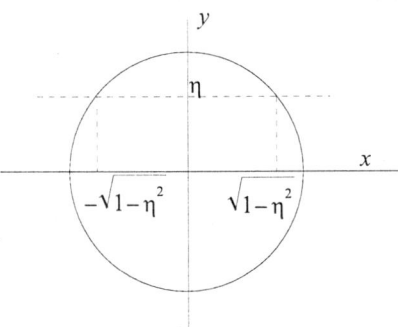

Fig. 1.11

For the relation between joint and marginal pdf's, we have

$$f_Y(\eta)=\int_{-\infty}^{\infty}f_{X,Y}(\xi,\eta)d\xi=\int_{-\sqrt{1-\eta^2}}^{\sqrt{1-\eta^2}}\frac{1}{\pi}d\xi=\frac{2}{\pi}\sqrt{1-\eta^2}$$

for $|\eta|\leq1$ and $f_Y(\eta)=0$ elsewhere. Bayes rule then leads to the result that

$$f_{X/Y}(\xi/\eta)=\frac{f_{X,Y}(\xi,\eta)}{f_Y(\eta)}=\frac{1}{2\sqrt{1-\eta^2}}$$

for $-\sqrt{1-\eta^2}<\xi<\sqrt{1-\eta^2}$ with $|\eta|\leq1$, and $f_{X/Y}(\xi/\eta)=0$ elsewhere.

The conditional expectation is given as

$$E\{X/Y(\omega)=b\}=\int_{-\infty}^{\infty}\xi\,f_{X/Y}(\xi/b)d\xi=\int_{-\sqrt{1-b^2}}^{\sqrt{1-b^2}}\xi\,\frac{d\xi}{2\sqrt{1-b^2}}=0$$

This result is obvious, since each $f_{X/Y}(\xi/\eta)$ is uniform on the interval $\left(-\sqrt{1-\eta^2},\sqrt{1-\eta^2}\right)$ and centered at the origin. The definition expression for the conditional covariance now takes the form

$$R_{X/Y} = \int_{-\infty}^{\infty} \left(\xi - E\{X/Y(\omega) = b\}\right)^2 f_{X/Y}(\xi/b)\,d\xi$$

$$= \int_{-\sqrt{1-b^2}}^{\sqrt{1-b^2}} \xi^2 \frac{1}{2\sqrt{1-b^2}}\,d\xi = \frac{1}{3}(1-b^2)$$

where $|b| \leq 1$. The conditional variance $R_{X/Y}$ is a function of b. Since X and Y are related, we feel that knowledge of the value $Y(\omega) = b$ should permit us to conclude something about the corresponding value of random variable X. We could pick up $E\{X/Y(\omega) = b\}$ as an estimate of X under the assumption that we are told the value of Y. Moreover, since $R_{X/Y}$ is a measure of the spread of X about its conditional mean $E\{X/Y(\omega) = b\}$, we could employ $R_{X/Y}$ as a measure of the quality of the estimate. For example, given $Y(\omega) = b = 1$ we say that the estimate $\hat{X} = E\{X/Y(\omega) = 1\} = 0$ with a variance of zero, or given $Y(\omega) = b = 0.5$, $\hat{X} = E\{X/Y(\omega) = 0.5\} = 0$ with a variance of 0.25.

Characteristic function: The characteristic function $\Phi_X(\omega)$ of a random n-dimensional vector X is defined as [36, 42]

$$\Phi_X(\omega) = E\{\exp(jX^T\omega)\} = \int_{-\infty}^{\infty} \cdots \int_{-\infty}^{\infty} \exp(j\xi^T\omega) f_X(\xi)\,d\xi_1 \cdots d\xi_n$$

where $j^2 = -1$ and ω is a n-dimensional vector. We note that $\Phi_X(\omega)$ is a scalar-valued function of the vector argument $\omega = [\omega_1 \ \ldots \ \omega_n]^T$. The inversion formula, which follows from the Fourier transformation theory, is

$$f_X(x) = \frac{1}{(2\pi)^n} \int_{-\infty}^{\infty} \cdots \int_{-\infty}^{\infty} \exp(-jx^T\omega) \Phi_X(\omega)\,d\omega_1 \cdots d\omega_n$$

In a similar way, the joint characteristic function of the random vectors X and Y is given as

$$\Phi_{XY}(\omega_1, \omega_2) = E\{\exp(jZ^T\omega)\} = \int_{-\infty}^{\infty} \cdots \int_{-\infty}^{\infty} \exp(jZ^T\omega) f_{XY}(x_1, \ldots, x_n, y_1, \ldots, y_m)\,dx_1 \cdots dx_n dy_1 \cdots dy_m$$

with

$$Z = \begin{bmatrix} X \\ Y \end{bmatrix} ; \ \omega = \begin{bmatrix} \omega_1 \\ \omega_2 \end{bmatrix}$$

where X and ω_1 are n-dimensional vectors, while Y and ω_2 are m-dimensional vectors, respectively. The inverse formula is

$$f_{XY}(x,y) = \frac{1}{(2\pi)^{n+m}} \int_{-\infty}^{\infty} \cdots \int_{-\infty}^{\infty} \exp(-jZ^T\omega) \Phi_{XY}(\omega_1, \omega_2)\,d\omega_{11} \cdots d\omega_{1n} d\omega_{21} \cdots d\omega_{2m}$$

where $\omega_1 = [\omega_{11}, \ldots, \omega_{1n}]^T$ and $\omega_2 = [\omega_{21}, \ldots, \omega_{2m}]^T$, while Z and ω are defined as before.

Finally, the conditional characteristic function of X given $Y(\omega) = y$ is written as

$$\Phi_{X/Y}(\omega) = E\{\exp(jX^T\omega)/Y(\omega) = y\} = \int_{-\infty}^{\infty} \cdots \int_{-\infty}^{\infty} \exp(jx^T\omega) f_{X/Y}(x/y)\,dx_1 \cdots dx_n$$

and its inverse is given by

$$f_{X/Y}(x/y) = \frac{1}{(2\pi)^n} \int_{-\infty}^{\infty} \cdots \int_{-\infty}^{\infty} \exp\left(-jx^T\omega\right) \Phi_{X/Y}(\omega) d\omega_1 \cdots d\omega_n$$

where $\omega_1 = [\omega_1, ..., \omega_n]^T$ and $x = [x_1, ..., x_n]^T$.

The characteristic function is useful in problems of determining the probability law of a function $g(\cdot)$ of a random variable X given the probability law of the latter. We illustrate its use with the following example.

Example 1.14 (characteristic and density functions): Let X be a Gaussian distributed scalar random variable with pdf

$$f_X(\xi) = \frac{1}{\sqrt{2\pi}\sigma} \exp\left(-\frac{\xi^2}{2\sigma^2}\right)$$

Assume that X is input into a square-law device whose output is $Y = \alpha X^2$; $\alpha > 0$. Determine the pdf $f_Y(\eta)$ of the device's output.

From the definition of characteristic function and the fact that $Y = \alpha X^2$, we have

$$\Phi_Y(\omega) = E\{\exp(jY\omega)\} = E\{\exp(j\alpha X^2\omega)\}$$

$$= \int_{-\infty}^{\infty} \exp(j\alpha\xi^2\omega) \frac{1}{\sqrt{2\pi}\sigma} \exp\left(-\frac{\xi^2}{2\sigma^2}\right) d\xi$$

$$= 2\int_0^{\infty} \exp(j\alpha\xi^2\omega) \frac{1}{\sqrt{2\pi}\sigma} \exp\left(-\frac{\xi^2}{2\sigma^2}\right) d\xi$$

However, $\eta = \alpha\xi^2$ and $d\eta = 2\alpha\xi d\xi = 2\sqrt{\alpha\eta}d\xi$. Hence

$$\Phi_Y(\omega) = \int_0^{\infty} \exp(j\eta\omega) \frac{1}{\sqrt{2\pi}\sigma} \exp\left(-\frac{\eta}{2\alpha\sigma^2}\right) \frac{d\eta}{\sqrt{\alpha\eta}}$$

$$= \int_0^{\infty} \exp(j\eta\omega) \frac{\exp\left(-\dfrac{\eta}{2\alpha\sigma^2}\right)}{\sigma\sqrt{2\pi\alpha\eta}} d\eta$$

Since, on the other hand

$$\Phi_Y(\omega) = \int_0^{\infty} \exp(j\eta\omega) f_Y(\eta) d\eta$$

one concludes

$$f_Y(\eta) = \frac{\exp\left(-\dfrac{\eta}{2\alpha\sigma^2}\right)}{\sigma\sqrt{2\pi\alpha\eta}}$$

1.7 Multivariable Gaussian distribution

The most important multivariable pdf is the normal or Gaussian one given by

$$f_X(\xi) = \frac{1}{\sqrt{(2\pi)^n |R_X|}} \exp\left(-\frac{1}{2}[\xi - \overline{X}]^T R_X^{-1}[\xi - \overline{X}]\right)$$

where $|R_X|$ is the determinant of the $n \times n$ matrix R_X, while n is the dimension of the random vector X and $\bar{X} = E\{X\}$. For the multivariable normal law we can compute the mean

$$E\{X\} = \bar{X}$$

and the covariance matrix

$$E\left\{\left[X - \bar{X}\right]\left[X - \bar{X}\right]^T\right\} = R_X$$

As mentioned before, there are basically two reasons why the Gaussian pdf has enjoyed the prominence in applications. First, it has been found through the experience that the Gaussian pdf provides a model which is a reasonable approximation to observed random behavior in certain physical systems. Secondly, by virtue of its mathematical form, the Gaussian pdf is analytically and computationally tractable. This is due to the fact that it is completely specified by its first and second order moments, namely, its mean $\bar{X}$ and covariance matrix R_X. In the expression for normal pdf we require the inverse of R_X, and have thus implicitly assumed that R_X is nonsingular matrix, i.e. $R_X > 0$. Thus, the normal pdf $f_X(\xi)$ does not exist if R_X is singular. For this reason, the Gaussian distribution is usually defined by its characteristic function. A random n-dimensional vector X is said to be Gaussian distributed if its characteristic function is

$$\Phi_X(\omega) = \exp\left(j\bar{X}^T\omega - \frac{1}{2}\omega^T R_X \omega \right)$$

where $\omega = [\omega_1, ..., \omega_n]^T$.

Central limit theorem: A partial justification for using the Gaussian distribution in practice is the central limit theorem [36, 42, 47]. The theorem which we state here without proof, is as follows. Let X_i ; $i = 1, ..., r$ be a set of independent, identically distributed random n-vectors with finite means $\bar{X}_i$ and covariance matrices R_i. Let Y_r and Z_r be the random n-vectors

$$Y_r = \sum_{i=1}^{r} X_i \; ; \; Z_r = P_r^{-1}\left(Y_r - \bar{Y}_r \right)$$

where

$$\bar{Y}_r = E\{Y_r\} = \sum_{i=1}^{r} \bar{X}_i \; ; \; P_r = \prod_{i=1}^{r} R_i$$

Then

$$\lim_{r \to \infty} f_{Z_r}(z) = \frac{1}{\sqrt{2\pi}} \exp\left(-\frac{1}{2} z^T z \right)$$

That is, as $r \to \infty$, Z_r becomes a zero-mean Gaussian n-vector whose covariance matrix is the identity matrix. This means that if the random phenomenon which we observe at the macroscopic level is the superposition of an arbitrarily large number of independent random phenomena which occur at the microscopic level, we are justified in describing the former phenomenon in terms of the Gaussian distribution.

1.8 Functions of two random variables

Consider now the special case of two random variables X and Y having the joint distribution $F_{X,Y}(\xi,\eta)$ and density $f_{X,Y}(\xi,\eta)$. We are also given a function $g(X,Y)$ of the random variables X and Y. If $g(X,Y)$ satisfies certain general conditions, then

$$Z = g(X,Y)$$

is a random variable whose distribution

$$F_Z(\zeta) = P\{\omega:Z(\omega) \le \zeta\}$$

and density $f_Z(\zeta)$ can be determined in terms of the function $g(X,Y)$ and the joint density $f_{X,Y}(\xi,\eta)$ of X and Y. To determine $F_Z(\zeta)$ for a given ζ, we must find the probability of the event $\{\omega:Z(\omega) \le \zeta\}$. We denote by D_ζ the region of the XY plane such that $g(X,Y) \le \zeta$. It is easy to see that

$$\{\omega:Z(\omega) \le \zeta\} = \{(X(\omega),Y(\omega)) \in D_\zeta\}$$

Therefore

$$F_Z(\zeta) = P\{(X(\omega),Y(\omega)) \in D_\zeta\} = \iint_{D_\zeta} f_{X,Y}(\xi,\eta)d\xi d\eta$$

Example 1.18 (distribution of the sum of random variables) : Let us take $g(X,Y) = X+Y$; this is the most important example of a function of two random variables. To determine $F_Z(\zeta)$, we note that the region D_ζ of the XY plane such that $X+Y \le \zeta$ is the half plane to the left of the line $X+Y = \zeta$, as in the Fig. 1.12 is depicted.

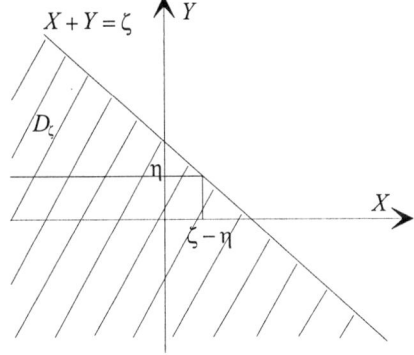

Fig. 1.12

We find, integrating over suitable strips

$$F_Z(\zeta) = \iint_{D_\zeta} f_{X,Y}(\xi,\eta)d\xi d\eta$$

$$= \int_{-\infty}^{\infty} \int_{-\infty}^{\zeta-\eta} f_{X,Y}(\xi,\eta)d\xi d\eta$$

Differentiating with respect to ζ we obtain

$$f_Z(\zeta) = \frac{dF_Z(\zeta)}{d\zeta} = \int_{-\infty}^{\infty} f_{X,Y}(\zeta-\eta,\eta)d\eta$$

If the random variables X and Y are independent, then

$$f_Z(\zeta) = \int_{-\infty}^{\infty} f_X(\zeta-\eta)f_Y(\eta)d\eta$$

$$= \int_{-\infty}^{\infty} f_X(\xi)f_Y(\zeta-\xi)d\xi = f_X \otimes f_Y$$

Thus, if the random variables X and Y are independent, then the density of their sum $Z = X+Y$ equals the convolution of their respective densities.

Given two functions $g(x,y)$ and $h(x,y)$ of the real variables x and y and two random variables X and Y, we form the random variables $Z = g(X,Y)$ and $W = h(X,Y)$. These random variables have a joint distribution and density $F_{Z,W}(\alpha,\beta)$ and $f_{Z,W}(\alpha,\beta)$, respectively. We shall determine these functions in terms of $g(X,Y)$, $h(X,Y)$ and the joint density $f_{X,Y}(\xi,\eta)$ of the random variables X and Y. With α and β two real numbers, we denote $D_{\alpha\beta}$ the region of the XY plane such that

$$g(X,Y) \le \alpha \text{ and } h(X,Y) \le \beta$$

Since

$$\{\omega : Z(\omega) \le \alpha, W(\omega) \le \beta\} = \{(X(\omega), Y(\omega)) \in D_{\alpha\beta}\}$$

we conclude that

$$F_{Z,W}(\alpha,\beta) = \iint\limits_{D_{\alpha\beta}} f_{X,Y}(\xi,\eta)\, d\xi\, d\eta.$$

If we denote by ($|\cdot|$ denotes the determinant)

$$J(\xi,\eta) = \begin{vmatrix} \dfrac{\partial g(\xi,\eta)}{\partial \xi} & \dfrac{\partial g(\xi,\eta)}{\partial \eta} \\[2mm] \dfrac{\partial h(\xi,\eta)}{\partial \xi} & \dfrac{\partial h(\xi,\eta)}{\partial \eta} \end{vmatrix}$$

the Jacobian of the transformation $Z = g(X,Y)$, $W = h(X,Y)$, we can express the joint density $f_{Z,W}(\alpha,\beta)$ of the random variables Z and W directly in terms of $f_{X,Y}(\xi,\eta)$ and $J(\xi,\eta)$. To determine $f_{Z,W}(\alpha,\beta)$, we solve the equations

$$g(\xi,\eta) = \alpha \; ; \; h(\xi,\eta) = \beta$$

for ξ and η in terms of α and β. If

$$(\xi_1,\eta_1),...,(\xi_n,\eta_n)$$

are all real solutions of these equations, then

$$f_{Z,W}(\alpha,\beta) = \frac{f_{X,Y}(\xi_1,\eta_1)}{|J(\xi_1,\eta_1)|} + \cdots + \frac{f_{X,Y}(\xi_n,\eta_n)}{|J(\xi_n,\eta_n)|}$$

where $|J|$ denotes the absolute value of J.

Proof: The proof is very similar to the case of a function of one random variable. Namely, let us denote by ΔD_{zw} the region of the points in the (x,y) plane such that

$$z < g(x,y) \le z + dz \; ; \; w < h(x,y) \le w + dw$$

Then, clearly

$$\{z < Z \le z + dz, w < W \le w + dw\} = \{(X,Y) \in \Delta D_{zw}\}$$

and

$$P\{z < Z \le z + dz, w < W \le w + dw\} = \iint\limits_{\Delta D_{zw}} f_{Z,W}(z,w) dz dw \approx f_{Z,W}(z,w) dz dw$$

Given z and w, the region ΔD_{zw} consists of differential parallelograms, one for each solution (x_i, y_i) of the system $g(x,y) = z$, $h(x,y) = w$.

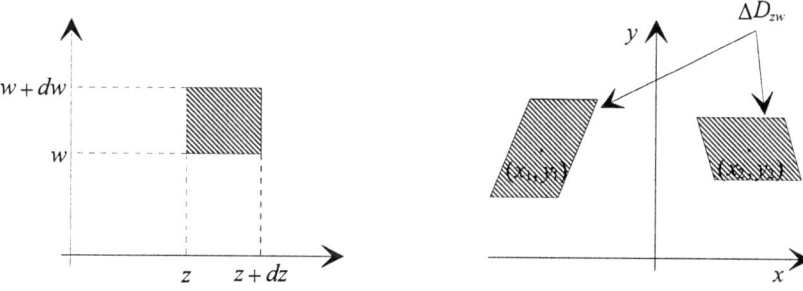

Fig. 1.13

The area of the i-th parallelogram equals

$$\Delta S_i = \frac{dz dw}{|J(x_i, y_i)|} \; ; \; i = 1, \dots, n$$

so that it contains the probability mass $f_{X,Y}(x_i, y_i) \Delta S_i$. Summing the probability mass in all these parallelograms, we obtain the probability mass in the parallelogram in the (z, w) plane (see, Fig. 1.13 representing the required probability $P\{z < Z \le z + dz, w < W \le w + dw\}$), i.e.

$$f_{Z,W}(z,w) dz dw = \iint\limits_{\Delta D_{zw}} f_{X,Y}(x,y) dx dy = \sum_i f_{X,Y}(x_i, y_i) \frac{dz dw}{|J(x_i, y_i)|}$$

which completes the proof.

Remark: We can easily generalize this result to determine the joint density of the k random variables

$$y_i = g_i(x_1, \dots, x_n) \; ; \; i = 1, \dots, k$$

in terms of the joint pdf $f(x_1, \dots, x_n)$ and the given functions $g_i(x_1, \dots, x_n)$

If $k > n$, then the unknown pdf is singular. In this case, one computes the statistics of $y_1, \dots, y_n$. If $k < n$, then, using auxiliary variables $y_{k+1} = x_{k+1}, \dots, y_n = x_n$, we increase the number of y's to n. Therefore, we can assume that $k=n$. To determine the joint density $f(y_1, \dots, y_n)$ for a given set of numbers $(y_1, \dots, y_n)$ we solve the system

$$g_i(x_1, \dots, x_n) = y_i \; ; \; i = 1, \dots, n$$

If this system has a single real solution $(x_1, \dots, x_n)$ (the values x_i are of course the functions of y_i), then

$$f_{Y_1,\dots,Y_n}\left(y_1,\dots,y_n\right)=\frac{f_{X_1,\dots,X_n}\left(x_1,\dots,x_n\right)}{\left|J\left(x_1,\dots,x_n\right)\right|}$$

where

$$J\left(x_1,\dots,x_n\right)=\begin{vmatrix}\dfrac{\partial g_1}{\partial x_1}&\cdots&\dfrac{\partial g_1}{\partial x_n}\\[2mm]\vdots&&\\[2mm]\dfrac{\partial g_n}{\partial x_1}&\cdots&\dfrac{\partial g_n}{\partial x_n}\end{vmatrix}$$

If the system has more than one solution, then we add in the right-hand side the corresponding expressions resulting from all solutions. If the above system has no real solutions, then $f_{Y_1,\dots,Y_n}\left(y_1,\dots,y_n\right)=0$.

The determination of the density of one function $Z=g\left(X,Y\right)$ of two random variables X and Y is sometimes facilitated by introducing an auxiliary variable, e.g. $W=X$ or $W=Y$. One determines first the joint density $f_{Z,W}\left(z,w\right)$ of Z and W. The unknown marginal density of Z is then found by integration

$$f_Z\left(\alpha\right)=\int_{-\infty}^{\infty}f_{Z,W}\left(\alpha,\beta\right)d\beta.$$

Example 1.15 (sum of two random variables): We shall rederive the density of $Z=X+Y$ using as auxiliary variable the random variable $W=Y$.

The system

$$\alpha=g\left(\xi,\eta\right)=\xi+\eta\ ;\ \beta=h\left(\xi,\eta\right)=\eta$$

has a unique solution

$$\xi_1=\alpha-\beta\ ;\ \eta_1=\beta$$

and since

$$J\left(\xi,\eta\right)=\begin{vmatrix}1&1\\0&1\end{vmatrix}=1$$

we obtain

$$f_{Z,W}\left(\alpha,\beta\right)=\frac{f_{X,Y}\left(\xi_1,\eta_1\right)}{\left|J\left(\xi_1,\eta_1\right)\right|}=f_{X,Y}\left(\alpha-\beta,\beta\right)$$

Therefore,

$$f_Z\left(\alpha\right)=\int_{-\infty}^{\infty}f_{X,Y}\left(\alpha-\beta,\beta\right)d\beta,$$

which is the same result we derived before on a quite different way.

Example 1.16 (a simple communication system): Let us determine the statistical characteristics of the amplitude and phase of a received signal, which is obtained by transmitting a periodical signal through a random media. The statistical model of a communication system is given in Fig. 1.14.

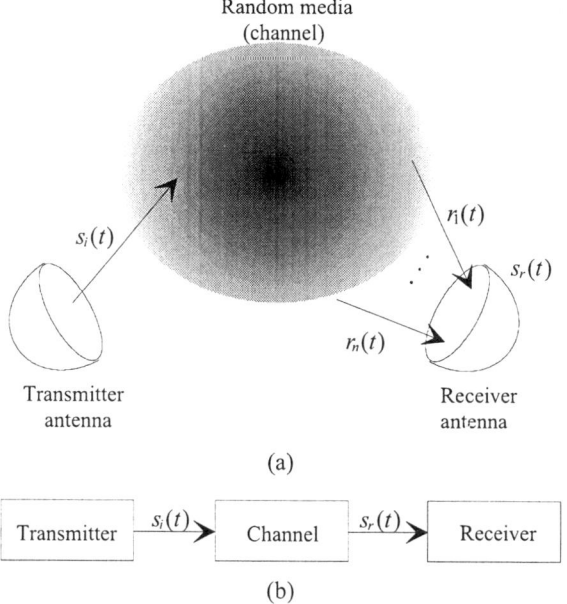

Random media
(channel)

$s_i(t)$

$r_1(t)$

$s_r(t)$

$r_n(t)$

Transmitter
antenna

Receiver
antenna

(a)

| Transmitter | $s_i(t)$ | Channel | $s_r(t)$ | Receiver |

(b)

Fig. 1.14 Block diagram for a simple communication system

The transmitted or input periodic signal $s_i(t)$ can be represented in a complex form as

$$s_i(t) = e^{j\omega_0 t}$$

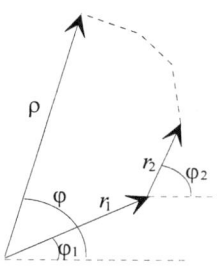

The received signal $s_r(t)$ is of the same shape, but with different amplitude and phase representing random variables, due to various disturbances across a random media (channel). Thus, the received signal takes the form (Fig. 1.14)

$$s_r(t) = \sum_k r_k e^{j(\omega_0 t + \varphi_k)} = \sum_k r_k e^{j\varphi_k} e^{j\omega_0 t}$$

where the elements of sum represent random reflections from the media. The resulting complex phasor, representing the influence of a random media, can be expressed as (see, Fig. 1.15)

Fig. 1.15

$$\rho e^{j\varphi} = \sum_k r_k e^{j\varphi_k}$$

Moreover, we can use the rectangular coordinates x and y, instead of the polar ones (ρ, φ), i.e.

$$\rho e^{j\varphi} = x + jy$$

where

$$x = \rho \cos(\varphi) \; ; \; y = \rho \sin(\varphi)$$

or equivalently

$$\rho = g(x,y) = \sqrt{x^2 + y^2} \; ; \; \varphi = h(x,y) = \arctan\left(\frac{y}{x}\right)$$

Thus, the complex form of the periodic received signal is given by

$$s_r(t) = \rho e^{j(\omega_0 t + \varphi)}$$

In this way, the joint probability density of ρ and φ is given by

$$f_{\rho,\varphi}(\rho,\varphi) = \frac{f_{X,Y}(x,y)}{|J(x,y)|}$$

where

$$J(x,y) = \begin{vmatrix} \dfrac{\partial g(x,y)}{\partial x} & \dfrac{\partial g(x,y)}{\partial y} \\[2mm] \dfrac{\partial h(x,y)}{\partial x} & \dfrac{\partial h(x,y)}{\partial y} \end{vmatrix} = \begin{vmatrix} \dfrac{x}{\sqrt{x^2+y^2}} & \dfrac{y}{\sqrt{x^2+y^2}} \\[2mm] -\dfrac{y}{\sqrt{x^2+y^2}} & \dfrac{x}{\sqrt{x^2+y^2}} \end{vmatrix} = \frac{1}{\sqrt{x^2+y^2}} = \frac{1}{\rho}$$

Furthermore, we can assume that X and Y are identically distributed, jointly normal and independent random variables, owing to the central limit theorem, so that the multivariate pdf of the 2-dimensional random vector $Z = \{X,Y\}$ is given by

$$f_Z(z) = \frac{1}{\sqrt{(2\pi)^n |R_Z|}} \exp\left\{-\frac{1}{2}(Z-\bar{Z})^T R_Z^{-1}(Z-\bar{Z})\right\}$$

where $n=2$, while the mean

$$\bar{Z} = \begin{bmatrix} E\{X\} \\ E\{Y\} \end{bmatrix} = \begin{bmatrix} 0 \\ 0 \end{bmatrix}$$

and the covariance matrix

$$R_Z = \begin{bmatrix} E\{(X-\bar{X})^2\} & E\{(X-\bar{X})(Y-\bar{Y})\} \\ E\{(X-\bar{X})(Y-\bar{Y})\} & E\{(Y-\bar{Y})(Y-\bar{Y})\} \end{bmatrix} = \begin{bmatrix} \sigma^2 & 0 \\ 0 & \sigma^2 \end{bmatrix} = \sigma^2 I$$

In this way, we obtain for the joint pdf

$$f_Z(z) = f_{X,Y}(x,y) = f_X(x) f_Y(y) = \frac{1}{2\pi\sigma^2} \exp\left\{-(x^2+y^2)/(2\sigma^2)\right\}$$

from which follows

$$f_{\rho,\varphi}(\rho,\varphi) = \frac{1}{2\pi}\frac{\rho}{\sigma^2} \exp\left(-\frac{\rho^2}{2\sigma^2}\right)$$

If we take

$$f_{\rho,\varphi}(\rho,\varphi) = f_\rho(\rho) f_\varphi(\varphi)$$

where

$$f_\varphi(\varphi) = \frac{1}{2\pi} \;\; ; \;\; f_\rho(\rho) = \frac{\rho}{\sigma^2}\exp\left(-\rho^2/(2\sigma^2)\right)$$

we conclude that the random phase φ of the received signal is uniformly distributed on the interval $(-\pi, \pi)$, while its amplitude ρ has the Reyleigh distribution (see Appendix C).

1.9 Worked examples

Example 1.17: One student is arbitrary chosen among the students attending the lecture of Probability theory. Let the event A denotes that the chosen student is a male. Let the event B denotes that the chosen student does not smoke , and let the event C means that the chosen student lives in a college apartment. 1) Describe event $AB\overline{C}$, where $\overline{C}$ denotes the complement event (set) of the event (set) C, i.e. $\overline{C}C = 0$ and $C + \overline{C} = \Omega$. 2) When the identity $ABC=A$ is valid? 3) In which case the relation $\overline{C} \subseteq B$ is satisfied? 4) Explain the situation when the identity $\overline{A} = B$ is valid.

Solution: 1) The event $AB\overline{C}$ means that an arbitrary chosen student is a boy, who does not smoke and that he lives in a college apartment.
2) The identity $ABC=A$ will be satisfied if all of the present boys do not smoke and live in the college apartments.
3) The relation $\overline{C} \subseteq B$ will be valid if all of the present students, who do not live in the college apartments, do not smoke.
4) The identity $\overline{A} = B$ will be satisfied if all the girls, present at the lecture, are not smokers, and all of the present boys are smokers.

Example 1.18: Let A, B and C are random events. Simplify the expression: 1) $(A+B)(B+C)$; 2) $(A + B)(A + \overline{B})$; 3) $(A + B)(A + \overline{B})(\overline{A} + B)$

Solution: If we use the next identities $AA=A$; $A+A=A$; $A\Omega = A$; $A+\Omega = \Omega$; $A0=0$; $A+0=A$; $A\overline{A} = 0$; $A + \overline{A} = \Omega$ then the expressions can be simplified on the following manner:

   1) $(A+B)(B+C)=AB+B+AC+BC=(AB+B+BC)+AC$
       $=B(A+\Omega +C)+AC=B\Omega +AC=B+AC$

   2) $(A + B)(A + \overline{B}) = A + A\overline{B} + BA + B\overline{B}$
       $= A + A(B + \overline{B}) + 0 = A + A\Omega = A + A = A$

   3) $(A + B)(A + \overline{B})(\overline{A} + B) = A(\overline{A} + B) = AB$

Example 1.19: In which situations the next identities are satisfied: 1) $A + B = \overline{A}$; 2) $AB = \overline{A}$; 3) $A + B = AB$?

Solution: 1) If there is only one outcome ω included in the event A, this outcome will belong to the event on the left hand side of the identity and will not belong to the event on the right hand side. As we want these two events to be equal, this means that must not exist any outcome ω within the event A, yielding that the event A is the impossible event (empty set). Moreover, one can conclude that the event B has to be certain event ($A = 0, B = \Omega$).

2) The event on the left hand side of the identity includes the outcomes which are common for both of the events A and B. On the other hand, the event on the right hand side does not include any of the outcomes from the event A. That means that the events A and B can not have any common element, so that the event on the right hand side has to be empty. Proceeding on a similar way, the event A is certain and the event B is impossible.

3) If there would be any outcome contained in the event A and not contained in the event B, a such outcome would occur on the left hand side of the identity, but would not occur on the right hand side. The same can be concluded for the outcomes contained in the event B and not contained in A. If we want the events AB and $A+B$ to be equal, there is no such outcomes, so then the events A and B have to be the same($A=B$).

Example 1.20: Determine the event X satisfying the next equations: 1) $A+X=A+B$; 2) $AB+X=(A+C)(B+C)$.

Solution: 1) As the event on the right hand side includes all elementary outcomes belonging the event A or event B, that means that the event X must include all outcomes contained in B and not contained in A; but it is also possible for some arbitrary outcomes from A to be included in X. In other words: $X = B \setminus A \cup AD$, where D is an arbitrary set, and $B \setminus A$ denotes a set consisting of the elements of B that are not in A, i.e. $B \setminus A$ is the difference $B \setminus A = B\overline{A}$.

2) The event on the right hand side is $AB+C$, so that, based on the previous solution, the event X is defined as $X = C \setminus AB + ABD$, where D is an arbitrary set.

Example 1.21: The box, whose all sides are painted, is divided into one thousand small boxes with the same dimensions. The obtained small boxes are mixed, and one of them is arbitrary chosen. Calculate the probability that the chosen box has two painted sides.

Solution: This problem requires the application of classical definition of probability. The number of possible outcomes is $n=1000$, while the number of favorable outcomes is $m=12 \cdot 8 = 96$ (because along each of 12 edges of the box there are 8 small boxes with two painted sides). So, the probability of the event that the chosen box has two painted sides is $P=m/n=96/1000=0.096$.

Example 1.22: Ten books on a shelf are arbitrary ordered. Calculate the probability of the event that three marked books will be together.

Solution: The number of possible outcomes is $n=10!$ because it represents the permutation of 10 elements without repeating. The number of favorable outcomes is $m = 8! \cdot 3!$ because seven books, which are not important for us, can be ordered in 7! possible ways, and in a such order we can put three marked books together on 8 different places among them. Also, three marked books can be ordered in 3! different ways, so that the number of desired outcomes is $m= 7! \cdot 8 \cdot 3! = 8! \cdot 3!$. Using the classical definition of probability, the probability of the concerned event is $P=m/n=1/15$.

Example 1.23: From the set of 52 cards, three cards are arbitrary taken. Calculate the probability that the taken cards will be three, seven and ace.

Solution: The number of possible outcomes is $n = C_{52}^3 = \begin{pmatrix} 52 \\ 3 \end{pmatrix}$, representing the combination of 3 among 52 elements without repeating. The number of desired outcomes is $m = C_4^1 C_4^1 C_4^1 = \begin{pmatrix} 4 \\ 1 \end{pmatrix}\begin{pmatrix} 4 \\ 1 \end{pmatrix}\begin{pmatrix} 4 \\ 1 \end{pmatrix} = 64$ since it is necessary to take one of the four cards with the same number, and to repeat this action three times (once for a card with the number three, the second time for a card with the number seven and, finally, it is necessary to take out one of four aces). So, the probability of the described event is equal to $P=m/n=0.0029$.

Example 1.24: At any moment of the time interval t, it is possible for the receiver, with the same probability, to detect two different signals. The receiver will be blocked if the time distance between them is less then τ. Calculate the probability of the event that the receiver will be blocked.

Solution: This problem can be solved by using the geometrical approach to probability. Namely, if t_1 denotes the moment of detecting the first signal, and if t_2 denotes the moment of detecting the second one, the point with the coordinates (t_1, t_2) placed within the square, shown in the Fig. 1.16, will represent the described event in a geometrical sense. The receiver will be blocked if the following inequality is satisfied $|t_1 - t_2| < \tau$. The points which satisfy this inequality are shown on the figure as marked surface.

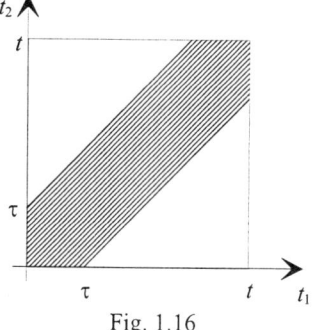

Fig. 1.16

So, the desired probability can be computed as the ratio of the areas of marked surface and whole square with side t: $P = (t^2 - (t-\tau)^2)/t^2 = (2t\tau - \tau^2)/t^2$

Example 1.25: The plane is divided by the equidistant parallel straight lines. The distance between them is d. Calculate the probability of the event that an arbitrary thrown needle of length l will intersect any line.

Solution: If h denotes the distance between the center of the needle and the nearest straight line (Fig. 1.17.) and Θ is the angle between the needle and the line orthogonal to the grid of lines, then the needle will intersect the nearest straight line if the next inequality is satisfied: $\dfrac{l}{2}\cos\theta > h$.

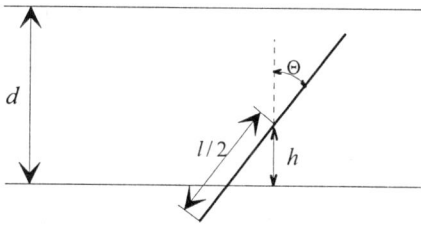

Fig. 1.17.

The problem can be solved by applying the geometrical approach to the probability. The experiment can be represented by the point with the coordinates h and Θ. The certain event is described by the rectangular whose one side is equal to $d/2$ (because h can take the values greater then or equal to zero and less then or equal to $d/2$), and the other side is equal to $\pi/2$ (because the angle Θ has to be nonnegative and less then or equal to $\pi/2$). The surface which represent the desired outcomes can be described by the points which satisfy the mentioned inequality ($\Theta \in [0, \pi/2]$; $0 < h < \dfrac{l}{2}\cos(\Theta)$). So, the required probability is equal to the ratio of the following areas:

$$P = \frac{\text{favorable area}}{\text{the whole area}} = \frac{\int_0^{\pi/2} \dfrac{l}{2}\cos\theta d\theta}{\dfrac{d\pi}{4}} = \frac{2l}{d\pi}.$$

Example 1.26: Two shooters are shooting independently. The probability of score for the first of the shooters is 0.8 and for the second one is 0.4. Both of them had one trial, and after that it is confirmed that the target is hit with one bullet. Calculate the probability of the event that the first shooter hit the target.

Solution: Let's form four mutually exclusive events whose sum (union) is equal to the certain event:

H_1 - the target is not hit;

H_2 - the target is hit by the first shooter only;

H_3 - the target is hit by the second shooter only;

H_4 - the target is hit by both of the shooters.

Let A denotes the event that the target is hit once. We want to compute the conditional probability $P(H_2/A)$. Using the formula for conditional probability we can write:

$$P(H_2/A) = \frac{P(H_2 A)}{P(A)} = \frac{P(A/H_2)P(H_2)}{P(A)},$$

where $P(A)$ can be computed using the formula for total probability:

$$\begin{aligned} P(A) = P(A\Omega) &= P(A(H_1 + H_2 + H_3 + H_4)) \\ &= P(AH_1) + P(AH_2) + P(AH_3) + P(AH_4) \\ &= P(A/H_1)P(H_1) + P(A/H_2)P(H_2) + P(A/H_3)P(H_3) + P(A/H_4)P(H_4) \end{aligned}$$

Finally, one obtains:

$$P(H_2/A) = \frac{P(A/H_2)P(H_2)}{P(A/H_1)P(H_1) + P(A/H_2)P(H_2) + P(A/H_3)P(H_3) + P(A/H_4)P(H_4)}$$

The last equation is Bayes formula. For our case the mentioned probabilities are equal to: $P(A/H_1) = P(A/H_4) = 0$; $P(A/H_2) = P(A/H_3) = 1$; $P(H_1) = 0.2 \cdot 0.6$; $P(H_2) = 0.8 \cdot 0.6$; $P(H_3) = 0.2 \cdot 0.4$ and $P(H_4) = 0.8 \cdot 0.4$, so the conditional probability $P(H_2/A)$ becomes:

$$P(H_2/A) = \frac{1 \cdot 0.8 \cdot 0.6}{0 \cdot 0.2 \cdot 0.6 + 1 \cdot 0.8 \cdot 0.6 + 1 \cdot 0.2 \cdot 0.4 + 0 \cdot 0.8 \cdot 0.4} = \frac{6}{7}$$

Example 1.27: There are ten boxes. Nine of them contain two black and two white balls, while the one contains five white balls and one black ball. One box is randomly chosen and one white ball is taken out from it. Calculate the probability of the event that the box with five white balls is chosen.

Solution: The events H_1 and H_2 given below are mutually exclusive and their sum (union) is equal to the certain event, where:

event H_1 - the chosen box is one of the nine boxes with the same content;

event H_2 - the chosen box is with five white balls.

Let A denotes the event that randomly taken ball from the randomly chosen box is a white ball. Thus, we want to calculate the conditional probability $P(H_2/A)$. Similarly as in the previous example, this probability can be computed in the next way:

$$P(H_2/A) = \frac{P(H_2 A)}{P(A)} = \frac{P(A/H_2)P(H_2)}{P(A/H_1)P(H_1) + P(A/H_2)P(H_2)} = \frac{\frac{5}{6} \cdot \frac{1}{10}}{\frac{2}{4} \cdot \frac{9}{10} + \frac{5}{6} \cdot \frac{1}{10}} = \frac{5}{32}$$

Example 1.28: Digital communication system operates in according to the next scheme (Fig. 1.18):

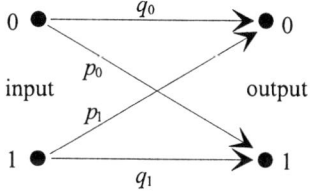

Fig. 1.18

Let us denote by π_0 the probability of occurring signal 0 and by π_1 the probability of the event that the signal 1 will occur at the system input X.

a) Calculate the probabilities γ_0 i γ_1 of occurring the signals 0 and 1, respectively, at the output Y of the communication system.

b) Calculate the conditional probabilities $P(X=1/Y=1)$, $P(X=1/Y=0)$, $P(X=0/Y=1)$ and $P(X=0/Y=0)$.

Solution: a) According to the formula of total probability, one can write:

$$\gamma_0 = P(Y = 0) = P(Y = 0/X = 0)P(X = 0) + P(Y = 0/X = 1)P(X = 1)$$
$$= q_0 \pi_0 + p_1 \pi_1$$

$$\gamma_1 = P(Y = 1) = P(Y = 1/X = 0)P(X = 0) + P(Y = 1/X = 1)P(X = 1)$$
$$= p_0 \pi_0 + q_1 \pi_1$$

b) Based on Bayes formula, the required conditional probabilities can be calculated as (see Fig. 1.18):

$$P(X = 1/Y = 1) = \frac{P(X = 1, Y = 1)}{P(Y = 1)} = \frac{P(Y = 1/X = 1)P(X = 1)}{\gamma_1} = \frac{q_1 \pi_1}{p_0 \pi_0 + q_1 \pi_1}$$

$$P(X = 0/Y = 1) = \frac{P(X = 0, Y = 1)}{P(Y = 1)} = \frac{P(Y = 1/X = 0)P(X = 0)}{\gamma_1} = \frac{p_0 \pi_0}{p_0 \pi_0 + q_1 \pi_1}$$

$$P(X = 1/Y = 0) = \frac{P(X = 1, Y = 0)}{P(Y = 0)} = \frac{P(Y = 0/X = 1)P(X = 1)}{\gamma_0} = \frac{p_1 \pi_1}{q_0 \pi_0 + p_1 \pi_1}$$

$$P(X = 0/Y = 0) = \frac{P(X = 0, Y = 0)}{P(Y = 0)} = \frac{P(Y = 0/X = 0)P(X = 0)}{\gamma_0} = \frac{q_0 \pi_0}{q_0 \pi_0 + p_1 \pi_1}$$

Example 1.29: Two basketball players are throwing the ball until one of them makes the score. The probability of making a score for the first player is 0.4 and for the second one is 0.6. Calculate the distribution of the random variable X which is equal to the number of throwing of the ball.

Solution: Let p_i denotes the probability of the event that the discrete random variable X will take the value i. Then, we can calculate:

$$p_1 = 0.4 \; ; \; p_2 = 0.6 \cdot 0.6 \; ; \; p_3 = 0.6 \cdot 0.4 \cdot 0.4 \; ; \; p_4 = 0.6 \cdot 0.4 \cdot 0.6 \cdot 0.6 ; \cdots$$
$$p_{2k+1} = 0.6^k \cdot 0.4^k \cdot 0.4 = 0.6^k \cdot 0.4^{k+1} \; ; \; p_{2k} = 0.6^k \cdot 0.4^{k-1} \cdot 0.6 = 0.6^{k+1} \cdot 0.4^{k-1}$$

These values form the so-called distribution of probabilities. On the other hand, it is possible to calculate the distribution function as:

$$\left(\forall x \in [j, j+1); j = 0,1,... \right); \; F_X(x) = \sum_{k=0}^{j} p_k h(x-k)$$

where $h(x)$ is the unit step function ($h(x)=0$ for $x<0$ and $h(x)=1$ for $x\geq 0$).

Example 1.30: Continuos random variable X is defined by its probability density function:

$$f(x)=\begin{cases} A\cos 2x \ ; & -\dfrac{\pi}{4}<x<\dfrac{\pi}{4} \\[2mm] 0 \ ; & \text{elsewhere} \end{cases}$$

a) Calculate the constant A;
b) Calculate the distribution function $F(x)$;
c) Calculate the probability $P(0<x<\pi/8)$.

Solution: a) Constant A can be calculated using the normalizing condition:

$$\int_{-\infty}^{\infty} f(x)dx = 1$$

In our case, one can obtain:

$$\int_{-\infty}^{\infty} f(x)dx = \int_{-\pi/4}^{\pi/4} A\cos(2x)dx = A\dfrac{\sin(2x)}{2}\Bigg|_{-\pi/4}^{\pi/4} = A = 1$$

b) The distribution function $F(x)$ can be calculated by integrating the corresponding pdf:

$$F(x) = \int_{-\infty}^{x} f(\tau)d\tau$$

In our case the distribution function takes the form:

$$\left(\forall x < -\pi/4\right): F(x) = 0;$$

$$\left(\forall x \in [-\pi/4, \pi/4)\right): F(x) = \int_{-\pi/4}^{x} f(\tau)d\tau = \dfrac{1}{2}[\sin(2x)+1]$$

$$\left(\forall x \geq \pi/4\right): F(x) = 1;$$

c) The probability of the event that the random variable X will take a value within some interval $[a,b]$ can be calculated using either the distribution function or probability density function; that is,

$$P(a<X\leq b) = F(b)-F(a) = \int_{a}^{b} f(x)dx$$

so, one can calculate:

$$P(0<X<\pi/8) = F(\dfrac{\pi}{8}) - F(0) - P\left(X = \dfrac{\pi}{8}\right)$$

$$= \dfrac{1}{2}\left[\sin\left(2\cdot\dfrac{\pi}{8}\right)+1\right] - \dfrac{1}{2}[\sin(0)+1] - 0 = \dfrac{1}{2}\sin\dfrac{\pi}{4} = \dfrac{\sqrt{2}}{4}$$

The last result is obtained taking into consideration the fact that for the continuous random variable probability of the event that the random variable will take some specific value is equal to zero.

Example 1.30: The probability density function of uniformly distributed random variable X is:

$$f(x)=\begin{cases} \dfrac{1}{b-a} \ ; & x\in[a,b] \\[2mm] 0 \ ; & \text{elsewhere} \end{cases}$$

Calculate the corresponding distribution function and the median of this random variable.
Solution: Based on the previous example, we can write:

$$F(x) = \int_{-\infty}^{x} f(\tau)d\tau \Rightarrow F(x) = \begin{cases} 0 \; ; \; x < a \\ \dfrac{x-a}{b-a} \; ; \; x \in [a,b] \\ 1 \; ; \; x \geq b \end{cases}$$

Median M_e of the random variable X is defined as the value of the argument x at which the distribution function $F(x)$ takes the value of 0.5:

$$F(x)\big|_{x=M_e} = 0.5 \Rightarrow \frac{M_e - a}{b-a} = 0.5 \Rightarrow M_e = \frac{a+b}{2}$$

Example 1.31: The distribution function of a random variable X is defined as:

$$F(x) = \begin{cases} 0 \; ; \; x \leq -a \\ A + B \arcsin\dfrac{x}{a} \; ; \; -a < x < a \\ 1 \; ; \; x \geq a \end{cases}$$

a) Determine the parameters A and B so that the distribution function be continuous, i.e. without break points; b) Calculate the probability ; c) Determine the corresponding probability density function; d) Calculate the quantile $x_{0.75}$ and the mode M_o of the random variable X.

Solution: a) The conditions for the point $\pm a$ not to be the break point are:

$$F(-a^-) = F(-a^+) = 0 = A + B\arcsin(-1) = A - B\frac{\pi}{2}$$

$$F(a^+) = F(a^-) = 1 = A + B\arcsin(1) = A + B\frac{\pi}{2}$$

Based on these two equations, it can be calculated:

$$A = \frac{1}{2} \; ; \; B = \frac{1}{\pi}$$

b) $$P(-a/2 < X < a/2) = F(a/2) - F(-a/2) = 1/3$$

c) The corresponding probability density function is given by:

$$f(x) = \frac{dF(x)}{dx} = \begin{cases} 0 \; ; \; |x| \geq a \\ \dfrac{1}{\pi\sqrt{a^2 - x^2}} \; ; \; \text{elsewhere} \end{cases}$$

Figure 1.20 presents this function for $a=1$ and it can be seen that there are two break points at ± 1.

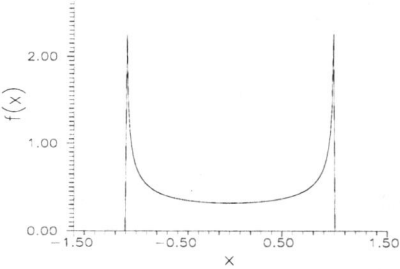

Fig. 1.20

d) The quantile $x_{0.75}$ is the value of argument x for which the distribution function $F(x)$ takes the value of 0.75:

$$F(x_{0.75}) = \frac{3}{4} = \frac{1}{2} + \frac{1}{\pi}\arcsin\left(\frac{x_{0.75}}{a}\right) \Rightarrow x_{0.75} = \frac{\sqrt{2}a}{2}$$

The mode of the random variable is the value of argument x for which the probability density function has a maximum. In our case, the mode does not exist (Fig. 1.20).

Example 1.32: Basketball player is throwing a ball up to the first score. The probability of score for each of the trials is the same and equal to p. Calculate the mean and the variance of the random variable X representing the number of trials.

Solution: X denotes discrete type random variable and can take the values from the set of positive integers. The probability of the event that the random variable X will take the value i is equal to:

$$p_i = P\{X = i\} = (1-p)^{i-1} p$$

since $(1-p)$ is the probability of the negative result of a trial. The mean value of discrete X can be calculated by applying the corresponding summation, instead of integration in the case of continuous variables:

$$m_X = E\{X\} = \sum_{i=1}^{\infty} i p_i = \sum_{i=1}^{\infty} i(1-p)^{i-1} p \qquad = p\sum_{i=1}^{\infty} i(1-p)^{i-1} = -p\sum_{i=1}^{\infty}\frac{d}{dp}(1-p)^i$$

$$= -p\frac{d}{dp}\sum_{i=1}^{\infty}(1-p)^i = -p\frac{d}{dp}\lim_{n\to\infty}(1-p)\frac{1-(1-p)^n}{1-(1-p)} = -p\frac{d}{dp}\left[\lim_{n\to\infty}\sum_{i=1}^{n}(1-p)^i\right]$$

$$= -p\frac{d}{dp}\lim_{n\to\infty}(1-p)\frac{1-(1-p)^n}{1-(1-p)}$$

$$= -p\frac{d}{dp}\frac{1-p}{p} = -p\frac{-p-1+p}{p^2} = \frac{1}{p}$$

The variance of X can be calculated in the same manner:

$$\sigma_X^2 = E\{(X-m_x)^2\} = E\{X^2\} - m_X^2 \qquad = \sum_{i=1}^{\infty} i^2 p_i - \frac{1}{p^2} = \sum_{i=1}^{\infty} i^2(1-p)^{i-1} p - \frac{1}{p^2}$$

$$= p\sum_{i=1}^{\infty}\frac{d}{dp} i(1-p)^i - \frac{1}{p^2} = p\sum_{i=1}^{\infty}\frac{d}{dp}(1-p)\frac{d}{dp}(1-p)^i - \frac{1}{p^2}$$

$$= p\frac{d}{dp}\left\{(1-p)\frac{d}{dp}\left[\sum_{i=1}^{\infty}(1-p)^i\right]\right\} - \frac{1}{p^2} = p\frac{d}{dp}\left\{(1-p)\frac{d}{dp}\left[\frac{1}{p}\right]\right\} - \frac{1}{p^2}$$

$$= p\frac{d}{dp}\left[-\frac{1-p}{p^2}\right] - \frac{1}{p^2} = -p\frac{p^2-2p}{p^4} - \frac{1}{p^2} = \frac{1-p}{p^2}$$

Example 1.33: The input of the operational amplifier, whose characteristic is given in the Fig 1.21.b, is a random variable X with uniform distribution within the interval [-2,2] (Fig. 1.21. a). Determine the distribution, the mean value and the variance of the random variable Y at the output of the amplifier.

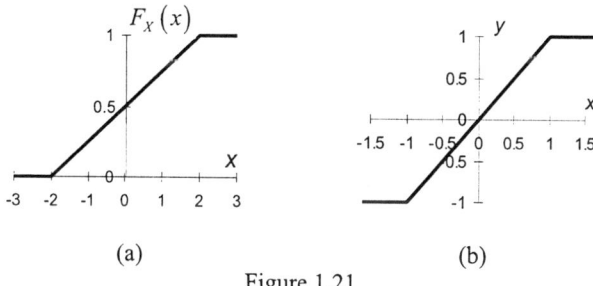

(a)   (b)

Figure 1.21

Solution: The random variable Y is neither the continuous nor the discrete one, because within the interval $(-1,1)$ it's nature is continuous (the distribution function is continuous without break points), while there are some non-zero probabilities for the random variable to take the values -1 and 1, so the distribution function will have the break points at -1 and 1, and that is the property of discrete random variable. This kind of random variable is called mixed type random variable. Let us, firstly, determine the density function:

$$y < -1 \Rightarrow F_Y(y) = P\{Y \le y\} = 0$$

$$F_Y(-1) = P\{Y \le -1\} = P\{Y = -1\} = P\{X \le -1\} = F_X(-1) = \frac{1}{4}$$

$$y \in (-1,1) \Rightarrow F_Y(y) = P\{Y \le y\} = P\{X \le y\} = F_X(y) = \frac{y+2}{4}$$

$$y \ge 1 \Rightarrow F_Y(y) = P\{Y \le y\} = P\{X \le 2\} = F_X(2) = 1$$

This function is shown in the Fig. 1.22. The probability density function can be calculated as the first derivative of the distribution function everywhere except at the break points. At the break points the probability density function can be calculated as for a discrete random variable:

$$f_Y(-1) = \left[F_Y(-1^+) - F_Y(-1^-)\right]\delta(y+1) = 0.25\delta(y+1)$$

$$f_Y(1) = \left[F_Y(1^+) - F_Y(1^-)\right]\delta(y-1) = 0.25\delta(y-1)$$

where $\delta(\cdot)$ is the unit delta impulse. For other values of y, we have $f_Y(y) = dF_Y(y)/dy$. Thus, the probability density function can be written in the form:

$$f_Y(y) = = \frac{1}{4}\delta(y+1) + \frac{1}{4}\delta(y-1) + \frac{1}{4}\left[h(y+1) - h(y-1)\right]$$

where $h(\cdot)$ denotes the unit step function. This function is shown in the Fig. 1.22.

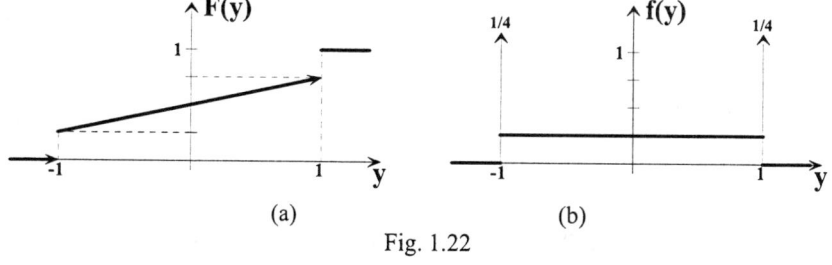

(a)   (b)

Fig. 1.22

The mean value of the mixed type random variable is a combination of the formulas for discrete and continuous type random variables, i.e.:

$$m_Y = (-1)\frac{1}{4} + 1 \cdot \frac{1}{4} + \int_{-1}^{1} y f_Y(y) dy = \int_{-1}^{1} \frac{1}{4} y \, dy = \left. \frac{y^2}{8} \right|_{-1}^{1} = 0$$

and in a similar way, the corresponding variance is:

$$\sigma_Y^2 = E\left\{ (Y - m_Y)^2 \right\} = \frac{1}{4}(-1-0)^2 + \frac{1}{4}(1-0)^2 + \int_{-1}^{1} y^2 \frac{1}{4} dy = \frac{1}{2} + \left. \frac{y^3}{12} \right|_{-1}^{1} = \frac{2}{3}$$

Example 1.34: Z is two-dimensional random vector whose components are the random variables X and Y. The joint distribution function of these random variables is:

$$F_{X,Y}(x,y) = \begin{cases} 1 \; ; \; 1 \le x \; ; \; 1 \le y \\ x \; ; \; 0 \le x < 1 \; ; \; 1 \le y \\ y \; ; \; 1 \le x \; ; \; 0 \le y < 1 \\ xy \; ; \; 0 \le x < 1 \; ; \; 0 \le y < 1 \\ 0 \; ; \; \text{elsewhere} \end{cases}$$

Calculate the marginal distribution functions $F_X(x)$ and $F_Y(y)$, as well as the joint and marginal probability density functions.

Solution: The marginal distribution function of the random variable X can be calculated by applying the next equations:

$$F_X(x) = P\{X \le x\} = P\{X \le x, Y \le \infty\} = F_{X,Y}(x,\infty),$$

so, in our case, this function takes the form:

$$F_X(x) = \begin{cases} 0 \; ; \; x < 0 \\ x \; ; \; 0 \le x < 1 \\ 1 \; ; \; 1 \le x \end{cases}$$

Similarly,

$$F_Y(y) = P\{Y \le y\} = P\{X \le \infty, Y \le y\} = F_{X,Y}(\infty,y)$$

yielding

$$F_Y(y) = \begin{cases} 0 \; ; \; y < 0 \\ y \; ; \; 0 \le y < 1 \\ 1 \; ; \; 1 \le y \end{cases}$$

Obviously, the random variables are uniformly distributed within the interval [0,1]. The joint probability density function can be obtained by differentiating twice the joint distribution function:

$$f_{X,Y}(x,y) = \frac{\partial^2 F_{X,Y}(x,y)}{\partial x \partial y} = \begin{cases} 1 \; ; \; 0 \le x < 1 \; ; \; 0 \le y < 1 \\ 0 \; ; \; \text{elsewhere} \end{cases}$$

The marginal probability density functions can be derived by integrating the joint probability density function:

$$f_X(x) = \frac{dF_X(x)}{dx} = \frac{dF_{X,Y}(x,\infty)}{dx} = \int_{-\infty}^{\infty} f_{X,Y}(x,y)dy = \begin{cases} 1 & ; \ x \in [0,1) \\ 0 & ; \ \text{elsewhere} \end{cases}$$

and, similarly for the random variable Y:

$$f_Y(y) = \frac{dF_Y(y)}{dy} = \frac{dF_{X,Y}(\infty,y)}{dy} = \int_{-\infty}^{\infty} f_{X,Y}(x,y)dx = \begin{cases} 1 & ; \ y \in [0,1) \\ 0 & ; \ \text{elsewhere} \end{cases}$$

Example 1.35: The product of the independent random variables Z and Y is a random variable X. The probability density function of the random variable Y is

$$f_Y(y) = \frac{1}{\sqrt{2\pi}} \exp\left(-\frac{1}{2}y^2\right)$$

and the discrete random variable Z is defined by the set of possible values $\{-1,1\}$ with the corresponding probabilities $(0.5, 0.5)$, i.e. $P\{Z = -1\} = P\{Z = 1\} = 0.5$. This can be represented as

$$Z : \begin{pmatrix} -1 & 1 \\ 0.5 & 0.5 \end{pmatrix}$$

Determine the next functions $F_{X/Y}(x/y)$, $F_X(x)$, $f_X(x)$, $f_{X/Y}(x/y)$, $f_{X,Y}(x,y)$ $F_{X/Y}(x/y)$, $F_X(x)$, $f_X(x)$, $f_{X/Y}(x/y)$, $f_{X,Y}(x,y)$.

Solution: The random variable $X=YZ$, assuming the value y of the random variable Y, is a discrete random variable, because it can take only two different values, y and $-y$ with the same probability. So, one can write:

$$X = YZ|_{Y=y} : \begin{pmatrix} -y & y \\ 0.5 & 0.5 \end{pmatrix} \Rightarrow F_{X/Y}(x/y) = \begin{cases} 0 & ; \ x < -|y| \\ 0.5 & ; \ x \in [-y, y) \\ 1 & ; \ |y| \le x \end{cases}$$

The distribution function of the random variable X can be calculated using the next equality:

$$F_X(x) = P(X \le x) = P\{ZY \le x\} = P\big((Y \le x, Z = 1) + (Y \ge -x, Z = -1)\big)$$

As the events on the right hand side of the last equation are mutually exclusive, and the random variables Y and Z are independent, we can proceed as:

$$F_X(x) = P(Y \le x)P(Z = 1) + P(Y \ge -x)P(Z = -1)$$

$$= P\{Y \le x\}P\{Z=1\} + \big[1 - P\{Y \le -x\}\big]P\{Z=-1\} = F_Y(x)\frac{1}{2} + \big(1 - F_Y(-x)\big)\frac{1}{2}$$

$$= \frac{1}{2}\left(1 + \int_{-\infty}^{x} \frac{1}{\sqrt{2\pi}} \exp\left(-\frac{1}{2}\lambda^2\right)d\lambda - \int_{-\infty}^{-x} \frac{1}{\sqrt{2\pi}} \exp\left(-\frac{1}{2}\lambda^2\right)d\lambda\right) = \frac{1}{2}\left(1 + \int_{-x}^{x} \frac{1}{\sqrt{2\pi}} \exp\left(-\frac{1}{2}\lambda^2\right)d\lambda\right)$$

The probability density function can be obtained by differentiating:

$$f_X(x) = \frac{dF_X(x)}{dx} = \frac{d}{dx}\left[\frac{1}{2}\big(1 + F_Y(x) - F_Y(-x)\big)\right]$$

$$= \frac{1}{2}\big(f_Y(x) + f_Y(-x)\big) = f_Y(x) = \frac{1}{\sqrt{2\pi}} \exp\left(-\frac{1}{2}x^2\right)$$

since $f_Y(x) = f_Y(-x)$. The conditional probability density function can be calculated by differentiating the corresponding distribution function:

$$f_{X/Y}(x/y) = \frac{dF_{X/Y}(x/y)}{dx} = 0.5\delta(x+y) + 0.5\delta(x-y).$$

The joint probability density function of the random variables X and Y can be determined from the conditional probability definition formula:

$$f_{X,Y}(x,y) = f_{X/Y}(x/y)f_Y(y) = \left[0.5\delta(x-y) + 0.5\delta(x+y)\right]\frac{1}{\sqrt{2\pi}}\exp\left(-\frac{1}{2}y^2\right)$$

Example 1.36: Prove the next statements:
   - two independent random variables are uncorrelated, but the opposite statement is not true;
   - for the Gaussian random variables independence is equivalent with the uncorrelation;

Solution: The random variables X and Y are independent if the next equation is satisfied:

$$f_{X,Y}(x,y) = f_X(x)f_Y(y)$$

On the other hand, the random variables X and Y are uncorrelated if the equation

$$E\left\{(X - m_x)(Y - m_y)\right\} = 0$$

is valid.
- Assume that the random variables X and Y are independent. Then, we can write:

$$f_{X,}(x,y) = f_X(x)f_Y(y) \Rightarrow$$

$$E\left\{(X - m_x)(Y - m_y)\right\} = \int_{-\infty}^{\infty}\int_{-\infty}^{\infty}(x - m_x)(y - m_y)f_{X,Y}(x,y)\,dxdy$$

$$= \int_{-\infty}^{\infty}\int_{-\infty}^{\infty}(x - m_x)(y - m_y)f_X(x)f_Y(y)\,dxdy = \int_{-\infty}^{\infty}(x - m_x)f_X(x)\,dx\int_{-\infty}^{\infty}(y - m_y)f_Y(y)\,dy$$

$$= \left[E\{X\} - m_X\right]\left[E\{Y\} - m_Y\right] = 0 \cdot 0 = 0$$

So, it is proven that they are uncorrelated. If we want to prove that the opposite statement is not true, it is enough to design a contra example. Assume some random variable X with uniform distribution within the interval $[-a,a]$, and let the random variable Y be a function of X: $Y = cX^2$. We can now write

$$E\left\{(X - m_X)(Y - m_Y)\right\} = E\left\{X\left(Y - \int_{-a}^{a}cx^2\frac{1}{2a}\,dx\right)\right\}$$

$$= E\left\{X\left(Y - \frac{c}{6a}x^3\Big|_{-a}^{a}\right)\right\} = E\left\{X\left(Y - \frac{ca^2}{3}\right)\right\}$$

$$= E\left\{X\left(cX^2 - \frac{ca^2}{3}\right)\right\} = \int_{-a}^{a}\left(cx^3 - \frac{ca^2}{3}x\right)\frac{1}{2a}\,dx = 0$$

So, they are uncorrelated but dependent, since $Y = cX^2$.

- Assume that the random variables X and Y are Gaussian and uncorrelated. The joint probability density function of X and Y is:

$$f_{X,}(x,y) = \frac{1}{\sqrt{2\pi|\Sigma|}}\exp\left(-\frac{1}{2}[x - m_x]\Sigma^{-1}\left[y - m_y\right]\right)$$

where Σ denotes the covariance matrix:

$$\Sigma = E\left\{\begin{bmatrix} X - m_X \\ Y - m_Y \end{bmatrix}[X - m_X \quad Y - m_Y]\right\}$$

$$= \begin{bmatrix} E\{(X - m_x)^2\} & E\{(X - m_x)(Y - m_Y)\} \\ E\{(X - m_x)(Y - m_Y)\} & E\{(Y - m_Y)^2\} \end{bmatrix} = \begin{bmatrix} \sigma_X^2 & \sigma_{XY} \\ \sigma_{XY} & \sigma_Y^2 \end{bmatrix}$$

Taking into account that the random variables X and Y are uncorrelated ($\sigma_{XY} = 0$) one can proceed as:

$$f_{X,Y}(x,y) = \frac{1}{\sqrt{(2\pi)^2|\Sigma|}}\exp\left(-\frac{1}{2}[x - m_X]\frac{\begin{bmatrix} \sigma_X^2 & 0 \\ 0 & \sigma_Y^2 \end{bmatrix}}{\sigma_X^2\sigma_Y^2}\begin{bmatrix} x - m_x \\ x - m_y \end{bmatrix}\right)$$

$$= \frac{1}{\sqrt{(2\pi)^2\,\sigma_X^2\sigma_Y^2}}\exp\left(-\frac{1}{2}\left(\frac{(x - m_X)^2}{\sigma_X^2} + \frac{(y - m_y)^2}{\sigma_Y^2}\right)\right)$$

$$= \frac{1}{\sqrt{2\pi}\sigma_X}\exp\left(-\frac{1}{2}\frac{(x - m_X)^2}{\sigma_X^2}\right)\frac{1}{\sqrt{2\pi}\sigma_Y}\exp\left(-\frac{1}{2}\frac{(y - m_Y)^2}{\sigma_Y^2}\right)$$

$$= f_X(x)f_Y(y)$$

Thus, the joint probability density function is equal to the product of the corresponding marginal density functions. So, it is proven that the uncorrelated Gaussian random variables are also independent.

Example 1.37: For the random variables X and Y, with the given joint probability density function

$$f_{X,Y}(x,y) = \begin{cases} c(x+y)^2 & ; \ x \in [-1,1] \ \text{and} \ y \in [-1,1] \\ 0 & ; \ \text{elsewhere} \end{cases}$$

investigate if they are independent, uncorrelated and orthogonal.

Solution: Taking into account the normalizing condition for the joint pdf, we can calculate the constant c:

$$\int_{-\infty}^{\infty}\int_{-\infty}^{\infty} f_{X,Y}(x,y)\,dxdy = 1 \Rightarrow c = \frac{3}{8}$$

Based on the joint pdf, it is possible to determine the corresponding marginal pdf's:

$$f_X(x) = \int_{-\infty}^{\infty} f_{X,Y}(x,y)\,dy = \int_{-1}^{1}\frac{3}{8}(x+y)^2\,dy\,(h(x+1) - h(x-1))$$

$$= \frac{3}{8}\left(x^2 y + xy^2 + \frac{y^3}{3}\right)\Bigg|_{-1}^{1}(h(x+1) - h(x-1))$$

$$= \frac{3}{8}\left(2x^2 + \frac{2}{3}\right)\left(h(x+1) - h(x-1)\right)$$

$$= \frac{1}{4}\left(3x^2 + 1\right)\left(h(x+1) - h(x-1)\right)$$

where $h(x)$ is the unit step function. Similarly, we also have:

$$f_Y(y) = \frac{1}{4}\left(3y^2 + 1\right)\left(h(y+1) - h(y-1)\right)$$

According to the inequality

$$f_{X,Y}(x,y) \neq f_X(x) f_Y(y)$$

we can conclude that the random variables are dependent.

If we want to investigate the correlation between the given random variables, it is necessary to calculate the expression

$$\text{cov}\{X,Y\} = E\{(X - m_X)(Y - m_Y)\}$$

and to check whether it is equal to zero. Firstly, let us calculate the mean values of the variables:

$$m_X = \int_{-\infty}^{\infty} x f_X(x)\, dx = \int_{-1}^{1} \frac{1}{4}\left(3x^3 + x\right) dx = 0$$

and, similarly,

$$m_Y = 0$$

So, the covariance will take the value:

$$\text{cov}\{X,Y\} = E\{(X - m_X)(Y - m_Y)\} = \int_{-\infty}^{\infty}\int_{-\infty}^{\infty} (x - m_X)(x - m_Y) f_{X,Y}(x,y)\, dxdy$$

$$= \int_{-1}^{1}\int_{-1}^{1} xy \frac{3}{8}(x+y)^2\, dxdy = \frac{8}{9}$$

Therefore, it can be concluded that the random variables are correlated.

Finally, the orthogonality of the random variables is satisfied if the next equation is satisfied:

$$E\{XY\} = 0$$

In our case, we can write:

$$E\{XY\} = \int_{-\infty}^{\infty}\int_{-\infty}^{\infty} xy f_{X,Y}(x,y)\, dxdy$$

$$= \frac{3}{8} \int_{-1}^{1}\int_{-1}^{1} xy(x+y)^2\, dxdy = \frac{8}{9}$$

so, the variables X and Y are not orthogonal.

Example 1.38: The random variable X is with uniform distribution within the interval [-1,2]. Calculate the characteristic function $\Phi_X(\omega)$ and, based on it, the first two uncentralized moments m_1 and m_2.

Solution: The characteristic function of the random variable is defined as

$$\Phi_X(\omega) = E\{\exp(j\omega X)\} = \int_{-\infty}^{\infty} \exp(j\omega x) f_X(x)\, dx$$

so, in our case, we can calculate:

$$\Phi_X(\omega) = \int_{-1}^{2} \frac{1}{3} \exp(j\omega x)\, dx$$

$$= \frac{1}{3} \frac{\exp(j\omega x)}{j\omega}\Big|_{-1}^{3} = \frac{1}{3j\omega}\left[\exp(2j\omega) - \exp(-j\omega)\right]$$

The relation between the characteristic function and uncentralized moments of a random variable can be obtained on the following way:

$$\Phi_X(\omega) = E\{\exp(j\omega X)\}$$

$$= E\left\{1 + j\omega X + \frac{(j\omega X)^2}{2!} + \frac{(j\omega X)^3}{3!} + \ldots\right\}$$

$$= E\left\{1 + j\omega X - \frac{(\omega X)^2}{2!} - j\frac{(\omega X)^3}{3!} + \ldots\right\}$$

$$= E\{1\} + E\{j\omega X\} - E\left\{\frac{(\omega X)^2}{2!}\right\} - E\left\{j\frac{(\omega X)^3}{3!}\right\} + \ldots$$

$$= 1 + j\omega m_1 - \frac{\omega^2}{2!} m_2 - j\frac{\omega^3}{3!} m_3 + \frac{\omega^4}{4!} m_4 + \ldots$$

where $m_i = E\{X^i\}$, $i = 1, 2, \ldots$ Based on the last equation, the moments can be calculated by differentiating the characteristic function:

$$\frac{d^n \Phi_X(0)}{d\omega^n} = j^n m_n \Rightarrow m_n = j^{-n} \frac{d^n \Phi_X(0)}{d\omega^n}$$

This expression is known as the moment theorem. In our case, the first two moments are equal to:

$$m_1 = E\{X\}$$

$$= \frac{1}{j} \frac{d}{d\omega}\left[\frac{1}{3j\omega}(\exp(2j\omega) - \exp(-j\omega))\right]_{\omega=0} = \frac{1}{2}$$

$$m_2 = E\{X^2\}$$

$$= -\frac{d^2}{d\omega^2}\left[\frac{1}{3j\omega}(\exp(2j\omega) - \exp(-j\omega))\right]_{\omega=0} = \frac{25}{27}$$

Example 1.39: The random variables X and Y are independent with uniform distributions: $X \sim U[0,1]$, $Y \sim [2,4]$;. a) Determine the probability density function of the random variable $Z = X + Y$; b) Calculate $var\{Z\}$, $var\{Z/X\}$ and $var\{Y/Z\}$.

Solution a) This task can be performed using the convolution theorem:

$$f_Z(z) = \int_{-\infty}^{\infty} f_X(x) f_Y(z-x)\, dx$$

$$= \int_0^1 f_Y(z-x)\,dx = -\int_z^{z-1} f_Y(t)\,dt = \int_{z-1}^z f_Y(t)\,dt$$

The value of the last integral is dependent on the argument z:

$$z < 2 \Rightarrow f_Z(z) = 0$$

$$z \in [2,3) \Rightarrow f_Z(z) = \int_2^z \frac{1}{2}\,dt = \frac{z}{2} - 1$$

$$z \in [3,4) \Rightarrow f_Z(z) = \int_{z-1}^z \frac{1}{2}\,dt = \frac{1}{2}$$

$$z \in [4,5) \Rightarrow f_Z(z) = \int_{z-1}^4 \frac{1}{2}\,dt = \frac{5-z}{2}$$

$$z \geq 5 \Rightarrow f_Z(z) = 0$$

The shape of the function $f_Z(z)$ is shown in the Fig. 1.23.

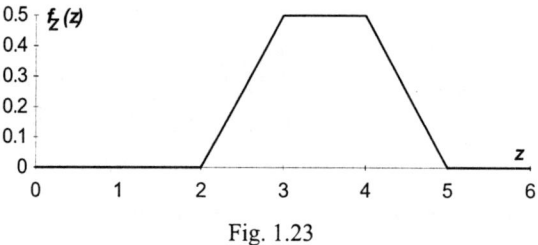

Fig. 1.23

b) As the random variables X and Y are independent, we can write:

$$\text{var}\{Z\} = \text{var}\{X+Y\} = \text{var}\{X\} + \text{var}\{Y\} = \frac{1}{12} + \frac{1}{3} = \frac{5}{12}$$

$$\text{var}\{Z/X\} = \text{var}\{X+Y/X\} = \text{var}\{Y/X\} = \text{var}\{Y\} = \frac{1}{3}$$

$$\text{var}\{Y/Z\} = \text{var}\{Z-X/Z\} = \text{var}\{X\} = \frac{1}{12}$$

Example 1.40: For the random variable X with uniform distribution within the interval $[-1,1]$, calculate the probability density function of the random variables obtained as the following:
a) $Y = 1 - X^3$; b) $Y = aX^2$; $a > 0$; c) $Y = arctg(X)$; d) $Y = \sin(X)$.

Solution: a) As the function $Y = g(X) = 1 - X^3$ is monotonous inside the interval $[-1,1]$, the probability density function of the random variable Y can be calculated using by the next formula:

$$f_Y(y) = \frac{f_X(x)}{|g'(x)|}\bigg|_{x=g^{-1}(y)} = \frac{1}{2}\frac{h(x+1)-h(x-1)}{3x^2}\bigg|_{x=g^{-1}(y)} = \frac{h(y)-h(y-2)}{6(1-y)^{2/3}} = \begin{cases} \dfrac{1}{6(1-y)^{2/3}} & ;\ y \in [0,2] \\[2mm] 0 & ;\ \text{elsewhere} \end{cases}$$

where $h(\cdot)$ is the unit step function.

b) The function $Y = g(X) = aX^2$ is not monotonous inside the interval $[-1,1]$, so the corresponding pdf $f_Y(y)$ can be calculated as the first derivative of the distribution function $F_Y(y)$. It should be noted that in this case we cannot use the procedure proposed in the previous example (1.41.a) (except if we divide the mentioned region in two monotonous regions as it is shown in example 1.11). However, the function $F_Y(y)$ has to be calculated using the definition expression:

$$F_Y(y) = P\{Y \le y\} = P\{aX^2 \le y\} = \begin{cases} 0 \; ; \; y < 0 \\ P\left\{X \in \left[-\sqrt{\dfrac{y}{a}}, \sqrt{\dfrac{y}{a}}\right]\right\}; \; y \in [0,a] \\ 1 \; ; \; y > a \end{cases}$$

or, equivalently

$$F_Y(y) = \begin{cases} 0 \; ; \; y < 0 \\ F_X\left(\sqrt{\dfrac{y}{a}}\right) - F_X\left(-\sqrt{\dfrac{y}{a}}\right) \; ; \; y \in [0,a] \\ 1 \; ; \; y > a \end{cases}$$

The first derivative of the distribution function represents the probability density function; i.e.

$$f(y) = \frac{dF_Y(y)}{dy} = \begin{cases} 0 \; ; \; y \notin [0,a] \\ f_X\left(\sqrt{y/a}\right)\dfrac{1}{2\sqrt{ay}} + f_X\left(-\sqrt{y/a}\right)\dfrac{1}{2\sqrt{ay}}; \; y \in [0,a] \end{cases}$$
$$= 1/2\sqrt{ay}\,(h(y) - h(y-a))$$

c) The function 'arctg' is a monotonous function, so that we can use the procedure from the example 1.41.a:

$$f_Y(y) = \frac{f_X(x)}{|g'(x)|}\bigg|_{x=g^{-1}(y)} = \frac{0.5[h(x+1) - h(x-1)]}{\dfrac{1}{1+x^2}}\bigg|_{x=\tan(y)} = \frac{1+tg^2(y)}{2}[h(y+\pi/4) - h(y-\pi/4)]$$

d) Using the procedure from the example 1.41.a or 1.41.c, we have

$$f_Y(y) = \frac{0.5[h(x+1) - h(x-1)]}{\cos(x)}\bigg|_{x=\arcsin(y)} = \frac{0.5[h(x+1) - h(x-1)]}{\sqrt{1 - \sin^2(x)}}\bigg|_{x=\arcsin(y)} = \frac{[h(y+\sin(1)) - h(y-\sin(1))]}{2\sqrt{1-y^2}}$$

Example 1.41: The random variable X is uniformly distributed within the interval $[0,1]$. Determine the function $Y = g(X)$, so that the corresponding pdf of the random variable Y has the form: $f_Y(y) = \lambda \exp(-\lambda y)h(y)$.

Solution: Assuming that the function $g(\cdot)$ is monotonous, we can write:

$$F_Y(y) = P\{Y \le y\} = P\{g(X) \le y\} = P\{X \le g^{-1}(y)\} = F_X(g^{-1}(y)) = \begin{cases} 0 \; ; \; g^{-1}(y) < 0 \\ g^{-1}(y) \; ; \; g^{-1}(y) \in [0,1] \\ 1 \; ; \; g^{-1}(y) > 1 \end{cases}$$

On the other hand, the density function of Y can be calculated as:

$$F_Y(y) = \int_{-\infty}^{y} f_Y(t)\,dt = \left(1 - \exp(-\lambda\,y)\right)h(y)$$

By equating the last two expressions, one obtains:

$$g^{-1}(y) = x = 1 - \exp(-\lambda\,y) \Rightarrow y = -\frac{1}{\lambda}\ln(1 - x)$$

or, equivalently

$$y = g(x) = -\frac{1}{\lambda}\ln(1 - x)$$

The obtained function is monotonous, so the adopted assumption is valid. The proposed procedure serves to determine a random variable Y with a given distribution $F_Y(y)$ by passing uniformly distributed random variable $X \sim U[0,1]$ through the nonlinearity $g(\cdot)$, satisfying the assumed monotonous condition (Fig. 1.24). The block diagram of the computer program for generating the random sequence $\{y(k)\}$ with the given exponential distribution function, is depicted in Fig. 1.25 MATLAB program for realizing the given block scheme 1.25 is given in Fig. 1.26, while Fig. 1.27 shows a realization of the sequence $\{y(k)\}$.

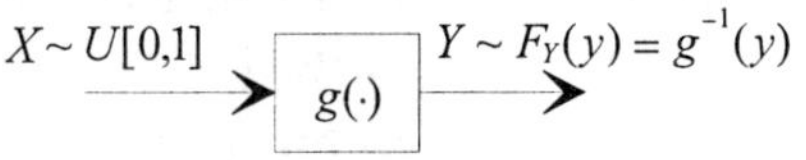

$$X \sim U[0,1] \quad \boxed{g(\cdot)} \quad Y \sim F_Y(y) = g^{-1}(y)$$

Fig. 1.24 Block scheme for generating a random variable with desired pdf

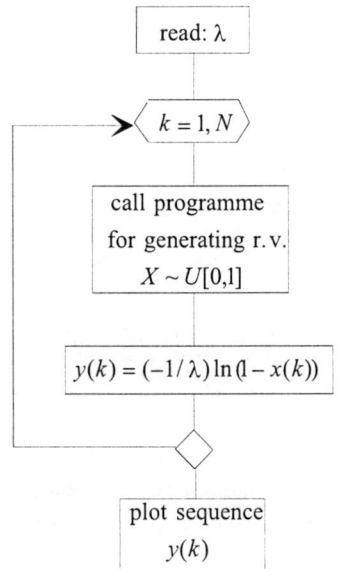

Fig. 1.25 Flow chart of the algorithm for generating the exponentially distributed random sequence

```
function example
lambda=input(' type the coefficient lambda:  ');
N=100;
for k=1:N
   x(k)=rand;
   y(k)=(-1/lambda)*log(1-x(k));
end
plot(y);title('the sequence {y(k)}');
end
```

Fig. 1.26 MATLAB code for the flow chart on Fig 1.25

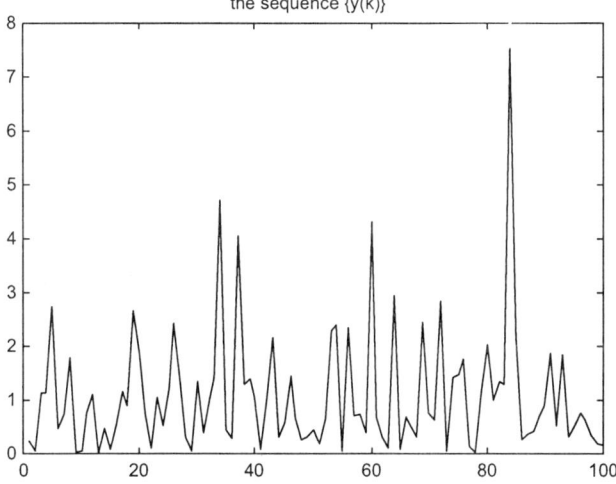

Fig. 1.27 The sequence obtained by the code given in Fig. 126

Example 1.42: The joint probability density function of the random variables X and Y is:
$f_{X,Y}(x,y) = \dfrac{1}{2\pi\sigma^2}\exp\left(-\dfrac{x^2+y^2}{2\sigma^2}\right)$. Calculate the joint pdf of the random variables P and Q, where

$P = \sqrt{X^2 + Y^2}\,;\ Q = X/Y$.

Solution: For the functions of two random variables, the joint probability density function can be calculated using the next formula:

$$T\left(\begin{bmatrix} X \\ Y \end{bmatrix}\right) = \begin{bmatrix} P \\ Q \end{bmatrix} \Rightarrow f_{P,Q}(p,q) = \frac{f_{X,Y}(x,y)}{|J(x,y)|}\Bigg|_{\begin{bmatrix} x \\ y \end{bmatrix} = T^{-1}\left(\begin{bmatrix} p \\ q \end{bmatrix}\right)}$$

where J denotes the Jacobian:

$$J = \begin{vmatrix} \partial p/\partial x & \partial p/\partial y \\ \partial q/\partial x & \partial q/\partial y \end{vmatrix}$$

In this case, the Jacobian takes the form:

$$J(x,y) = \begin{vmatrix} \dfrac{x}{\sqrt{x^2+y^2}} & \dfrac{y}{\sqrt{x^2+y^2}} \\ \dfrac{1}{y} & -\dfrac{x}{y^2} \end{vmatrix} = -\dfrac{\sqrt{x^2+y^2}}{y^2}.$$

If we want to represent the joint pdf $f_{P,Q}(p,q)$ as the function of p and q, it is necessary to substitute the expressions

$$p^2 = x^2 + y^2 \ ; \ y^2 = p^2(1+q^2)$$

into the Jacobian, yielding:

$$f_{P,Q}(p,q) = \frac{p^2}{p(1+q^2)} \frac{1}{2\pi\sigma^2} \exp\left(-\frac{p^2}{2\sigma^2}\right) h(p)$$

where $h(\cdot)$ is the unit step function. In the expression for $f_{P,Q}(p,q)$, the step function $h(p)$ appeared, because P can take only nonnegative values, owing to the definition $P = \sqrt{X^2 + Y^2}$.

Example 1.43: The random variables X and Y are independent, with corresponding pdf's $f_X(x) = \exp(-x)h(x)$ and $f_Y(y) = \exp(-y)h(y)$. Investigate the dependence of the random variables P and Q, where $P = X + Y$; $Q = X/(X+Y)$.

Solution: Similarly, as it is done in the previous example, we can calculate the joint probability density functions of the random variables P and Q as:

$$f_{X,Y}(x,y) = f_X(x)f_Y(y)$$

and

$$f_{P,Q}(p,q) = \frac{f_{X,Y}(x,y)}{|J(x,y)|}\bigg|_{\left[\begin{smallmatrix} x \\ y \end{smallmatrix}\right] = T^{-1}\left(\left[\begin{smallmatrix} p \\ q \end{smallmatrix}\right]\right)}$$

$$= \frac{\exp(-x-y)h(x)h(y)}{|x+y|^{-1}}\bigg|_{\left[\begin{smallmatrix} x \\ y \end{smallmatrix}\right] = \left[\begin{smallmatrix} pq \\ p-pq \end{smallmatrix}\right]} = p\exp(-pq)h(p)\big[h(q)-h(q-1)\big]$$

The step functions in the last expressions are introduced since the random variable P has to be nonnegative and the random variable Q has to take a value within the interval $[0,1]$. Based on the calculated joint pdf, it is possible to calculate the corresponding marginal probability density functions $f_P(p)$ and $f_Q(q)$ as:

$$f_P(p) = \int_{-\infty}^{\infty} f_{P,Q}(p,q)dq = p\exp(-p)h(p)$$

$$f_Q(q) = \int_{-\infty}^{\infty} f_{P,Q}(p,q)dp = h(q)-h(q-1)$$

As the equation

$$f_{P,Q}(p,q) = f_P(p)f_Q(q)$$

is valid, we conclude that the random variables P and Q are independent.

CHAPTER 2

FUNDAMENTALS OF STOCHASTIC PROCESSES

The concept of a stochastic process is complex. A simple presentation of the fundamental ideas is given here. Our prime purposes are to define notation and to motivate the reader to review a detailed text. However, note at this stage that noise models in terms of stochastic processes can be used in various ways. One possibility is to let the noise models describe random disturbances that are assumed to be present. Sensor noise and the effects of unmodelled sources in dynamic systems are typical interpretations of such descriptions. Another possibility is to use the noise model as a mean of expressing model uncertainties. If the model description relating input and output variables of a dynamic system is uncertain, such uncertainties may be incorporated in a noise model. Some ideas along this line will be presented in the next chapter. A third possibility is to regard the noise as a tuning variable when constructing a filter. By changing the noise model parameters, a user can change the frequency properties of an associated optimal filter. The idea will be illustrated in Sections VII and VIII. Finally, a noise model can be used as means of achieving certain good properties of a feedback system designed by stochastic control theory. The last idea will not be discussed in this book, because the book is devoted to modeling and analysis of stochastic signals and systems, but not to the control of a such system.

2.1. Basic concepts

Recall that a random variable X is defined as a function mapping the outcomes ω of an experiment σ to the real numbers, that is $X(\omega)$. If the random variable is also a function of time as well, then it is called a stochastic process [2, 17, 36]. More formally, a stochastic process is a function of two variables t and ω:

$$X(t,\omega) \; ; \; t \in T \; ; \; \omega \in \Omega$$

where T is a set of index parameters (continuous or discrete) and Ω is the probability (sample) space. If T is the set of discrete-time instants $T = \{t_k \; ; \; k = 0,1,...\}$ we have a discrete-time stochastic process. In the other hand, if T involves intervals on the time axis as $T = \{t : t \geq t_0\}$ or $T = \{t : 0 \leq t \leq T\}$ the process is called a continuous-time stochastic process. There are four possibilities of $X(\cdot,\cdot)$ above [23, 48]:

1) for any t and ω, $X(t,\omega)$ is a family of random time functions named realizations of the process;

2) for t varying and ω fixed, $\omega = \omega_i$; $X(t,\omega_i)$ is a time function or a realization (sample) of the process;

3) for ω varying and t fixed, $t = t_i$, $X(t_i,\omega)$ is a random variable;

4) for t and ω fixed, that is $t = t_i$ and $\omega = \omega_i$, $X(t_i,\omega_i)$ is a number.

Before continuing, let us consider the following example of a well known process named a random walk.

Example 2.1 (random walk): Toss a coin at each time instant t, that is, $t \in T = \{0, 1, ..., N-1\}$ and $\Omega = \{\text{head}, \text{non}-\text{head}\} = \{h, nh\}$. The random variable is

$$X(t, \omega) = \begin{cases} k & \text{for } \omega = h \\ -k & \text{for } \omega = nh \end{cases} \; ; \; t = 0, 1, ..., N-1$$

where k is a positive constant. For a fair coin, we have

$$P\{X(t, \omega) = k\} = \frac{1}{2} \; ; \; P\{X(t, \omega) = -k\} = \frac{1}{2}$$

Let $Y(t, \omega_i)$ be the position score at time instant t for the i-th realization ω_i. Then

$$Y(N, \omega_i) = \sum_{j=0}^{N-1} X(t_j, \omega_i) \; ; \; i = 1, 2, ...$$

where we assume $Y(0, \cdot) = 0$. For each t, Y is a sum of discrete random variables $X(\cdot, \omega)$ and, for each i, $X(t, \omega_i)$ is the i-th realization of the process. Let, for example, $N=6$; $\omega_1 = \{h\ h\ nh\ h\ nh\ nh\}$ and $\omega_2 = \{nh\ h\ h\ nh\ nh\ nh\}$. Then

$$Y(5, \omega_1) = \sum_{t=0}^{5} X(t, \omega_1) = X(0, \omega_1) + \cdots + X(5, \omega_1) = k + k - k + k - k - k = 0$$

$$Y(5, \omega_2) = \sum_{t=0}^{5} X(t, \omega_2) = X(0, \omega_2) + \cdots X(5, \omega_2) = -k + k + k - k - k - k = -2k$$

Note that $Y(t, \omega)$, or more simply $Y(t)$, can be generated by the difference equation

$$Y(t+1) = Y(t) + X(t) \; ; \; t = 0, 1, ..., N-1 \; ; \; Y(0) = 0.$$

which represents a mathematical model of the process.

Some important concepts for random processes follow.

The values of a random process at m-distinct time instants define the m-dimensional random vector. The function

$$F_X(a_1, ..., a_m; t_1, ..., t_m) = P\{X(\omega, t_1) \le a_1, ..., X(\omega, t_m) \le a_m\} \tag{2.1}$$

where $P(\cdot)$ denotes probability, is called the finite dimensional *(joint) distribution function* of the random process. In equation (2.1) it is sometimes convenient to include the notation

$$F_X(a_1, ..., a_m; t_1, ..., t_m) = F_X(a_1(t_1), ..., a_m(t_m)).$$

An illustration is given in the figure bellow.

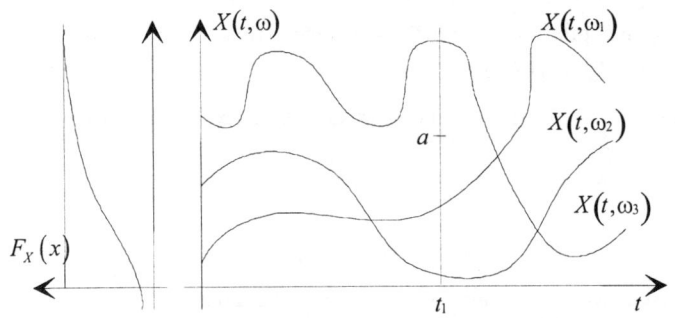

Figure 2.1: A stochastic process and its one-dimensional distribution function
$$F_X(a, t_1) = P\{X(\omega, t_1) \le a\}$$

Equivalent description could also be given in terms of the joint density

$$f_X(a_1,...,a_m) = \frac{\partial^n F(a_1,...,a_m;t_1,...,t_m)}{\partial a_1 \cdots \partial a_m} \tag{2.2}$$

Of course, the above expressions are valid for vector processes $X(t,\omega)$ as well, where $a_i, i = 1,...,m$ are n-dimensional vectors. Then the joint density function is described by

$$f_X(a_1,...,a_m) = \frac{\partial^{mn} F_X(a_1,...,a_m;t_1,...,t_m)}{\partial a_1^1 \cdots \partial a_1^n \cdots \partial a_m^1 \cdots \partial a_m^n} \tag{2.3}$$

where a_i^j ; $j = 1,...,n$ and $i = 1,...,m$ is the j-th component of the n-vector a_i. Thus, the process is completely described if we specify the joint distribution, or density function, of the m random variables, or vectors, $X(\omega,t_1),...,X(\omega,t_m)$ for all integers m and time points $t_1,...,t_m$ in the index set T.

The approach given above, and defined by either joint distribution or density function, is the most natural way of describing the probability law that governs a stochastic process. In practice, however, it has been found more expedient to describe certain stochastic processes in terms of their *first* and *second order* moments, which are called the *mean value* and *covariance function*, respectively. The standard definitions of expectation in Chapter 1 still approach to stochastic processes.

The *mean-value* function of a random process X is defined by

$$m_X(t) = \overline{X}(t) = E\{X(t)\} = \int_{-\infty}^{\infty} a f_X(a,t) da \tag{2.4}$$

The mean value function is an ordinary function of time t. High order moments are defined similarly, as in Chapter 1.

The *covariance function* of a n-vector process X is defined by

$$R_X(t,s) = \text{cov}\{X(t),X(s)\} = E\{[X(t)-\overline{X}(t)][X(s)-\overline{X}(s)]^T\}$$
$$= \int_{-\infty}^{\infty} \cdots \int_{-\infty}^{\infty} [\xi_1 - \overline{X}(t)][\xi_2 - \overline{X}(s)]^T f_X(\xi_1,\xi_2,t,s) d\xi_1 d\xi_2 \tag{2.5}$$

where $\xi_1^T = [\xi_1^1,...,\xi_1^n]$, $\xi_2^T = [\xi_2^1,...,\xi_2^n]$ and $d\xi_i = d\xi_i^1 \cdots d\xi_i^n$.

Similarly, the *mean square* or *correlation function* of a process is defined by

$$\Psi_X(t,s) = E\{X(t)X^T(s)\} = \int_{-\infty}^{\infty} \cdots \int_{-\infty}^{\infty} \xi_1\xi_2 f_X(\xi_1,\xi_2,t,s) d\xi_1 d\xi_2 \tag{2.6}$$

where

$$R_X(t,s) = E\{[X(t)-\overline{X}(t)][X(s)-\overline{X}(s)]^T\} = \Psi_X(t,s) - \overline{X}(t)\overline{X}^T(s) \tag{2.7}$$

We observe that for $t \neq s$ the correlation and covariance functions (2.5) and (2.6) , respectively, give a measure of "time-correlation" for the stochastic process, as distinguished from the "space-correlation" that results when $t = s$. We now say that the stochastic process $X(t)$ is *uncorrelated* if

$$\Psi_X(t,s) = E\{X(t)\}E\{X^T(s)\} = \overline{X}(t)\overline{X}^T(s) \text{ or } R_X(t,s) = 0 \tag{2.8}$$

Furthermore, we say that a stochastic process $X(t)$ is *independent* if, for any time points $t_1,...,t_m$ in T and any integer m, the joint distribution function

$$F_X(a_1,...,a_m;t_1,...,t_m) = \prod_{i=1}^{n} F_X(a_i,t_i) \tag{2.9}$$

where $F_X(a_i,t_i) = P\{X(t_i) \le a_i\}$ is the marginal distribution of the random variable $X(t_i)$. If a stochastic process is independent, then for $t,s \in T, t \ne s$, $f_X(\xi_1,\xi_2;t,s) = f_X(\xi_1,t) f_X(\xi_2,s)$ and it follows

$$\Psi_X(t,s) = \int_{-\infty}^{\infty} \cdots \int_{-\infty}^{\infty} \xi_1 \xi_2^T f_X(\xi_1,t) f_X(\xi_2,s) d\xi_1 d\xi_2$$
$$= \int_{-\infty}^{\infty} \cdots \int_{-\infty}^{\infty} \xi_1 f_X(\xi_1,t) d\xi_1 \int_{-\infty}^{\infty} \cdots \int_{-\infty}^{\infty} \xi_2 f_X(\xi_2,s) d\xi_2 = \bar{X}(t)\bar{X}^T(s) \tag{2.10}$$

Hence, an independent stochastic process is also an uncorrelated one, but an uncorrelated stochastic process has not to be and independent one. If, in addition, the process has zero-mean value, then

$$R_X(t,s) = \Psi_X(t,s) = 0 \quad \text{for all} \quad t,s \in T \; ; t \ne s$$

A process with the property $\Psi_X(t,s) = 0$ is known as an *orthogonal process*.

The *cross-covariance function* of n-vector stochastic processes $X(t)$ and $Y(t)$ is defined by

$$R_{X,Y}(t,s) = \text{cov}\{X(t),Y(s)\} = E\left\{\left[X(t)-\bar{X}(t)\right]\left[Y(s)-\bar{Y}(s)\right]^T\right\}$$
$$= \int_{-\infty}^{\infty} \cdots \int_{-\infty}^{\infty} \left[\xi-\bar{X}(t)\right]\left[\eta-\bar{Y}(s)\right]^T f_{X,Y}(\xi,\eta;t,s) d\xi d\eta \tag{2.11}$$

where $\xi^T = [\xi_1,...,\xi_n], \eta^T = [\eta_1,...,\eta_m]$ and $d\xi = d\xi_1 \cdots d\xi_n \; ; d\eta = d\eta_1 \cdots d\eta_m$.

Example 2.2 (mean value and covariance function): Let $X(t) = a+bt$ where a and b are random variables. Then

$$\bar{X}(t) = m_X(t) = E\{X(t)\} = E\{a\} + E\{b\}t = \bar{a} + \bar{b}t$$

and

$$R_X(t,s) = E\{(a+bt)(a+bs)\} - \bar{X}(t)\bar{X}(s)$$
$$= E\{a^2\} + E\{ab\}(t+s) + E\{b^2\}ts - \bar{X}(t)\bar{X}(s)$$

Suppose that $\bar{a} = \bar{b} = 0$ and $\sigma_a^2 = \sigma_b^2 = \sigma^2$, as well that a and b are uncorrelated, i.e. $E\{ab\} = \bar{a}\bar{b}$. Then $\bar{X}(t) = 0$ for all t, and $R_X(t,s) = \sigma^2(1+ts)$. So, we see that dealing with stochastic processes is similar to dealing with random variables, except that we must account corresponding functions for the time index.

Example 2.3. (independence and uncorrelatedness): Consider the scalar stochastic process $\{X(t), t \in T = \{0,1\}\}$ with $X(1) = \alpha X^2(0)$, where $\alpha > 0$. Obviously, this is not an independent process. Suppose that $X(0)$ is uniformly distributed between -1 and 1, as shown in the Fig. 2.2.

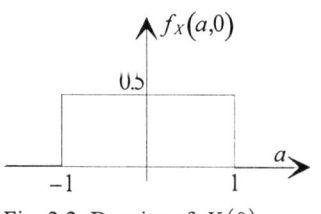

Fig. 2.2: Density of $X(0)$

We have first that

$$\Psi_X(1,0) = E\{X(1)X(0)\} = E\{\alpha X^3(0)\} = \alpha \int_{-1}^{1} \frac{1}{2} \xi^3 d\xi = 0$$

On the other hand, it is clear that $\overline{X}(0) = 0$, owing to symmetry of $f_X(a,0)$, and

$$E\{X(1)\} = E\{\alpha X^2(0)\} = \alpha \int_{-1}^{1} \frac{1}{2} \xi^2 d\xi = \frac{\alpha}{3}$$

from which it follows that $\overline{X}(1)\overline{X}(0) = 0$. Thus, $\Psi_X(1,0) = \overline{X}(1)\overline{X}(0) = 0$, and we have a case of an uncorrelated process which is not independent.

2.2. Gauss-Markov stochastic process

In dealing with stochastic processes, certain assumptions must be made in order to increase their usefulness in modeling. A common class of useful processes is called Markov processes, with loosely means that the probability of reaching future state depends only on the present and not the past one; i.e. the future is independent on the past. More precisely, let t_i ($i = 1,...,n$) and t be elements of the index set T such that $t_1 < t_2 < \cdots < t_n < t$. A stochastic process $\{X(\omega,t), t \in T\}$ is called a Markov process if [23, 36]

$$P\{X(t) \le a | X(t_1) = a_1,..., X(t_n) = a_n\} = P\{X(t) \le a | X(t_n) = a_n\} = F_X(a,t|a_n,t_n) \quad (2.12)$$

where $P\{\cdot | X(t_1) = a_1,..., X(t_n) = a_n\}$ denotes the conditional probability given $X(t_1) = a_1,..., X(t_n) = a_n$. That is, the future state $X(t)$ depends only on the present one, $X(t_n)$, and not the past ones $\{X(t_1),..., X(t_{n-1})\}$. In terms of the conditional probability density function, the above definition means

$$f_X(a,t | X(t_1) = a_1,..., X(t_n) = a_n) = f_X(a,t | X(t_n) = a_n) = f_X(a,t|a_n,t_n) \quad (2.13)$$

We now exhibit a specific way of describing a Markov process. As mentioned before, one way of describing the process is to specify the joint probability density of m variables (or vectors) , $X(t_1),..., X(t_m)$ for all integers m and m-time points $t_1 < t_2 < \cdots < t_m$ in T and all scalars (or vectors) $a_1,..., a_m$. Then, from Bayes rule

$$f_X(X(t_1),..., X(t_m)) = f_X(a_1,..., a_m; t_1,..., t_m)$$
$$= f_X(a_m; t_m | a_{m-1},..., a_1; t_{m-1},..., t_1) f_X(a_{m-1},..., a_1; t_{m-1},..., t_1) \quad (2.14)$$

However, since the process is Markov

$$f_X(a_m, t_m | a_{m-1},..., a_1, t_{m-1},...,^., t_1) = f_X(a_m, t_m | a_{m-1}, t_{m-1}) \quad (2.15)$$

which means that

$$f_X(a_1,..., a_m, t_1,..., t_m) = f_X(a_1,..., a_{m-1}, t_1,..., t_{m-1}) f_X(a_m, t_m | a_{m-1}, t_{m-1}) \quad (2.16)$$

Again, from the Bayes rule and the fact that the process is Markov

$$f_X\left(a_1,...,a_{m-1},t_1,...,t_{m-1}\right) = f_X\left(a_1,...,a_{m-2},t_1,...,t_{m-2}\right)f_X\left(a_{m-1},t_{m-1}\mid a_{m-2},t_{m-2}\right) \qquad (2.17)$$

Substituting this result into the preceding equation, and continuing in this fashion by reapplying the Bayes' rule (the so-called chain rule) and the Markov property of the process, we are let finally to the result that

$$f_X\left(a_1,...,a_m,t_1,...,t_m\right) = f_X\left(a_m,t_m\mid a_{m-1},t_{m-1}\right)f_X\left(a_{m-1},t_{m-1}\mid a_{m-2},t_{m-2}\right)\cdots f_X\left(a_2,t_2\mid a_1,t_1\right)f_X\left(a_1,t_1\right) (2.18)$$

Hence, a Markov process is completely determined by the initial density $f_X\left(a_1;t_1\right)$ and the transient density $f_X\left(a_i,t_i\mid a_j,t_j\right) = f_X\left(X(t_i)\mid X(t_j)\right);t_i>t_j$ and $t_i,t_j\in T$, assuming that all the indicated probability density functions in (2.18) exist. Of course, this does not mean that the conditional densities are the same for different time indices.

Another important process is the *Gaussian or normal process* [36, 48, 59]. A stochastic process $\{X(\omega,t),t\in T\}$ is defined to be Gaussian if for any m time points $t_1,...,t_m$ in T, where m is any integer, the set of m n-random vectors $X(t_1),...,X(t_m)$ ($n=1$ if $X(t)$ is a scalar random variable) is jointly Gaussian distributed. In terms of the joint density function

$$f_X(x) = \frac{1}{\sqrt{(2\pi)^{nm}\det(R_X)}}\exp\left(-\frac{1}{2}(x-\bar{X})^T R_X^{-1}(x-\bar{X})\right) \qquad (2.19)$$

where $X = \{X^T(t_1)...X^T(t_m)\}^T$, while $\bar{X}=E\{X\}$ and $R_X = E\{(X-\bar{X})(X-\bar{X})^T\}$ are the respective mean value and covariance matrix of X. Note that a Gaussian vector is completely characterized by its mean vector and covariance matrix, which reduces to variance in the scalar case $(m=1)$. The normal process is usually symbolized as $X\sim N(\bar{X},R_X)$. Particularly, if $X(t_1),...,X(t_m)$ are independent Gaussian n-random vectors, a stochastic process $\{X(\omega,T),t\in T\}$ is said to be a Gaussian *white process*, or Gaussian *white noise*. A more detailed concept of white noise is given at the end of this chapter.

Probably the most important aspect of Gaussian distributions is that in the limit, as the number of samples gets large, many distributions have the Gaussian one as their limiting distribution. Also, the well-known central limit theorem makes Gaussian distribution important. Recall that this theorem states simply that the sum of a large number of independent random variables tends to be Gaussian. Thus, primarily for these reasons, many processes are modeled as Gaussian.

The following properties of Gaussian random variables or processes hold:

- *Linear transformation:* Linear transformations of Gaussian variables are Gaussian; that is, if X is a random vector with Gaussian distribution and mean value m_X and covariance matrix R_X, and $Y = aX + b$, then Y is Gaussian with the mean value $m_Y = E\{Y\} = am_X + b$ and the covariance matrix $R_Y = E\{(Y-\bar{Y})(Y-\bar{Y})^T\} = aR_Xa^T$.

- *Sums of Gaussian variables:* Sums of independent Gaussian variables yield a Gaussian-distributed variable with mean and covariance (in the scalar case the covariance matrix reduces to the variance of the process) equal to the sums of the respective means and covariances; that is, if X_i are

Gaussian variables with mean $m_X(i)$ and covariance $R_X(i)$ and $Y = \sum_i k_i X_i$, then Y is Gaussian with the mean $m_Y = \sum_i k_i m_X(i)$ and the covariance $R_Y = \sum_i k_i R_X(i) k_i^T$.

- *Uncorrelated Gaussian variables:* Uncorrelated Gaussian variables are independent.

- *Conditional Gaussian variables:* Conditional Gaussian variable are Gaussian distributed; that is, if X and Y are jointly Gaussian, then the conditional density $f_{X/Y}(x|y)$ is Gaussian (normal) with the conditional mean

$$m_{X/Y} = E\{X|Y=y\} = m_X + R_{XY} R_Y^{-1}(y - m_Y)$$

and the conditional covariance

$$R_{X/Y} = E\{(X - m_{X/Y})(X - m_{X/Y})^T | Y = y\} = R_X - R_{XY} R_Y^{-1} R_{YX}$$

where $m_X = E\{X\}, m_Y = E\{Y\}, R_X = \text{cov}\{X\}, R_Y = \text{cov}\{Y\}$ and $R_{XY} = \text{cov}\{X,Y\}$. Also note that Y and $\tilde{X} = X - E\{X|Y\}$ are orthogonal, that is

$$E_{X,Y}(\tilde{X}Y^T) = E_{X,Y}\{(X - E\{X|Y\})Y^T\} = 0$$

where $E_{X,Y}\{\cdot\}$ denotes the expectation with respect to the vector random variables X and Y. The proof of these statements is given in the next example.

It should be noted that these expressions are valid for scalar and vector processes, so that we use R^{-1} notation wherever is possible for generality.

Example 2.4. (Gaussian conditional expectation): To prove the last statements we let X and Y be jointly Gaussian n and m dimensional random vectors, respectively, with the joint density function

$$f_{X,Y}(x,y) = f_Z(z) = \frac{1}{\sqrt{(2\pi)^{n+m}|P_Z|}} \exp\left\{-\frac{1}{2}(z - m_z)^T P_Z^{-1}(z - m_z)\right\}$$

where

$$Z = \begin{bmatrix} X \\ Y \end{bmatrix}; \quad m_z = \begin{bmatrix} E\{X\} \\ E\{Y\} \end{bmatrix} = \begin{bmatrix} m_X \\ m_Y \end{bmatrix}$$

$$P_Z = E\{(Z - m_z)(Z - m_z)^T\} = \begin{bmatrix} E\{(X - m_X)(X - m_X)^T\} & E\{(X - m_X)(Y - m_Y)^T\} \\ E\{(Y - m_Y)(X - m_X)^T\} & E\{(Y - m_Y)(Y - m_Y)^T\} \end{bmatrix}$$

$$= \begin{bmatrix} P_X & P_{XY} \\ P_{YX} & P_Y \end{bmatrix} = \begin{bmatrix} P_X & P_{XY} \\ P_{XY}^T & P_Y \end{bmatrix}$$

and $|\cdot|$ denotes the determinant. When dealing with the density function of two jointly Gaussian random vectors, it is often convenient to have an explicit expression for P_Z^{-1} in terms of P_X, P_{XY} and P_Y. This can be obtained by defining

$$P_Z^{-1} = \begin{bmatrix} A & B \\ B^T & C \end{bmatrix}$$

where the matrices A, B and C are determined so that $P^{-1}P = I$. The results are

$$A = \left(P_X - P_{XY}P_Y^{-1}P_{YX}\right)^{-1} = P_X^{-1} + P_X^{-1}P_{XY}CP_{YX}P_X^{-1}$$

$$B = -AP_{XY}P_Y^{-1} = -P_X^{-1}P_{XY}C$$

$$C = \left(P_Y - P_{YX}P_X^{-1}P_{XY}\right)^{-1} = P_Y^{-1} + P_Y^{-1}P_{YX}AP_{XY}P_Y^{-1}$$

We now utilize these relations to develop an expression for the conditional density $f(X|Y)$. Then, from the Bayes rule

$$f(X/Y) = \frac{f(X,Y)}{f(Y)} = \frac{f(Z)}{f(Y)}$$

where

$$f(Y) = \frac{1}{\sqrt{(2\pi)^m |P_Y|}} \exp\left\{-\frac{1}{2}(Y-m_Y)^T P_Y^{-1}(Y-m_Y)\right\}$$

Substituting the expressions for $f(Z)$ and $f(Y)$, we obtain

$$f(X|Y) = \frac{1}{\sqrt{(2\pi)^n \frac{|P_Z|}{|P_Y|}}} \exp\left\{-\frac{1}{2}\begin{bmatrix} X-m_X \\ Y-m_Y \end{bmatrix}^T \begin{bmatrix} A & B \\ B^T & C-P_Y^{-1} \end{bmatrix} \begin{bmatrix} X-m_X \\ Y-m_Y \end{bmatrix}\right\}$$

Expanding the quadratic form in the exponential, and substituting the expressions for A, B and C, as well rearranging the terms, we have

$$\begin{bmatrix} X-m_X \\ Y-m_Y \end{bmatrix}^T \begin{bmatrix} A & B \\ B^T & C-P_Y^{-1} \end{bmatrix} \begin{bmatrix} X-m_X \\ Y-m_Y \end{bmatrix}$$

$$= (X-m_X)^T A(X-m_X) + 2(X-m_X)^T B(Y-m_Y) + (Y-m_Y)^T (C-P_Y^{-1})(Y-m_Y)$$

$$= (X-m_X)^T A(X-m_X) + 2(X-m_X)^T AP_{XY}P_Y^{-1}(Y-m_Y) + (Y-m_Y)^T P_Y^{-1}P_{YX}AP_{XY}P_Y^{-1}(Y-m_Y)$$

$$= (X-m)^T Q(X-m)$$

where

$$m = m_X + P_{XY}P_Y^{-1}P_{YX}(Y-m_Y)$$

Noting that

$$P_Z = \begin{bmatrix} P_X & P_{XY} \\ P_{YX} & P_Y \end{bmatrix} = \begin{bmatrix} P_X - P_X P_Y^{-1}P_{YX} & P_{XY} \\ 0 & P_Y \end{bmatrix}\begin{bmatrix} I_n & 0 \\ P_Y^{-1}P_{XY} & I_m \end{bmatrix}$$

where I_n is the $n \times n$ identity matrix and I_m is the $m \times m$ identity matrix, we have

$$|P_Z| = |P_X - P_{XY}P_Y^{-1}P_{YX}||P_Y|$$

Hence

$$\frac{|P_Z|}{|P_Y|} = |P_X - P_{XY}P_Y^{-1}P_{YX}| = |Q|$$

The expression for $f(X|Y)$ can now be written as

$$f(X|Y) = \frac{1}{\sqrt{(2\pi)^n |Q|}} \exp\left\{-\frac{1}{2}(X-m)^T Q^{-1}(X-m)\right\}$$

We see that $f(X|Y)$ is a Gaussian with the conditional mean $m = E\{X|Y\}$ and the conditional covariance $Q = E\{(X-m)(X-m)^T |Y\} = \text{cov}\{X|Y\}$, which completes the proof of the first part of the property 4. Moreover, to prove that $\tilde{X} = X - m$ and Y are orthogonal, we form

$$E_{XY}\{(X-m)Y^T\} = E_Y\{E\{(X-m)Y^T |Y\}\} = E_Y\{E\{(X-m)|Y\}Y^T\} = 0$$

since

$$E\{(X-m)|Y\} = E\{X|Y\} - m = m - m = 0$$

Finally, a stochastic process is Gauss-Markov if and only if it is both Gaussian and Markov. It is clear that the Gaussian nature of these processes dictates the amplitude distribution, while the Markov nature governs the process's evolution in time. The Gauss-Markov process is a natural concept to use when extending the notion of system state model to the stochastic case. This idea will be discussed in sections 3 and 4 of the book. Additionally, we note that a Gaussian white process can be viewed as a Gauss-Markov process for which the transition density is $f(X(t)|X(\tau)) = f(X(t))$ for all $t, \tau \in T$ with $t > \tau$, since the amplitudes $X(t)$ and $X(\tau)$ are independent random variables.

Example 2.5. (scalar Gauss-Markov process): Consider the scalar process $\{X(t), t \geq 0\}$ which is defined by the differential equation

$$\frac{dX(t)}{dt} = -\frac{1}{t+1}X(t)$$

where $X(0)$ is a Gaussian random variable with mean value zero and variance $\sigma_0^2 > 0$. By direct integration, the solution is given by

$$X(t) = \frac{X(0)}{t+1} \; ; \; t \geq 0$$

Since $X(0)$ is Gaussian distributed, it follows that, for any m time points $t_1, ..., t_m$ in $T = [0, \infty)$, the joint density $f(X(t_1), ..., X(t_m))$ is the Gaussian one. In addition, for some $t_2 > t_1 \geq 0$, it is easily shown that

$$X(t_2) = \frac{t_1 + 1}{t_2 + 1} X(t_1)$$

As a result, for any ordered set of time points $t_1 < t_2 < \cdots < t_m$

$$f(X(t_m)|X(t_{m-1}), ..., X(t_1)) = f(X(t_m)|X(t_{m-1}))$$

Consequently, the process is Gauss-Markov. We know that

$$f(X(0)) = \frac{1}{\sqrt{2\pi}\sigma_0} \exp\left(-\frac{1}{2}\frac{X^2(0)}{\sigma_0^2}\right)$$

Let us determine the transition density $f\big(X(t)\big|X(\tau)\big)$ for $t>\tau$ and $t,\tau \in T =[0,\infty)$. Since $\{X(t),t\ge 0\}$ is Gaussian, $f\big(X(t)\big)$ and $f\big(X(\tau)\big)$ are Gaussian densities, so that the conditional density

$$f\big(X(t)\big|X(\tau)\big)=\frac{f\big(X(t),X(\tau)\big)}{f\big(X(\tau)\big)}$$

is also Gaussian, and is completely specified in terms of its conditional mean value and variance. Since

$$X(t)=\frac{\tau+1}{t+1}X(\tau)\ ;t>\tau$$

then, obviously, the conditional mean value

$$E\big\{X(t)\big|X(\tau)\big\}=\frac{\tau+1}{t+1}X(\tau)$$

and the conditional variance

$$E\left\{\Big[X(t)-E\big\{X(t)\big|X(\tau)\big\}\Big]^{2}\right\}=0 .$$

This simply means that $f\big(X(t)\big|X(\tau)\big)$ is a unit delta function. The result is not surprising, since a knowledge of $X(\tau)$ permits us to determine $X(t)$ exactly for all $t\ge \tau \ge 0$. Furthermore, since $X(t)$ is a Gaussian random variable, we can characterize it in terms of its mean value and variance. In particular, the mean value

$$\bar{X}(t)=E\big\{X(t)\big\}=E\left\{\frac{X(0)}{t+1}\right\}=\frac{1}{t+1}E\big\{X(0)\big\}=0$$

and the variance

$$\sigma^{2}(t)=E\left\{\big(X(t)-\bar{X}(t)\big)^{2}\right\}=E\left\{\frac{X^{2}(0)}{(t+1)^{2}}\right\}=\frac{\sigma_{0}^{2}}{(t+1)^{2}}$$

from which it follows

$$f(a,t)=\frac{t+1}{\sqrt{2\pi}\sigma_{0}}\exp\left(-\frac{(t+1)^{2}a^{2}}{2\sigma_{0}^{2}}\right)$$

or, in slightly different notation

$$f\big(X(t)\big)=\frac{t+1}{\sqrt{2\pi}\sigma_{0}}\exp\left(-(t+1)^{2}\frac{X^{2}(t)}{2\sigma_{0}^{2}}\right).$$

2.3. Properties of stochastic processes

In this section we discuss properties of stochastic processes that are used frequently in practice. Simply stated, a stochastic process is called *stationary* if the finite-dimensional joint distribution of $X(t_{1}),...,X(t_{m})$ is identical to the distribution of $X(t_{1}+\tau),...,X(t_{m}+\tau)$ for all $\tau,m,t_{1},...,t_{m}$. In terms of density function, a process $\{X(\omega,t), t \in T, \omega \in \Omega\}$ is stationary if the joint density function is time-invariant, that is [23, 59]

$$f\left(X(t_1+\tau),...,X(t_m+\tau)\right) = f\left(X(t_1),...,X(t_m)\right) \tag{2.20}$$

or, in slightly different notation

$$f_X\left(a_1,...,a_m; t_1+\tau,...,t_m+\tau\right) = f_X\left(a_1,...,a_m; t_1,...,t_m\right) \tag{2.21}$$

for all m, $\tau, t_1,...,t_m, a_1,...,a_m$. The process is *stationary of order M* if the preceding equation holds for $m \le M$ only. Special cases are:

1) For $M = 1$, $f\left(X(t)\right) = f\left(X(t+\tau)\right) = f(x)$ for all t, i.e. it does not depend on t. This implies that the mean value $\overline{X}(t) = E\{X(t)\}$ is constant, and such a process is called *mean-stationary*.

2) For $M = 2$, $f_X\left(X(t_1), X(t_2)\right) = f_X\left(a_1, a_2; t_1, t_2\right)$ is equal to $f_X\left(X(t_1+\tau), X(t_2+\tau)\right) = f_X\left(a_1, a_2; t_1+\tau, t_2+\tau\right)$, so that for $\tau = t_2 - t_1$ one obtains $f_X\left(a_1, a_2; t_1, t_2\right) = f_X\left(a_1, a_2; t_1+\tau, t_2+\tau\right) = f_X\left(a_1, a_2; t_2, t_2+\tau\right)$ for all t_2, τ. Since this is also true for $t_2 = 0$, we have that the joint distribution $f\left(X(t_1), X(t_2)\right)$ is only a function of the difference τ. In this way covariance function $R_X\left(t_1, t_2\right) = R_X\left(t_2 - t_1\right) = R_X(\tau)$, and such a process is called *covariance-stationary*.

3) A process that is both mean-stationary and covariance-stationary is called *wide-sense* or *weakly stationary*. Thus, the mean-value function of a weakly stationary process is constant, while its covariance function $R_X\left(t_2, t_1\right)$ is a function of the difference $t_2 - t_1$ of the arguments only.

The *covariance function* of a *weakly stationary process* $X(t)$ has the following properties:

$R_X(\tau) = R_X(-\tau)$;

$\left|R_X(\tau)\right| \le R_X(0)$;

$R_X(0) = \sigma_X^2$, i.e. the value $R_X(0)$ of the covariance function at the origin is the variance of the process; it tells how large the fluctuations of the process are ;

The quadratic form in Z_i, $\displaystyle\sum_{i,j=1}^{n} Z_i Z_j R_X\left(t_i - t_j\right)$ is non-negative definite for all integer n and every choice of time points t_i, $i = 1, 2, ..., n$;

If $R_X(\tau)$ is continuos for $\tau = 0$, then $R_X(\tau)$ is continuos for all τ.

The first and third statements follow directly from the definition of covariance function

$$R_X\left(t_2, t_1\right) = R_X\left(t_2 - t_1\right) = \text{cov}\{X(t_2), X(t_1)\} = \text{cov}\{X(t_1), X(t_2)\} = R_X\left(t_1, t_2\right) = R_X\left(t_1 - t_2\right)$$

and

$$R_X(t,t) = R_X(0) = \text{cov}\{X(t), X(t)\} = \sigma_X^2$$

Statement two follows from the Schwartz inequality

$$E\{|XY|\} \le \sqrt{E\{X^2\} E\{Y^2\}}$$

To prove this inequality we let α be a real constant and consider the inequality

$$\left(|X| + \alpha |Y| \right)^2 \geq 0$$

Taking mean values of both sides, we get

$$E\{X^2\} + 2\alpha E\{|XY|\} + \alpha^2 E\{Y^2\} = E\{Y^2\}\left[\alpha + \frac{E\{|XY|\}}{E\{Y^2\}} \right]^2 + E\{X^2\} - \frac{E^2\{|XY|\}}{E\{Y^2\}} \geq 0$$

If the left-hand side should remain non-negative for all α, we get

$$E^2\{|XY|\} \leq E\{X^2\} E\{Y^2\}$$

which is identical to Schwartz inequality. If we assume now that $m_X = E\{X(t)\} = 0$, then

$$R_X(\tau) = \text{cov}\{X(t_1 + \tau), X(t_1)\} = E\{X(t_1 + \tau) X(t_1)\} ; \; R_X(0) = E\{X^2(t_1)\} = \sigma_X^2$$

and, if we apply the Schwartz inequality, we obtain

$$E\{X(t_1 + \tau) X(t_1)\} \leq \sqrt{E\{X^2(t_1 + \tau) E\{X^2(t_1)\}\}} = \sqrt{\sigma_X^2 \sigma_X^2} = \sigma_X^2 = R_X(0)$$

which completes the proof of the second statement.

To prove the fourth statement, we assume again that $E\{X\} = 0$ and form

$$E\left\{ \left[\sum_{i=1}^n Z_i X(t_i) \right]^2 \right\} = \sum_{i,j=1}^n Z_i Z_j E\{X(t_i) X(t_j)\} = \sum_{i,j=1}^n Z_i Z_j R_X(t_i - t_j)$$

Since the left-hand side is the expectation of non-negative quantity, it is non-negative and the statement 4 is proven. Finally, to prove the property 5, we form

$$\left| R_X(s+h, t+k) - R_X(t,s) \right| =$$

$$= \left| E\{X(s+h) X(t+k)\} - E\{X(s) X(t)\} + E\{X(s+h) X(t)\} - E\{X(s+h) X(t)\} \right|$$

$$= \left| E\{X(s+h)[X(t+k) - X(t)]\} + E\{[X(s+h) - X(s)] X(t)\} \right|$$

$$= \left| \text{cov}\{X(s+h), X(t+k) - X(t)\} + \text{cov}\{X(s+h) - X(s), X(t)\} \right|$$

$$\leq \left\{ R_X(s+h, s+h)[R_X(t+k, t+k) - 2R_X(t+k, t) + R_X(t,t)] \right\}^{1/2} +$$

$$+ \left\{ [R_X(s+h, s+h) - 2R_X(s+h, s) + R_X(s,s)] R_X(t,t) \right\}^{1/2}$$

where the last inequality follows from the Schwartz inequality [2]. Now let $h, k \to 0$; then $R_X(t+k, t) \to R_X(t,t)$ and $R_X(t+k, t+k) \to R_X(t,t)$, because $R_X(s,t) = R_X(s-t)$ is continuos for $s=t$. The right member of the last inequality will thus converge to zero, and the statement is proven.

If the covariance function is normalized by $R_X(0)$, the *correlation function*, which is defined by

$$\rho_X(\tau) = \frac{R_X(\tau)}{R_X(0)}$$

is obtained. It follows from the property 2 that $|R_X(\tau)| \le R_X(0)$. The correlation function is therefore less than one in magnitude. The values $\rho_X(\tau)$ gives the correlation between values of the process with a spacing τ. Values close to one mean that there are strong correlations, zero values indicate no correlation; negative values indicate negative correlation. An investigation of the shape of the correlation function, thus, indicates the temporal interdependencies of the process. Therefore, it is very useful to study realizations of stochastic processes and their covariance functions, in order to develop insight into their relationships. The idea will be discussed further through the worked examples at the end of this chapter.

Example 2.6. (wide-sense stationarity): Determine if the following process $X(t) = a\cos(t) + b\sin(t)$, for a and b zero-mean uncorrelated with the variance σ^2, is wide-sense stationary.

The mean value of the process is

$$\bar{X}(t) = E\{X(t)\} = E\{a\}\cos(t) + E\{b\}\sin(t) = 0$$

while the covariance function

$$R_X(t+\tau,t) = E\{[X(t+\tau) - \bar{X}][X(t) - \bar{X}]\}$$
$$= E\{[a\cos(t+\tau) + b\sin(t+\tau)][a\cos(t) + b\sin(t)]\}$$
$$= E\{a^2\}\cos(t+\tau)\cos(t) + E\{b^2\}\sin(t+\tau)\sin(t) + E\{ab\}\sin((t+\tau)+t)$$

Since $E\{a^2\} = E\{b^2\} = \sigma^2$ and $E\{ab\} = E\{a\}E\{b\} = 0$, one obtains

$$R_X(t+\tau,t) = \sigma^2\left[\cos(t)\cos(t+\tau) + \sin(t)\sin(t+\tau)\right] = \sigma^2\cos(t-(t+\tau))$$

or, equivalently

$$R_X(t+\tau,t) = R_X(\tau) = \sigma^2\cos(\tau).$$

Thus, the process is wide-sense stationary.

Since a stochastic process is both random and time dependent, we can imagine averages which are computed over the time variable, as well as by the expectation. Processes for which, for any function $g(X(t))$, the time average of $g(X(t))$ gives the same limit as the distribution (ensemble) average (or expectation) of $g(X(t))$, are called *ergodic* [23, 39, 42]. In symbol form this states that for an ergodic discrete process with arbitrary $g(\cdot)$

$$\lim_{N\to\infty}\frac{1}{N+1}\sum_{k=0}^{N}g(X^i(k)) = E\{g(X)\} \qquad (2.22)$$

or, equivalently, for continuous process

$$\lim_{I\to\infty}\frac{1}{I}\int_0^I g(X^i(t))dt = E\{g(X)\} \qquad (2.23)$$

where we use $X^i(t)$ as a shortened notation of $X(t,\omega_i)$, with ω_i being an outcome of the chance experiment. Basically, this means that all the statistical information in the ensemble can be obtained from one realization of the process. In practice, when a single measurement of a process is available, it is usually assumed to be ergodic. Thus, a stochastic process will be *ergodic in the mean* $(g(X) = X)$ if the finite-time average in continuous case, or finite-sample average in discrete case

$$\overline{X}_T = \lim_{I \to \infty} \frac{1}{I} \int_0^I X^i(t)\,dt \; ; \; \overline{X}_T = \lim_{N \to \infty} \frac{1}{N+1} \sum_{k=0}^{N} X^i(k), \tag{2.24}$$

respectively, converge to the expectation $\overline{X} = E_X\{X\}$, that is $\overline{X}_T = \overline{X}$. In an entirely similar way, we may define *correlation and covariance ergodicity*. A discrete stochastic process is said to be *ergodic in the covariance* $(g(X) = (X - \overline{X})(X - \overline{X})^T)$ if the second order sample-average

$$E_T\left\{\left[X(t_2) - \overline{X}_T\right]\left[X(t_1) - \overline{X}_T\right]^T\right\} = \lim_{N \to \infty} \sum_{i=0}^{N}\left[X^i(t_2) - \overline{X}_T\right]\left[X^i(t_1) - \overline{X}_T\right]^T / (N+1) \tag{2.25}$$

converges to the distribution average (covariance function) $R_X(t_2, t_1) = R_X(t_2 - t_1)$. It is necessary for a process to be stationary, in order for it to be ergodic. Thus, all ergodic processes are stationary. However, not all stationary processes are ergodic. Let us give a simple example.

Example 2.7. (stationarity and ergodicity): Let us consider an ensemble of random constant voltages, depicted in Fig. 2.3.

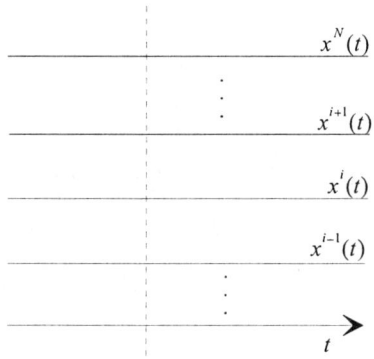

Fig.2.3. An ensemble of random constant voltages $X(t)$

This ensemble is stationary, since the distribution average or mean value

$$\overline{X}(t) = E\{X(t)\} \approx \frac{1}{N} \sum_{i=1}^{N} X^i(t)$$

is independent of the time index t, i.e. $\overline{X}(t) = \overline{X}$. However, the amplitudes of members of the given ensemble $\{X^i(t)\}$ are quite different (Fig. 2.3.), and this means that the time-averages at the particular random voltage realizations

$$\overline{X}_T^i = \frac{1}{I} \int_0^T X^i(t)\,dt = X^i(t)\frac{1}{I}\int_0^I dt = X^i(t) \; ; \; i = 1, 2, ..., N$$

are not the same; that is $\overline{X}_T \neq \overline{X}$. Therefore, the given process is stationary, but it is not ergodic.

The relationship between stationary and ergodic processes is illustrated in Fig. 2.4.

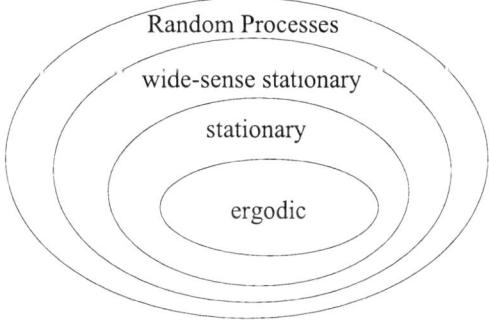

Fig. 2.4. The relationship between stationary and ergodic stochastic processes

2.4. Spectral representation of stochastic processes

In the preceding section we have been concerned with methods for the characterization of random processes. An alternative method of characterizing random processes will be discussed in this section. It is concerned with signal decomposition into simple elementary components. Particularly, spectral density concepts result from a decomposition of the signal power in a random wave [2, 13, 25, 28, 43].

The average power in the i-th record of a random scalar ergodic waveform $X(t, \omega_i)$ is

$$P_{av} = \lim_{T \to \infty} \frac{1}{T} \int_{-T/2}^{T/2} X^2(t, \omega_i) \, dt \qquad (2.26)$$

In order to resolve existence problems associated with the Fourier transform of a waveform $X(t, \omega_i)$, which may have infinite energy, we define a truncated zero-mean signal

$$X_T(t, \omega_i) = \begin{cases} X(t, \omega) - \bar{X} \;;\; |t| \leq T/2 \\ 0 \;;\; |t| > T/2 \end{cases} \qquad (2.27)$$

where $\bar{X}$ is the mean value, such that the Fourier transform of the truncated signal is

$$X_T^i(s) = \int_{-\infty}^{\infty} X_T(t, \omega_i) \exp(-st) dt = \int_{-T/2}^{T/2} \left[X(t, \omega_i) - \bar{X} \right] \exp(-st) dt \qquad (2.28)$$

The average power in the truncated waveform can be expressed by using the Parceval's theorem, which states that

$$\int_{-\infty}^{\infty} f_1(t) f_2(t) \, dt = \frac{1}{2\pi j} \int_{-j\infty}^{j\infty} F_1(\pm s) F_2(\mp s) \, ds \qquad (2.29)$$

In this way, we obtain for the average power in the truncated waveform

$$\frac{1}{T} \int_{-T/2}^{T/2} \left[X_T(t, \omega_i) \right]^2 dt = \int_{-\infty}^{\infty} \frac{\left[X_T(t, \omega_i) \right]^2}{T} dt = \frac{1}{2\pi j} \int_{-j\infty}^{j\infty} \frac{X_T^i(s) X_T^i(-s)}{T} ds \qquad (2.30)$$

We will define the term

$$S_X^i(T,s) = \frac{X_T^i(s) X_T^i(-s)}{T} \tag{2.31}$$

as the spectral density of the i-th record $X(t,\omega_i)$ in the time interval $(-T/2, T/2)$. We define the *power spectral density* of the random waveform $X(t,\omega)$ as the limit of the ensemble average (expectation) of the power spectral density of the records; that is

$$S_X(s) = \lim_{T\to\infty} E\{S_X^i(s,T)\} = \lim_{T\to\infty} \frac{1}{T} E\{X_T^i(s) X_T^i(-s)\} \tag{2.32}$$

The Fourier transform of $X_T(t,\omega_i)$ in (2.28) may be now used to obtain

$$\begin{aligned}
E\{S_X^i(T,s)\} &= E\left\{ \frac{1}{T} \int_{-T/2}^{T/2} \int_{-T/2}^{T/2} X_T(\lambda_1,\omega_i) X_T(\lambda_2,\omega_i) \exp(-s\lambda_1) \exp(-s\lambda_2) d\lambda_1 d\lambda_2 \right\} \\
&= \frac{1}{T} \int_{-T/2}^{T/2} \int_{-T/2}^{T/2} R_X(\lambda_1-\lambda_2) \exp(-s\lambda_1) \exp(-s\lambda_2) d\lambda_1 d\lambda_2
\end{aligned} \tag{2.33}$$

where $R_X(\lambda_1-\lambda_2) = E\{X_T(\lambda_1,\omega_i) X_T(\lambda_2,\omega_i)\}$ is the covariance function. Now, if we make the change of variable $\tau = \lambda_1 - \lambda_2$ and $\eta = \lambda_1 + \lambda_2$, it is possible to show that

$$\begin{aligned}
E\{S_X^i(T,s)\} &= \int_{-T/2}^{T/2} \left(1 - \frac{|\tau|}{T}\right) R_X(\tau) \exp(-s\tau) d\tau \\
&= \int_{-T/2}^{T/2} R_X(\tau) \exp(-s\tau) d\tau - \int_{-T/2}^{T/2} \frac{|\tau|}{T} R_X(\tau) \exp(-s\tau) d\tau
\end{aligned} \tag{2.34}$$

The second expression will go to zero as $T \to \infty$ if

$$\int_{-\infty}^{\infty} |\tau R_X(\tau)| d\tau < \infty \tag{2.35}$$

If the above condition is satisfied, we have

$$S_X(s) = \lim_{T\to\infty} \frac{1}{T} E\{X_T^i(s) X_T^i(-s)\} = \lim_{T\to\infty} E\{S_X^i(T,s)\} = \int_{-\infty}^{\infty} R_X(\tau) \exp(-s\tau) d\tau \tag{2.36}$$

which shows that the *power spectral density* (for an *ergodic process*) is the Fourier transform of the *covariance function*. The fact that the covariance function and spectral density constitute a Fourier transform pair is known as the *Wiener-Khintchine relationship (theorem)*. Thus, we could have simply defined spectral density as the Fourier, or bilateral Laplace, transform of the covariance function of the stationary random process

$$S_X(s) = \int_{-\infty}^{\infty} R_X(\tau) \exp(-s\tau) d\tau \ ; \ R_X(\tau) = \frac{1}{2\pi j} \int_{-j\infty}^{j\infty} S_X(s) \exp(s\tau) ds \tag{2.37}$$

The Wiener-Khintchine relationship may also be expressed by use of the Fourier cosine transformation. We may easily show that covariance functions and spectral densities are symmetric in that $R_X(\tau) = R_X(-\tau)$ and $S_X(s) = S_X(-s)$. By using the Euler's identity

$$\exp(s\tau) = \cos(s\tau) + j\sin(s\tau)$$

and substituting $s = j\omega$, the above expressions become

$$S_X(\omega) = \int_{-\infty}^{\infty} R_X(\tau)\cos(\omega\tau)d\tau = 2\int_{0}^{\infty} R_X(\tau)\cos(\omega\tau)d\tau$$

$$R_X(\tau) = \frac{1}{2\pi}\int_{-\infty}^{\infty} S_X(\omega)\cos(\omega\tau)d\omega = \frac{1}{\pi}\int_{0}^{\infty} S_X(\omega)\cos(\omega\tau)d\omega$$

(2.38)

It follows from the last expressions that in order to establish $R(\tau)$ by a given $S(\omega)$, and vice versa, it is useful to implement the cosine-transformation tables. Some of them are given in Table 2.1. Additionally, Table 2.2. shows the correspondence between a process $X(t)$, autocovariance $R_X(\tau)$ and power spectrum $S_X(\omega)$. The justification follows easily from definitions (2.38) and the elementary properties of Fourier transforms.

Table 2.1. The Fourier cosine-transformation expressions

$f(t)$	$F(\omega) = \int_{0}^{\infty} f(\tau)\cos(\omega\tau)d\tau$
$\exp(-a\tau)$	$\dfrac{a}{a^2+\omega^2}$
$\exp(-a\tau)\cos(\omega_0\tau)$	$\dfrac{a}{2}\left[\dfrac{1}{a^2+(\omega-\omega_0)^2}+\dfrac{1}{a^2+(\omega+\omega_0)^2}\right]$
$\exp(-\beta^2\tau^2)$	$\dfrac{\sqrt{\pi}}{2\beta}\exp\left\{-\left(\dfrac{\omega}{2\beta}\right)^2\right\}$
$\exp(-\beta^2\tau^2)\cos(\omega_0\tau)$	$\dfrac{\sqrt{\pi}}{2\beta}\exp\left\{-\dfrac{\omega^2+\omega_0^2}{4\beta^2}\right\}\cosh\left(\dfrac{\omega_0\omega}{2\beta^2}\right)$;
$\dfrac{\sin(\omega_0\tau)}{\tau}$	$\dfrac{\pi}{2}$; $0<\omega<\omega_0$ 0 ; $\omega>\omega_0$
1 ; $0<\tau<a$ 0 ; $\tau>a$	$\dfrac{\sin(\omega a)}{\omega}$
$1/(\tau^2+a^2)$	$\dfrac{\pi}{2a}\exp\{-a|\omega|\}$
$\exp(-a|\tau|)\left(\cos\left(\sqrt{\omega_0^2-a^2}\,\tau\right)+\dfrac{a}{\sqrt{\omega_0^2-a^2}}\sin\left(\sqrt{\omega_0^2-a^2}\,|\tau|\right)\right)$	$\dfrac{a\omega_0^2}{\left(\omega^2-\omega_0^2\right)^2+4a^2\omega^2}$

Comment: While using Table 2.1. it is necessary to keep in mind the multipliers corresponding to eqs. (2.38).

Table 2.2. The correspondence between stochastic signal $X(t)$, its covariance function $R_X(\tau)$ and power spectrum $S_X(\omega)$

$X(t)$	$R_X(\tau)$	$S_X(\omega)$
$aX(t)$	$\|a\|^2 R_X(\tau)$	$\|a\|^2 S_X(\omega)$
$\dfrac{dX(t)}{dt}$	$-\dfrac{d^2 R_X(\tau)}{d\tau^2}$	$\omega^2 S_X(\omega)$
$\dfrac{d^n X(t)}{dt^n}$	$(-1)^n \dfrac{d^{2n} R_X(\tau)}{d\tau^{2n}}$	$\omega^{2n} S_X(\omega)$
$X(t)\exp(\pm j\omega_0 t)$	$R_X(\tau)\exp(\pm j\omega_0\tau)$	$S_X(\omega \mp \omega_0)$
$X(t)\cos(\omega_0 t)$	$R_X(\tau)\cos(\omega_0\tau)$	$\dfrac{1}{2}\left[S_X(\omega+\omega_0)+S_X(\omega-\omega_0)\right]$

For discrete-time-stationary random processes we define the power spectral density as being the discrete Fourier, or bilateral Z, transform of the covariance function, which is sometimes called the covariance sequence

$$S_X(z) = T\sum_{n=-\infty}^{\infty} R_X(nT)z^{-n} \; ; \; R_X(nT) = \frac{1}{2\pi Tj}\int_C S_X(z)z^{n-1}dz \qquad (2.39)$$

where the contour C is a circle in the region of convergence of $S_X(z)$. The properties of Laplace and Z transformations are given in the appendix A.

The spectral-density function has a good physical interpretation. The integral $2\int_{\omega_1}^{\omega_2} S_X(\omega)d\omega$ represents the power of the signal in the frequency band (ω_1,ω_2). The area under the spectral density curve, thus, represents the signal power in a certain frequency band. The presence of peaks in the spectrum indicates that there are almost periodic components. The total area under the curve is

$$R_X(0) = \sigma_X^2 = \frac{1}{\pi}\int_0^\infty S_X(\omega)d\omega \qquad (2.40)$$

Notice that the mean-value function, the covariance function and the spectral density are characterized by the first two moments of the distribution only. Therefore, signals whose realizations are very different may have the same first two moments. In this way, the random telegraph wave that switches between the values 0 and 1 may have the same spectrum as the noise from a simple RC circuit.

2.5. The concept of white noise

Having introduced the power spectral density $S_X(\omega)$, we will now use it to introduce special stochastic processes which are called *white noise*. We will thus consider a weakly stationary stochastic process $\{X(t), t \in T\}$. Without loosing generality, we can assume that the mean value $E\{X(t)\} = 0$. Recalling that the spectral density function expressed how the variance of the process was distributed on different frequencies (the quantity $\int_{-\infty}^{\infty} S_X(\omega)d\omega$ equals the variance σ_X^2

of the process, eq. (2.40), while the quantity $2\int_{\omega_1}^{\omega_2} S_X(\omega)d\omega$ can be interpreted as the contribution

to the variance from frequencies in the interval (ω_1,ω_2)), we introduce the following definition: *A weakly stationary stochastic process with $S_X(\omega)=const.$ is called white noise* [1, 2, 59].

The analogy with the spectral properties of white light explains the name given to the process. We will now investigate the implications of the definition. To do this, we will consider the discrete and continuos time processes separately.

To further analyze the properties of discrete-time white noise, we will first calculate its covariance function. Introducing $S_X(\omega)=const.=c$ into (2.37), we find

$$R_X(\tau)=\frac{1}{2\pi}\int_{-\pi}^{\pi}\exp(j\omega\tau)c\,d\omega=\frac{c}{\pi\tau}\sin(\pi\tau) \qquad (2.41)$$

It should be noted that the integration in (2.41) is performed from $-\pi$ to π instead of $-\infty$ to ∞ as it is done in (2.37) for the continuous-time case. This can be explained by the effects of sampling a continuous-time signal. Namely, the spectrum of the discrete-time signal is periodic in ω with period $2\pi/T$ (here $T=1$ is used). Therefore, the path along the imaginary axis between ($\pm\pi$) is only used (see the relation between s-plane and z-plane in appendix A).

Discrete time white noise thus has the property

$$R_X(\tau)=\begin{cases} c \; ; \; \tau=0 \\ 0 \; ; \; \tau=\pm1,\pm2,... \end{cases} \; ; \; R_X(0)=\sigma_X^2=c \qquad (2.42)$$

This implies that the values of the process at different times are uncorrelated, and if we consider Gaussian or normal white noise also independent. White noise in the discrete time case is thus a process which consists of a sequence of uncorrelated (in the normal case also independent) stochastic variables. Discrete-time white noise is therefore sometimes referred to as a completely uncorrelated process or a pure random process.

The continuos time processes are much more difficult to analyze than discrete time processes. The introduction of continuos time white noise is a good example of the sort of difficulties that will be encountered in the sequel. In analogy with the discrete time case, the definition implies

$$S_X(\omega)=const.=c \qquad (2.43)$$

As the variance of the process

$$\sigma_X^2=\frac{1}{2\pi}\int_{-\infty}^{\infty}S_X(\omega)d\omega=\frac{1}{2\pi}\int_{-\infty}^{\infty}c\,d\omega \qquad (2.44)$$

we find immediately that continuos-time white noise does not have finite variance. Continuos-time white noise, thus, is not a second order random process; that is, it has not a finite energy and is not physically realizable. As the Fourier transform of a constant equals the Dirac delta function, we find that formally white noise has the covariance function

$$R_X(\tau)=\frac{1}{2\pi}\int_{-\infty}^{\infty}\exp(j\omega\tau)c\,d\omega=c\delta(\tau) \qquad (2.45)$$

Hence, we find that continuous-time white noise has also the property that $X(t)$ and $X(s)$ are uncorrelated if $t\neq s$, in complete analogy with the discrete time case. Notice, however, that in the continuos-time case we have the complication that white noise does not have finite variance; i.e. finite energy.

Since white noise has infinite variance, we could try to introduce other processes which have essentially constant spectral densities, but finite variance. One possibility is the so-called band limited white noise, characterized by the spectral density:

$$S_X(\omega) = \begin{cases} c & ; \ |\omega| < \Omega \\ 0 & ; \ |\omega| \geq \Omega \end{cases}$$

(2.46)

This process has the covariance function

$$R_X(\tau) = \frac{1}{2\pi} \int_{-\Omega}^{\Omega} c \exp(j\omega\tau) d\omega = \frac{c}{\pi\tau} \sin(\Omega\tau)$$

(2.47)

The correlation of two values of the process $X(t)$ and $X(s)$ separated by a given interval $\tau = |t - s|$ can, thus, be made arbitrarily small, by choosing Ω sufficiently large. Notice, however, that for given Ω the values of the process at time t and s will always be correlated, if t and s are chosen sufficiently close together. We will now analyze what happens when $\Omega \to \infty$. For $\tau \neq 0$, the value of the function will tend to zero. As $R_X(0) = c\Omega/\pi$, we find that $R_X(0)$ will tend to infinity. It is easier analytically to consider the integral

$$I(\tau) = \int_{-\infty}^{\tau} R_X(s) ds = \frac{c}{\pi} \int_{-\infty}^{\tau} \frac{\sin(\Omega s)}{s} ds = \frac{c}{\pi} \int_{-\infty}^{\Omega\tau} \frac{\sin(x)}{x} dx$$

Because

$$\int_{0}^{\infty} \frac{\sin(x)}{x} dx = \pi$$

we find that

$$\lim_{\Omega \to \infty} I(\tau) = \begin{cases} 0 & ; \ \tau < 0 \\ \dfrac{c}{2} & ; \ \tau = 0 \\ c & ; \ \tau > 0 \end{cases}$$

The integral of the covariance function is thus equal a step function, so that the limiting covariance function then becomes a Dirac delta function, i.e.

$$R_X(\tau) = \frac{dI(\tau)}{d\tau} \to c\delta(\tau)$$

Apart from band limited noise, the random process with the covariance function

$$R_X(\tau) = \frac{a}{2} \exp(-a|\tau|)$$

(2.48)

and the spectral density

$$S_X(\omega) = \int_{-\infty}^{\infty} R_X(\tau) d\tau = \frac{a^2}{a^2 + \omega^2}$$

(2.49)

is frequently used. In this case, we have

$$\lim_{a \to \infty} S_X(\omega) = 1 \ ; \ \lim_{a \to \infty} R_X(\tau) = \delta(\tau)$$

(2.50)

The concept of white noise is very important in the theory and application of stochastic processes. All stochastic processes that are needed will be generated simply by filtering white noise.

White noise is, thus, the equivalent of the Dirac delta function for deterministic systems. This idea will be elaborated in the next chapter.

2.6 Worked examples

Example 2.8: Let $\{X(t), t \geq 0\}$ be a stochastic process defined by the relation $X(t) = V \cos(\omega t)$, where ω is a constant and V is a uniformly distributed random variable within the interval $[0,1]$, i.e. $\omega = const.$; $f_V(v) = [h(v) - h(v-1)]$. a) Sketch few samples of this process; b) Calculate the probability density function of the random variable $X(t)$ for $t = 0$; $t = \pi/4\omega$; and $t = \pi/2\omega$; c) Calculate the mean value of the process $m_X(t) = E\{X(t)\}$; d) Investigate the stationarity of the process.

Solution: a) In order to sketch some samples of the process it is necessary to fix some possible values of the random variable V and, according to that, to sketch the corresponding deterministic functions $x(t)$. The samples of the process for different time instants can be obtained by choosing the values from different deterministic functions $x(t)$. This can be done very easy using some programming package. For example, using the programming package MATLAB, we made the program in Fig. 2.5.a by which the deterministic functions $x(t)$ for $V=0$, $V=0.2$, $V=0.5$, $V=0.8$ and $V=1$ are plotted and, after that, for the time instants from the set $\{0, 0.1, 0.2, ..., 3.1\}$ the values from the arbitrary chosen functions $x(t)$ are taken to be the samples of the process. The obtained realizations are shown in the Fig. 2.5.b.

```
function example
w=1; % definition of the constant w
t=0:0.05:3.1; %definition of time instants
V=[0 0.2 0.5 0.8 1]; % possible values of random variable V
% calculation of deterministic functions
f0=V(1)*cos(w*t); f2=V(2)*cos(w*t); f5=V(3)*cos(w*t);
f8=V(4)*cos(w*t); f1=V(5)*cos(w*t);
% simulation of random process
% 'round' is a function which assigns the closest integer to a decimal
% number. In that way, choosing one of five possible values among
%  V(1), V(2),...,V(5), we choose one of the deterministic functions
for i=1:length(t)
   x(i)=V(round(rand*5+0.5))*cos(w*t(i));
   % 'rand' is a function which generates the random number uniformly
   % distributed within the interval [0,1]
end
% Now we will plot six different functions on the same graph. All of
% them will have the same horizontal x axes defined by vector t. The
% corresponding vertical axes y are defined by corresponding vectors
% f0, f2, ...
plot(t,f0,t,f2,t,f5,t,f8,t,f1,t,x);
end
```

(a)

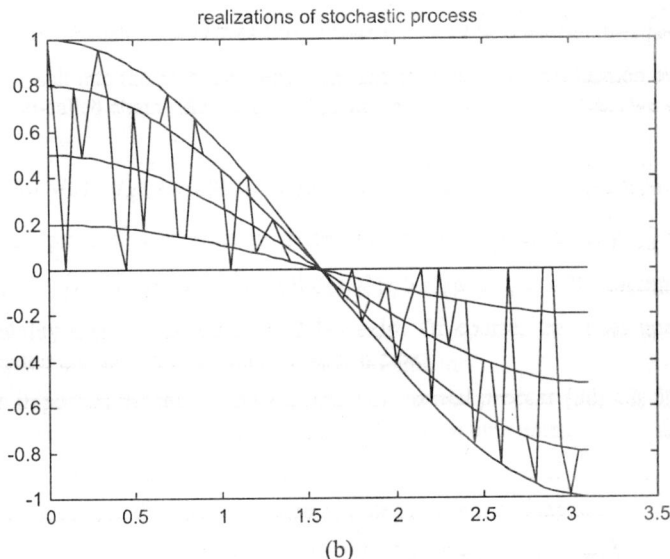

(b)

Fig. 2.5. (a) MATLAB program for simulating a stochastic process; (b) Realizations of a stochastic process

b) If we fix the time instant the stochastic process reduces to a random variable, for which it is possible to determine the corresponding probability density function:

$$t = 0 \Rightarrow X(0) = V \Rightarrow f_{X(0)}(x) = h(x) - h(x-1)$$

$$t = \frac{\pi}{4\omega} \Rightarrow X\left(\frac{\pi}{4\omega}\right) = \frac{\sqrt{2}}{2}V \Rightarrow f_{X\left(\frac{\pi}{4\omega}\right)}(x) = \sqrt{2}\left[h(x) - h\left(x - \frac{\sqrt{2}}{2}\right)\right]$$

$$t = \frac{\pi}{2\omega} \Rightarrow X\left(\frac{\pi}{2\omega}\right) = 0 \Rightarrow f_{X\left(\frac{\pi}{2\omega}\right)}(x) = \delta(x)$$

where $h(x)$ is the unit step function and $\delta(x)$ is the unit delta impulse.

c) The mean value of the process is given by:

$$E\{X_t\} = E\{V\cos(\omega t)\} = E\{V\}\cos(\omega t) = \frac{\cos(\omega t)}{2} = m_X(t)$$

d) Two conditions have to be satisfied in order that the process is wide sense stationary. The first one is that the mean value is constant, i.e. $m_X(t) = const.$, while the second one is that the auto-correlation function is only a function of one variable (the difference of the time instants τ); that is $R_X(t+\tau,t) = R_X(\tau)$. Taking into account the result from c) it is obvious that the first condition is not satisfied, so the process is not wide-sense stationary. In the light of this result, the process is also not strictly stationary.

Example 2.9: The stochastic process is defined by the relation $Z(t) = Y\cos(t) + X\sin(t)$ where X and Y are independent discrete random variables with the identical distributions: $X: \begin{pmatrix} -1 & 2 \\ 2/3 & 1/3 \end{pmatrix}$; $Y: \begin{pmatrix} -1 & 2 \\ 2/3 & 1/3 \end{pmatrix}$ a) Calculate mean value, auto-correlation and auto-covariance functions of the process; b) Analyze the stationarity of the process.

Solution:

a) The mean value of the process is

$$E\{Z(t)\} = E\{Y\cos t + X\sin t\} = m_Y\cos t + m_X\sin t = 0$$

since $m_X = (-1)\dfrac{2}{3} + 2\dfrac{1}{3} = 0 = m_Y$, while its auto-correlation function (2.8) is given by

$$R_Z(t+\tau,t) = E\{Z(t+\tau)Z(t)\} = E\{(Y\cos(t+\tau) + X\sin(t+\tau))(Y\cos(t) + X\sin(t))\}$$
$$= E\{Y^2\}\cos(t)\cos(t+\tau) + E\{XY\}(\sin(t)\cos(t+\tau) + \cos(t)\sin(t+\tau)) +$$
$$+ E\{X^2\}\sin(t)\sin(t+\tau)$$

Taking into account that:

$$E\{X^2\} = E\{Y^2\} = (-1)^2\dfrac{2}{3} + (2)^2\dfrac{1}{3} = 2 \ ; \ E\{XY\} = E\{X\}E\{Y\} = m_X m_Y = 0$$

we have

$$R_X(t+\tau,t) = 2\cos(t)\cos(t+\tau) + 0\cdot(\sin(t)\cos(t+\tau) + \cos(t)\sin(t+\tau)) + 2\sin(t)\sin(t+\tau)$$
$$= 2\cos(\tau)$$

On the other hand, if we want to calculate the auto-covariance function (2.7), it is necessary to take into account the mean values of the process:

$$C_Z(t+\tau,t) = E\{(Z(t+\tau) - m_Z(t+\tau))(Z(t) - m_Z(t))\}$$
$$= E\{Z(t+\tau)Z(t)\} - m_Z(t+\tau)E\{Z\} - m_Z(t+\tau)E\{Z(t)\} + m_Z(t)m_Z(t+\tau)$$
$$= R_Z(t+\tau,t) - m_Z(t+\tau)m_Z(t)$$

However, as the mean value of the process is equal to zero, in this case the auto-covariance and auto-correlation functions are equal:

$$C_Z(t+\tau,t) = R_Z(t+\tau,t) = 2\cos(\tau)$$

b) The stochastic process $Z(t)$ satisfies the following conditions:

$$m_Z(t) = 0$$
$$R_Z(t+\tau,t) = 2\cos(\tau) = R_Z(\tau)$$

so we can conclude that it is a weakly-stationary process. However, this process is not strictly-stationary, and this can be checked very easy. Namely, it is enough to find two time instants t_1 and t_2 for which the random variables $Z(t_1)$ and $Z(t_2)$ have different distributions. In our case, let us take the moments $t_1 = 0$ and $t_2 = \pi$. The corresponding random variables are with different distributions; that is,

$$Z(0): \begin{pmatrix} -1 & 2 \\ 2/3 & 1/3 \end{pmatrix}; \ Z(\pi): \begin{pmatrix} 1 & -2 \\ 2/3 & 1/3 \end{pmatrix}$$

so, we can conclude that the process $Z(t)$ is not strictly sense stationary.

Example 2.10: Stochastic process $N_t; t \in [0,\infty)$ is called Poisson process if it satisfies the following conditions:

i) N_t can take only nonnegative integer values and $N_0 = 0$;

ii) N_t has stationary and independent increments;

iii) $\Pr\{N_{t+\Delta t} - N_t = 1\} = \lambda \Delta t + o(\Delta t)$ ($o(\Delta t)$ is infinitely small with respect to Δt, that is

$\lim\limits_{\Delta t \to 0} \dfrac{o(\Delta t)}{\Delta t} = 0$);

iv) $\Pr\{N_{t+\Delta t} - N_t > 1\} = o(\Delta t)$.

a) Calculate the probability $\Pr\{N_t = k\}$; b) Determine the expression for $o(\Box t)$; c) Prove that the sum of two Poisson processes is also a Poisson process.

Solution: a) Let us introduce the notation $\Pr\{N_t = k\} = p_k(t)$. Firstly, we are going to calculate $p_0(t)$:

$$p_0(t+\Delta t) = \Pr\{N_{t+\Delta t} = 0\} = \Pr\{N_t = 0\}\Pr\{N_{t+\Delta t} - N_t = 0\}$$

$$= p_0(t)\left[1 - \Pr\{N_{t+\Delta t} - N_t = 1\} - \Pr\{N_{t+\Delta t} - N_t > 1\}\right] \Rightarrow$$

$$p_0(t+\Delta t) = p_0(t)\left[1 - \lambda \Delta t - o(\Delta t)\right] \Rightarrow \frac{p_0(t+\Delta t) - p_0(t)}{\Delta t} = -\lambda p_0(t) - p_0(t)\frac{o(\Delta t)}{\Delta t} \Rightarrow$$

$$\lim\limits_{\Delta t \to 0} \frac{p_0(t+\Delta t) - p_0(t)}{\Delta t} = \frac{dp_0(t)}{dt} = -\lambda p_0(t) \Rightarrow p_0(t) = \Pr\{N_t = 0\} = p_0(0)\exp(-\lambda t) = \exp(-\lambda t)$$

In the next step, we are going to form the differential equation for $p_k(t)$:

$$p_k(t+\Delta t) = \Pr\{N_{t+\Delta t} = k\}$$

$$= \Pr\{N_t = k\}\Pr\{N_{t+\Delta t} - N_t = 0\} + \Pr\{N_t = k-1\}\Pr\{N_{t+\Delta t} - N_t = 1\}$$

$$+ \Pr\{N_t < k-1\}\Pr\{N_{t+\Delta t} - N_t > 1\}$$

$$= p_k(t)\left[1 - \lambda \Delta t - o(\Delta t)\right] + p_{k-1}(t)\left[\lambda \Delta t + o(\Delta t)\right] + \Pr\{N_t < k-1\}o(\Delta t) \Rightarrow$$

$$\frac{p_k(t+\Delta t) - p_k(t)}{\Delta t} = -\lambda p_k(t) + \lambda p_{k-1}(t) + \left[\Pr\{N_t < k-1\} + p_{k-1}(t) - p_k(t)\right]o(\Delta t) \Rightarrow$$

$$\lim\limits_{\Delta t \to 0} \frac{p_k(t+\Delta t) - p_k(t)}{\Delta t} = \frac{dp_k(t)}{dt} = -\lambda p_k(t) + \lambda p_{k-1}(t)$$

If we apply the Laplace transformation on the last equation, the corresponding algebraic equation will be obtained:

$$sP_k(s) - p_k(0) + \lambda P_k(s) = \lambda P_{k-1}(s)$$

where the initial condition is:

$$p_k(0) = \Pr\{N_0 = k\} = 0; \text{ for } k \geq 1;$$

so

$$(\forall k \geq 1)\ (s+\lambda)P_k(s) = \lambda P_{k-1}(s)$$

The last equation is recurrent algebraic relation:

$$(\forall k \geq 1)\ P_k(s) = \frac{\lambda}{s+\lambda}P_{k-1}(s)$$

where the Laplace transform of $p_0(k)$ can be calculated and substituted in the last equation:

$$P_0(s) = L\{p_0(t)\} = L\{\exp(-\lambda t)\} = \frac{1}{s+\lambda} \Rightarrow P_k(s) = \frac{\lambda^k}{(s+\lambda)^{k+1}}$$

Applying the inverse Laplace transformation on the last expression, it is possible to calculate the desired function $p_k(t)$ (see the table of Laplace transform in the appendix A)

$$(\forall k \geq 1)\ p_k(t) = \Pr\{N_t = k\} = \frac{\lambda^k t^k}{k!}\exp(-\lambda t)$$

b) The expression for $o(\Delta t)$ can be calculated taking into account $p_1(\Delta t)$:

$$\Pr\{N_{\Delta t} = 1\} = p_1(\Delta t) = \frac{\lambda \Delta t}{1!}e^{-\lambda \Delta t} = \Pr\{N_{\Delta t} - N_0 = 1\} = \lambda \Delta t + o(\Delta t)$$

Solving the last equation for $o(\Delta t)$, we obtain:

$$o(\Delta t) = \lambda \Delta t\left(\exp(-\lambda \Delta t)-1\right)$$

c) If we assume that N_t and M_t are Poisson processes with the parameters $\square$ and $\square$, respectively, it is easy to prove that the process $Q_t = M_t + N_t$ will be Poisson process with the parameter $\lambda + \mu$. To prove this it is enough to check the conditions i)-iv) mentioned above.

Example 2.11: The process $\{Y(t);\ -\infty < t < \infty\}$ is a stationary zero-mean Gaussian process with auto-covariance function (2.7) given by

$$C_Y(\tau) = \begin{cases} 1 - \dfrac{|\tau|}{T}; & -T \leq \tau \leq T \\ 0\ ; & \text{otherwise} \end{cases}.$$

This process is sampled at n equidistant time instants $t_j = jT/2;\ j = 1, 2, ..., n$.
a) Determine the expectation and auto-covariance function of the sample-mean (arithmetic mean of the samples): $\hat{m}_n = \dfrac{1}{n}\sum_{j=1}^{n} Y_j$;
b) Determine the probability density function of the random variable $\hat{m}_n$.

Solution: A stochastic process is said to be Gaussian if the joint pdf of its samples $\{Y_1, Y_2, ..., Y_n\}$ has a form:

$$f_{Y_1,...,Y_n}(y_1,...,y_n) = \frac{1}{(2\pi)^{n/2}|\Lambda|^{1/2}}\exp\left\{-\frac{1}{2}(Y - E\{Y\})'\Lambda^{-1}(Y - E\{Y\})\right\}$$

where:

$$Y = [y_1 \cdots y_n]'\ ;\ \Lambda = E\left\{(Y - E\{Y\})(Y - E\{Y\})'\right\}$$

a) The mean value of the sample mean $\hat{m}_n$ is

$$\hat{m}_n = \frac{1}{n}\sum_{j=1}^{n} Y_j \Rightarrow E\{\hat{m}_n\} = \frac{1}{n}\sum_{j=1}^{n} E\{Y_j\} = 0$$

while its variance

$$\text{var}\{\hat{m}_n\} = E\{\hat{m}_n^2\} = E\left\{\left(\frac{1}{n}\sum_{j=1}^{n}Y_j\right)^2\right\}$$

$$= \frac{1}{n^2}E\left\{\left(\sum_{j=1}^{n}Y_j\right)^2\right\} = \frac{1}{n^2}E\left\{\sum_{j=1}^{n}Y_j^2\right\} + \frac{1}{n^2}2E\left\{\sum_{\substack{j,i=1\\i\neq j}}^{n}Y_iY_j\right\} =$$

$$= \frac{1}{n^2}\sum_{i=1}^{n}\sum_{j=1}^{n}R_Y\left((i-j)\frac{T}{2}\right)$$

$$= \frac{1}{n^2}\sum_{i=1}^{n}\sum_{j=1}^{n}C_Y\left((i-j)\frac{T}{2}\right) = \frac{2n-1}{n^2}$$

Where $R(\cdot)$ denotes the autocorrelation function (2.8).

b) The random variable $\hat{m}_n$ is the linear combination of n Gaussian zero-mean random variables, so we can conclude that it is also Gaussian with the pdf:

$$f_{\hat{m}_n}(x) = \frac{1}{\sqrt{2\pi\,\text{var}\{\hat{m}_n\}}}\exp\left(-\frac{1}{2}\frac{(x-E\{\hat{m}_n\})^2}{\text{var}\{\hat{m}_n\}}\right) = \frac{1}{\sqrt{2\pi}\sqrt{\frac{2n-1}{n^2}}}\exp\left(-\frac{1}{2}\frac{x^2}{\frac{2n-1}{n^2}}\right)$$

CHAPTER 3

LINEAR DISCRETE-TIME STOCHASTIC SYSTEMS

A very useful aid to understanding the properties of stationary stochastic processes is found by considering the response of a linear stationary system to a stationary input process. The purpose of this chapter is to investigate how the properties of stochastic processes change when they are filtered by linear dynamic systems. Tools for analyzing the properties of the linear discrete-time systems with random inputs are given. By such an approach it is possible to describe a wide class of disturbances in a control system, as their effects on a system. This is important task, since the nature of the disturbances determines the quality of regulation in a process-control system. Some ways to eliminate the disturbances will be given in the next chapters.

3.1 Transfer function models

To generate stochastic processes with specified statistics, we choose to use two models that are equivalent: the input-output or transfer function models and state space models. Each model has its own advantages: the input-output, or transfer function, model is easy to use, while the state-space model is easily generalized [1, 15, 17, 29].

3.1.1 Linear filtering

Consider a linear discrete-time system as show in fig. 3.1. Suppose we let the input to be u, and suppose we take the output to be y.

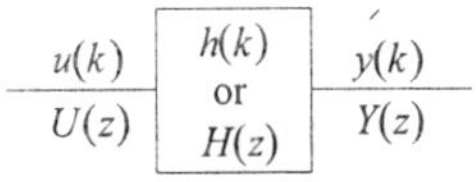

Fig. 1. Transfer function model of linear discrete-time system

Now suppose we deliberately select u to be the unit discrete pulse defined by

$$u(k) = \begin{cases} 1 & ; \ k = 0 \\ 0 & ; \ k \neq 0 \end{cases} = \delta(k) \tag{3.1}$$

Let the response of the system to a unit pulse input $\delta(k)$, the so-called impulse response, be $y(k) = h(k)$, $k = 0,1,...$ The Z-transform of this sequence is (see, appendix A)

$$H(z) = \sum_{k=-\infty}^{\infty} h(k) z^{-k} \tag{3.2}$$

and it is known as the transfer function of a system. Exciting a causal linear system, for which $h(k) = 0$ if $k < 0$, with an input sequence $\{u(k), k = 0,1,...\}$, we obtain the output at time k (for simplicity it is assumed that the sampling period is chosen as the time unit) as

$$y(k) = \sum_{n=-\infty}^{k} h(k-n)u(n) = \sum_{n=0}^{\infty} h(n)u(k-n) = h(k) \otimes u(k) \qquad (3.3)$$

where $\otimes$ denotes the convolution sum. We can derive the convolution sum (3.3) from the properties of linearity and stationarity. First we need more formal definitions of "linear" and "stationary". A system with input u and output y is linear if superposition applies. Thus, if $y_1(k)$ is the response to $u_1(k)$, and $y_2(k)$ is the response to $u_2(k)$, then the system is linear if and only if, for every scalars α and β, the response to $\alpha u_1(k) + \beta u_2(k)$ is $\alpha y_1(k) + \beta y_2(k)$.

A system is stationary, or time-invariant, if the properties of the system do not change with time. For example, if we put the system at rest (no internal energy in the system) and apply a certain signal $u(k)$, we observe a response $y(k)$. If we repeat this experiment at a later time (N periods later) and the system is again at rest and we apply $u(k-N)$, we should see $y(k-N)$. Thus, if the system is linear and response to a unit pulse $\delta(k)$ is $h(k)$, then response to a pulse of intensity u_0 is $u_0 h(k)$. Furthermore, if the system is stationary, or time-invariant, then a delay of the input will delay the response. Thus, if $u = u(l)$ for $k = l$ and $u = 0$ for $k \neq l$, then the response will be $u(l)h(k-l)$. Finally, the total response at time k to a sequence of these pulses is the sum of the responses, i.e.

$$y(k) = \sum_{l=0}^{k} u(l)h(k-l)$$

Now, note that if the input sequence began in the distant past, we must include terms for $l < 0$, perhaps back to $l = -\infty$, from which one obtains (3.3), which completes the proof.

Assume now that the input u is a stationary stochastic process with a given mean-value m_u and a given covariance function R_u. Taking mean-value of the input-output relationship (3.3), we obtain

$$m_y(k) = E\{y(k)\} = E\left\{\sum_{n=0}^{\infty} h(n)u(k-n)\right\}$$

$$= \sum_{n=0}^{\infty} h(n)E\{u(k-n)\} = \sum_{n=0}^{\infty} h(n)m_u(k-n) \qquad (3.4)$$

The mean value of the output is thus obtained by sending the mean-value of the input through the system.

To determine the covariance, first observe that a subtraction of (3.4) from (3.3) gives

$$y(k) - m_y(k) = \sum_{n=0}^{\infty} h(n)\left[u(k-n) - m_u(k-n)\right]$$

The difference between the input signal and its mean value thus propagates through the system in the same way as the input signal itself. When calculating the covariance, it can be assumed that the mean values are zero. The definition of the covariance function gives

$$R_Y(\tau) = E\{y(k+\tau)y(\tau)\}$$

$$= E\left\{\sum_{n=0}^{\infty} h(n)u(k+\tau-n)\left[\sum_{l=0}^{\infty} h(l)u(k-l)\right]\right\} \qquad (3.5)$$

Since the system unit pulse response $h(k)$ is not random, and $h(k)=0$ for $k<0$, both $h(n)$ and $h(l)$ may be removed from the integral implied by the $E\{\cdot\}$ operation, with the result

$$R_Y(\tau) = \sum_{n=-\infty}^{\infty} h(n) \sum_{l=-\infty}^{\infty} h(l) E\{u(k+\tau-n)u(k-l)\} \tag{3.6}$$

The expectation in (3.6) is now recognized as $R_u((k+\tau-n)-(k-l)) = R_u(\tau+l-n)$, and substituting this expression in (3.6), we find

$$R_Y(\tau) = \sum_{n=-\infty}^{\infty} h(n) \sum_{l=-\infty}^{\infty} h(l) R_u(\tau+l-n) \tag{3.7}$$

Equation (3.6) is not especially enlightening, but the Z-transform of it is. We proceed with several simple steps as follows

$$Z\{R_Y(\tau)\} = \sum_{\tau=-\infty}^{\infty} R_Y(\tau) z^{-\tau} = \sum_{\tau=-\infty}^{\infty} \sum_{n=-\infty}^{\infty} h(n) \sum_{l=-\infty}^{\infty} h(l) R_u(\tau+l-n) z^{-\tau}$$

Exchanging the order, since $h(n)$ and $h(l)$ do not depend on τ, we have

$$Z(R_Y(\tau)) = \sum_{n=-\infty}^{\infty} h(n) \sum_{l=-\infty}^{\infty} h(l) \sum_{\tau=-\infty}^{\infty} R_u(\tau+l-n) z^{-\tau}$$

Now we let $m = \tau+l-n$ in the last sum, leading to

$$Z\{R_Y(\tau)\} = \sum_{n=-\infty}^{\infty} h(n) \sum_{l=-\infty}^{\infty} h(l) \sum_{m=-\infty}^{\infty} R_u(m) z^{-(m+n-l)}$$

Finally, we use the fact that $z^{-(m+n-l)} = z^{-m} z^{-n} z^{l}$ and distribute these terms to the corresponding sums, with the result

$$Z\{R_Y(\tau)\} = \sum_{n=-\infty}^{\infty} h(n) z^{-n} \sum_{l=-\infty}^{\infty} h(l) z^{l} \sum_{m=-\infty}^{\infty} R_u(m) z^{-m} = S_Y(z) \tag{3.8}$$

For reasons soon to be clear, we call the Z-transform of $R_Y(\tau)$ for $z = \exp(j\omega)$ the spectrum, or spectral density, of y and use the symbol $S_Y(z)$, and similarly for u and $S_u(z)$. With these symbols and recognition that the Z-transform of the unit-pulse response $h(k)$ is the system-transfer function $H(z)$ in (3.2), the relation (3.8) becomes

$$S_Y(z) = H(z) H(z^{-1}) S_u(z) \tag{3.9}$$

To give an interpretation of (3.9) we make two observations. First note that $R_Y(0) = E\{y^2(k)\}$ is the mean-square value, or power in the y-process. By the inverse Z-transform integral, we have

$$E\{y^2(k)\} = R_Y(0) = \frac{1}{2\pi j} \oint_C S_Y(z) z^{-1} dz = \frac{1}{2\pi j} \oint_C H(z) H(z^{-1}) S_u(z) z^{-1} dz \tag{3.10}$$

where the contour C is a circle in the region of convergence of $S_y(z)$. Now, as a second step, we suppose that $H(z)$ is the transfer function of a very narrow bandpass filter centered at ω_0, so that $H(z)H(z^{-1})$ for $z = \exp(j\omega)$ is $|H(\exp(j\omega))|^2$ and is nearly zero except at ω_0. Then the integral (3.10) may be approximated by assuming that $S_u(\exp(j\omega))$ is nearly constant at the value

$S_u\left(\exp(j\omega_0)\right)$, where $\left|H\left(\exp(j\omega)\right)\right|$ is nonzero, and may thus be removed from the integral. The result is

$$E\{y^2(k)\} = S_u\left(\exp(j\omega_0)\right)\frac{1}{2\pi j}\oint_C H(z)H(z^{-1})z^{-1}dz = S_u\left(\exp(j\omega_0)\right)K \tag{3.11}$$

In (3.11) we have defined the integral as a positive constant K dependent on the exact area of the narrow band characteristic of $H(\cdot)$. But now we can give good intuitive meaning to (3.11). The mean square of the output of a very narrow band filter is proportioned to the S_u. If S_u is constant for all $z = \exp(j\omega)$, we say the process is white (after the spectrum of white light which has equal intensity at all frequencies). Hence we call $S_u\left(\exp(j\omega)\right)$ the power spectral density of the u-process. Thus, the result (3.9) has a following physical interpretation. The number $\left|H\left(\exp(j\omega)\right)\right|$ is the steady-state amplitude of the response of the system to a sine wave with frequency ω. The value of the spectral density of the output is then the product of the power gain $\left|H\left(\exp(j\omega)\right)\right|^2$ and the spectral density of the input $S_u\left(\exp(j\omega)\right)$.

A similar calculation gives the following formula for the cross-covariance of the input and the output:

$$R_{yu}(\tau) = E\{y(k+\tau)u(k)\}$$

$$= E\left\{\sum_{n=0}^{\infty}h(n)u(k+\tau-n)u(k)\right\} = \sum_{n=0}^{\infty}h(n)E\{u(k+\tau-n)u(k)\} \tag{3.12}$$

$$= \sum_{n=0}^{\infty}h(n)R_u(\tau-n) = h(n)\otimes R_u(n)$$

where the symbol $\otimes$ denotes the convolution sum. Notice that it has been assumed that all finite sums exist and that the operations of infinite summation and mathematical expectation have been freely exchanged in these calculations. Introducing the definition of cross-spectral density $S_{yu}(z) = Z\{R_{yu}(\tau)\}$, we proceed as

$$S_{yu}(z) = \sum_{\tau=-\infty}^{\infty}R_{yu}(\tau)z^{-\tau} = \sum_{\tau=-\infty}^{\infty}z^{-\tau}\sum_{n=0}^{\infty}h(n)R_u(\tau-n)$$

$$= \sum_{n=-\infty}^{\infty}h(n)\sum_{\tau=-\infty}^{\infty}z^{-\tau}R_u(\tau-n)$$

Now let $\tau-n=l$ in the second sum

$$S_{yu}(z) = \sum_{n=-\infty}^{\infty}h(n)\sum_{l=-\infty}^{\infty}R_u(l)z^{-(n+l)}$$

but $z^{-(n+l)} = z^{-n}z^{-l}$, which leads to

$$S_{yu}(z) = \sum_{n=-\infty}^{\infty}h(n)z^{-n}\sum_{l=-\infty}^{\infty}R_u(l)z^{-l}$$

Introducing the pulse-transfer function (3.2) of the system, and the definition of the power-spectral density of the input, we recognize these two separate sums as

$$S_{yu}(z) = H(z)S_u(z) \tag{3.13}$$

It follows from (3.13) that the cross-spectral density is equal to the transfer function of the system if the input is white noise with unit spectral density, i.e. $S_u\left(\exp\left(j\omega\right)\right)=1$ for all ω. This fact can be used to determine the pulse-transfer function of a system $H(z)$.

We summarize this discussion with the *Theorem of Filtering of Stationary Process*, which states that given a stable stationary discrete-time dynamic system with unit sampling period and the pulse-transfer function H in (3.2) with the input signal u being a stationary stochastic process with mean m_u and spectral density S_u , then the output y is also a stationary process with the mean value in (3.4)

$$m_y = m_u \sum_{i=0}^{n} h(n) = m_u H(1)$$

(3.14)

and the spectral density in (3.9)

$$S_y\left(\exp\left(j\omega\right)\right)= H\left(\exp\left(j\omega\right)\right)H\left(\exp\left(-j\omega\right)\right)S_u\left(\exp\left(j\omega\right)\right)$$

(3.15)

The cross-spectral density between the input and the output is given by (3.13)

$$S_{yu}\left(\exp\left(j\omega\right)\right)= H\left(\exp\left(j\omega\right)\right)S_u\left(\exp\left(j\omega\right)\right)$$

(3.16)

3.1.2. Stability

A very important qualitative property of dynamic systems is stability. A system may be said to be stable if its response is appropriate for the given stimulus. For input-output models described by (3.3), the most common definition of "appropriate response" is that for every Bounded Input u, we should have a Bounded Output y. If this is true, we say the system is BIBO stable. A test for BIBO stability may be given directly in terms of the unit pulse response h. Suppose the input u is bounded, i.e. there is an M such that $\left|u(k)\right|\leq M < \infty$ for all k. If we consider the magnitude of the response given by (3.3), it is easy to see that [15, 24]

$$\left|y(k)\right|\leq\left|\sum_{n=-\infty}^{\infty} h(n)u(k-n)\right|\leq \sum_{n=-\infty}^{\infty}\left|h(n)\right|\left|u(k-n)\right|\leq M \sum_{n=-\infty}^{\infty}\left|h(n)\right|$$

Thus, the output will be bounded for every bounded input if

$$\sum_{n=-\infty}^{\infty}\left|h(n)\right|< \infty$$

(3.17)

The definition of BIBO stability requires that all poles of the rational pulse-transfer function (3.2)

$$H(z)=\frac{B\left(z^{-1}\right)}{A\left(z^{-1}\right)},$$

(3.18)

where A and B are polynomials in z or z^{-1} with real constant coefficients (the roots of the system characteristic equation $A\left(z^{-1}\right)=0$ are poles of the system described by (3.18), while the roots of the equation $B\left(z^{-1}\right)=0$ are called system zeros) lie inside the unit circle in the z-plane. For the system (3.18) , where p_i and z_i are the poles and zeros, respectively, we can express $H(z)$ as

$$H(z)=\sum_{i}\frac{K_i z}{z-p_i}$$

(3.19)

Since the impulse response $h(k)= Z^{-1}\left\{H(z)\right\}$, if the inverse Z-transform, Z^{-1}, of these terms tend to zero as time increases the system is BIBO stable. The inverse Z-transform of the i-th term of the partial-fraction expansion (3.19) at discrete time k yields

$$Z^{-1}\left\{\frac{K_i z}{z-p_i}\right\} = K_i p_i^k$$

Thus, if the magnitude of p_i is less then unity, i.e. $|p_i| < 1$, this term approaches zero as k approaches infinity. Then, the system of (3.18) is stable if the magnitude of each p_i is less then unity. The factors $(z - p_i)$ originate in the term $A(z^{-1})$ of (3.18). Therefore, the system is BIBO stable provided that all roots of the characteristic equation $A(z^{-1}) = 0$ must lie inside the unit circle in the z-plane.

3.1.3 Spectral factorization

Filtering Theorem gives an important result that is fundamental for modeling of stochastic processes. It follows from this theorem that the random process generated from a linear system with a white noise input has the spectral density given by (3.15), where $S_u(\exp(j\omega))$ is constant and equal to $R_u(0) = \sigma_u^2$, with σ_u^2 being the input variance. If the system is finite dimensional, i.e. the polynomials B and A in (3.18) have a finite degree, the pulse-transfer function H is then a rational function in $\exp(j\omega)$, and the spectral density S_y will also be rational in $\exp(j\omega)$, or equivalently in $\cos(\omega)$. Such a spectral density is called rational. Introducing $z = \exp(j\omega)$ and $S_U(\exp(j\omega)) = 1$, the right hand side of (3.15) can be written as [1, 2, 30]

$$F(z) = H(z)H(z^{-1}) \tag{3.20}$$

If z_i is a zero of $H(z)$, then z_i^{-1} is a zero of $H(z^{-1})$. The zeros of the function $F(z)$ are thus symmetric with respect to the real axis and mirrored in the unit circle. If the coefficients of $H(z)$ are real, the zeros of $F(z)$ will also be symmetric with respect to the real axis. The same arguments hold for the poles of $H(z)$. The poles and zeros of $F(z)$ will thus have the pattern shown in the figure 3.2.

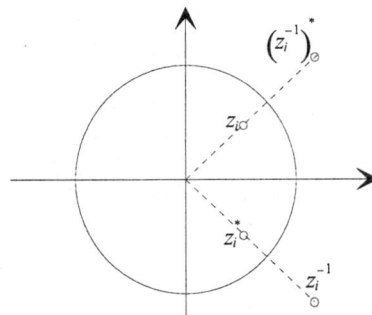

Fig. 3.2.: Symmetry of the poles and zeros of the rational spectral density

It is now straightforward to find a transfer function H that corresponds to a given rational spectral density as follows:

1. Determine the poles p_i and the zeros z_i of the function F in (3.20) associated with the spectral density, i.e. $S(z) = F(z)$.

2. It follows from the symmetry of the poles and zeros (see, fig. 3.2) that the poles and zeros always appear in pairs such that $z_i z_j = 1$ and $p_i p_j = 1$. In each pair choose the pole, or the zero, that is less or equal to one in magnitude; then form the desired transfer function from the chosen poles and zeros as

$$H(z) = k \frac{\prod_i (z - z_i)}{\prod_i (z - p_i)} = \frac{B(z)}{A(z)}$$

Since the stochastic process is stationary, the chosen poles p_i will all be strictly less than one in magnitude. There may, however, be zeros that have unit magnitude. We summarize this discussion with the *Spectral factorization theorem* which states that given a spectral density $S(\exp(j\omega))$, which is rational in $\cos(\omega)$, there exists a linear system with the pulse-transfer function $H(z) = B(z)/A(z)$, such that the output obtained when the system is driven by white noise is a stationary random process with the spectral density $S(\exp(j\omega))$. The polynomial A has all its zeros inside the unit disc. The polynomial B has all its zeros inside or on the unit disc.

The spectral factorization theorem is very important, and it implies that all stationary random processes can be thought of as being generated by stable linear system driven by white noise. This means a considerable simplification, both in theory and practice, since it is sufficient to understand how systems behave when excited by white noise. Thus, it is sufficient to be able to simulate white noise, and all other stationary processes with rational spectral density can then be formed by filtering. Summarizing, the spectral simulation procedure is as follows:

1. Calculate $S_y(z)$, $z = \exp(j\omega T)$ (usually it is adopted that the sampling period $T = 1$);

2. Perform the spectral factorization of eq. (3.20), where $F(z) = S_Y(z)$;

3. Generate a white noise sequence of variance $R_u(0) = \sigma_u^2$;

4. Use the white-noise sequence to excite the system $H(z)$.

The most difficult part of this procedure is to perform the spectral factorization of the step 2. For simple systems, the factorization can be performed by equating the coefficients of the known $S_y(z)$, obtained in the step 1, with the unknown coefficients of the spectral factors and solving the resulting nonlinear algebraic equations. Thus,

$$S_Y(z) = \begin{cases} \dfrac{B_s(z, z^{-1})}{A_s(z, z^{-1})} \quad ; \; (\text{step 1}) \\[4mm] H(z) H(z^{-1}) \sigma_u^2 = \dfrac{B(z)}{A(z)} \dfrac{B(z^{-1})}{A(z^{-1})} \sigma_u^2 \quad ; \; (\text{step 2}) \end{cases} \qquad (3.21)$$

where $B_s(z, z^{-1}) = B(z) B(z^{-1})$ and $A_s(z, z^{-1}) = A(z) A(z^{-1})$, with $A(z)$ and $B(z)$ being the stable polynomials (see, fig. 3.2). For high-order systems, more efficient iterative techniques exist, including multi input-multi output systems. Moreover, to construct the power spectrum from the step 1, we can use the *sum decomposition* as follows

$$S_Y(z) = Z\{R_Y(n)\} = \sum_{n=-\infty}^{\infty} R_Y(n) z^{-n}$$

$$= \sum_{n=0}^{\infty} R_Y(n) z^{-n} + \sum_{n=-\infty}^{\infty} R_Y(n) z^{-n} - R_Y(0) = S_Y^+(z) + S_Y^-(z) - R_Y(0) \qquad (3.22)$$

where S^+ and S^- signify positive and negative time. Furthermore, it can be shown that S_Y^+ has only poles inside the unit circle, while S_Y^- has only those outside. We illustrate this discussion with a simple example.

Example 3.1 (modeling of a stochastic process): Suppose we would like to generate a random sequence with autocovariance

$$R_Y(nT) = a^{|nT|}, \quad \text{for } 0 < a < 1 \text{ and } T = 1$$

From the sum decomposition of eq. (3.22), we obtain

$$S_Y(z) = S_Y^+(z) + S_Y^-(z) - R_Y(0)$$

$$S_Y^+(z) = \sum_{n=0}^{\infty} R_Y(n) z^{-n} = \sum_{n=0}^{\infty} (az^{-1})^n = \frac{1}{1 - az^{-1}}$$

$$S_Y^-(z) = \sum_{n=-\infty}^{0} R_Y(n) z^{-n} = \sum_{n=-\infty}^{0} (az)^n = \frac{1}{1 - az}$$

from which one concludes

$$S_Y(z) = \frac{1}{1 - az^{-1}} + \frac{1}{1 - az} - 1 = \frac{1 - a^2}{(1 - az^{-1})(1 - az)} = H(z) H(z^{-1}) \sigma_u^2$$

Since S_Y is already in the factorized form, we identify H and σ_u^2 by inspection

$$B_s(z, z^{-1}) = B(z) B(z^{-1}) = 1$$

$$A_s(z, z^{-1}) = A(z) A(z^{-1}) = (1 - az^{-1})(1 - az)$$

$$S_u(z) = \sigma_u^2 = 1 - a^2$$

so that

$$H(z) = \frac{B(z)}{A(z)} = \frac{1}{1 - az^{-1}}$$

It is worthwhile noting that $S_Y^-(z) = S_Y^+(z^{-1})$.

An important consequence of the spectral factorization theorem is that for systems with one output, it is always possible to represent the net effect of all disturbances with one equivalent disturbance. This disturbance is obtained by calculating the total spectral density of the output signal and applying the spectral factorization theorem. Furthermore, since a continuos function can be approximated on a compact interval with a rational function, it follows that the input-output model (3.3) and the spectral factorization theorem can give signals whose spectra are arbitrarily close to any continuos function. Notice, however, that these are models with nonrational spectral densities (for example, in turbulence theory there are spectral densities that decay as fractional powers of ω for large ω). Finally, as it is mentioned in relation to eq. (3.21), it is often assumed that the polynomial B has all its zeros inside the unit disc. This means that the inverse filter $1/H$ of the system H is stable.

3.1.4 Polynomial representation: ARMAX models

The input-output, or transfer function, model is represented by a pulse transfer function H, and it is given by [13, 32, 33]

$$H(z) = B(z^{-1}) / A(z^{-1}) \tag{3.23}$$

where A and B are polynomials in z or z^{-1}, i.e.

$$A(z^{-1}) = 1 + \sum_{i=1}^{n} a_i z^{-i} \quad ; \quad B(z^{-1}) = \sum_{i=0}^{m} b_i z^{-i} \tag{3.24}$$

If we consider the equivalent time domain representation, then we have a difference equation relating the output sequence $\{y(k)\}$ to the input sequence $\{u(k)\}$. Here, we use the physical interpretation of z^{-1} as the backward shift operator, which states that $z^{-1}y(k) = y(k-1)$. In this way, we obtain

$$y(k) = \frac{B(z^{-1})}{A(z^{-1})} u(k) = \frac{\sum_{i=0}^{m} b_i z^{-i}}{1 + \sum_{i=1}^{n} a_i z^{-i}} u(k) \tag{3.25}$$

or equivalently

$$A(z^{-1}) y(k) = B(z^{-1}) u(k) \tag{3.26a}$$

from which one concludes

$$y(k) - \sum_{i=1}^{n} a_i y(k-i) = \sum_{i=0}^{m} b_i u(k-i) \tag{3.26b}$$

When the system is excited by a random input sequence, $\{e(k)\}$, and a deterministic input sequence, the so-called exogenous input $\{u(k)\}$, the model (3.26) takes a slightly different form

$$A(z^{-1}) y(k) = B(z^{-1}) u(k) + C(z^{-1}) e(k) \tag{3.27a}$$

where A and B are polynomials defined by (3.24) and

$$C(z^{-1}) = 1 + \sum_{i=0}^{l} c_i z^{-i} \tag{3.27b}$$

The model (3.27) is known in the literature as the autoregressive moving-average model with exogenous input, or ARMAX model. The ARMAX model represents the general form for popular time-series and digital-filter models, that is, it represents:

1. for $C=0$, the pulse transfer function or infinite impulse response (IIR) model;
2. for $A=1$ and $C=0$, the finite impulse response (FIR) model;
3. for $B=0$ and $C=1$, the autoregressive (AR) model;
4. for $A=1$ and $B=0$, the moving-average (MA) model;
5. for $B=0$, the autoregressive moving-average (ARMA) model;
6. for $C=1$, the autoregressive model with exogenous input (ARX).

The ARMAX model is shown in fig. 3.3.

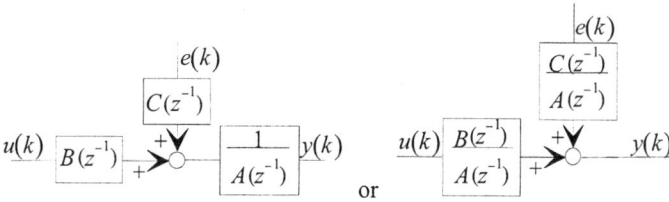

Fig. 3.3 ARMAX model

ARMAX models can easily be used for signal-processing purposes, since they are basically digital filters, or linear systems, with two inputs: known input $u(k)$ and random input $e(k)$. Consider the following example of spectral shaping using the ARMAX model.

Example 3.2 (spectral shaping using polynomial representation): Let $a = 0.5$ in the previous example and random input $\{e(k)\}$ is a zero-mean white stochastic process with unit variance $\sigma_e^2 = 1$. Now generate the output sequence $\{y(k)\}$ using an AR model with the pulse-transfer function

$$H(z) = \frac{Z\{y(k)\}}{Z\{e(k)\}} = \frac{Y(z)}{E(z)} = \frac{1}{1 - 0.5z^{-1}} = \frac{C(z^{-1})}{A(z^{-1})}$$

Cross-multiplying and using the backward-shift property of z^{-1}, we obtain the difference equation

$$\frac{Y(z)}{E(z)} = \frac{1}{1 - 0.5z^{-1}} \Rightarrow y(k) = 0.5y(k-1) + e(k)$$

From the preceding example, the covariance function is given by $R_y(\tau) = 0.5^{|\tau|}$, while the corresponding power spectrum is

$$S_y(z) = \frac{1 - a^2}{(1 - az^{-1})(1 - az)}$$

or equivalently, for $a=0.5$ and $z = \exp(j\omega)$

$$S_y(\exp(j\omega)) = \frac{0.75}{1 - 0.5(\exp(j\omega) + \exp(-j\omega)) + 0.5^2} = \frac{0.75}{1 - \cos(\omega) + 0.25}$$

MATLAB program code, given by the Fig. 3.4, generates the stochastic signal with obtained power spectral density, and uses the periodogram method for estimating the spectral density based on the signal samples. Also, this program makes the comparison between the estimated and true spectrum.

```
function probe
% generation of white stochastic process samples e(k)
N=500; the length of the sequence
for i=1:N
  e(i)=randn;
end
% generation of the coloured stochastic process samples y(k) with given spectrum
y(1)=e(1);
for i=2:N
  y(i)=0.5*y(i-1)+e(i);
end
% the shape of the theoretical spectrum
for i=1:512
  w(i)=(i-1)*pi/(2*512);
  PY1(i)=1/(1.25-cos(w(i)));
end
% the estimation of power spectral density using periodogram method
Y=fft(y,1024);PY2=Y.*conj(Y);PY2=real(PY2)/1024;
plot(w,PY1,'--',w,PY2(1:512));legend('theoreitcal','estimated'); pause;
```

Figure 3.4. MATLAB program code

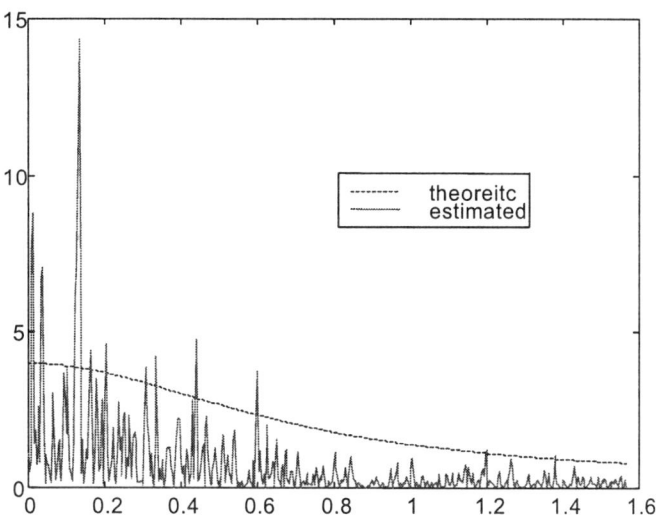

Fig. 3.5. Estimated and theoretical spectrum

3.2 State-space models

An alternative model that has been quite popular and has had to many intcresting estimation results is the state-space model, or internal representation of a discrete time-invariant process. The concept of states has its roots in cause-and-effect relationship in classical mechanics. The motion of a system of particles is uniquely determined for all future times by the present positions and moments of the particles and the future forces. How the present positions and moments were achieved is not important. The state is an abstraction of this property; it is the minimal information about the history of a system required to predict its future motion. For stochastic systems, it cannot be required that the future motion be determined exactly. A natural extension of the motion, or state, for stochastic systems is to require that the probability distribution of future states be uniquely given by the current states. As mentioned before, stochastic processes with this property are called Markov processes. Markov processes are thus the stochastic equivalent of state-space models [1, 2, 15, 25, 30, 35, 36].

3.2.1 Analysis

Consider a discrete-time system where the sampling period is chosen as the time unit ($T = 1s$). Let the state at time k be given by $x(k)$. If the mean value is linear in $x(k)$ and the distribution around the mean is independent of $x(k)$, then $x(k+1)$ can be represented as

$$x(k+1) = Ax(k) + Gw(k) \qquad (3.28)$$

where $w(k)$ is a random variable with zero-mean that is independent of $x(k)$, and independent of all past values of x. This implies that $w(k)$ is also independent of all past w's. The sequence $\{w(k), k = \cdots, -1, 0, 1, \cdots\}$ is a sequence of independent, equally distributed random variables. The stochastic process $\{w(k)\}$ is thus discrete-time white noise. The covariance of the random variable $w(k)$ is denoted by Q, i.e. $E\{w(k)w^T(k)\} = Q$. Equation (3.28) is called a linear stochastic difference equation. To define the random process $\{x(k)\}$ completely, it is necessary to specify the initial condition $x(0)$. It is assumed that initial state has the mean $m_x(0) = E\{x(0)\} = m_0$ and the covariance matrix $P_x(0) = E\{[x(0)-m_0][x(0)-m_0]^T\} = P_0$. The block diagram of the model is shown in Fig. 3.6.

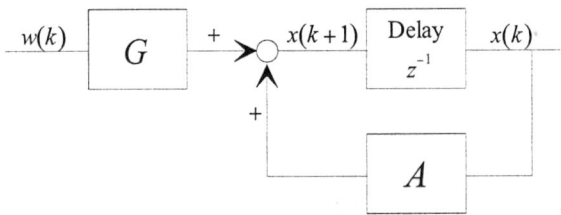

Fig. 3.6: Discrete linear system dynamics model: state-space model

Here it is understood that on any one "run" of the system the input $w(k)$, $k = ..., -1, 0, 1, ...$ is any sample function, or realization, of the white noise sequence $\{w(k)\}$, and the initial condition vector $x(0)$ is any sample, or realization, drawn from the given distribution of initial condition random vector $x(0)$.

The solution to the state-difference equation (3.28) is the following. From eq. (3.28) it is clear that

$$x(k+1) = A(k)x(k) + G(k)w(k)$$
$$x(k+2) = A(k+1)x(k+1) + G(k+1)w(k+1)$$

Substituting the former into the latter and rearranging terms, we have

$$x(k+2) = A(k+1)\left[A(k)x(k) + G(k)w(k)\right] + G(k+1)w(k+1)$$
$$= \Phi(k+2,k)x(k) + \Phi(k+2,k+1)G(k)w(k) + G(k+1)w(k+1)$$
$$= \Phi(k+2,k)x(k) + \sum_{i=k+1}^{k+2} \Phi(k+2,i)G(i-1)w(i-1)$$

where $\Phi(k+2,k) = A(k+1)A(k)$. Continuing in this manner, we obtain the general relationship

$$x(k+n) = \Phi(k+n,k)x(k) + \sum_{i=k+1}^{k+n} \Phi(k+n,i)G(i-1)w(i-1) \qquad (3.29)$$

where $n = 1, 2, ...$, with

$$\Phi(k+n,k) = A(k+n-1) \cdot A(k+n-2) \cdots A(k) \qquad (3.30a)$$

Here is assumed that the system model (3.28) is time-varying, i.e. $A = A(k)$ and $G = G(k)$ in (3.28). Thus, for a time-invariant system, where A and G in (3.28) are constant matrices, we have

$$\Phi(k+n,k) = A^{(k+n)-k} = A^n \quad \text{for} \quad n > 0 \qquad (3.30b)$$

For $k+n = j$, it is obvious that

$$x(j) = \Phi(j,k)x(k) + \sum_{i=k+1}^{j} \Phi(j,i)G(i-1)w(i-1) \qquad (3.31)$$

Setting $k = 0$ in (3.31), one obtains

$$x(j) = \Phi(j,0)x(0) + \sum_{i=1}^{j} \Phi(j,i)G(i-1)w(i-1) \qquad (3.32)$$

We show now that the process $\{x(k), k \in I\}$, with I being a set of integers, is Markov. Let $t_1 < t_2 < \cdots < t_m$ be any m time points in some interval I, where m is any integer. Also, let j and k be the integers in I that correspond to the time points t_m and t_{m-1}, respectively. The situation may be considered as shown in fig. 3.7. It is clear that there will, in general, be points in I that lie between the time points t_i, $i = 1,...m$, as well as outside the interval of time spanned by the points t_i.

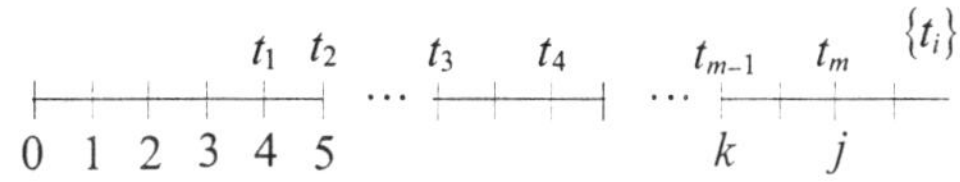

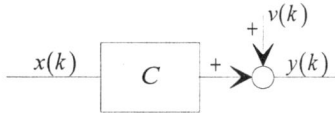

Fig. 3. 7: Index set I and arbitrary set of order time points $\{t_i\}$

Recalling that j corresponds to t_m and k to t_{m-1}, it is clear from (3.31) that the conditional probability distribution of $x(t_m)$ for any given set of values of $x(t_{m-1}),...,x(t_1)$ depends only upon $x(t_{m-1})$,; that is the conditioning is dependent only on $x(t_{m-1})$. Since this is true for any m time points $t_1 < t_2 < \cdots < t_m$, where m is any integer, $\{x(k), k \in I\}$ is obviously Markov process.

It is trivial to show when the process is Gaussian, or normal. If $x(0)$ and each $w(i-1)$, $i = 1,...,j$, are Gaussian, then $x(j)$ is a linear transformation of Gaussian random variables, and therefore Gaussian. Thus, we can represent a Gauss-Markov process easily using the state-space models (28).

When we include the measurement model as well, then the Gauss-Markov model evolves, based on the state-space representation (3.28) and the corresponding discrete measurement model

$$y(k) = Cx(k) + v(k) \qquad (3.33)$$

where y is the measurement vector, C is a matrix that relates the state vector to the measurement vector, and v is the measurement error vector or measurement noise. We assume that the measurement error process can be modeled as a Gaussian white discrete-time process, whose mean $E\{v(k)\} = 0$ and covariance $E\{v(k)v^T(k)\} = R$, where R is a given matrix. We also assume that $\{v(k)\}$ is independent of $x(0)$ for all $k = ...,-1,0,1,...$, i.e. $E\{[x(0) - m_0]v^T(k)\} = 0$, as well as that $w(k)$ and $v(j)$ are uncorrelated with each other for all $k, j = ...,-1,0,1,...$; that is, $E\{w(k)v^T(j)\} = 0$. The measurement model has the block diagram shown in fig. 3.8.

Fig. 3.8: Discrete linear system measurement model

Our measurement model presupposes that measurement errors $v(k)$ made at one time are independent of those at any other time. The justification for the assumed independence of $x(0)$ and $\{v(k)\}$ is that, physically, we expect the mechanism from which the measurement errors $\{v(k)\}$ arise to be independent of the one leading to the initial state $x(0)$. Generally, the measurement system for a given system is peripheral to the system dynamics, so that the hardware and electronic component imperfections, which cause measurement errors, function independently of the initial state $x(0)$. Also, from basic implementation considerations, it is plausible to expect that mechanisms producing system disturbances $\{w(k)\}$ are unrelated to those that generate measurement errors $\{v(k)\}$ in many cases [51].

Since the Gauss-Markov model of eq. (3.28) is stationary, it is completely characterized statistically by its mean and covariance. To obtain the mean-value function

$$m_x(k) = E\{x(k)\}$$

simply take the mean values of both sides of (3.28). Because $w(k)$ has zero mean, the following difference equation is obtained

$$m_x(k+1) = Am_x(k) \tag{3.34a}$$

The initial condition is

$$m_x(0) = m_0 = E\{x(0)\} \tag{3.34b}$$

The mean value will thus propagate in the same way as the unperturbed system $(w(k)=0)$. To calculate the covariance matrix, introduce

$$P_x(k) = \text{cov}\{x(k)\} = E\{\tilde{x}(k)\tilde{x}^T(k)\} \tag{3.35a}$$

where

$$\tilde{x}(k) = x(k) - m_x(k) \tag{3.35b}$$

It follows from eqs. (3.28) and (3.34a) that $\tilde{x}$ satisfies

$$\tilde{x}(k+1) = x(k+1) - m_x(k+1) = Ax(k) + Gw(k) - Am_x(k) = A\tilde{x}(k) + Gw(k) \tag{3.36a}$$

with initial condition

$$E\{\tilde{x}(0)\} = E\{x(0) - m_0\} = 0 \tag{3.36b}$$

To calculate the covariance, form the expression

$$\tilde{x}(k+1)\tilde{x}^T(k+1) = \left[A\tilde{x}(k) + Gw(k)\right]\left[A\tilde{x}(k) + Gw(k)\right]^T \tag{3.37a}$$
$$= A\tilde{x}(k)\tilde{x}^T(k)A^T + A\tilde{x}(k)w^T(k)G^T + Gw(k)\tilde{x}^T(k)A^T + Gw(k)w^T(k)G^T$$

Taking mean values gives

$$P_x(k+1) = AP_x(k)A^T + GQG^T \tag{3.37b}$$

because $w(k)$ and $\tilde{x}(k)$ are independent, i.e. $E\{\tilde{x}(k)w^T(k)\} = 0$ (from eqs. (3.32), (3.34) and (3.35) $\tilde{x}(k)$ is combination of $w(0), w(1), ..., w(k-1)$, all of which are uncorrelated with $w(k)$). The initial conditions are

$$P_x(0) = E\left\{\tilde{x}(0)\tilde{x}^T(0)\right\} = P_0 \tag{3.37c}$$

The recursive equation (3.37b) for P_x tells how the covariance propagates. This equation is fundamental equation for the time-domain analysis of discrete-systems with stochastic inputs. Note that (3.37) represents a nonstationary situation, since the covariance $P_x(k)$ depends on the time k of the occurrence of x. However, if the system is time-invariant and stable, then we have A, G, C, Q and R are the constant matrices, as well the characteristic roots (eigenvalues) of A are inside the unit circle (the characteristic roots of A are the solutions of the characteristic equation $\det(zI - A) = 0$, which defines the poles of the state-error dynamics system in (3.36)). In this case, the effects of the initial condition $P_x(0)$ gradually diminish, and $P_x(k)$ approaches a stationary value. This value is given by the solution to the Lyapunov equation (in steady state $(k \to \infty)$, we have $P_x(k+1) = P_x(k) = P_x$ in (3.37))

$$P_x = AP_x A^T + GQG^T \tag{3.38}$$

Moreover, the different terms of (3.37) also have good physical interpretations. The covariance P_x may represent the uncertainty in the state; the term $AP_x(k)A^T$ tells how the uncertainty at time k propagates due to the system dynamics, and the term GQG^T describes the increase of uncertainty due to the disturbance w.

To calculate the covariance function of the state, observe that

$$\tilde{x}(k+1)\tilde{x}^T(k) = \left[A\tilde{x}(k) + Gw(k)\right]\tilde{x}^T(k)$$

Because $w(k)$ and $\tilde{x}(k)$ are independent, and $w(k)$ has zero mean

$$R_x(k+1,k) = \text{cov}\left\{x(k+1), x(k)\right\} = E\left\{\tilde{x}(k+1)\tilde{x}^T(k)\right\} = AP_x(k)$$

Repeating this discussion,

$$R_x(k+\tau,k) = R_x(\tau) = \text{cov}\left\{x(k+\tau), x(k)\right\} = A^\tau P_x(k) ; \ \tau \geq 0 \tag{3.39a}$$

Because the covariance function is an even function, i.e. $R_x(-\tau) = R_x(\tau)$, one concludes

$$R_x(\tau) = \text{cov}\left\{x(k+\tau), x(k)\right\} = E\left\{\tilde{x}(k+\tau)\tilde{x}^T(k)\right\} = A^{|\tau|}P_x(k) \tag{3.39b}$$

The mean-value function of the system output y in (3.33) is given by

$$m_y = E\left\{y(k)\right\} = CE\left\{x(k)\right\} + E\left\{v(k)\right\}$$

and since $E\left\{v(k)\right\} = 0$, one obtains

$$m_y = Cm_x \tag{3.40}$$

The covariance function of the output is given by

$$R_y(k+\tau,k) = R_y(\tau) = R\left\{\tilde{y}(k+\tau)\tilde{y}^T(k)\right\} \tag{3.41a}$$

where

$$\tilde{y}(k) = y(k) - m_y = C\tilde{x}(k) + v(k) ; \ \tilde{x}(k) = x(k) - m_x \tag{3.41b}$$

Because $v(k)$ and $\tilde{x}(k)$ are independent by assumption, it follows from eqs. (3.39a) and (3.41)

$$R_y(\tau) = CR_x(\tau)C^T \; ; \; \tau > 0 \tag{3.42a}$$

and it is also possible to show that for zero lag $\tau = 0$

$$R_y(0) = E\{\tilde{y}(k)\tilde{y}^T(k)\} = CR_x(0)C^T + R \tag{3.42b}$$

The cross-covariance between y and x is given by

$$R_{yx}(k+\tau,k) = E\{\tilde{y}(k+\tau)\tilde{x}^T(k)\} = E\{[C\tilde{x}(k+\tau)+v(k+\tau)]\tilde{x}^T(k)\} = CR_X(\tau) \tag{3.43}$$

The obtained results are so important that they deserve to be summarized in the following theorem:

Theorem: Consider a random process $\{x(k)\}$ defined by the stochastic difference equation (3.28), where $\{w(k)\}$ is a white-noise process with zero-mean and covariance Q. Let the initial state $x(0)$ has mean m_0 and covariance P_0. The mean-value function of the process is then given by

$$m_x(k+1) = Am_x(k) \; ; \; m(0) = m_0 \tag{3.44}$$

and the covariance function by

$$R_x(k+\tau,\tau) = R_x(\tau) = A^{|\tau|}P_x(k) \tag{3.45}$$

where $P_x(k) = \text{cov}\{x(k), x(k)\}$ is given by

$$P_x(k+1) = AP_x(k)A^T + GQG^T \; ; \; P_x(0) = P_0 \tag{3.46}$$

If the system has an output y defined by (3.33), where $\{v(k)\}$ is a white-noise process with zero-mean and covariance R, then the mean-value function of y is given by

$$m_y = Cm_x \tag{3.47}$$

and its covariance for nonzero lags τ is given by

$$R_y(k+\tau,k) = R_y(\tau) = CR_x(k+\tau,k)C^T = CR_x(\tau)C^T \tag{3.48}$$

while for zero lag $\tau = 0$

$$R_y(k,k) = R_y(0) = CR_x(0)C^T + R \tag{3.49}$$

The cross-covariance between x and y is given by

$$R_{yx}(k+\tau,k) = R_y(\tau) = CR_x(\tau) \tag{3.50}$$

3.2.2. Equivalence of linear models: transformation of state-space model to transfer function model

The measurement power spectral density is easily obtained by taking the Z-transform of eq. (3.33) for the time-invariant case, i.e. [1, 30, 48]

$$Y(z) = CX(z) + V(z) \tag{3.51}$$

The power spectral density of the process $\{y(k)\}$ may then be represented in terms of the Fourier, or Z, transform of the covariance of a truncated signal $\{y_T(k)\}$ with truncating interval T

$$S_y(\exp(j\omega)) = \lim_{T \to \infty} E\left\{ \frac{Y_T(\exp(j\omega))Y_T^T(\exp(-j\omega))}{T} \right\} = Z\{R_y(\tau)\}\Big|_{z=\exp(j\omega)} \tag{3.52}$$

Using (3.51), one obtains

$$S_y\left(\exp(j\omega)\right) = C \lim_{T\to\infty} E\left\{\frac{X_T\left(\exp(j\omega)\right)X_T^T\left(\exp(-j\omega)\right)}{T}\right\}C^T + \lim_{T\to\infty} E\left\{\frac{V\left(\exp(j\omega)\right)V^T\left(\exp(-j\omega)\right)}{T}\right\}$$

$$= CS_x\left(\exp(j\omega)\right)C^T + S_v\left(\exp(j\omega)\right) = CS_x\left(\exp(j\omega)\right)C^T + R$$

$$(3.53)$$

Here is used the fact that

$$E\left\{\frac{X_T\left(\exp(j\omega)\right)V^T\left(\exp(-j\omega)\right)}{T}\right\} = 0,$$

since the process $\{x(k)\}$ is uncorrelated with the white measurement noise $\{v(k)\}$ by assumption, and $S_v(z) = R$.

The expression for $X(z)$ may be obtained by taking the Z-transform, or Fourier transform, of eq. (3.28) as

$$zX(z) = AX(z) + GW(z) \tag{3.54a}$$

or

$$X(z) = T(z)W(z) \quad \text{for} \quad T(z) = (zI - A)^{-1}G \tag{3.54b}$$

The spectral density $S_x(z)$ may then be written as

$$S_x(z) = \lim_{T\to\infty} E\left\{\frac{X_T(z)X_T^T(z^{-1})}{T}\right\}$$

$$= T(z)\lim_{T\to\infty} E\left\{\frac{W(z)W^T(z^{-1})}{T}\right\}T^T(z^{-1}) = T(z)S_w(z)T^T(z^{-1}) \tag{3.55}$$

$$= T(z)QT^T(z^{-1})$$

Thus, we see that using formal deterministic concept of eqs. (3.28) and (3.33)

$$Y(z) = H(z)W(z) + V(z) \quad \text{for} \quad H(z) = CT(z) \tag{3.56}$$

so that the power spectral density of the output process $\{y(k)\}$ is given by

$$S_Y(z) = H(z)S_w(z)H^T(z^{-1}) + S_v(z) \quad \text{for} \quad S_w(z) = Q \text{ and } S_v(z) = R \tag{3.57}$$

Here is used the fact that the noise processes $\{w(k)\}$ and $\{v(k)\}$ are uncorrelated by assumption. The block diagram of the transfer function model is given in fig. 3.9.

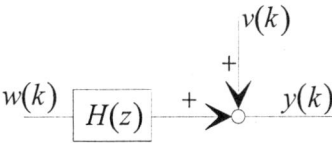

Fig. 3.9: Transfer function representation of linear state-space model with random input

Before we leave this subject, let us analyze a simple example.

Example 3.3 (analysis of the first order state-space model):

Consider the first order system

$$x(k+1) = ax(k) + w(k) + u(k)$$
$$y(k) = x(k) + v(k)$$

where w and v are sequences of uncorrelated random variables with zero-mean values and variances $Q = 1$ and $R = 0.5$, respectively. Let $u(k)$ be the unit step, i.e. $u(k) = 1$ for $k \geq 0$ and $u(k) = 0$ for $k < 0$, and let the state $x(0)$ at time $k = 0$ have the mean $m_0 = 0$ and variance $P_0 = 1$. It follows from (3.34) that the mean value of the state is given by

$$m_x(k+1) = am_x(k) + 1 \; ; \; m_x(0) = m_0 = 0$$

which has the solution

$$m_x(k) = a^k m_0 + \sum_{i=0}^{k-1} a^{k-1-i} = a^{k-1} \sum_{i=0}^{k-1} (a^{-1})^i = a^{k-1} \frac{1-(a^{-1})^k}{1-a^{-1}} = \frac{a^k - 1}{a-1}$$

This has a limiting value $m_x(\infty) = \lim_{k \to \infty} m_x(k)$ if and only if $|a| < 1$; that is, if the system is stable. It should be noted that the corresponding transfer function $H(z)$ in (3.56) reduces to $H(z) = \dfrac{1}{z-a}$, so that the system has a pole $z = a$. In that case, a limiting value is

$$m_x(\infty) = \frac{1}{1-a}$$

To find the covariance of the state use (3.37)

$$P_x(k+1) = a^2 P_x(k) + Q \; ; \; P_x(0) = P_0 = 1 \; ; \; Q = 1$$

Thus,

$$P_x(k) = (a^2)^k + \sum_{i=0}^{k-1} (a^2)^{k-1-i} = a^{2k} + \frac{a^{2k} - 1}{a^2 - 1} = \frac{a^{2(k+1)} - 1}{a^2 - 1}$$

There is a limiting solution $P_x(\infty)$ if and only if $|a| < 1$, and then

$$P_x(\infty) = \frac{1}{1-a^2}$$

If w, v and $x(0)$ are normal variables then $x(k)$ and $y(k)$ are also normal, and the state probability density function (pdf) $f_{x(k)}(x)$ is given by

$$f_{x(k)}(x) = \frac{1}{\sqrt{2\pi P_x(k)}} \exp\left(-\frac{1}{2} \frac{(x - m_x(k))^2}{P_x(k)} \right)$$

Note that this is actually a conditional pdf given $m_x(0)$, $P_x(0)$ and inputs $u(0), u(1), ..., u(k-1)$. The conditional pdf of the state is commonly defined as *hyperstate*, and we sketch it as a function of k. This is done in fig. 3.10 for $a = 0.5$ and $a = 2$.

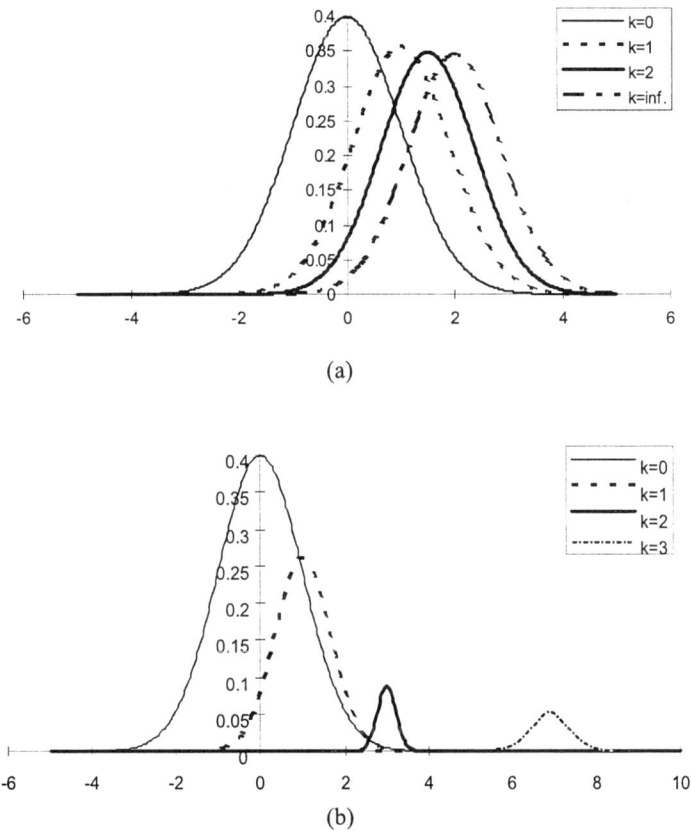

(a)

(b)

Fig. 3.10: Propagation of the pdf of state: a) stable system (a=0.5); b) unstable system (a=2)

It is evident that the effect of the process, or state, noise w is to "flatten" the pdf for $x(k)$ as k increases, thus decreasing our confidence that $x(k)$ is near $m_x(k)$. There are two cases. If the system is stable a limiting pdf $f_x(x) = \lim_{k \to \infty} f_{x(k)}(x)$ is reached and subsequent propagation does not inject further uncertainty (fig. 3.10.a). This limiting case is known as statistical steady state. In this case, we can say that $\lim_{k \to \infty} x(k)$ is near $m_x(\infty)$ to some tolerance described by $P_x(\infty)$. On the other hand, if a system is unstable, there is no limiting pdf and, as k increases, both the value of $m_x(k)$ and the probable spread of $x(k)$ about $m_x(k)$, $P_x(k)$, go to infinity (fig.3.10.b).

The correlation function of the output process $\{y(k)\}$ is given by (3.48) and (3.49), i.e.

$$R_y(\tau) = a^{|\tau|} P_x(\infty) = \frac{a^{|\tau|}}{1 - a^2} \; ; \; R_y(0) = P_x(\infty) + R = \frac{1}{1 - a^2} + 0.5$$

and its power spectral density is given by (3.56) and (3.57)

$$S_y(z) = H(z) H(z^{-1}) + 0.5$$

where, from (3.54b) and (3.56),

$$H(z) = (z-a)^{-1}$$

Thus

$$S_y(z)\Big|_{z=\exp(j\omega)} = \left[0.5 + \frac{1}{1-a(z+z^{-1})+a^2}\right]_{z=\exp(j\omega)} = 0.5 + \frac{1}{1-2a\cos(\omega)+a^2}$$

So we conclude that for stationary processes the input-output and state space representation are equivalent in the stochastic, as well as the deterministic cases (see Ex. 3.6).

3.2.3. Equivalence of linear models: Transformation of ARMAX to State Space Model

In this section we show the equivalence between the ARMAX, or polynomial, and state-space models for scalar processes. That is, we choose a particular coordinate system in the state space, the so-called *observer canonical form*, and obtain a relationship between entries of the state space system to coefficients of the ARMAX model. An example is presented that shows how these models can be applied to realize a stochastic process [25].

The general difference equation form of the ARMAX model is given by

$$y(k) = -\sum_{i=1}^{n} a_i y(k-i) + \sum_{i=0}^{m} b_i u(k-i) + \sum_{i=0}^{l} c_i e(k-i) \tag{3.58}$$

where $n \geq m$, l and $\{e(k)\}$ is a zero-mean white noise sequence with spectrum given by R_e,

It is straight forward to show that the ARMAX model can be represented in the observer *canonical form*

$$x(k) = A_0 x(k-1) + B_0 u(k-1) + G_0 e(k-1) \tag{3.59a}$$

$$y(k) = C_0^T x(k) + b_0 u(k) + c_0 e(k) \tag{3.59b}$$

where x, u, e and y are the n-state vector, scalar input, noise and output, respectively, with

$$A_0 = \begin{bmatrix} 0 & \vdots & -a_n \\ \text{-----} & & -a_{n-1} \\ I_{n-1} & \vdots & \vdots \\ & \vdots & -a_1 \end{bmatrix} ; \quad B_0 = \begin{bmatrix} -a_n b_0 \\ \vdots \\ -a_{m+1} b_0 \\ \text{-----} \\ b_m - a_m b_0 \\ \vdots \\ b_1 - a_1 b_0 \end{bmatrix} ; \quad G_0 = \begin{bmatrix} -a_n c_0 \\ \vdots \\ -a_{l+1} c_0 \\ \text{-----} \\ c_l - a_l c_0 \\ \vdots \\ c_1 - a_1 c_0 \end{bmatrix} \tag{3.59c}$$

$$C_0^T = [0 \cdots 0 \ 1]$$

where I_{n-1} denotes the $(n-1) \times (n-1)$ identity matrix.

Example 3.4 (transformation of ARMAX model to observer canonical state-space form): Let us take for simplicity that $n=3$, $m=1$ and $l=1$ in (3.58). Then input-output model (3.58) can be written as

$$y(k) = z^{-1}\left\{\left(b_1 u(k) - a_1 y(k) + c_1 e(k)\right) + z^{-1}\left[-a_2 y(k) + z^{-1}\left(-a_3 y(k)\right)\right]\right\} + c_0 e(k) + b_0 u(k)$$

where z^{-1} is the backward-shift operator, i.e. $z^{-1}\{x(k)\} = x(k-1)$.

By introducing the state coordinates as

$$x_1(k) = z^{-1}(-a_3y(k))$$
$$x_2(k) = z^{-1}\left[-a_2y(k) + z^{-1}(-a_3y(k))\right]$$
$$x_3(k) = z^{-1}\left\{(b_1u(k) - a_1y(k) + c_1e(k)) + z^{-1}\left[-a_2y(k) + z^{-1}(-a_3y(k))\right]\right\}$$

one obtains

$$y(k) = x_3(k) + c_0e(k) + b_0u(k)$$
$$x_1(k+1) = zx_1(k) = -a_3y(k)$$
$$\qquad = -a_3x_3(k) - a_3b_0u(k) - a_3c_0e(k)$$
$$x_2(k+1) = zx_2(k) = -a_2y(k) + x_1(k)$$
$$\qquad = -a_2x_3(k) - a_2b_0u(k) - a_2c_0e(k) + x_1(k)$$
$$x_3(k+1) = zx_3(k) = b_1u(k) - a_1y(k) + c_1e(k) + x_2(k)$$
$$\qquad = b_1u(k) - a_1x_3(k) - a_1b_0u(k) - a_1c_0e(k) + c_1e(k) + x_2(k)$$

Finally, the obtained system of linear equations can be represented in the matrix form

$$\begin{bmatrix} x_1(k+1) \\ x_2(k+1) \\ x_3(k+1) \end{bmatrix} = \begin{bmatrix} 0 & 0 & -a_3 \\ 1 & 0 & -a_2 \\ 0 & 1 & -a_1 \end{bmatrix} \begin{bmatrix} x_1(k) \\ x_2(k) \\ x_3(k) \end{bmatrix} + \begin{bmatrix} -a_3b_0 \\ -a_2b_0 \\ b_1 - a_1b_0 \end{bmatrix} u(k) + \begin{bmatrix} -a_3c_0 \\ -a_2c_0 \\ c_1 - a_1c_0 \end{bmatrix} e(k)$$
$$y(k) = \begin{bmatrix} 0 & 0 & 1 \end{bmatrix} x(k) + b_0u(k) + c_0e(k)$$

which completes the conversion.

If we assume that $\{e(k)\}$ is zero-mean Gaussian, then a particular Gauss-Markov model of eqs. (3.28) and (3.33) evolves from eq. (3.59) and it is called the *innovations model*

$$x(k) = Ax(k-1) + Bu(k-1) + Ge(k-1) \qquad (3.60a)$$

$$y(k) = Cx(k) + b_0u(k) + c_0e(k) \qquad (3.60b)$$

where the covariance matrix

$$R_e^* = \mathrm{cov}\left\{\begin{bmatrix} Ge(k) \\ c_0e(k) \end{bmatrix}, \begin{bmatrix} Ge(k) \\ c_0e(k) \end{bmatrix}\right\} = E\left\{\begin{bmatrix} Ge(k) \\ c_0e(k) \end{bmatrix}\begin{bmatrix} Ge(k) \\ c_0e(k) \end{bmatrix}^T\right\} = \begin{bmatrix} GR_eG' & c_0GR_e \\ c_0R_eG^T & c_0R_ec_0 \end{bmatrix} ; \quad R_e = E\left\{e(k)e^T(k)\right\}$$
$$(3.61)$$

The covariance matrix R_e^* in eq. (3.61) can be represented in the "factorized" form

$$R_e^* = \begin{bmatrix} G\sqrt{R_e} \\ c_0\sqrt{R_e} \end{bmatrix}\begin{bmatrix} \sqrt{R_e}G^T & \sqrt{R_e}c_0 \end{bmatrix} \qquad (3.62)$$

where $\sqrt{R_e}$ denotes the square root of the matrix R_e. Using the innovations model of eqs. (3.60) and (3.61), it is quite easy to simulate correlated state and measurement noises, because of the factorization property of eq. (3.62), as it is shown in fig. 3.11.

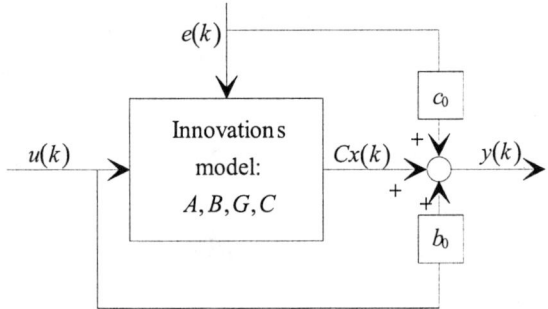

Fig. 3.11: Correlated noise simulation using the innovations state-space model

Comparing the innovations model of eqs. (3.59) to the Gauss-Markov model of eqs. (3.28) and (3.33), we see that they are equivalent, except that eq. (3.59a) corresponds to the case when w and v are correlated. That is, the standard Gauss-Markov model of eq. (3.28) for correlated process (state) and measurement noise is given by

$$x(k) = Ax(k-1) + Bu(k-1) + Gw^*(k-1) \tag{3.63a}$$

$$y(k) = Cx(k) + b_0 u(k) + v^*(k) \tag{3.63b}$$

where the covariance matrix

$$R^* = \text{cov}\left\{ \begin{bmatrix} Gw^* \\ v^* \end{bmatrix}, \begin{bmatrix} Gw^* \\ v^* \end{bmatrix} \right\} = E\left\{ \begin{bmatrix} Gw^* \\ v^* \end{bmatrix} \begin{bmatrix} w^{*T}G^T & v^{*T} \end{bmatrix} \right\} = \begin{bmatrix} GR_w G^T & GR_{wv} \\ R_{vw}G^T & R_v \end{bmatrix} \tag{3.64}$$

with $\quad R_w = E\{w^*(k)w^{*T}(k)\}$; $R_{wv} = E\{w^*(k)v^{*T}(k)\}$; $R_{vw} = E\{v^*(k)w^{*T}(k)\}$ $\quad$ and $R_v = E\{v^*(k)v^{*T}(k)\}$. To simulate a system with correlated w and v, it is more complicated to use the Gauss-Markov model (3.63) then the innovations model (3.60), because R^* must be factorized as

$$R^* = \begin{bmatrix} R_1^* \\ R_2^* \end{bmatrix} \begin{bmatrix} R_1^{*T} & R_2^{*T} \end{bmatrix}$$

where R_i^*, $i = 1,2$ are the matrix square roots. Once the factorization is performed, the correlated noise is simulated using

$$\begin{bmatrix} w^* \\ v^* \end{bmatrix} = \begin{bmatrix} R_1^* w \\ R_2^* v \end{bmatrix}$$

where w and v are uncorrelated. The correlated noise simulation is shown in fig. 3.12.

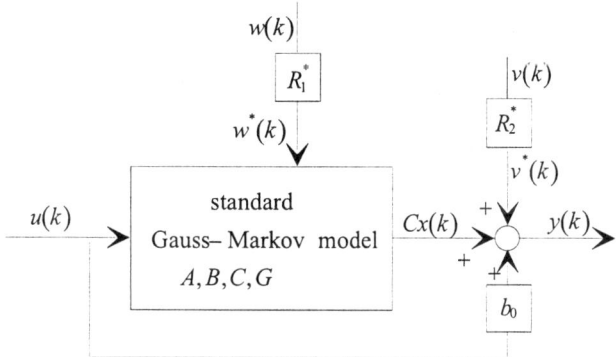

Fig. 3.12: Correlated noise simulation using standard Gauss-Markov model

Note that we need only factor R_e in the innovations model, whereas in the standard Gauss-Markov model we must factor R^*, which is of larger dimension. We conclude this section by showing how to simulate a process with given covariance function using both ARMAX and Gauss-Markov models.

Example 3.5. (Simulation of a stochastic process using ARMAX and state-space models): Suppose we are asked to simulate a process $\{y(k)\}$ with covariance $R_y(\tau) = \exp(-a|\tau|)\cos(\omega\tau)$; $\tau = nT$ using ARMAX and state-space simulators. We use the spectrum simulation procedure; that is, we use the sum decomposition

$$S_y(z) = S_y^+(z) + S_y^-(z) - R_y(0)$$

From tables of Z -transforms (see appendix A), we have

$$S_y^+(z) = Z\{\exp(-anT)\cos(\omega nT)\} = \frac{z(z - \exp(-aT)\cos(\omega T))}{z^2 - 2z\exp(-aT)\cos(\omega T) + \exp(-2aT)}$$

or equivalently

$$S_y^+(z) = \frac{\beta_0 z^2 + \beta_1 z + \beta_2}{z^2 + \alpha_1 z + \alpha_2}$$

where $\beta_0 = 1$, $\beta_1 = \exp(-aT)\cos(\omega T), \beta_2 = 0, \alpha_1 = -2\exp(-aT)\cos(\omega T)$ and $\alpha_2 = \exp(-2aT)$. Next, we must perform the spectral factorization by equating the coefficients in $S_y(z)$, formed from the sum decomposition, and the coefficients in $S_y(z)$ obtained from the adopted spectral factors.. Firstly, from the sum decomposition, we obtain

$$S_y(z) = \frac{B_s(z, z^{-1})}{A_s(z, z^{-1})}$$

where

$$B_s(z, z^{-1}) = \left[\beta_2 + \alpha_2(\beta_0 - 1)\right](z^2 + z^{-2}) + \left[\alpha_1\beta_2 + \beta_1 + \alpha_2\beta_1 - \alpha_1\alpha_2 + \alpha_1(\beta_0 - 1)\right](z + z^{-1})$$
$$+ 2\beta_0 - 1 - \alpha_1^2 - \alpha_2^2 + 2\alpha_1\beta_1 + 2\alpha_2\beta_2$$

and

$$A_s(z, z^{-1}) = (z^2 + \alpha_1 z + \alpha_2)(z^{-2} + \alpha_1 z^{-1} + \alpha_2)$$

Next, we adopt the spectral factors; that is

$$S_y(z) = H(z)H(z^{-1})R_e = \frac{\left(c_0z^2+c_1z+c_2\right)\left(c_0z^{-2}+c_1z^{-1}+c_2\right)R_e}{\left(z^2+a_1z+a_2\right)\left(z^{-2}+a_1z^{-1}+a_2\right)} = \frac{B_H(z)}{A_H(z)}\frac{B_H(z^{-1})}{A_H(z^{-1})}R_e$$

Comparing the factored form of $S_y(z)$ with the sum decomposition, we see that the denominator is found by inspection as

$$A_s(z,z^{-1}) = \alpha_2\left(z^2+z^{-2}\right)+\left(\alpha_1+\alpha_1\alpha_2\right)\left(z+z^{-1}\right)+\left(\alpha_1^2+\alpha_2^2+1\right)$$

$$= \left(z^2+\alpha_1z+\alpha_2\right)\left(z^{-2}+\alpha_1z^{-1}+\alpha_2\right)$$

$$A_H(z) = z^2+a_1z+a_2$$

Therefore,

$$a_1 = \alpha_1 = -2e^{-aT}\cos\omega T \; ; \; a_2 = \alpha_2 = e^{-2aT}$$

It remains to factor the numerators. Multiplication of the polynomials in the numerator of the preceding factorization gives

$$B_H(z)B_H(z^{-1}) = c_0c_2\left(z^2+z^{-2}\right)+\left(c_0c_1+c_1c_2\right)\left(z+z^{-1}\right)+\left(c_0^2+c_1^2+c_2^2\right)$$

Equating numerator coefficients of $B_s(z,z^{-1})$, we have

$$\beta_2+\alpha_2\left(\beta_0-1\right) = c_0c_2$$

$$\alpha_1\beta_2+\beta_1+\alpha_2\beta_1-\alpha_1\alpha_2+\alpha_1\left(\beta_0-1\right) = c_0c_1+c_1c_2$$

$$2\beta_0-1-\alpha_1^2-\alpha_2^2+2\alpha_1\beta_1+2\alpha_2\beta_2 = c_0^2+c_1^2+c_2^2$$

Substituting the values for β_i, α_i in our problem, we have

$$c_0c_2 = \beta_2 = 0 \quad \text{implying} \quad c_2 = 0$$

$$c_0c_1 = \beta_1\left(1+\alpha_2\right)-\alpha_1\alpha_2 = -\left(1+e^{-2aT}\right)e^{-aT}\cos\omega T$$

$$c_0^2+c_1^2+c_2^2 = 1-\alpha_1^2-\alpha_2^2+2\alpha_1\beta_1$$

which can be solved for c_0 and c_1 using the quadratic formula. Thus we have

$$H(z) = \frac{z(c_0z+c_1)}{z^2+a_1z+a_2} = \frac{c_0+c_1z^{-1}}{1+a_1z^{-1}+a_2z^{-2}}$$

By inspection of the coefficients, we have the ARMA model in backward-shift operator z^{-1} notation

$$A(z^{-1})y(k) = C(z^{-1})e(k)$$

where

$$A(z^{-1}) = 1+a_1z^{-1}+a_2z^{-2} \; ; \; C(z^{-1}) = c_0+c_1z^{-1}$$

and $e(k)$ is zero-mean normal with variance R_e.

The state-space model is easily found using the observer canonical form of eq. (3.58), which gives

$$x(k) = \begin{bmatrix} 0 & -a_2 \\ 1 & -a_1 \end{bmatrix}x(k-1)+\begin{bmatrix} -a_2c_0 \\ c_1-a_1c_0 \end{bmatrix}e(k-1); \quad y(k) = \begin{bmatrix} 0 & 1 \end{bmatrix}x(k)+c_0e(k)$$

We note that the algebra can become quite complicated even in this simple case. Iterative algorithms should be used to obtain the spectral factors.

Based on these results, it is possible to make the simulation of the considered process. The simulation is made using by programming package MATLAB and corresponding code is given by the Fig. 3.13. Also, after the generation of process samples, its autocorrelation function is estimated and compared with the desired one. Fig. 3.14. represents the generated samples of the white noise

$\{e(k)\}$ and the colored noise $\{x(k)\}$. Fig. 3.15. shows the corresponding estimations of the autocorrelation functions $R_e(k)$ and $R_x(k)$, respectively

```
function example1

% definition of the parameters a and w, and sampling time T
a=input('define the parameter a:   '); w=input('define the parameter w:   '); T=input('sampling time   T:   ');

% calculation of the transfer function parameters
a1=-exp(-a*T)*cos(w*T); a2=exp(-2*a*T); b1=((a1*a2)-(1+exp(-2*a*T))*exp(-a*T)*cos(w*T))/(1+a2);
d1=-(1+exp(-2*a*T))*exp(-a*T)*cos(w*T); d2=1-a1^2-a2^2+2*a1*b1;
c0=sqrt((d2+sqrt(d2^2-4*d1^2))/2); c1=d1/c0;
if abs(c0)>abs(c1)
    tmp=c1;c1=c0;c0=tmp;
end

% generation of samples of white process  e(k)
N=input('how many samples would you like to generate:   ');

for i=1:N
   e(i)=randn;
end

% initial conditions
x(1)=e(1); x(2)=e(2);

% difference equation
for i=3:N
   x(i)=-a1*x(i-1)-a2*x(i-2)+c0*e(i)+c1*e(i-1);
end

% plotting of the sequences
figure(1);plot((1:N)*T,e);title('sequence of white noise');pause;
figure(2);plot((1:N)*T,x);title('coloured noise');pause;

% estimation of the autocorelation function

for k=1:13
   re(k)=0;
   for j=1:N-k
      re(k)=re(k)+e(j+k-1)*e(j);
   end
   re(k)=re(k)/(N-k+1);
end
figure(3);plot((-12:12)*T,[re(13:-1:2) re]);
title('autocorelation funciont of white noise'); pause;

for k=1:13
   rx(k)=0;
   for j=1:N-k
      rx(k)=rx(k)+x(j+k-1)*x(j);
   end
   rx(k)=rx(k)/(N-k+1);
end

% comparing of obtained and desired autocorrelation function
figure(4);
plot((-12:12)*T,[rx(13:-1:2) rx],(-12:12),exp(-abs(a*[-12:12])).*cos(w*[-12:12]));
end
```

Fig. 3.13. Program code for generation of colored noise with the desired autocorrelation function

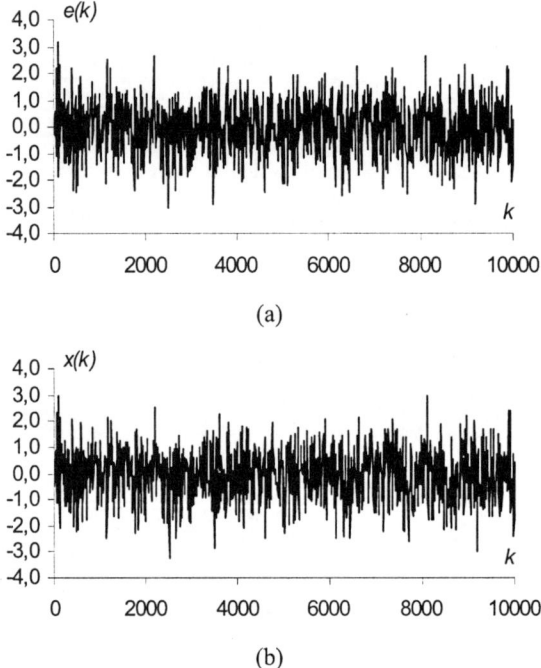

Fig. 3.14. Samples of the generated (a) white noise; (b) colored noise

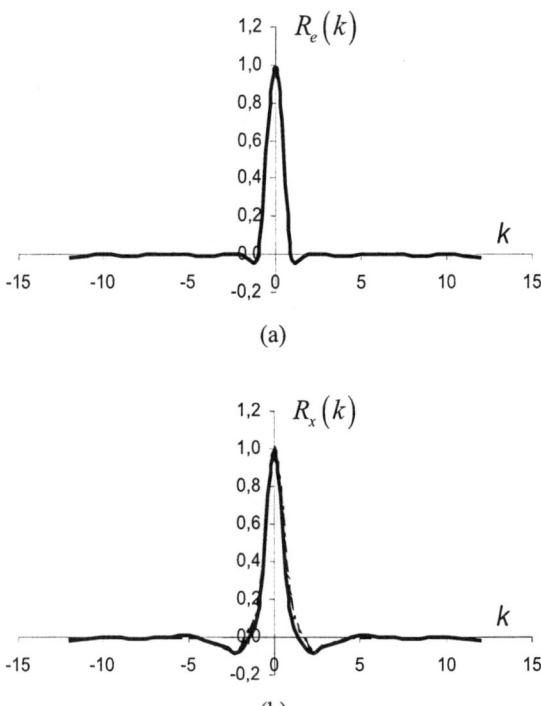

Fig. 3.15. The estimations of the autocorrelation functions (a) white noise; (b) colored noise

3.2.4. Observability and Controllability

Next we discuss the concepts of observability and controllability. These concepts are theoretical in nature but actually provide information about a system that is fundamental in analyzing its performance. We will define the basic concepts and then show how these system properties can be used [17, 25, 48].

Observability of a system means that the initial states of a system can be determined from a suitable measurement of its output. From a practical point of view, it is desirable to know that measurement of the system output will reveal all information about the system states and that one or more of the system modes (states) is not hidden from observation. Loosely, observability is the ability to choose measurement variables that enable the construction of the state variables that cannot be measured directly. More precisely, the system is observable if for any $x(0)$ there is a finite N such that $x(0)$ can be computed from observation of $y(0), y(1), \ldots, y(N-1)$. We will consider the development of a test for observability according to this definition. The system is described by (3.63) with $u = 0$, $w^* = 0$ and $v^* = 0$. Clearly the deterministic input is irrelevant here if we assume that all values of $u(k)$ are available in the computation of $x(0)$. If some inputs, such as a disturbance w^*, are not available, we have a very different problem. The successive outputs from $k=0$ are

$$
\begin{aligned}
y(0) &= Cx(0) \\
y(1) &= Cx(1) = CAx(0) \\
&\vdots \\
y(N-1) &= CA^{N-1}x(0)
\end{aligned}
$$

In matrix form, these equations are

$$
\begin{bmatrix} y(0) \\ y(1) \\ \vdots \\ y(N-1) \end{bmatrix} = \begin{bmatrix} C \\ CA \\ \vdots \\ CA^{N-1} \end{bmatrix} x(0)
$$

We have assumed that the number of states and hence the number of columns of the coefficients matrix of these equations is n; the number of rows is N. If N is less than n, we cannot possibly find a solution for every $x(0)$. If, on the other hand, N is greater than n, we will add a row CA^n, and so on. But, by the Cayley Hamilton theorem, A^n is a linear combination of lower powers of A, and the new rows add no new rank. Therefore we have a solution if and only if the so-called observability matrix

$$
Q_o = \begin{bmatrix} C \\ \hline CA \\ \hline \vdots \\ \hline CA^{n-1} \end{bmatrix}
$$

is nonsingular, i.e. that it is of rank n.

The concept of complete controllability is analogous to observability in that it examines the requirements of a system to be controlled by an input u. A system is said to be completely controllable if for every $x(0)$ and x_f there is a finite N and a sequence of controls

$u(0),...,u(N-1)$ such that if the system has the initial state $x(0)$ at $k=0$, it is forced to the "final" state x_f at $k = N$. In other word, any initial state $x(0)$ can be transferred to any final state x_f in a finite time N by an input signal u.

Let us develop a test for controllability for this definition. the system is described by (3.63a) with $w^* = 0$. Solving for a few steps, we find that

$$x(1) = Ax(0) + Bu(0)$$
$$x(2) = Ax(1) + Bu(1)$$
$$= A^2 x(0) + ABu(0) + Bu(1)$$
$$\vdots$$

$$x(N) = A^N x(0) + \sum_{j=0}^{N-1} A^{N-1-j} Bu(j)$$

The last equation can be written in the matrix form:

$$x(N) = A^N x(0) + \left[B \mid AB \mid \cdots \mid A^{N-1}B \right] \begin{bmatrix} u(N-1) \\ u(N-2) \\ \vdots \\ u(0) \end{bmatrix}$$

If $x(N)$ is equal to x_f, then we must be able to solve the equations

$$\left[B \mid AB \mid \cdots \mid A^{N-1}B \right] \begin{bmatrix} u(N-1) \\ u(N-2) \\ \vdots \\ u(0) \end{bmatrix} = x_f - A^N x(0)$$

As we saw in the discussion of observability, new columns in these equations cannot be independent of previous columns $B, AB, ..., A^{n-1}B$ if $N>n$, because of the Cayley-Hamilton theorem. Thus, the test for controllability is that the so-called controllability matrix

$$Q_c = \left[B \mid AB \mid \cdots \mid A^{n-1}B \right]$$

must be nonsingular, i.e. that it is of rank n.

If we take the transpose of Q_c and let $B^T = C$ and $A^T = A$ then we find the observability matrix Q_o. This property was first observed by Kalman, who termed it *duality*. Thus, observability and controllability are dual properties of linear systems. The exploration of the many ramifications of duality is beyond the scope of this book, and no attempt will be made to pursue this topic further. We also observe that controllability and observability are the properties of the system and not of the particular coordinate system (choice of state variables) in which the system is represented.

Controllability and observability yield important information about the system under investigation. If the system is "overmodeled", i.e. the number of state variables selected to represent the system is larger than the required or minimal number necessary, then the system will not be completely controllable or observable. In fact, a necessary and sufficient condition for a system to be of minimal dimension is that it be completely controllable and observable. Nonminimal systems lead to pole-zero cancellations in the corresponding pulse transfer function (input-output model).

It is possible to show that there exists a unique representation of a system that can be obtained from finite-difference equation, or input-output model. It is called observer canonical form and is given by (3.59).

Another popular representation of digital systems is a controllable canonical form. Given the following state space model and finite-difference equation representations for a digital system

$$
\begin{bmatrix} x_1(k+1) \\ x_2(k+1) \\ \vdots \\ x_n(k+1) \end{bmatrix} = \begin{bmatrix} 0 & \vdots & I_{n-1} \\ \cdots & \cdots & \cdots & \cdots \\ -a_n & \vdots & -a_{n-1} & \cdots & -a_1 \end{bmatrix} \begin{bmatrix} x_1(k) \\ x_2(k) \\ \vdots \\ x_n(k) \end{bmatrix} + \begin{bmatrix} b_1^* \\ b_2^* \\ \vdots \\ b_n^* \end{bmatrix} u(k) + \begin{bmatrix} c_1^* \\ c_2^* \\ \vdots \\ c_n^* \end{bmatrix} e(k) \tag{3.65}
$$

and

$$
\begin{aligned}
y(k+n) + a_1 y(k+n-1) + \cdots + a_n y(k) &= b_n u(k+n-1) + b_{n-1} u(k+n-2) + \cdots + b_1 u(k) \\
&\quad + c_n e(k+n-1) + c_{n-1} e(k+n-2) + \cdots + c_1 e(k)
\end{aligned} \tag{3.66}
$$

where symbol $u(k)$ represents a controllable signal (e.g., test signal or feedback control signal) whereas $e(k)$ represents an uncontrollable signal (e.g. random disturbance), the representations in Eqs. (3.65) and (3.66) are equivalent when

$$
\begin{aligned}
x_1(k) &= y(k) \\
x_2(k) &= y(k+1) - b_1^* u(k) - c_1^* e(k) \\
&\vdots \\
x_n(k) &= y(k+n-1) - b_1^* u(k+n-2) - b_2^* u(k+n-3) - \cdots - b_{n-1}^* u(k) \\
&\quad - c_1^* e(k+n-2) - c_2^* e(k+n-3) - \cdots - c_{n-1}^* e(k)
\end{aligned} \tag{3.67}
$$

where

$$
b^* = \left(b_1^*, b_2^*, \ldots, b_n^* \right)^T = T^{-1} b \; ; \; b = \left(b_1, b_2, \ldots, b_n \right)^T \tag{3.68}
$$

$$
c^* = \left(c_1^*, c_2^*, \ldots, c_n^* \right)^T = T^{-1} c \; ; \; c = \left(c_1, c_2, \ldots, c_n \right)^T \tag{3.69}
$$

$$
T = \begin{bmatrix}
1 & 0 & 0 & \cdots & 0 \\
a_1 & 1 & 0 & \cdots & 0 \\
a_2 & a_1 & 1 & \cdots & 0 \\
\vdots & \vdots & \vdots & \ddots & \vdots \\
a_{n-2} & a_{n-3} & a_{n-4} & \cdots & 0 \\
a_{n-1} & a_{n-2} & a_{n-3} & \cdots & 1
\end{bmatrix} \tag{3.70}
$$

We give the proof of this result for the case $e = 0$ for all k. The proof for the case when $e \neq 0$ parallels our development that follows. Our approach is to derive Eq. (3.66) from Eq. (3.65). To begin, let $k = k+1$ in the last equation in Eq. (3.67), and solve the resulting equation or $y(k+n)$

$$
y(k+n) = x_n(k+1) + b_1^* u(k+n-1) + b_2^* u(k+n-2) + \cdots + b_{n-1}^* u(k+1) \tag{3.71}
$$

We also know from the last equation in matrix Eq. (3.65) that

$$
x_n(k+1) = -a_n x_1(k) - a_{n-1} x_2(k) - \cdots - a_1 x_n(k) + b_n^* u(k) \tag{3.72}
$$

Hence, substituting this expression into Eq. (3.71) and making use of the first $(n-1)$ equations in Eq. (3.67) as collecting common terms, we find

$$y(k+n)+a_1 y(k+n-1)+\cdots+a_n y(k)=b_1 u(k+n-1)+b_2 u(k+n-2)+\cdots+b_{n-1}u(k+1)+b_n u(k)$$

where

$$
\begin{aligned}
b_1 &= b_1^* \\
b_2 &= a_1 b_1^* + b_2^* \\
&\vdots \\
b_{n-1} &= a_{n-2}b_1^* +\cdots+ a_2 b_{n-3}^* + a_1 b_{n-2}^* + b_{n-1}^* \\
b_n &= a_{n-1}b_1^* + a_{n-2}b_2^* +\cdots+ a_1 b_{n-1}^* + b_n^*
\end{aligned}
\tag{3.73}
$$

Clearly, Eq. (3.73) can also be written as

$$b = Tb^* \; ; \; b^T = (b_1,...,b_n)$$

where T is given by (3.70), which completes the proof.

Example 3.6 (comparison of transfer function and state space model): The zero-mean white noise $e(k)$ with unit variance is at the input of the discrete-time system with transfer function $H(z) = \dfrac{0.4}{(z-0.2)(z-0.5)}$. Calculate the autocorrelation function of the process $x(k)$ at the output of this system.

Solution: This problem can be solved by using two different methods.

I method is based on using the theorem of spectral factorization. Namely, knowing the power spectral density of the input process and transfer function of the system, it is possible to calculate the power spectral density of the output process as:

$$
\begin{aligned}
S_x(z) &= H(z)H(z^{-1})S_e(z) \\
&= \frac{0.4}{(z-0.2)(z-0.5)}\frac{0.4}{(z^{-1}-0.2)(z^{-1}-0.5)}\cdot 1
\end{aligned}
$$

Taking into account that the power spectral density of the process $x(k)$ is defined as double-side Z transform of the corresponding autocorrelation function, and keeping in mind the properties of the autocorrelation function, we can write (this is known as spectral factorization)

$$S_x(z) = S_x^+(z) - R_x(0) + S_x^-(z)$$

where $S_x^+(z)$ represents one-side Z transform of the causal part of the function $R_x(k)$ (it corresponds to the positive time indices k and the stable poles of S_x) and $S_x^-(z) = S_x^+(z^{-1})$. So, in order to divide the power spectral density $S_x(z)$ in three mentioned parts, one can write:

$$S_x(z) = \frac{0.16}{(z-0.2)(z-0.5)(z^{-1}-0.5)(z^{-1}-0.2)}$$

$$= \frac{az^2+bz+c}{(z-0.2)(z-0.5)} - a + \frac{az^{-2}+bz^{-1}+c}{(z^{-1}-0.2)(z^{-1}-0.5)}$$

$$= \frac{A(z)}{(z-0.2)(z-0.5)(z^{-1}-0.5)(z^{-1}-0.2)};$$

$$A(z) = (az^2+bz+c)(z^{-2}-0.7z^{-1}+0.1) - a(z^2-0.7z+0.1)(z^{-2}-0.7z^{-1}+0.1)$$
$$+(az^{-2}+bz^{-1}+c)(z^2-0.7z+0.1)$$

Equating the coefficients of the polynomials in the corresponding nominators, it is possible to calculate the values of the unknown coefficients a, b and c:

$$a = 0.2716 \; ; \; b = -0.0173 \; ; \; c = 0$$

So, the causal part of the autocorrelation function $R_x(k)$ can be obtained as an inverse one-side Z transform of the function $S_x^+(z)$ (see, the appendix A):

$$R_x^+(k) = Z^{-1}\{S_x^+(z)\} = Z^{-1}\left\{\frac{0.2716z^2-0.0173z}{(z-0.2)(z-0.5)}\right\}$$

$$= \begin{cases} 0.2716 \; ; \; k = 0 \\ 0.1975 \cdot (0.5)^{k-1} - 0.024 \cdot (0.2)^{k-1} \; ; \; k \geq 1 \end{cases}$$

Finally, the autocorrelation function of the process $x(k)$ becomes:

$$R_x(k) = \begin{cases} 0.2716 \; ; \; k = 0 \\ 0.1975 \cdot (0.5)^{k-1} - 0.024 \cdot (0.2)^{k-1} \; ; \; k \geq 1 \\ R_x(-k) \; ; \; k \leq -1 \end{cases}$$

II method is based on the state-space system model. If we represent the transfer function $H(z)$ in diagonal state space form, we obtain

$$X(z) = \left[-\frac{4/3}{z-0.2} \quad \frac{4/3}{z-0.5}\right]^T E(z)$$

We have in the z-domain

$$z\xi_1(z) = 0.2\xi_1(z) + E(z)$$
$$z\xi_2(z) = 0.5\xi_2(z) + E(z)$$
$$X(z) = -\frac{4}{3}\xi_1(z) + \frac{4}{3}\xi_2(z)$$

or equivalently in the time-domain

$$\xi_1(k+1) = 0.2\xi_1(k) + e(k)$$
$$\xi_2(k+1) = 0.5\xi_2(k) + e(k)$$
$$x(k) = -\frac{4}{3}\xi_1(k) + \frac{4}{3}\xi_2(k)$$

The last relations can be put in the corresponding state-space matrix form:

$$\xi(k+1) = \Phi\xi(k) + \Gamma e(k) = \begin{bmatrix} 0.2 & 0 \\ 0 & 0.5 \end{bmatrix}\xi(k) + \begin{bmatrix} 1 \\ 1 \end{bmatrix}e(k)$$

$$x(k) = H\xi(k) = \begin{bmatrix} -\dfrac{4}{3} & \dfrac{4}{3} \end{bmatrix}\xi(k)$$

Let us start from the definition of the autocorrelation function:

$$R_x(i+k,i) = E\{x(i+k)x(i)\}$$

On the other hand, the output of the system $x(i)$ can be calculated from the state-space model as:

$$x(1) = H\xi(1) = H\Phi\xi(0) + H\Gamma e(0)$$
$$x(2) = H\xi(2) = H(\Phi\xi(1) + \Gamma e(1)) = H\Phi^2\xi(0) + H\Phi\Gamma e(0) + H\Gamma e(1)$$
$$\vdots$$
$$x(i) = H\Phi^i\xi(0) + H\Phi^{i-1}\Gamma e(0) + \cdots + H\Phi\Gamma e(i-2) + H\Gamma e(i-1)$$
$$= H\Phi^i\xi(0) + H\sum_{j=0}^{i-1}\Phi^{i-1-j}\Gamma e(j)$$

Based on the last relation, the autocorrelation function can be represented as:

$$R_x(i+k,i) = E\left\{\left[H\Phi^{i+k}\xi(0) + H\sum_{j=0}^{i+k-1}\Phi^{i+k-1-j}\Gamma e(j)\right]\left[H\Phi^i\xi(0) + H\sum_{l=0}^{i-1}\Phi^{i-1-l}\Gamma e(l)\right]\right\}$$

assuming that $k \geq 0$ without loosing generality. Now, we can suppose that the covariance matrix of the initial condition $\xi(0)$ is known and equal to Σ_0, and that the initial condition is not correlated to the samples of the white noise $e(j)$. Also, if we take into account that the operators of expectation and summation can be exchanged and that the input $e(j)$ has the mean value equal to zero, the last relation can be simplified:

$$R_x(i+k,i) = H\Phi^{i+k}E\{\xi(0)\xi^T(0)\}(\Phi^i)^T H^T + H\sum_{l=0}^{i-1}\Phi^{i-1-l}\Gamma H\Phi^{i+k}E\{\xi(0)e(l)\}$$
$$+ H\sum_{j=0}^{i+k-1}\Phi^{i+k-1-j}\Gamma H\Phi^i E\{\xi(0)e(j)\} + H\sum_{j=0}^{i+k-1}\sum_{l=0}^{i-1}\Phi^{i+k-1-j}\Gamma E\{e(j)e(l)\}\Gamma^T(\Phi^{i-1-l})^T H^T$$
$$= H\Phi^{i+k}\Sigma_0(\Phi^i)^T H^T + H\sum_{j=0}^{i+k-1}\sum_{l=0}^{i-1}\Phi^{i+k-1-j}\Gamma E\{e(j)e(l)\}\Gamma^T(\Phi^{i-1-l})^T H^T$$

As the process $e(j)$ is zero-mean, with unit variance, white process, the last relation can be written in the next form:

$$R_x(i+k,i) = H\Phi^{i+k}\Sigma_0 \left(\Phi^i\right)^T H^T + H\sum_{l=0}^{i-1}\Phi^{i+k-1-l}\Gamma\Gamma^T \left(\Phi^{i-1-l}\right)^T H^T \; ; \; k \geq 0$$

If we assume the matrix Σ_0 in the form

$$\Sigma_0 = \begin{bmatrix} \sigma_1^2 & \sigma_{12} \\ \sigma_{12} & \sigma_2^2 \end{bmatrix}$$

and substitute the values of the matrices Φ, Γ and H, the autocorrelation function becomes:

$$R_x(i+k,i) = \frac{16}{9}\left[0.2^{2i+k}\sigma_1^2 - 0.2^i 0.5^{i+k}\sigma_{12} - 0.2^{i+k}0.5^i\sigma_{12} + 0.5^{2i+k}\sigma_2^2\right]$$

$$+\frac{16}{9}\left[\frac{1}{24}\left(0.2^{k-2} - 0.2^{2i+k-2}\right) - \frac{1}{9}\left(0.5^{k-1}0.2^{-1} - 0.5^{i+k-1}0.2^{i-1}\right)\right] \; ; \quad k \geq 0$$

$$+\frac{16}{9}\left[-\frac{1}{9}\left(0.2^{k-1}0.5^{-1} - 0.2^{i+k-1}0.5^{i-1}\right) + \frac{1}{3}\left(0.5^{k-2} - 0.5^{2i+k-2}\right)\right]$$

Obviously, the autocorrelation function depends on the time indices k and i, and it is not possible to express it as a function of the time difference k. Thus, the process $x(i)$ is not wide sense stationary and we can not apply the spectral factorization theorem. This is the reason that the results obtained by the first and second method are not identical. However, if we take that the time index i tends to infinity in the last expression, we obtain

$$\lim_{i\to\infty} R_x(i+k,i) = \frac{16}{9}\left(\frac{1}{18}0.5^{k-2} - \frac{1}{360}0.2^{k-2}\right) = R_x(k); \; k \geq 0$$

This result is identical with the solution obtained by using the spectral factorization theorem (the first method). So, it can be generally concluded that the results obtained by using the spectral factorization theorem and the state-space model are equivalent in the steady state, provided that the considered system is stable.

CHAPTER 4

LINEAR CONTINUOUS TIME STOCHASTIC SYSTEMS

We will now turn to continuous time systems and continuous time processes. The analysis is completely analogous to the discrete time case treaded in the chapter 3.

4.1 Input-output models

Let us consider a linear time invariant (LTI) dynamical system with one input u and one output y, as is shown in Fig. 4.1.

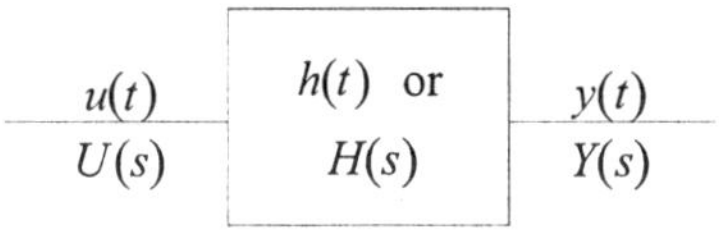

$u(t)$	$h(t)$ or	$y(t)$
$U(s)$	$H(s)$	$Y(s)$

Fig. 4.1 Linear continuous time single input single output (SISO) system

The transfer function model is termed the input-output description because it primarily operates on the input and output variables $\{u(t)\}$ and $\{y(t)\}$; that is, the transfer function is defined by [2, 23]

$$H(s) = L\{h(t)\} = \frac{L\{y(t)\}}{L\{u(t)\}} = \frac{Y(s)}{U(s)} \tag{4.1}$$

where $h(t)$ is the impulse response of the system, i.e. the response of the system to the delta or impulse function $u(t) = \delta(t)$ (the function $\delta(t)$ has the properties $\int_{-\infty}^{\infty} \delta(t)dt = 1$ and $L\{\delta(t)\} = 1$), $L\{\cdot\}$ is the Laplace transform, and s is a complex variable in the Laplace transform domain. Here it is assumed that there is no internal energy in the system, i.e. initial conditions are equal to zero. A system which has a property that its response is equal to zero before an excitation; that is $h(t) = 0$ for $t < 0$, is termed a *causal system*. The transfer function of a general time invariant system can be written as

$$H(s) = \frac{B(s)}{A(s)} = \frac{b_m s^m + \cdots + b_1 s + b_0}{a_n s^n + \cdots + a_1 s + a_0}; \ m \leq n \tag{4.2}$$

where $B(\cdot)$ and $A(\cdot)$ are polynomials with fixed coefficients (the constant n is called the system order). The roots of the equation $B(s) = 0$ are termed zeros of system, while the roots of the equation $A(s) = 0$, the so-called characteristic equation, are termed poles of system.

4.1.1 Stability

While there is considerable difficulty to determine the stability of a physical system under all conditions, the stability question is not difficult for LTI system models (4.2). For these systems, we use the bounded-input, bounded-output (BIBO) definition of stability (see the chapter 3).

A system is BIBO stable if, for every bounded input, the output remains bounded for all time. We now develop the criterion for BIBO stability of LTI systems. Let the characteristic equation $A(s) = 0$ of eq. (4.2) be represented in factored form [2, 25]

$$A(s) = a_n \prod_{i=1}^{n} (s - p_i) = 0 \tag{4.3}$$

From eq. (4.1) we can express the output as

$$Y(s) = \frac{B(s)}{a_n \prod_{i=1}^{n} (s - p_i)} U(s) = \sum_{i=1}^{n} \frac{k_i}{s - p_i} + Y_u(s) \tag{4.4}$$

where $Y_u(s)$ is the sum of the terms, in the partial fraction expansion, that originate in the poles of $U(s)$. The inverse Laplace transform of $Y(s)$ yields

$$y(t) = \sum_{i=1}^{n} k_i \exp(p_i t) + c_u(t) = c_n(t) + c_u(t) \tag{4.5}$$

where $c_u(t)$ is the forced response and $c_n(t)$ is the natural response, since the terms in $c_n(t)$ originate in the poles of the transfer function and the functional forms are independent of the input. If $u(t)$ is bounded, all terms in $c_u(t)$ will remain bounded, since $c_u(t)$ is of the functional form of $u(t)$. Thus, if the output becomes unbounded, it is because at least one of the natural-response terms, $k_i \exp(p_i t)$, has become unbounded. This unboundness cannot occur if the real part of each of the poles p_i is negative. We see from this discussion that the requirement for a LTI system to be stable is that all poles of the transfer function must lie in the left half of the s-plane. If a system has poles on the imaginary axis ($j\omega$- axis) of the s-plane, with all other poles in the left half-plane, the steady-state output will be sustained oscillations for a bounded input, unless the input is a sinusoid (which is bounded) whose frequency is equal to the magnitude of the $j\omega$-axis poles. For this case the output becomes unbounded. Such a system is called *marginally stable*, since only certain bounded inputs (sinusoids of the frequency of the poles) will cause the output to become unbounded. For an unstable LTI system, there is at least one pole in the right half of the s-plane, for this case the output will become unbounded for any input.

4.1.2 Linear filtering

Taking inverse Laplace transform of eq. (4.1); that is $Y(s) = H(s)U(s)$, the input-output relation, then becomes

$$y(t) = h(t) \otimes u(t) = \int_{-\infty}^{\infty} h(t-s)u(s) ds = \int_{0}^{\infty} h(s)u(t-s) ds \tag{4.6}$$

where $\otimes$ denotes the convolution.

Now let the input signal be a stochastic process of *second order*. A stochastic process $\{u(t), t \in T\}$ is said to be of second order if $E\{u^2(t)\} < \infty$ for all $t \in T$. For such process the mean value $m_u(t) = E\{u(t)\}$ and the covariance function $R_u(s,t) = E\{[u(s) - m_u(s)][u(t) - m_u(t)]\}$ exist, and the second order properties of the distribution can thus be expressed by these two functions.

We must then first ensure that the integral in (4.6) has a meaning. To do so we will first introduce integrals of stochastic processes. Consider an interval $[a,b] \in T$. Let $a = t_0 < t < \cdots < t_n = b$ be a subdivision of $[a,b]$. Consider the sum

$$I_n = \sum_{k=1}^{m} u(\tau_k)[t_k - t_{k-1}]$$

where $t_{k-1} \leq \tau_k \leq t_k$ and $\{u(t), t \in T\}$ is a second order stochastic process. The process $\{u(t), t \in T\}$ is said to be Riemann integrable if I_n converges, in some sense, as $n \to \infty$ in such a way that $\max_{1 \leq k \leq n} |t_k - t_{k-1}| \to 0$.

We will now discuss what we should mean by the limit of a sequence of stochastic variables $\{I_n(\omega), n = 1, 2, \ldots; \omega \in \Omega\}$, with Ω being a probability space. This can be formulated in many ways. In the following we will discuss the most common limit concepts. The sequence $\{I_n(\omega)\}$ converges *with probability one* to the stochastic variable $I(\omega)$ if $I_n(\omega) \to I(\omega)$ for all ω, except possibly for a set of random outcomes ω having zero probability, i.e.

$$P\{\omega : I_n(\omega) \to I(\omega)\} = 1 \tag{4.7}$$

The sequence $\{I_n(\omega)\}$ converges to $I(\omega)$ *in probability* if for every $\varepsilon > 0$

$$\lim_{n \to \infty} P\{\omega : |I_n(\omega) - I(\omega)| \geq \varepsilon\} = 0 \tag{4.8}$$

The sequence $\{I_n(\omega)\}$ converges to $I(\omega)$ *in the mean square* if

$$\lim_{n \to \infty} E\{|I_n(\omega) - I(\omega)|^2\} = 0 \tag{4.9}$$

The different convergence concepts are related. We have, for example, that convergence with probability one implies convergence in probability, and convergence in the mean square implies convergence in probability. The criteria for convergence with probability one are often difficult to establish. We will therefore in the sequel use convergence in the mean square for the simple reason that it leads to very simple analysis (to investigate mean square convergence we use the Cauchy criterion).

Therefore, the process $\{u(t), t \in T\}$ is said to be Riemann integrable if I_n converges to a limit in the mean square as $n \to \infty$, and the limit is called the mean square Riemann integral of u over $[a,b]$ and is denoted by [2, 48, 59]

$$I = \int_a^b u(t) dt \tag{4.10}$$

We have the following result: The second order process $\{u(t), t \in T\}$ with mean value $m_u(t)$ and covariance function $R_u(s,t)$ is Riemann integrable if the integrals

$$\int_a^b m_u(t) dt \quad \text{and} \quad \int_a^b \int_a^b R_u(s,t) ds dt$$

exist. In that case we have that the operations $E\{\cdot\}$ and $\int(\cdot)$ can be interchanged; that is

$$E\left\{\int_a^b u(t)dt\right\} = \int_a^b E\{u(t)\}\,dt = \int_a^b m_u(t)\,dt \tag{4.11a}$$

$$E\left\{\int_a^b \int_a^b u(t)u(s)\,dtds\right\} = \int_a^b \int_a^b E\{u(t)u(s)\}\,dtds = \left[\int_a^b m_u(t)\,dt\right]^2 + \int_a^b \int_a^b R_u(s,t)\,dsdt \tag{4.11b}$$

Let us now analyze a meaning of the convolution integral (4.6). To do this, we will first consider a finite integration interval. It then follows, from (4.11), if the integral is interpreted as a mean square limit of Riemann sums, it will exist if the covariance $R_u(s,t)$ of the input is continuos in both its arguments. Thus, the expression

$$\int_0^a h(s)u(t-s)\,ds$$

has a meaning for finite a. To find out if the limit

$$\lim_{a\to\infty}\int_0^a h(s)u(t-s)\,ds$$

also exists, we form the sequence

$$I(a,b) = \int_a^b h(s)u(t-s)\,ds$$

Therefore, we have

$$E\{I^2(a,b)\} = E\left\{\int_a^b \int_a^b h(s)u(t-s)h(s')u(t-s')\,dsds'\right\}$$

$$= \int_a^b \int_a^b h(s)h(s')R_u(t-s,t-s')\,dsds'$$

If the system is asymptotically stable, we find from (4.5) ($u(t)=\delta(t)$ so that $h(t)=c_n(t)$)

$$|h(s)| \le const.\exp(-\alpha t)\ ;\ \alpha > 0$$

As $\{u(t)\}$ is a stochastic process of second order, we get

$$R_u(s,t) \le E\{u^2(t)\} \le \infty$$

Hence

$$E\{I^2(a,b)\} \to 0\ \text{ as }\ a,b\to\infty$$

The limit thus exists and we find

$$E\{y^2(t)\} \le const < \infty$$

The output y of the dynamical system is thus a stochastic process of second order. We will now determine its mean value and its covariance function. We have

$$m_y(t) = E\{y(t)\} = E\left\{\int_0^\infty h(s)u(t-s)\,ds\right\} = \int_0^\infty h(s)m_u(t-s)\,ds \tag{4.12}$$

We thus find that the mean value of the input propagates through the system as a deterministic signal.

The covariance function of the output will now be determined. We have

$$R_y(s,t) = \text{cov}\{y(s), y(t)\}$$
$$= \text{cov}\left\{\int_0^\infty h(s')u(s-s')ds', \int_0^\infty h(s'')u(t-s'')ds''\right\}$$
$$= \int_0^\infty \int_0^\infty h(s')h(s'')\text{cov}(u(s-s'), u(t-s:))ds'ds''$$
$$= \int_0^\infty \int_0^\infty h(s')h(s'')R_u(s-s', t-s'')ds'ds'' \tag{4.13}$$

We also have the cross-covariance function of the input and of the output is given by

$$R_{uy}(s,t) = \text{cov}\{u(s), y(t)\}$$
$$= \text{cov}\left\{u(s), \int_0^\infty h(s')u(t-s')ds'\right\}$$
$$= \int_0^\infty h(s')\text{cov}\{u(s), u(t-s')\}ds'$$
$$= \int_0^\infty h(s')R_u(s, t-s')ds' \tag{4.14}$$

Summing up we get the following theorem: Consider a continuous linear time invariant dynamical system which has the impulse response $h(t)$. Let the input signal u be a stochastic process of second order with mean value $m_u(t)$ and the covariance function $R_u(s,t)$. If the dynamical system is stable (asymptotically) and if the function $R_u(s,t)$ is continuous, the integral

$$J(t) = \int_0^\infty h(s)u(t-s)ds \tag{4.15}$$

exists as a mean square limit of Reimann sums. The output y is a stochastic process with the mean value

$$m_y(t) = \int_0^\infty h(s)m_u(t-s)ds \tag{4.16}$$

and the covariance function

$$R_y(s,t) = \int_0^\infty \int_0^\infty h(s')h(s'')R_u(s-s', t-s'')ds'ds'' \tag{4.17}$$

The covariance function of the input and of the output is

$$R_{uy}(s,t) = \int_0^\infty h(s')R_u(s, t-s')ds' \tag{4.18}$$

We will now concentrate to stationary process. If the input is a weakly stationary process we have

$$m_u(t) = m_u = const \tag{4.19a}$$

$$R_y(s,t) = \int_0^\infty \int_0^\infty h(s')h(s'')R_u(s-t-s'+s'')ds'ds'' \tag{4.19b}$$

$$R_{uy}(s,t) = \int_0^\infty h(s')R_u(s-t+s')ds' \tag{4.19c}$$

As these equations are convolutions, they can be simplified if we introduce Fourier or Laplace transforms.

Let S_u denote the spectral density of the input, and H transfer function of the dynamical system

$$S_u(\omega) = F\{R_u(\tau)\} = \frac{1}{2\pi} \int_{-\infty}^{\infty} \exp(-j\omega\tau) R_u(\tau) d\tau \qquad (4.20)$$

$$H(s) = L\{h(t)\} = \int_0^{\infty} \exp(-st) h(t) dt \qquad (4.21)$$

Equations (4.19) then give

$$m_y = m_u H(0) \qquad (4.22a)$$

$$S_y(\omega) = \frac{1}{2\pi} \int_{-\infty}^{\infty} \exp(-j\omega\tau) R_y(\tau) d\tau = \frac{1}{2\pi} \int_{-\infty}^{\infty} \exp(-j\omega\tau) \int_0^{\infty} \int_0^{\infty} h(s') h(s'') R_u(\tau - s' + s'') ds' ds'' d\tau$$

$$= \frac{1}{2\pi} \int_{-\infty}^{\infty} d\tau \int_0^{\infty} ds' \int_0^{\infty} ds'' h(s') \exp(-j\omega s') h(s'') \exp(-j\omega s'') R_u(\tau - s' + s'') \exp(-j\omega(\tau - s' + s''))$$

$$= H(j\omega) H(-j\omega) S_u(\omega)$$

$$(4.22b)$$

$$S_{uy}(\omega) = \frac{1}{2\pi} \int_{-\infty}^{\infty} \exp(-j\omega\tau) R_{uy}(\tau) d\tau = \frac{1}{2\pi} \int_{-\infty}^{\infty} \exp(-j\omega\tau) \int_0^{\infty} h(s') R_u(\tau + s') ds' d\tau$$

$$= \frac{1}{2\pi} \int_{-\infty}^{\infty} d\tau \int_0^{\infty} ds' \exp(j\omega s') h(s') \exp(-j\omega(\tau + s')) R_u(\tau + s') = H(-j\omega) S_u(\omega)$$

Summing up we find the following theorem, which is known as *Linear Filtering Theorem for stationary processes.*. Let a time invariant dynamics system has the transfer function H and the input signal is a weakly stationary stochastic process with the mean value m_u and the spectral density $S_u(\omega)$. If the dynamical system is (asymptotically) stable and if

$$R_u(0) = \text{var}\{u(t)\} = \int_{-\infty}^{\infty} S_u(\omega) d\omega \le const < \infty \qquad (4.23)$$

then the output signal is a weakly stationary process with mean value

$$m_y = H(0) m_u \qquad (4.24)$$

and spectral density

$$S_y(\omega) = H(j\omega) H(-j\omega) S_u(\omega) = |H(j\omega)|^2 S_u(\omega) \qquad (4.25)$$

The input-output cross-spectral density is

$$S_{uy}(\omega) = H(-j\omega) S_u(\omega) \qquad (4.26)$$

This theorem is analogous to the corresponding theorem related to the discrete time systems (see preceding section), and the physical interpretations are identical. The condition (4.23), which has no correspondence in discrete-time case, ensures that the input signal has finite variance (it should be noted that $|R_u(\tau)| \le |R_u(0)|$ and that $R_u(\tau)$ is continuous for all τ if it is continuous for $\tau = 0$).
This is a fundamental difference between continuous time and discrete time processes.

The spectral density relations of eqs. (4.25) and (4.26) can also be derived by a simple procedure. We may consider the Fourier transform of a single signal record and use the limiting arguments as applied to the truncated signals with the truncating interval T. The spectral densities $S_y(s)$ and $S_{uy}(s)$ may then be written

$$S_y(s) = \lim_{T \to \infty} \frac{E\{Y_T(s)Y_T(-s)\}}{T}$$

and

$$S_{uy}(s) = \lim_{T \to \infty} \frac{E\{U_T(s)Y_T(-s)\}}{T}$$

However, we recognize from (4.1) $H(s)$ as the transfer function relating y and u; that is

$$Y_T(s) = H(s)U_T(s) \quad \text{and} \quad Y_T(-s) = H(-s)U_T(-s)$$

Therefore, we have simply

$$S_y(s) = H(s)H(-s)\lim_{T \to \infty}\frac{E\{U_T(s)U_T(-s)\}}{T} = H(s)H(-s)S_u(s)$$

and

$$S_{uy}(s) = H(-s)\lim_{T \to \infty}\frac{E\{U_T(s)U_T(-s)\}}{T} = H(-s)S_u(s)$$

where $s = j\omega$.

It follows from (98) that if the input u is white noise with $S_u(\omega) = 1$, then the spectral density of the output is given by

$$S_y(s)\big|_{s=j\omega} = H(s)H(-s)\big|_{s=j\omega} = |H(j\omega)|^2 \tag{4.27}$$

This means that any disturbance whose spectral density can be written in this form may be generated by sending continuous-time white noise through a linear system (filter) with the transfer function H.

4.1.3 Spectral factorization

Since finite dimensional linear systems have rational transfer functions of the form (4.2), it follows that signal with arbitrary rational spectral densities can be generated from linear finite-dimensional systems. The covariance function $R(\tau)$ is nonnegative and symmetric. It then follows that the spectral density S is also symmetric. If S is rational, than it follows from (4.27) that its poles and zeros are symmetric with respect to the real and imaginary axis of the s-plane (the poles and zeros of the transfer function $H(s)$ are symmetric with respect to the real axis, and if s_i is a zero (pole) of $H(s)$ then $-s_i$ is a zero (pole) of $H(-s)$; that is, the poles and zeros of $H(s)H(-s)$ are thus symmetric with respect to the real axis and mirrored in the imaginary axis). The transfer function H in (4.27) can then be chosen so that all its poles are in the left half-plane and all its zeros in the left half-plane or on the imaginary axis. The following analog of the spectral factorization theorem in the preceding section is thus obtain, which is known as *Spectral factorization theorem for continuous time processes:* Given a rational spectral density $S(\omega)$, there exists a finite-dimensional linear systems with the rational transfer function [1, 2, 30]

$$H(s) = \frac{B(s)}{A(s)} \tag{4.28}$$

such that the output obtained when the system is driven by white noise is a stationary stochastic process with the given spectral density S. The polynomial A has all its roots in the left half-plane. The polynomial B has no roots in the right half-plane.

It should be noted that the condition (4.23) is not fulfilled when the input signal u is white noise, so that the above filtering theorem does not hold. Also notice that the integral (4.6) does not have a meaning when u is white noise, since continuous-time white noise does not have a bounded variation due to its infinite variance. However, the condition (4.23) will be fulfilled and the integral (4.6) will exist if the input signal u is band limited white noise.

4.1.4 The concept of stochastic process with uncorrelated increments (Wiener process)

These difficulties with continuous-time white noise can be avoided by introducing a stochastic process that formally is the time integral [2, 44, 59]

$$v(t) = \int_0^t u(t)\,dt \qquad (4.29)$$

of white noise u. The stochastic process v has mean value

$$m_v(t) = E\{v(t)\} = \int_0^t E\{u(t)\}\,dt = tm_u \qquad (4.30)$$

where m_u is mean-value of a stationary white noise process u. If the covariance function of u is

$$\text{cov}\{u(t),u(s)\} = E\{[u(t)-m_u][u(s)-m_u]\} = r_0\delta(t-s) \qquad (4.31)$$

it is easy to show that for $s,t > 0$

$$\begin{aligned} \text{cov}\{v(t),v(s)\} &= E\{[v(t)-m_v(t)][v(s)-m_v(s)]\} \\ &= E\left\{\int_0^t\int_0^s [u(\lambda_1)-m_u][u(\lambda_2)-m_u]\,d\lambda_1 d\lambda_2\right\} = r_0\min(s,t) \end{aligned} \qquad (4.32)$$

In addition, we note that the increment $v(s)-v(t)$ is given by

$$v(s)-v(t) = \int_0^s u(\lambda)\,d\lambda - \int_0^t u(\lambda)\,d\lambda = \int_t^s u(\lambda)\,d\lambda \qquad (4.33)$$

and has mean value

$$E\{v(s)-v(t)\} = E\left\{\int_t^s u(\lambda)\,d\lambda\right\} = |s-t|m_u \qquad (4.34)$$

and variance

$$\text{var}\{v(s)-v(t)\} = E\left\{\int_t^s\int_t^s [u(\lambda_1)-m_u][u(\lambda_2)-m_u]\,d\lambda_1 d\lambda_2\right\} = r_0|s-t| \qquad (4.35)$$

The stochastic process $\{v(t), t \in T\}$ is called a Wiener process if it also is zero-mean Gaussian. The Wiener process was devised by Wiener as a simple model for Brownian motion. We let $v(t)$ denote the displacement from the origin at time t of a particle submerged in a fluid. The motion of the particle over a time interval which is long enough is the result of the change of impulse due to many collisions. It is thus reasonable to assume that the displacement of the particle over the time interval (s,t), which is long compared white the time between impacts, is the sum of a large number of small disturbances, and therefore subject to application of the central limit theorem. Thus, the distribution of the increment $v(s)-v(t)$ will approximate a Gaussian one.

The Wiener process, also called the Brownian motion process, has many interesting properties such as:

- $v(0) = 0$

- $\{v(t)\}$ is normal process

- $E\{v(t)\} = 0$ for all $t > 0$

- The Wiener process $v(t)$ has independent increments, that is $v(0) = 0$ and the random variable $v(t_1) - v(t_0), v(t_2) - v(t_1), ..., v(t_k) - v(t_{k-1})$ are independent for $t_k > t_{k-1} > \cdots > t_0 \geq 0$. Since the process $v(t_k + T) - v(t_{k-1} + T)$ has the same probability function as $v(t_k) - v(t_{k-1})$ the Wiener process has stationary independent increments.

- The Wiener process is a Markov process, since

$$f_{v(t_3)/v(t_1),v(t_2)}(\alpha_3 / \alpha_1, \alpha_2) = f_{v(t_3)/v(t_2)}(\alpha_3 / \alpha_2) \; ; \; t_1 \leq t_2 \leq t_3 \tag{4.36}$$

We may show this in a simple way by noting that

$$v(t_3) = \int_0^{t_3} u(\lambda) d\lambda = \int_0^{t_2} u(\lambda) d\lambda + \int_{t_2}^{t_3} u(\lambda) d\lambda = v(t_2) + \int_{t_2}^{t_3} u(\lambda) d\lambda$$

such that

$$E\{v(t_3)/v(t_2)\} = v(t_2)$$

$$\text{var}\{v(t_3)/v(t_2)\} = E\left\{\left[v(t_3) - E\{v(t_2)\}\right]^2 / v(t_2)\right\} = E\left\{\int_{t_2}^{t_3} \int_{t_2}^{t_3} u(\lambda_1) u(\lambda_2) d\lambda_1 d\lambda_2\right\}$$

or using (4.35)

$$\text{var}\{v(t_3)/v(t_2)\} = r_0(t_3 - t_2)$$

Additionally, since the Wiener process $v(t)$ is normal its probability function is completely characterized by the mean value and variance.

-The Wiener process is a martingale process in that the conditional expectation of $v(t_k)$, given the values $v(t_0), v(t_1), ..., v(t_{k-1})$, is equal to the most recently observed value $v(t_{k-1})$, where $t_k > t_{k-1} > \cdots > t_0$; that is

$$E\{v(t_k)/v(t_{k-1}), v(t_{k-2}), ..., v(t_0)\} = v(t_{k-1})$$

However, it is not necessary for the process to be a martingale process in order for it to be a Markov process.

-The sample functions (realizations) of a Wiener process are continuous functions, but they do not have a bounded variation, since $\text{var}\{v(t)\} = r_0 t$, and they are almost nowhere differentiable. A nonrigorous explanation for the observation that $v(t)$ is almost nowhere differentiable follows from eq. (4.24); that is

$$\text{var}\{v(t+\tau) - v(t)\} = r_0\tau \; ; \; \tau \geq 0$$

such that the increment $v(t+\tau) - v(t)$ has a value somewhere near $\sqrt{r_0\tau}$; the derivative of the increment is then of the order $\sqrt{\tau}/\tau$, which diverges as τ becomes zero.

Taking into account the above discussion, we also found that a case of white noise input u could be modeled by

$$y(t) = \int_{-\infty}^{t} h(t-s)\,dv(s) = \int_{0}^{\infty} h(s)\,dv(t-s) \tag{4.37}$$

where $\{v(t), t \in T\}$ is a stationary stochastic process with uncorrelated increments $dv = v(t+dt) - v(t) = \int_{t}^{t+dt} u(\lambda)\,d\lambda$ which has the incremental mean and the incremental covariance given by eqs. (4.30) and (4.35), respectively; that is $E\{dv\} = m_u dt$ and $E\{dv^2\} = r_0 dt$. To show that the infinite integral in (4.35) exists we form

$$E\left\{\left[\int_{a}^{b} h(t-s)\,dv(s)\right]^2\right\} \leq \max_{a \leq s \leq b} h^2(t-s) E\left\{\left[\int_{a}^{b} dv(s)\right]^2\right\}$$

$$= \max_{a \leq s \leq b}\{h^2(t-s)\} \int_{a}^{b} E\{dv^2\}\,dt$$

$$= \max_{a \leq s \leq b}\{h^2(t-s)\} r_0 (b-a)$$

Here is used the Schwartz inequality. As the dynamical system is (asymptotically) stable we have, owing to eq. (4.5), that impulse response $h(t)$ is bounded; that is

$$|h(t)| \leq const.\ \exp(-\alpha t)\ ;\ \alpha > 0$$

We thus find that

$$E\left\{\left[\int_{a}^{b} h(t-s)\,dv(s)\right]^2\right\} \to 0 \quad \text{as} \quad \max(a,b) \to \infty$$

The infinite integral (4.37) then exists according to the Cauchy criterion, and the stochastic process $\{y(t)\}$ thus is a process of second order. We will now determine the mean value and the covariance function

$$m_y(t) = E\{y(t)\} = E\left\{\int_{-\infty}^{t} h(t-s)\,dv(s)\right\} = \int_{-\infty}^{t} h(t-s) E\{dv(s)\}$$

$$= \int_{-\infty}^{t} h(t-s) m_u(s)\,ds$$

If m_u is constant, the mean value of y is also constant. It follows that the covariance function of y is given by (here is assumed that $m_u = 0$ so that $m_y = 0$)

$$R_y(s,t) = E\{y(s)y(t)\} = E\left\{\int_{-\infty}^{s}\int_{-\infty}^{t} h(s-\lambda_1)h(t-\lambda_2)\,dv(\lambda_1)\,dv(\lambda_2)\right\}$$

$$= \int_{-\infty}^{s}\int_{-\infty}^{t} h(s-\lambda_1)h(t-\lambda_2) E\{u(\lambda_1)u(\lambda_2)\}\,d\lambda_1\,d\lambda_2$$

$$= \int_{-\infty}^{s}\int_{-\infty}^{t} h(s-\lambda_1)h(t-\lambda_2) r_0\delta(\lambda_1-\lambda_2)\,d\lambda_1\,d\lambda_2$$

$$= \int_{-\infty}^{s} h(s-\lambda_1)\int_{-\infty}^{t} h(t-\lambda_2) r_0\delta(\lambda_1-\lambda_2)\,d\lambda_1\,d\lambda_2$$

$$= \int_{-\infty}^{s} h(s-\lambda_1)h(t-\lambda_1) r_0\,d\lambda_1 = r_0 \int_{0}^{\infty} h(s-t+\lambda_1)h(\lambda_1)\,d\lambda_1 = R_y(s-t)$$

Thus, the covariance function $R_y(s,t) = R_y(s-t)$ and we find that the stochastic process $\{y(t)\}$ is weakly stationary. The spectral density of y is given by

$$S_y(\omega) = \int_{-\infty}^{\infty} \exp(-j\omega\tau) R_y(\tau) d\tau$$

$$= r_0 \int_{-\infty}^{\infty} \exp(j\omega\tau) \int_0^{\infty} h(\tau + \lambda) h(\lambda) d\lambda d\tau$$

$$= r_0 \int_0^{\infty} \exp(j\omega\lambda) h(\lambda) d\lambda \int_{-\infty}^{\infty} \exp(-j\omega(\tau+\lambda)) h(\tau+\lambda) d\tau$$

As system is causal, we have that the impulse response $h(t) = 0$ for $t \leq 0$. Hence

$$S_y(\omega) = r_0 \int_0^{\infty} \exp(j\omega\lambda) h(\lambda) d\lambda \int_0^{\infty} \exp(-j\omega\lambda) h(\lambda) d\lambda = r_0 H(-j\omega) H(j\omega)$$

where $H(s) = L\{h(t)\}$ is the transfer function of the system, which is equal to the Laplace transform of the impulse response h. Summing up we find the *Representation theorem* which states the following: Consider a rational spectral density function $S(\omega)$. Then there exists an (asymptotically) stable time invariant dynamical system with the impulse response h such that the stochastic process derived by $y(t) = \int_{-\infty}^{\infty} h(t-s) dv(s)$, where v is a stationary process with uncorrelated increments, has the spectral density $S(\omega)$.

Example 4.1 (electrical circuit with random excitation): In the circuit of fig. 4.2 the generator $i(t)$ is a random white process with the power spectral density $S_i(\omega) = S_0 \left[A^2 / Hz \right]$. We shall determine the power spectral $S_u(\omega)$ of the voltage $u(t)$, its standard deviation σ_u and correlation function $R_u(\tau)$.

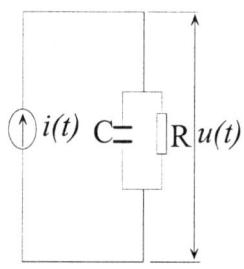

Fig. 4.2.

This voltage can be considered as output of the linear system with input $i(t)$. The corresponding system transfer function is

$$H(s) = \frac{U(s)}{I(s)} = \frac{R}{RCs+1}$$

The corresponding power spectral density $S_u(\omega)$ can be calculated using the theorem of spectral factorization:

$$S_u(\omega) = |H(j\omega)|^2 S_i(\omega) = \frac{R^2 S_0}{1+(\omega RC)^2}$$

We obtain from eq. (2.38.) that the covariance function is given by

$$R_u(\tau) = \frac{1}{\pi} \int_0^{\infty} S_u(\omega) \cos(\omega\tau) d\omega = \frac{1}{\pi} \int_0^{\infty} \frac{S_0 R^2}{1+(\omega CR)^2} \cos(\omega\tau) d\omega$$

Substituting ωRC with x, we have

$$R_u(\tau) = \frac{2 S_0 R}{\pi C} \int_0^{\infty} \frac{1}{1+x^2} \cos\left(\frac{\tau}{CR} x\right) dx$$

The last integral cannot be solved in the closed form. If we assume that $CR = 1$, the numerical integration can be used. Fig. 4.3. presents the MATLAB functions prepared for this purpose.

```
function program1
global T;  % the joint variable for both of the programs 'program1' and 'program2'
for i=1:51
   t(i)=(i-1)*0.1;
   T=t(i);
   r(i)=quad8('program2',0,10000);
end
plot(t,r);title('normalized autocorrelation function for positive values of time');
end
```

```
function y=program2(x)
% this program describes the integrand of the integral which defines the auto-correlation function
global T;
y=cos(T*x)./(1+x.^2);
end
```

Fig. 4.3. The MATLAB functions for numerical calculation of the integral

The obtained result is presented in Fig. 4.4.

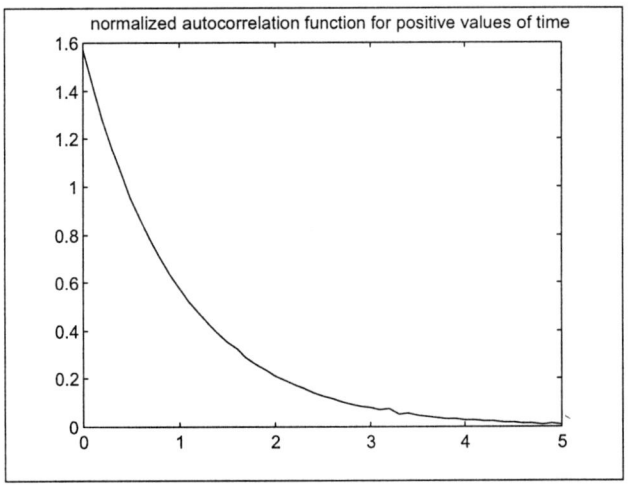

Fig. 4.4. The numerically computed auto-correlation function for $t \geq 0$

Using the last simulation (and assumption $CR=1$), variance of the process $u(t)$ can be calculated:

$$\sigma_u^2 = R_u(0) = 1.5698\frac{2S_0R}{\pi C}$$

and the standard deviation is equal to

$$\sigma_u = \sqrt{\sigma_u^2} = \sqrt{1.5698\frac{S_0R}{2C}}$$

4.2. State-space models

State models for continuos-time processes can be obtained by a formal generalization of eq. (3.28) to

$$\frac{dx}{dt} = Ax + Gw^*$$ (4.38)

where w^* is a vector whose elements are zero-mean white-noise stochastic processes. Since w^* has infinite variance, that is $E\{w^*(t)w^{*T}(s)\} = Q(t)\delta(t-s)$ with $\delta(\cdot)$ being the Dirac delta function, it is customary to introduce a stochastic process w that formally is the time integral of white noise w^* [1, 2, 30, 35, 36]. The signal w is thus assumed to have zero-mean, uncorrelated increments and the covariance matrix (see eqs. (4.29)-(4.35))

$$\text{cov}\{w(t), w(t)\} = E\{w(t)w^T(t)\} = Qt$$ (4.39)

In this way, the eq. (4.38) can be rewritten in terms of differentials

$$dx = Axdt + Gdw$$ (4.40)

where $dw = w^* dt$, i.e. w is the integral of white-noise w^*. However, a precise meaning can be given to (4.40) without any reference to white noise. Therefore, this form is common in mathematically oriented texts. The form is also useful as a reminder that dw has a magnitude (amplitude) proportional to $\sqrt{dt}$. It is also assumed that dw is uncorrelated with the state x.

Equation (4.40) is called a stochastic differential equation [23, 48]. To specify it fully, it is also necessary to give the initial probability distribution of x at the starting time $t = t_0$ (it is commonly used $t_0 = 0$). A continuous-time analog of the theorem for discrete-time systems in the section 3.2.1 is then obtained. To do this, we must solve eq. (4.38).

The homogeneous (no input, i.e. $w^* = 0$) solution to (4.38) is

$$x_h(t) = \exp(A(t-t_0))x(t_0)$$ (4.41)

where, by definition, the matrix exponential is

$$\exp(A(t-t_0)) = \sum_{k=0}^{\infty} A^k \frac{(t-t_0)^k}{k!}$$ (4.42)

This is proved by differentiating (4.41) with respect to t, which can be done by differentiating (4.42) term by term, yielding

$$\dot{x}_h(t) = A\left[I + A(t-t_0) + A^2 \frac{(t-t_0)^2}{2!} + \cdots \right] x(t_0)$$

or

$$\dot{x}_h(t) = A\exp(A(t-t_0))x(t_0) = Ax_h(t)$$ (4.43)

which is the unforced portion of (4.38).

It can be shown that the solution given by (4.41) is unique, which leads to very interesting properties of the matrix exponential. For example, consider two values of $t: t_1$ and t_2. We have

$$x(t_1) = \exp(A(t_1-t_0))x(t_0)$$

and

$$x(t_2) = \exp\left(A(t_2 - t_0)\right)x(t_0)$$

Since t_0 is arbitrary also, we can express $x(t_2)$ as if the equation solution began at t_1, for which

$$x(t_2) = \exp\left(A(t_2 - t_1)\right)x(t_1)$$

Substituting for $x(t_1)$ gives

$$x(t_2) = \exp\left(A(t_2 - t_1)\right)\exp\left(A(t_1 - t_0)\right)x(t_0')$$

We now have two separate expressions for $x(t_2)$ and if the solution is unique, these must be the same. Hence we conclude that

$$\exp\left(A(t_2 - t_0)\right) = \exp\left(A(t_2 - t_1)\right)\exp\left(A(t_1 - t_0)\right) \tag{4.44}$$

for all t_2, t_1 and t_0. Note especially that if $t_2 = t_0$, then

$$I = \exp\left(-A(t_1 - t_0)\right)\exp\left(A(t_1 - t_0)\right)$$

Thus, we can obtain the inverse of $\exp(At)$ by merely changing the sign of t. We will use this result in computing the particular solution of (4.38).

The particular solution, when w^* is not zero, is obtained by using the method of variation of parameters (due to the French mathematician Josep Louis Lagrange). We guess the solution in the form

$$x_p(t) = \exp\left(A(t - t_0)\right)\lambda(t) \tag{4.45}$$

where $\lambda(t)$ is a vector of variable parameters to be determined (as contrasted to the constant parameters $x(t_0)$ in eq. (4.41)). Substituting (4.45) in (4.38), we obtain

$$A\exp\left(A(t - t_0)\right)\lambda(t) + \exp\left(A(t - t_0)\right)\dot{\lambda}(t) = A\exp\left(A(t - t_0)\right)\lambda(t) + Gw^*$$

and, using the fact that the inverse is found by changing the sign of the exponent, we can solve for $\dot{\lambda}(t) = d\lambda(t)/dt$ as

$$\dot{\lambda}(t) = \exp\left(-A(t - t_0)\right)Gw^*$$

Assuming that the input w^* is zero for $t < t_0$, we can integrate $\dot{\lambda}(t)$ from t_0 to t to obtain

$$\lambda(t) = \int_{t_0}^{t} \exp\left(-A(t - t_0 - \tau)\right)Gw^*(\tau)d\tau$$

Hence, from (4.45)

$$x_p(t) = \exp\left(A(t - t_0)\right)\int_{t_0}^{t} \exp\left(-A(\tau - t_0)\right)Gdw(\tau)$$

where $dw = w^* d\tau$. Simplifying, using the results of (4.44), we obtain the particular solution in the form of

$$x_p(t) = \int_{t_0}^{t} \exp\left(A(t - \tau)\right)Gdw(\tau) \tag{4.46}$$

The total solution for $w^* = 0$ and $w^* \neq 0$ is the sum of (4.41) and (4.46)

$$x(t) = \exp\left(A(t - t_0)\right)x(t_0) + \int_{t_0}^{t} \exp\left(A(t - \tau)\right)Gdw(\tau) \tag{4.47a}$$

Since t_0 is arbitrarily, one can rewrite (4.47a) as

$$x(s) = \exp\big(A(s-t)\big)x(t) + \int^s \exp\big(A(s-\tau)\big)Gdw(\tau) \tag{4.47b}$$

for any $s \geq t$.

The matrix

$$\Phi(t,\tau) = \exp\big(A(t-\tau)\big) \tag{4.48}$$

is called *the state transition matrix* of the system characterized by the state differential equation (4.38). Thus, the solution (4.47a) to the state differential equation (4.38) can be also written as

$$x(t) = \Phi(t,t_0)x(t_0) + \int_{t_0} \phi(t,\tau)Gdw(\tau) \tag{4.49}$$

The state-transition matrix can be found by solving the homogeneous linear differential equation

$$\frac{d\Phi(t,\tau)}{dt} = A\Phi(t,\tau) \text{ for } \Phi(t,t) = I \tag{4.50}$$

The eq. (4.50) is a direct consequence of eqs. (4.41) and (4.43), since from (4.41) and (4.48) one obtains

$$x_h(t) = \Phi(t,t_0)x(t_0)$$

and

$$\dot{x}_h(t) = \frac{d\Phi(t,t_0)}{dt}x(t_0)$$

while eq. (4.43) reduces to

$$\dot{x}_h(t) = A\Phi(t,t_0)x(t_0)$$

Thus, comparing the last two equations, one obtains eq. (4.50). It is also important to note that if we keep eqs. (4.47) and (4.49) in mind, then the following definition of the state variable makes sense:

A state variable is the minimal set of variables such that knowledge of these variables at any time, say t_0, plus information about the input w^* is sufficient to specify the state at any time t.

As in the discrete-time case, the model is easily extended to include a control input, say $u(t)$, by writing

$$\dot{x}(t) = Ax(t) + Gw^*(t) + Bu(t) \tag{4.51}$$

where u is a known control input to the system, and w^* is a disturbance input. The solution to the state differential equation (4.51) is given by

$$x(t) = \Phi(t,t_0)x(t_0) + \int_{t_0} \Phi(t,\tau)Bu(\tau)d\tau + \int_{t_0} \Phi(t,\tau)Gdw(\tau) \tag{4.52}$$

where $dw = w^*d\tau$. The block diagram for the process $\{x(t), t \geq t_0\}$ is shown in fig. 4.5.

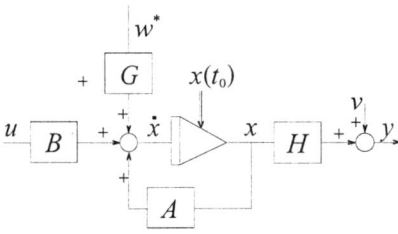

Fig.4.5: Block diagram for system of eqs. (4.51) and (4.52)

When the system order, that is the dimension of the state x, is small, then it is possible to solve the state-space equation using Laplace transform technique by hand. If the system order is not small, then computational techniques must be employed. Thus, taking Laplace transforms of eqs. (4.51) (we assume a disturbance input $w^* = 0$, but since the equations are linear, the effect of w^* can be added later)

$$sX(s) - x(0) = AX(s) + BU(s)$$

or solving for X, we have

$$(sI - A)X(s) = x(0) + BU(s)$$
$$X(s) = (sI - A)^{-1} x(0) + (sI - A)^{-1} BU(s)$$

(4.53)

We define the resolving matrix as

$$\Phi(s) = L\{\Phi(t)\} = (sI - A)^{-1} = \frac{\text{adj}(sI - A)}{\det(sI - A)}$$

(4.54)

where $\det(sI - A) = 0$, with $\det(A)$ denoting the determinant of a matrix A, is called the characteristic equation of the system. So to find the state transition matrix, we can find the resolving matrix and merely take inverse Laplace transform, that is $\Phi(t) = L^{-1}\{\Phi(s)\}$. This is by far the most direct way of obtaining closed form solutions to linear differential equations, that is

$$x(t) = \Phi(t)x(0) + \int_0^t \Phi(t - \tau)Bu(\tau)d\tau$$

(4.55)

where

$$\Phi(t) = \exp(At) = L^{-1}\{\Phi(s)\}; \quad \Phi(s) = (sI - A)^{-1}$$

(4.56)

is the fundamental system matrix. Here the starting time $t_0 = 0$ is used.

We assume that the collection of sensors which are attached to the system (4.51) to monitor its behavior has an output (collection of measured variables) which can be modeled by the relation

$$y(t) = Hx(t) + v(t)$$

(4.57a)

In eq. (4.57), y is a vector termed the measurement vector, v is also a vector called the measurement error vector or measurement noise, and H is a matrix which relates the states to the measurements. Also, y is sometimes called the system output. The block diagram for the complete model of eqs. (4.51) and (4.57) is the one which we gave in fig. 4.5. We assume that the measurement error process $\{v(t), t \geq t_0\}$ is white noise with zero-mean and covariance

$$E\{v(t)v^T(s)\} = R\delta(t - s)$$

(4.57b)

for all $t, s \geq t_0$. We assume further that $\{v(t)\}$ is independent of $x(t_0)$ and $\{w^*(t), t \geq t_0\}$. Our justification for the above model is the same as for the discrete-time one, as discussed earlier.

Analogous to the discrete-time model of section 3.2.1, the following theorem can be defined:

Consider a stochastic process defined by the linear stochastic differential equation (4.40), where the process w has zero-mean and incremental covariance $\text{cov}\{dw\} = Qdt$. Let the initial state $x(0)$ have mean m_0 and covariance P_0. The mean-value function of the process x is then given by

$$\frac{dm(t)}{dt} = Am(t) \; ; \; m(0) = m_0 \tag{4.58}$$

and the covariance function is given by

$$\text{cov}\{x(s), x(t)\} = \exp\left(A(s-t)\right)P(t) = \Phi(s,t)P(t) \; ; \; s \geq t \tag{4.59}$$

where $P(t) = \text{cov}\{x(t), x(t)\}$ is given by

$$\frac{dP(t)}{dt} = AP(t) + P(t)A^T + GQG^T \; ; \; P(0) = P_0 \tag{4.60}$$

Proof: The formula (4.58) for the mean value is obtained simply by taking the mean value of (4.40)

$$E\{dx\} = AE\{x\}\, dt + GE\{dw\}$$

from which one concludes

$$\frac{dE\{x(t)\}}{dt} = AE\{x(t)\}$$

or equivalently

$$\frac{dm(t)}{dt} = Am(t)$$

where $m(t) = E\{x(t)\}$. Here is used the fact that $E\{dw\} = 0$. A derivation of eq. (4.60) can be carried out by working directly with the solution (4.49) of eq. (4.40) and the definition of $P(t)$. First we see that

$$x(t) = \Phi(t,t_0)x(t_0) + \int_{t_0}^t \Phi(t,\tau)Gdw(\tau)$$

and

$$m(t) = \Phi(t,t_0)m_0 + \int_{t_0}^t \Phi(t,\tau)GE\{dw(\tau)\} = \Phi(t,t_0)m_0$$

from which it follows trivially that

$$x(t) - m(t) = \Phi(t,t_0)\left[x(t_0) - m_0\right] + \int_{t_0}^t \Phi(t,\tau)Gdw(\tau)$$

Then from the definition of $P(t)$

$$P(t) = E\left\{\left(\Phi(t,t_0)\left[x(t_0) - m_0\right] + \int_{t_0}^t \Phi(t,\tau)Gdw(\tau)\right)\left(\Phi(t,t_0)\left[x(t_0) - m_0\right] + \int_{t_0}^t \Phi(t,\tau)Gdw(\tau)\right)^T\right\}$$

$$= \Phi(t,t_0)E\left\{\left[x(t_0) - m_0\right]\left[x(t_0) - m_0\right]^T\right\}\Phi^T(t,t_0) +$$

$$+ \Phi(t,t_0)\int_{t_0}^t E\left\{\left[x(t_0) - m_0\right]\left[dw(s)\right]^T\right\}G^T\Phi^T(t,s) +$$

$$+ \int_{t_0}^t \Phi(t,\tau)GE\left\{dw(\tau)\left[x(t_0) - m_0\right]^T\right\}\Phi^T(t,t_0) +$$

$$+ \int_{t_0}^t \int_{t_0}^t \Phi(t,\tau)GE\left\{dw(\tau)\left[dw(s)\right]^T\right\}G^T\Phi^T(t,s)$$

Since $x(t_0)$ is independent of $\{dw(t), t \geq t_0\}$, the middle two terms on the right-hand side vanish. Also, since $dw(t) = w^*(t)dt$, where w^* is white noise with $\text{cov}\{w^*(t), w^*(s)\} = Q\delta(t-s)$,

$$E\{dw(\tau)[dw(s)]^T\} = E\{w^*(\tau)w^{*T}(s)\} d\tau ds = Qd\tau ds\delta(\tau-s),$$

so that we carry out the integration in the last term with respect to τ. Consequently,

$$P(t) = \Phi(t,t_0)P_0\Phi^T(t,t_0) + \int_{t_0}^{t} \Phi(t,\tau)GQG^T\Phi^T(t,\tau)d\tau \qquad (4.61)$$

Equation (4.61) gives the desired covariance matrix

$$P(t) = E\{[x(t)-m(t)][x(t)-m(t)]^T\}.$$

To reconcile this result with eq. (4.60), we differentiate it with respect to t to obtain

$$\dot{P}(t) = \dot{\Phi}(t,t_0)P_0\Phi^T(t,t_0) + \Phi(t,t_0)P_0\dot{\Phi}(t,t_0)$$
$$+ \Phi(t,t)GQG^T\Phi^T(t,t) + \int_{t_0}^{t} \dot{\Phi}(t,\tau)GQG^T\Phi^T(t,\tau)d\tau + \int_{t_0}^{t} \Phi(t,\tau)GQG^T\dot{\Phi}^T(t,\tau)d\tau$$

Taking into account (4.50), further follows

$$\dot{P}(t) = \Phi(t,t_0)P_0\dot{\Phi}^T(t,t_0)A^T + A\Phi(t,t_0)P_0\Phi^T(t,t_0)$$
$$+ GQG^T + \int_{t_0}^{t} \Phi(t,\tau)GQG^T\Phi^T(t,\tau)d\tau A^T + A\int_{t_0}^{t} \Phi(t,\tau)GQG^Td\tau$$
$$= A\left[\Phi(t,t_0)P_0\Phi^T(t,t_0) + \int_{t_0}^{t} \Phi(t,\tau)GQG^Td\tau\right] \qquad (4.62)$$
$$+ \left[\Phi(t,t_0)P_0\Phi^T(t,t_0) + \int_{t_0}^{t} \Phi(t,\tau)GQG^Td\tau\right]A^T + GQG^t$$

Finally, substituting (4.61) into the first two terms on the right-hand side of eq. (4.62), we have
$$\dot{P} = \dot{\Phi} = AP + PA^T + GQG^T$$
which is identically eq. (4.60).

Equation (4.61) is obviously the solution of eq. (4.60). However, it is of little use for numerical computation of $P(t)$, in general, because it requires that $\Phi(t,\tau)$ be determined first. For computational purposes, it is more expedient to determine $P(t)$ by numerical integration of eq. (4.60).

A derivation of eq. (4.60) can be also carried out by working directly with the eq. (4.40). To do this, notice that

$$d(xx^T) = (x+dx)(x+dx)^T - xx^T$$
$$= xdx^T + dxx^T + dxdx^T$$

Equation (4.40) then gives

$$d(xx^T) = x[Axdt + Gdw]^T + [Axdt + Gdw]x^T$$
$$+ [Axdt + Gdw][Axdt + Gdw]^T$$

Taking mean values gives

$$dE\{xx^T\} = E\{xx^T\}A^Tdt + AE\{xx^T\}dt + GE\{dwdw^T\}G^T + AE\{xx^T\}A^Tdt^2$$

because dw is uncorrelated with x. Furthermore, since

$$E\{dw dw^T\} = Q dt$$

it follows that

$$dP = PA^T dt + AP dt + Q dt + APA^T dt^2$$

Dividing by dt and taking the limit as dt goes to zero gives the differential equation in (4.40).

To obtain eq. (4.59), notice that for $s \ge t$

$$x(s) = \exp(A(s-t))x(t) + \int_t^s \exp(A(s-\tau))G dw(\tau)$$

so that

$$m(s) = E\{x(s)\} = \exp(A(s-t))m(t)$$

since $E\{dw(\tau)\} = 0$. Hence

$$\tilde{x}(s) = x(s) - m(s) = \exp(A(s-t))\tilde{x}(t) + \int_t^s \exp(A(s-\tau))G dw(\tau)$$

Multiplying by $\tilde{x}^T(t) = [x(t) - m(t)]^T$ from the right and taking mathematical expectation gives

$$E\{\tilde{x}(s)\tilde{x}(t)\} = \exp(A(s-t))E\{\tilde{x}(t)\tilde{x}^T(t)\} + \int_t^s \exp(A(s-\tau))GE\{dw(\tau)\tilde{x}^T(t)\}$$

and since $dw(\tau)$ is uncorrelated with $dw(\tau)$ for $\tau \ge t$, one obtains

$$E\{\tilde{x}(s)\tilde{x}^T(t)\} = \exp(A(s-t))E\{\tilde{x}(t)\tilde{x}^T(t)\}$$

which is identical to eq. (4.59).

If the model is extended to included control input, as in eq. (4.51), the only modification which is necessary in the process description is that

$$E\{dx\} = AE\{x\} dt + GE\{dw\} + Bu$$

from which one obtains the differential equation for the mean

$$\dot{m}(t) = Am(t) + Bu(t) \tag{4.63}$$

where u is a known control input and B is a given matrix. Furthermore, we see that eq. (4.60) is unchanged. Moreover, it is linear. Hence, its solution consists of the homogenous solution which depends upon the initial condition $P(t_0)$ and the particular solution, which depends upon the forcing function Q. This permits us to determine the contribution to the total uncertainty $P(t)$ of the uncertainty associated with the initial condition $x(t_0)$ and that associated with the state noise w^*, respectively. Once eqs. (4.58) or (4.63) and (4.60) are solved, the state process $\{x(t), t \ge t_0\}$ can be characterized in terms of its mean value $m(t)$ and covariance function $\text{cov}\{x(s), x(t)\}$.

Having obtained this result we might ask whether it was really worthwhile to go through the procedure of developing the stochastic differential equation in the linear case. Would it not be possible to obtain the same result by formal manipulation of the equality

$$\frac{dx}{dt} = Ax + w^*$$

where $\{w^*\}$ is continuos time white noise; that is, a stationary stochastic process with zero-mean and covariance function

$$\text{cov}\{w^*(t), w^*(s)\} = E\{w^*(t)w^{*T}(s)\} = Q\delta(t-s)$$

where Q is a constant spectral density matrix.

To demonstrate that erroneous results are easily obtained, consider, for example, the evaluation of the state covariance matrix

$$P(t) = E\left\{\left(x(t) - m_x(t)\right)\left(x(t) - m_x(t)\right)^T\right\} = E\left\{x(t)x^T(t)\right\} - m_x(t)m_x^T(t)$$

We get

$$\frac{dP(t)}{dt} = E\left\{\frac{dx}{dt}x^T\right\} + E\left\{x\frac{dx^T}{dt}\right\} - \frac{dm_x}{dt}m_x^T - m_x\frac{dm_x^T}{dt}$$

$$= E\left\{\left[Ax + w^{\bullet}\right]x^T\right\} + E\left\{x\left[Ax + w^{\bullet}\right]^T\right\} - Am_xm_x^T - m_x\left(Am_x\right)^T$$

$$= AE\left\{xx^T\right\} + E\left\{xx^T\right\}A^T - Am_xm_x^T - m_xm_x^TA^T$$

$$= A\left(E\left\{xx^T\right\} - m_xm_x^T\right) + \left(E\left\{xx^T\right\} - m_xm_x^T\right)A^T$$

$$= AP + PA^T$$

This is apparently wrong. We only get the first two terms of eq. (4.60). Notice that the result would be the same as if $w^{\bullet} = 0$. The erroneous result originates from the fact that the derivative dx/dt does not exist. Taking into account that dx is of magnitude $\sqrt{dt}$, we find that the ordinary rules for differentiation do not apply.

An alternative way to find a correct differential equation for the state covariance matrix $P(t)$ should be based on formal manipulations of delta functions. To do this, write

$$\frac{dP}{dt} = E\left\{\frac{d\tilde{x}}{dt}\tilde{x}^T\right\} + E\left\{\tilde{x}\frac{d\tilde{x}^T}{dt}\right\} \quad ; \quad \tilde{x} = x - m_x$$

where

$$\frac{d\tilde{x}}{dt} = A\tilde{x} + Gw^{\bullet}$$

Let $\Phi(t,t_0)$ be the state-transition matrix for A. Then the solution to the last differential equation is given by

$$\tilde{x}(t) = \Phi(t,t_0)\tilde{x}(t_0) + \int_{t_0} \Phi(t,\tau)Gw^{\bullet}(\tau)d\tau$$

We shall soon need the cross-correlation $R_{\tilde{x}w^{\bullet}}(t)$. To find it, use the uncorrelatedness assumption between $\tilde{x}(t_0)$ and $w^{\bullet}$. Thus

$$R_{\tilde{x}w^{\bullet}}(t) = \text{cov}\left\{\tilde{x}(t), w^{\bullet}(t)\right\} = E\left\{\tilde{x}(t)w^{\bullet T}(t)\right\}$$

$$= E\left\{\left[\Phi(t,t_0)\tilde{x}(t_0) + \int_{t_0}\Phi(t,\tau)Gw^{\bullet}(\tau)d\tau\right]w^{\bullet T}(t)\right\}$$

$$= \Phi(t,t_0)E\left\{\tilde{x}(t_0)w^{\bullet T}(t)\right\} + \int_{t_0}\Phi(t,\tau)GE\left\{w^{\bullet}(\tau)w^{\bullet}(t)\right\}d\tau$$

$$= \int_{t_0}\Phi(t,\tau)GQ\delta(t-\tau)d\tau$$

However, only half of the area of Dirac delta function should be considered as being to the left of $t_0 = 0$. Hence

$$\int_{t_0} \Phi(t,\tau) GQ\delta(t-\tau)d\tau = \Phi(t,t)GQ\int_{t_0} \delta(t-\tau)d\tau$$

and

$$R_{\tilde{x}w^{\cdot}}(t) = \frac{1}{2}\Phi(t,t)GQ = \frac{1}{2}GQ$$

since $\Phi(t,t) = I$. To find a differential equation for $P(t)$, write

$$\frac{dP}{dt} = E\left\{\left(A\tilde{x} + Gw^{\cdot}\right)\tilde{x}^{T}\right\} + E\left\{\tilde{x}\left(A\tilde{x} + Gw^{\cdot}\right)^{T}\right\}$$

$$= AE\left\{\tilde{x}\tilde{x}^{T}\right\} + GE\left\{w^{\cdot}\tilde{x}^{T}\right\} + E\left\{\tilde{x}\tilde{x}^{T}\right\}A^{T} + E\left\{\tilde{x}w^{\cdot T}\right\}G^{T}$$

$$= AP + PA^{T} + GR_{w^{\cdot}\tilde{x}}(t)G^{T}$$

Since

$$R_{w^{\cdot}\tilde{x}}(t) = R_{\tilde{x}w^{\cdot}}^{T}(t) = \left(\frac{1}{2}GQ\right)^{T} = \frac{1}{2}QG^{T}$$

one obtains

$$\frac{dP}{dt} = AP + PA^{T} + GQG^{T}$$

which is identical to eq. (4.60).

Finally, it should be noted the following properties of the symmetric Dirac delta function

$$\int_{a}^{b} f(\tau)\delta(t-\tau)d\tau = \begin{cases} 0 \; ; \; t \in (-\infty, a) \cup (b, \infty) \\ \dfrac{f(a)}{2} \; ; \; t = a \\ \dfrac{f(b)}{2} \; ; \; t = b \\ f(t) \; ; \; t \in (a,b) \end{cases}$$

where $f(\tau)$ is any function continuous on the closed interval $[a,b]$.

Example 4.2 (first order state space model): Consider the scalar stochastic differential equation

$$dx = -axdt + dw$$

where

$$E\{x(t_0)\} = m_0 \; ; \; \text{var}\{x(t_0)\} = \sigma_0^2$$

and the process $\{w(t), t \geq t_0\}$ has incremental variance $\text{var}\{dw\} = Qdt$. It follows from (4.58) that the mean-value is given by

$$\frac{dm}{dt} = -am \; , \; m(t_0) = m_0$$

The equation has the solution

$$m(t) = m_0 \exp\left(-a(t-t_0)\right)$$

The covariance function is given by Eq. (4.59)

$$R(s,t)=\text{cov}\left(x(s),x(t)\right)=\exp\left(-a(s-t)\right)P(t)\ ;\ s\geq t$$

and

$$R(s,t)=\exp\left(-a(t-s)\right)P(s)\ ;\ s\leq t$$

Equation (4.60) gives the following differential equation for P.

$$\frac{dP(t)}{dt}=-2aP(t)+Q\ ;\ P(t_0)=\sigma_0^2$$

This differential equation has the solution (see Eqs. (4.48) and (4.49))

$$P(t)=\exp\left(-2a(t-t_0)\right)\sigma_0^2+\int_{t_0}^{t}\exp\left(-2a(t-\tau)\right)Qd\tau$$

$$=\exp\left(-2a(t-t_0)\right)\sigma_0^2+\frac{Q}{2a}\left[1-\exp\left(-2a(t-t_0)\right)\right]$$

As $t_0\to-\infty$, the mean value goes to zero and the covariance function goes to

$$R(s,t)=\frac{Q}{2a}\exp\left(-a|s-t|\right)$$

Since the limiting covariance function depends only on the argument difference $\tau=s-t$, the limiting process is weakly stationary and its covariance function can be written as

$$R(\tau)=\frac{Q}{2a}\exp\left(-|\tau|\right)$$

The corresponding spectral density is given by (see the table of Fourier transforms in section 2)

$$S(\omega)=\text{F}\{R(\tau)\}=\frac{Q}{2\pi}\frac{1}{\omega^2+a^2}$$

Example 4.3 (first order state space model with control input): Consider linear stochastic system defined by a scalar stochastic differential equation

$$dx=ax+u+dw$$

where $x(0)$ is zero-mean random variable with variance σ_0^2, u is the unit step and the process $\left(w(t),t\geq0\right)$ has zero-mean and incremental variance $\text{var}\{dw\}=Qdt$ with $Q{=}1$. It follows from (4.63) that the mean value of the scalar state x is given by

$$\frac{dm}{dt}=am+u$$

Instead to use the usual state equation solution as in Eq. (4.49), we shall use the Laplace transform techniques to solve the above differential equation for the mean value. Thus, taking Laplace transform

$$sM(s)-m(0)=aM(s)+U(s)$$

and solving for $M(s)$

$$M(s)=\frac{1}{s-a}U(s)=\frac{1}{s-a}\frac{1}{s}=\frac{1/a}{s-a}+\frac{-1/a}{s}$$

Here is used the fact that $m(0)=0$ and $U(s)=\text{L}\{u(t)\}=1/s$. Using the Table of Laplace transforms (see appendix A), one obtains

$$m(t) = \left[\frac{1}{a}\exp(at) - 1\right]h(t)$$

where $h(t)$ is the unit step. This reaches a finite steady-state value only if $a < 0$; that is, if the system is stable. Namely, the transfer function of the system

$$\frac{M(s)}{U(s)} = \frac{1}{s-a}$$

has the pole $s = a$, and it is stable if and only if the pole lies in the left-half part of the s-plane, i.e. if $\mathrm{Re}\{a\} < 0$. In this case $m(t)$ behaves as in Fig. 4.6.a.

Equation (4.60) becomes

$$\dot{P}(t) = 2aP(t) + 1$$

with $P(0) = \sigma_0^2 = P_0$. This may be solved by using the equation similar to Eq. (4.49), or by separation of variables. Thus,

$$\int_{P_0}^{P} \frac{dp}{2ap+1} = \int_0^t d\tau$$

or

$$P(t) = \left(P_0 + \frac{1}{2a}\right)\exp(2at) - \frac{1}{2a}$$

There is a bounded limiting solution if and only if $a < 0$; that is, if the system is stable. In this case $P(t)$ behaves as in Fig.4.6.b.

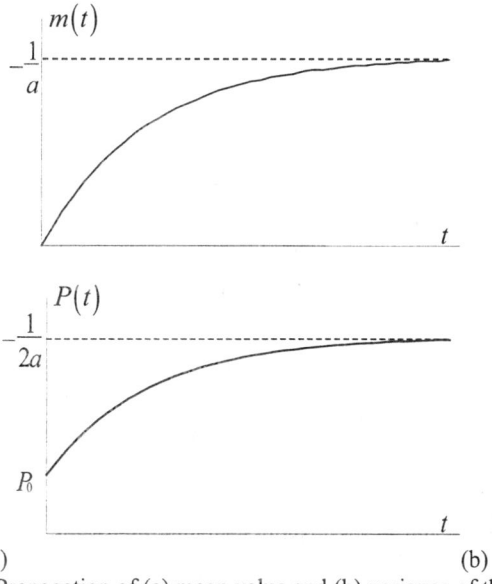

(a)   (b)

Fig. 4.6. Propagation of (a) mean value and (b) variance of the state

The mean value function of the system output y in (4.57) is given by

$$m_y(t) = E\{y(t)\} = HE\{x(t)\} + E\{v(k)\}$$

and since $E\{v(k)\} = 0$, one obtains

$$m_y(t) = Hm_x(t) \tag{4.64}$$

where matrix $m_x(t)$ is defined by Eq. (4.58). The covariance function of the output is given by

$$R_y(t,s) = E\left\{\left[y(t) - m_y(t)\right]\left[y(s) - m_y(s)\right]^T\right\}$$

where

$$\tilde{y}(t) = y(t) - m_y(t) = H\left[x(t) - m_x(t)\right] + v(t) = H\tilde{x}(t) + v(t)$$

so that

$$R_y(t,s) = HR_x(t,s)H^T + R\delta(t-s) \tag{4.65}$$

since $v(t)$ is uncorrelated with $\tilde{x}(t)$ for any t and $E\{v(t)v^T(s)\} = R\delta(t-s)$, with $\delta(t)$ being the Dirac delta function. The covariance function of the state $R_x(t,s)$ is defined by Eq. (4.59).

The covariance matrix of the output is given by

$$\begin{aligned} P_y(t) &= E\{\tilde{y}(t)\tilde{y}^T(t)\} = E\left\{\left[H\tilde{x}(t) + v(t)\right]\left[H\tilde{x}(t) + v(t)\right]^T\right\} \\ &= HP_x(t)H^T + E\{v(t)v^T(t)\} = HP_x(t)H^T + \mathrm{cov}\{v(t)\} \end{aligned} \tag{4.66}$$

where the covariance matrix $P_x(t)$ is defined by Eq. (4.60), and $\mathrm{cov}\{v(t)\}$ is given by (4.57.b).

The cross-covariance function between the output y and the state x is given by

$$\begin{aligned} R_{yx}(t,s) &= E\{\tilde{y}(t)\tilde{x}^T(s)\} = E\left\{\left[H\tilde{x}(t) + v(t)\right]\tilde{x}^T(s)\right\} \\ &= HR_x(t,s) \end{aligned} \tag{4.67}$$

where the covariance function $R_x(t,s)$ is defined by Eq. (4.59).

A very important subclass of the foregoing development is the stationary (at least wide-sense-stationary) problem, in which the means are constant and the covariance function is only a function of the variable $\tau = t - s$. The system coefficient matrix A and disturbance (noise) matrix G are constant, so that Eq. (4.58) gives

$$\dot{m}_x = 0 = Am_x + Gm_w.$$

or

$$m_x = A^{-1}Gm_w.$$

Particularly, if $m_w = 0$ one obtains $m_x = 0$. Furthermore, if A is singular, then there will be no stationary value of m_x, since it will be necessary for m_x to become unbounded. In simple physical terms, a singular A implies that the system transfer function has integrating capability (one or more poles at the origin), and a constant input will produce an unbounded output.

The spectral-density relations can be derived by a simple procedure, similarly as in the discrete-time case. Let us think of $w^*(t)$ and $x(t)$ in Eq. (4.38) as deterministic stationary zero-mean signals with usual Fourier transforms $W(s)$ and $X(s)$. To be more rigorous we may consider the Fourier transform of a single record and use the limiting arguments of section 2, as applied to the truncated signals. The spectral densities $S_{xw}(s)$ and $S_x(s)$ may then be written

$$S_{xw}(s) = E\{X(s)W^T(-s)\} \tag{4.68}$$

and

$$S_x(s) = E\{X(s)X^T(-s)\}$$

The expression for $X(s)$ may be obtained by taking the Fourier transform of Eq. (4.38) as

$$sX(s) = AX(s) + GW(s) \tag{4.69}$$

or

$$X(s) = (sI - A)^{-1}GW(s) = \Phi(s)GW(s) \tag{4.70}$$

Therefore $S_{xw}(s)$, as given by Eq. (4.68), becomes

$$S_{xw}(s) = E\{\Phi(s)GW(s)W^T(-s)\} = \Phi(s)GS_x(s) \tag{4.71}$$

and

$$\begin{aligned} S_x(s) &= E\{\Phi(s)GW(s)[\Phi(-s)GW(-s)]^T\} \\ &= \Phi(s)GE\{W(s)W^T(-s)\}G^T\Phi^T(-s) \\ &= \Phi(s)GS_w(s)G^T\Phi^T(-s) \end{aligned} \tag{4.72}$$

Although the proposed procedure is formal and one should hence be careful in its application in nonstandard situations, the procedure is nonetheless a very useful one.

Example 4.4 (single input single output state space system model): Let us use spectral-density input-output relations, Eqs. (4.71) and (4.72), to find the spectral densities $S_{yw}(s)$ and $S_y(s)$ for the simple stationary single-input single-output system given by the state space model

$$\dot{x}(t) = Ax(t) + gw^*(t)$$
$$y(t) = hx(t)$$

We will assume that the spectral density $S_w(s)$ of w^* is known. Equations (4.71) and (4.72) may be used directly to obtain the spectral densities $S_{xw}(s)$ and $S_x(s)$ as

$$S_{xw}(s) = \Phi(s)S_w(s)$$
$$S_x(s) = \Phi(s)gS_w(s)g^T\Phi^T(-s) \; ; \; \Phi(s) = (sI - A)^{-1}$$

The desired spectral density $S_{yw}(s)$ can be obtained from $S_{xw}(s)$ by using the formal procedure discussed above, i.e.

$$S_{yw}(s) = E\{Y(s)W^T(-s)\} = hE\{X(s)W(-s)\} = hS_{xw}(s) = h\Phi(s)gS_w(s)$$

However, we recognize $h\Phi(s)g$ as the transfer function relating y and w^*, or

$$\frac{Y(s)}{W(s)} = H(s) = h\Phi(s)g \ ; \ \Phi(s) = (sI - A)^{-1}.$$

Therefore, $S_{yw}(s)$ becomes

$$S_{yw}(s) = H(s)S_w(s)$$

In a completely similar fashion, it is easy to show that $S_y(s)$ is given by

$$S_y(s) = E\{Y(s)Y(-s)\} = H(s)H(-s)E\{W(s)W(-s)\} = H(s)H(-s)S_w(s).$$

Let us now examine the nature of the state $\{x(t), t \geq t_0\}$ defined by the model (4.38) or (4.40). The process is obviously Markov, since the solution (4.49) of Eq. (4.38) can be written

$$x(t_m) = \Phi(t_m, t_{m-1})x(t_{m-1}) + \int_{t_{m-1}}^{t_m} \Phi(t_m, \tau)Gdw(\tau)$$

where $t_m > t_{m-1} \geq t_0$.

Also, for any $t \geq t_0$ let $x^{(n)}(t)$ be given by

$$x^{(n)}(t) = \Phi(t,t_0)x(t_0) + \int_{t_0}^{t} \Phi(t,\tau)Gdw^{(n)}(\tau) = \Phi(t,t_0)x(t_0) + \int_{t_0}^{t} \Phi(t,\tau)Gw^{*(n)}(\tau)d\tau$$

$$= \Phi(t,t_0)x(t_0) + \sum_{i=0}^{n-1} \Phi(t,t_0 + i\Delta t)Gw^{*(n)}(t_0 + i\Delta t)\Delta t \qquad (4.73)$$

where $w^{*(n)}(t)$ denote the piecewise constant white sequence, and the interval $[t_0, t]$ has been subdivided into n subintervals each of length $\Delta t = (t - t_0)/n$. The one component, say $w_i^*(n)$, of the vector process $\{w^{*(n)}(t)\}$ is indicated in Fig. 4.7.

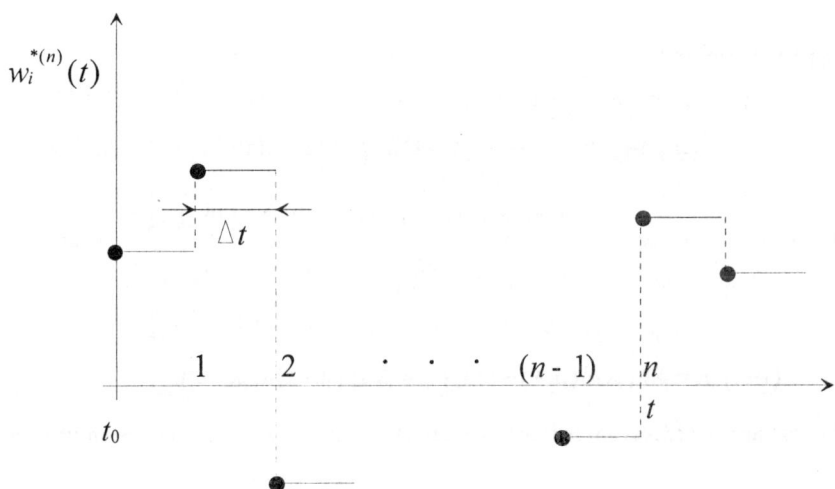

Fig. 4.7.: Component of piecewise constant white sequence sample function

For $\{x(t), t \geq t_0\}$ we know that

$$x(t) = \Phi(t,t_0)x(t_0) + \int_{t_0}^{t} \Phi(t,\tau)Gw^*(\tau)d\tau$$

Since

$$\lim_{n\to\infty}\left\{w^{*(n)}(\tau), t_0 \leq \tau \leq t\right\} = \left\{w^*(\tau), t_0 \leq \tau \leq t\right\}$$

it follows that

$$\lim_{n\to\infty}\left\{x^{(n)}(t), t \geq t_0\right\} = \left\{x(t), t \geq t_0\right\}$$

Referring now to eq. (4.73), it is clear that the term $\Phi(t,t_0)x(t_0)$ is a Gaussian random vector for all $t \geq t_0$ if $x(t_0)$ is a Gaussian random vector. Moreover, for each $n = 1,2,...$ it is clear that if $\{w^*(t)\}$ is a Gaussian white noise then

$$\sum_{i=0}^{n-1}\Phi(t,t_0 + i\Delta t)Gw^{*(n)}(t_0 + i\Delta t)\Delta t$$

is also a Gaussian random vector, being simply sum of n independent Gaussian random vectors $w^{*(n)}(t_0 + i\Delta t)$, $i = 0,1,...,n-1$. It follows immediately that for any $n = 1,2,...,$ $x^{(n)}(t)$ in eq. (4.73) is a Gaussian random vector for all $t \geq 0$. Hence, $\{x(t), t \geq t_0\}$ is a Gauss-Markov process. Its mean follows by Eq. (4.58), while the evolution of its covariance matrix is given by Eq. (4.60). Once Eqs. (4.58) and (4.60) are solved to obtain $m(t)$ and $P(t)$ for $t \geq 0$, the Gauss-Markov process $\{x(t), t \geq t_0\}$ can be characterized in terms of its probability density function

$$f(x,t) = \frac{1}{\sqrt{(2\pi)^n \det(P(t))}}\exp\left\{-\frac{1}{2}\left[x - m(t)\right]^T P^{-1}(t)\left[x - m(t)\right]\right\} \qquad (4.74)$$

where n denotes the dimension of the state vector x, or in terms of its characteristic function

$$\Phi(s,t) = \exp\left\{jm^T(t)s - \frac{1}{2}s^T P(t)s\right\} \qquad (4.75)$$

where s is a real n-dimensional vector.

Finally, let us denote that the system model which we have developed in this chapter is a generalization of the system models which were introduced in classical communication and control theory. In the classical work, the model was that of a time-invariant linear system whose input was a stationary white noise and in which the effect of initial conditions was assumed negligible. The time index set was chosen to be $-\infty < t < \infty$, and the resulting stochastic process was stationary. These simplifications permitted analysis to be conducted in the frequency domain in a straightforward manner. The extensions here are apparent: the system is, in general, time-varying (the matrices A, B and G in Eq. (4.51) can be functions of time t, i.e. $A = A(t), B = B(t)$ and $G = G(t)$); the noise inputs need not be stationary (that is the mean value of w^* in Eq. (4.51) is a function of time, i.e. $E\{w^*(t)\} = m_{w^*}(t)$, and its covariance function $\text{cov}\{w^*(t), w^*(s)\} = Q(t)\delta(t-s)$; similarly the mean-value of the measurement noise in eq. (4.57) $E\{v(t)\} = m_v(t)$ and $\text{cov}\{v(t), v(s)\} = R(t)\delta(t-s)$; and the effects of uncertainty in the initial conditions are included.

Example 4.5 (state space model with colored noise): Let us consider the state-space system model

$$\dot{x}(t) = A(t)x(t) + G(t)w^*(t)$$

where $w^*(t)$ is nonwhite such that it can in general be modeled as

$$\dot{\xi}(t) = \Sigma(t)\xi(t) + C(t)\gamma(t)$$
$$w^*(t) = H(t)\xi(t) + v(t)$$

where $\gamma(t)$ and $v(t)$ are white and assumed uncorrelated, that is $\mathrm{cov}\{\gamma(t),\gamma(s)\} = \Gamma(t)\delta(t-s)$, $\mathrm{cov}\{v(t),v(s)\} = V(t)\delta(t-s)$ and $\mathrm{cov}\{\gamma(t),v(t)\} = 0$. We may augment the state by the definition of a new state vector

$$q(t) = \begin{bmatrix} x(t) \\ \xi(t) \end{bmatrix}$$

such that the augmented differential equation

$$\dot{q}(t) = F(t)q(t) + D(t)w(t)$$

where

$$F(t) = \begin{bmatrix} A(t) & G(t)H(t) \\ 0 & \Sigma(t) \end{bmatrix} ; \quad D(t) = \begin{bmatrix} G(t) & 0 \\ 0 & C(t) \end{bmatrix} ; \quad w(t) = \begin{bmatrix} v(t) \\ \gamma(t) \end{bmatrix}$$

is now in the form of Eq. (4.38) which is driven by white noise. Thus, Eqs. (4.59) and (4.60) may be used to obtain $P_q(t)$ and $R_q(t,s)$ and Eq. (4.58) may be used to obtain $m_q(t)$. We need the initial conditions

$$m_q(t_0) = \begin{bmatrix} m_x(t_0) \\ m_\xi(t_0) \end{bmatrix} ; \quad P_q(t_0) = \begin{bmatrix} P_x(t_0) & 0 \\ 0 & P_\xi(t_0) \end{bmatrix}$$

Thus, we see that the previously developed mean and covariance equations are indeed quite general.

Example 4.6 (first order state space model with nonstationary noise): The dynamic system model is given by a scalar differential equation

$$\dot{x}(t) = -x(t) + w^*(t)$$

and we shall assume that the state of a system is known at $t = t_0 = 0$, $x(0) = 0$, and thus the prior mean and variance are $m_0 = 0$ and $P_0 = \mathrm{var}\{x(0)\} = 0$. The input noise w^* is white with unit mean and a sinusoidal variance parameter, that is $m_w = 1$ and $\mathrm{var}\{w^*\} = E\{(w^*(t) - m_w)^2\} = Q(t)\delta(t)$; $Q(t) = 1 + \cos t$.

The mean value of $x(t)$ may now be easily determined by taking the expected value of both sides of the above differential equation, to obtain (see eq. (4.58))

$$\frac{dm_x(t)}{dt} = -m_x(t) + m_w = -m_x(t) + 1$$

This may be solved by using the separation of variables

$$\int_{m_0}^{m_x} \frac{dm_x}{-m_x+1} = \int_0^t dt$$

or

$$m_x(t) = 1 - \exp(-t)$$

Since the solution of the state differential equation (4.38) is given by eq. (4.49), the above results can be obtained by the use of the integral equation

$$m_x(t) = \Phi(t,t_0)m_0 + \int_{t_0}^t \Phi(t,\tau)GE\{w^*(\tau)\}d\tau$$

where the fundamental matrix Φ is the solution of the differential equation (4.50). In our case $G=1, A=-1, m_0=0, m_w = E\{w^*(\tau)\}=1$ so that $\Phi(t) = \exp(-t)$ and

$$m_x(t) = \int_0^t \exp(-(t-\tau))d\tau = \exp(-t)\int_0^t \exp(\tau)d\tau = 1 - \exp(-t)$$

The integral expression for m_x is difficult to use, in general, because of the convolution integral

$$\int_0^t \Phi(t,\tau)Gm_w(\tau)d\tau$$

and the need to determine the state-transition matrix Φ by solving the Eq. (4.50). The convolution integral is particularly troublesome because it is not well suited for either hand or machine evaluation. Therefore it is more desirable to write the expression for $m_x(\tau)$ in the form of differential equation (4.58).

The variance equation is given by (4.60)

$$\dot{P}_x(t) = -2P_x(t) + 1 + \cos t \; ; \; P_x(0) = P_0 = 0$$

The solution is given by eq. (4.61), where $G=1$, $P_0=0$, $Q(t)=\cos t+1$, $\Phi(t)=\exp(At)=\exp(-t)$, $\Phi(t,\tau)=\Phi(t-\tau)=\exp(-(t-\tau))$, so that

$$P(t) = \Phi(t)P_0\Phi^T(t) + \int_0^t \Phi(t-\tau)GQ(\tau)G^T d\tau$$

$$= \int_0^t \exp(-2(t-\tau))[1+\cos\tau]d\tau = \frac{4\cos t + 2\sin t - 9\exp(-2t) + 5}{10}$$

Similarly as in the case of mean value function, eq. (4.61) is not well suited for computation of P, in general, because it requires that the state transition matrix Φ be determined first and then the corresponding convolution integral

$$\int_0^t \Phi(t,\tau)G(\tau)Q(\tau)G^T(\tau)d\tau$$

has to be solved. Therefore, it is more desirable to write the expression for $P(t)$ in the form of the covariance equation (4.60) and to determine $P(t)$ by direct (numerical) integration of eq. (4.60).

Since the state-noise variance parameter $Q(t)=1+\cos t$ is a function of time, the state-variance $P_x(t)$ will not reach a constant steady state. The stochastic steady-state solution occurs theoretically after infinite time $(t \to \infty)$ and is given by (in reality the steady state solution of $P_x(t)$ occurs after 3 or 4 sec)

$$P_x(t) = \frac{4\cos t + 2\sin t + 5}{10} = \frac{1}{2} + \frac{1}{\sqrt{10}}\sin(t+63.4)$$

Thus we see that the state variance is sinusoidal, although it never becomes zero, as does the noise variance.

4.3 Discretization of continuous stochastic systems

With the increasing sophistication of microprocessors, more and more estimation and control schemes are being implemented digitally. In a digital implementation the state of the plant is required at discrete-time instants. It therefore behooves us to discuss state estimation for continuous-time systems using discrete, or sampled, data [2, 3, 30].

Suppose the system model is described by

$$\dot{x}(t) = Ax(t) + Bu(t) + Gw^*(t) \tag{4.76}$$

with measurements

$$y(t) = Hx(t) + v(t) \tag{4.77}$$

Let $x(t_0)$ is a random vector with mean m_0 and covariance matrix P_0, and assume further that $\{w^*(t), t \geq t_0\}$ and $\{v(t), t \geq t_0\}$ are zero mean white noise processes uncorrelated with each other and $x(t_0)$. The covariances kernels of $w(t)$ and $v(t)$ are given by

$$\text{cov}\{w^*(t), w^*(\tau)\} = Q\delta(t-\tau)$$
$$\text{cov}\{v(t), v(\tau)\} = R\delta(t-\tau) \tag{4.78}$$

where the matrices Q and R are the covariance parameters, which are known in the literature as the spectral densities. Suppose the microprocessor samples control input $u(t)$ and measurement $y(t)$ every T sec. We can apply the relations derived in the previous section for discrete systems to the continuous plant by first discretizing it. To do this, we begin with the solution (4.52) to (4.76), that is

$$x(t) = \exp\left(A(t-t_0)\right)x(t_0) + \int_{t_0}^{t} \exp\left(A(t-\tau)\right)Bu(\tau)d\tau + \int_{t_0}^{t} \exp\left(A(t-\tau)\right)Gw^*(\tau)d\tau \tag{4.79}$$

To describe the state propagation between samples, let $t_0 = kT, t = (k+1)T$ for an integer k. Defining the sampled state function as $x_k = x(kT)$, we can write

$$x_{k+1} = \exp(AT)x_k + \int_{kT}^{(k+1)T} \exp\left(A\big[(k+1)T - \tau\big]\right)Bu(\tau)d\tau$$
$$+ \int_{kT}^{(k+1)T} \exp\left(A\big[(k+1)T - \tau\big]\right)Gw^*(\tau)d\tau \tag{4.80}$$

Assuming that control input $u(\tau)$ is reconstructed from the discrete control sequence u_k by using a zero-order hold, $u(\tau)$ has a constant value of $u(kT) = u_k$ over the integration interval T. The third term is a smoothed, i.e. low-pass filtered, version of the continuous white process noise w^* weighted by the state transition matrix Φ and the noise input matrix G. This term describes a discrete zero-mean white noise sequence, since the random variable

$$w_k = \int_{kT}^{kT+T} \exp\left(A\left[(k+1)T-\tau\right]\right)Gw^*(\tau)d\tau$$
$$= \int_{kT}^{(k+1)T} \exp\left(A\left[(k+1)T-\tau\right]\right)Gdw(\tau)$$

(4.81)

has zero-mean (because w^* has zero-mean) and the random variables w_k and w_l are uncorrelated for $k \neq l$ (because the increments dw over disjoint intervals are uncorrelated). Now (4.80) becomes

$$x_{k+1} = \exp(AT)x_k + \int_{kT}^{kT+T} \exp\left(A\left[(k+1)T-\tau\right]\right)Bd\tau u_k + w_k$$

On changing variables twice ($\lambda = \tau - kT$ and then $\tau = T - \lambda$), we obtain

$$x_{k+1} = \exp(AT)x_k + \int_0^T \exp(A\tau)Bd\tau u_k + w_k$$

(4.82)

This is the sampled version of (4.76), which can be written as

$$x_{k+1} = A_d x_k + B_d u_k + w_k$$

(4.83)

with

$$A_d = \exp(AT) \ ; \ B_d = \int_0^T \exp(A\tau)Bd\tau$$

(4.84)

To find the covariance Q_d of the new noise sequence w_k (similarly as in Fig.4.7, a sample function of a white noise sequence w_k can be thought of as a time function which is composed of the superposition of an arbitrarily large number of independent pulses of duration T which have random amplitudes) write

$$Q_d = E\{w_k w_k^T\}$$
$$= \int_{kT}^{(k+1)T} \int_{kT}^{(k+1)T} \exp\left(A\left[(k+1)T-\tau\right]\right)GE\{w^*(\tau)w^{*T}(s)\}G^T \exp\left(A^T\left[(k+1)T-s\right]\right)d\tau ds$$

But, owing to (4.78)

$$E\{w^*(\tau)w^{*T}(s)\} = Q\delta(\tau-s)$$

so that

$$Q_d = \int_{kT}^{(k+1)T} \exp\left(A\left[(k+1)T-\tau\right]\right)GQG^T \exp\left(A^T\left[(k+1)T-\tau\right]\right)d\tau$$

By changing variables twice, as above, we have

$$Q_d = \int_0^T \exp(A\tau)GQG^T \exp(A^T\tau)d\tau$$

(4.85)

It is worth noting that even Q is diagonal, Q_d need not be. Thus, sampling can destroy independence among the components of the process (state) noise.

Discretizing the measurement equation (4.77) is easy since it has no dynamics, i.e.

$$y_k = Hx_k + v_k$$

(4.86)

To find the covariance R_d of v_k in terms of the given R in (4.78) we need to think a little. Define the unit rectangle as

$$\Pi(t) = \begin{cases} 1 \ ; \ |t| \leq 0.5 \\ 0 \ ; \ \text{otherwise} \end{cases}$$

(4.87)

and note that $\lim_{T \to 0} \Pi(t/T) = \delta(k)$, the Kronecker delta ($\delta(k)$ has a value of one at $k = 0$). We can write the covariance of v_k in terms of its spectral density R_d as

$$R_v(k) = R_d \delta(t-k) \tag{4.88}$$

Since $\delta(t-k)$ has a value of one at $k = t$, in this discrete case the covariance is equal to R_d, a finite matrix. The covariance of $v(t)$, on the other hand, is given by (4.78), i.e.

$$\text{cov}\{v(t), v(\tau)\} = E\{v(t)v^T(\tau)\} = R\delta(t-\tau) \tag{4.89}$$

where R is a spectral density, that is a finite matrix and. Since $\lim_{T \to 0} \frac{1}{T}\Pi\left(\frac{t}{T}\right) = \delta(t)$, to make (4.88) approaches (4.89) in the limit we must write

$$R\delta(t) = \lim_{T \to 0} (R_d T) \frac{1}{T}\Pi\left(\frac{t}{T}\right) \tag{4.90}$$

so that

$$R_d = \frac{R}{T} \tag{4.91}$$

We note that

$$\lim_{T \to 0} \frac{R}{T} \tag{4.92}$$

does not make sense. However, the quantity with which we are dealing is defined over an interval of width T and we can envision the situation as depicted in fig. 4.8.

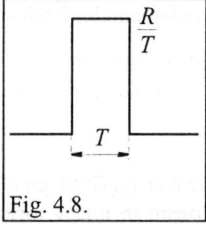

Fig. 4.8.

Fig. 4.8 represents the covariance matrix for piecewise constant discrete white sequence v_k (a sample function of v_k can be thought of as a time function which is composed of the superposition of independent pulses of duration T which have random amplitudes).

Then in the limit as $T \to 0$, the function which is $1/T$ over the interval T and zero elsewhere becomes the Dirac delta function. By examing (4.91), it is evident that eq. (4.91) is the relation required between R and R_d, provided that the noise processes $v(t)$ and v_k have the same spectral densities. Note also that if R is diagonal then so is R_d. Furthermore, sampling does not affect the a priori information m_0, P_0.

It is worthwhile to write down the first few terms of the infinite series expansions for the sampled matrices

$$A_d = \exp(AT) = I + AT + \frac{A^2 T^2}{2!} + \cdots$$

$$B_d = \int_0^T \exp(A\tau) B \, d\tau = BT + \frac{ABT^2}{2!} + \cdots \tag{4.93}$$

$$Q_d = \int_0^T \exp(A\tau) GQG^T \exp(A^T \tau) \, d\tau = GQG^T T + \frac{(AGQG^T + GQG^T A^T)T^2}{2!} + \cdots$$

If terms of order T^2 are disregarded, then the result is Euler's approximation to the sampled system

$$A_d = I + AT \; ; \; B_d = BT \; ; \; Q_d = GQG^T T \; ; \; R_d = R/T \tag{4.94}$$

In Euler's approximation note that process noise covariance Q_d results from multiplication by T, and measurement noise covariance R_d results from division by T.

Example 4.7 (sampling motion governed by Newton's laws): Consider the continuous system described by:

$$\dot{x}(t) = \begin{bmatrix} 0 & 1 \\ 0 & 0 \end{bmatrix} x(t) + \begin{bmatrix} 0 \\ 1/m \end{bmatrix} u(t) + w^*(t) = Ax(t) + Bu(t) + w^*(t) \tag{I}$$

If $u(t) = mg$ is a constant, and $w^*(t) = 0$, then by an easy integration components of two dimensional state vector $x = \begin{bmatrix} x_1 & x_2 \end{bmatrix}^T$ satisfy

$$x_1(t) = x_1(0) + gt \tag{II}$$

$$x_2(t) = x_2(0) + x_1(0)t + \frac{1}{2}gt^2 \tag{III}$$

For an input force $u(t) = F = mg$, therefore, eq. (I) is just a formulation of Newton's law $F = ma$, with x_1 the position and x_2 the velocity; that is, $\dot{x}_1 = x_2, \dot{x}_2 = g$.

Let $x(0)$ be a random vector with mean value m_0 and covariance matrix P_0, and w^* is the zero-mean white noise process with spectral density Q, that is $\operatorname{cov}\{w^*(t)\} = Q$. Additionally, measurements of position $x_1(t)$ are made so that

$$y(t) = \begin{bmatrix} 1 & 0 \end{bmatrix} x(t) + v(t) = Hx(t) + v(t) \tag{IV}$$

where measurement noise v is a zero-mean white noise process with spectral density R. It is also assumed that $\{v(t), t \geq 0\}$, $\{w^*(t), t \geq 0\}$ and $x(0)$ are uncorrelated. If measurements are made at intervals of T units, the discretized system can quickly be written down by noting that, the sequences for the sampled system matrices are all finite, so

$$A_d = \exp(AT) = I + AT = \begin{bmatrix} 1 & T \\ 0 & 1 \end{bmatrix} \tag{V}$$

$$B_d = BT + \frac{ABT^2}{2} = \frac{1}{m} \begin{bmatrix} \frac{T^2}{2} \\ T \end{bmatrix} \tag{VI}$$

Here is used the fact that A is nilpotent, that is $A^2 = 0$. Letting

$$Q = \begin{bmatrix} q_1 & q_2 \\ q_3 & q_4 \end{bmatrix}$$

we get

$$Q_d = QT + \begin{bmatrix} q_2 & q_4/2 \\ q_4/2 & 0 \end{bmatrix} T^2 + \begin{bmatrix} q_4/3 & 0 \\ 0 & 0 \end{bmatrix} T^3 \tag{VII}$$

and

$$R_d = R/T \tag{VIII}$$

The sampled version of (I), (IV) is thus

$$x_{k+1} = \begin{bmatrix} 1 & T \\ 0 & 1 \end{bmatrix} x_k + \frac{1}{m} \begin{bmatrix} T^2/2 \\ T \end{bmatrix} u_k + w_k \tag{IX}$$

$$y_k = \begin{bmatrix} 1 & 0 \end{bmatrix} x_k + v_k \tag{X}$$

with noise covariances Q_d and R_d given above. Note that even if Q is diagonal, Q_d may not be.

We are now ready to develop the mean value function and covariance expressions in eqs. (4.60) and (4.63), for continuous stochastic system by considering the limiting behavior of the corresponding expressions in eqs. (3.44) and (3.46) for its discrete-time equivalent. We repeat here for convenience eqs. (3.44) and (3.46)

$$m_x(k+1) = A_d m_x(k) + B_d u(k)$$
$$P_x(k+1) = A_d P_x(k) A_d^T + Q_d$$

Here is supposed that the discrete white process noise w_k has zero mean value, as that the control input u_k exists (the presence of known control input and non-zero mean value function of noise changes only the mean-value function of the state in eq. (3.44), but not the covariance expression (3.46)). Note also that in the discrete-time equivalent (4.83) of the continuous system (4.76) the noise-input matrix $G_d = I$, so that $G_d Q_d G_d = Q_d$ in eq. (3.46). Then, making use of the Euler's approximations in eqs. (4.75), one obtains

$$m_x(k+1) = (I + AT) m_x(k) + BTu_k$$
$$P_x(k+1) = [I + AT] P_x(k) [I + AT]^T + GQG^T T$$

so that

$$\frac{m_x(k+1) - m_x(k)}{T} = Am_x(k) + Bu_k$$
$$\frac{P_x(k+1) - P_x(k)}{T} = AP_x(k) + P_x(k) A^T + GQG^T + AP_x(k) A^T T$$

Taking the limit as $T \to 0$, we have the matrix differential equations

$$\dot{m}(t) = Am(t) + Bu(t)$$
$$\dot{P}(t) = AP(t) + P(t) A^T + GQG^T$$

which are identically eqs. (4.63) and (4.60), respectively.

Finally, it is straightforward to find equations for discretizing time-varying continuous systems, for which $A = A(t)$, $B = B(t)$, $G = G(t)$ in eq. (4.76) and $H = H(t)$ in eq. (4.77), with the spectral matrices $Q = Q(t)$ and $R = R(t)$ and the usual whiteness and uncorrelatedness assumptions. Let $\Phi(t,t_0)$ be the state-transition matrix of $x(t)$ defined by eq. (4.50), which we repeat here for convenience

$$\frac{d}{dt} \Phi(t,t_0) = A(t) \Phi(t,t_0) \;;\; \Phi(t_0,t_0) = I \tag{4.95}$$

It is not difficult to show that the discrete system and noise covariance matrices are time-varying matrices given by

$$A_d(k) = \Phi\big((k+1)T, kT\big) \tag{4.96}$$

$$B_d(k) = \int_{kT}^{(k+1)T} \Phi\big((k+1)T, \tau\big) B(\tau) d\tau \tag{4.97}$$

$$G_d(k) = I \tag{4.98}$$

$$Q_d(k) = \int_{kT}^{(k+1)T} \Phi\big((k+1)T, \tau\big) G(\tau) Q(\tau) G^T(\tau) \Phi^T\big((k+1)T, \tau\big) d\tau \tag{4.99}$$

$$H_d(k) = H(kT) \tag{4.100}$$

$$R_d(k) = \frac{R(kT)}{T} \tag{4.101}$$

4.4 State-space and transfer function models equivalence

The state-space and transfer function models are equivalent for liner time-invariant (LTI) systems in the sense that one can be obtained from the other. For singe input-single output (SISO) systems, this is easy, but for multi input-multi output (MIMO) systems, this leads to the problem of minimal realization [13, 15, 25].

It is easy to see from eqs. (4.53) and (4.54) that the frequency-domain versions of the state equations, when the disturbance input $w^* = 0$, are given by

$$X(s) = \Phi(s)BU(s) + \Phi(s)x(0) \quad ; \quad \Phi(s) = (sI - A)^{-1} \tag{4.102}$$

Assuming that the measurement noise $v = 0$, the output equation (4.77a) can be written in terms of the state equation as

$$Y(s) = HX(s) = H\Phi(s)BU(s) + H\Phi(s)x(0) \tag{4.103}$$

Now the transfer function $G(s)$ is defined as

$$G(s) = \left.\frac{Y(s)}{U(s)}\right|_{x(0)=0} = H\Phi(s)B = H(sI - A)^{-1}B \tag{4.104}$$

Taking the inverse Laplace transform of the equation leads to the impulse response of an LTI system as

$$h(t,\tau) = H\Phi(t,\tau)B = H\exp\big(A(t-\tau)\big)B = h(t-\tau) \quad \text{for } t > \tau \tag{4.105}$$

So we see that state-space models can be used to produce the transfer function output as well.

Furthermore, we discuss a technique to convert transfer function models to state-space models and vice versa. We are able to do this because the models are equivalent. The underlying theory is that of similarity transformations and beyond the scope of this discussion. Here we present two techniques that always work for SISO systems. It is possible to show that there exists a unique representation of a system that can always be obtained from a transfer function model. It is called *observer canonical form*, and if we are given the transfer function model

$$G(s) = \frac{B(s)}{A(s)} = \frac{b_0 s^n + b_1 s^{n-1} + \cdots + b_n}{s^n + a_1 s^{n-1} + \cdots + a_n} \tag{4.106}$$

we obtain the state-space model by inspection of the coefficients and vice versa:

$$\frac{dx(t)}{dt} = A_o x(t) + B_o u(t) \tag{4.107}$$

$$y(t) = H_o x(t) + D_o u(t) \tag{4.108}$$

where the quadruple (A_o, B_o, C_o, D_o) is given by

$$A_o = \begin{bmatrix} 0 \cdots 0 & \vdots & -a_n \\ \hline & \vdots & -a_{n-1} \\ I_{n-1} & \vdots & \vdots \\ & \vdots & -a_1 \end{bmatrix} \; ; \; B_o = \begin{bmatrix} b_n - a_n b_0 \\ \vdots \\ b_1 - a_1 b_0 \end{bmatrix} \; ; \; H_o^T = \begin{bmatrix} 0 \\ \vdots \\ 0 \\ 1 \end{bmatrix} \; ; \; D_o = b_0 \tag{4.109}$$

with I_{n-1} being $(n-1) \times (n-1)$ dimensional identity matrix.

Example 4.8 (observer canonical form): Let us take for simplicity that $n = 3$ in eq. (4.106). Then the transfer function model can be written as

$$Y(s) = G(s)U(s)$$

$$= \frac{b_0 s^3 + b_1 s^2 + b_2 s + b_3}{s^3 + a_1 s^2 + a_2 s + a_3} U(s) = \frac{b_0 + b_1 s^{-1} + b_2 s^{-2} + b_3 s^{-3}}{1 + a_1 s^{-1} + a_2 s^{-2} + a_3 s^{-3}} U(s)$$

from which one obtains

$$Y(s) = s^{-1}\left\{ \left(b_1 U(s) - a_1 Y(s)\right) + s^{-1}\left[\left(b_2 U(s) - a_2 Y(s)\right) + s^{-1}\left(b_3 U(s) - a_3 Y(s)\right)\right]\right\} + b_0 U(s)$$

By introducing the state coordinates as

$$X_1(s) = s^{-1}\left(b_3 U(s) - a_3 Y(s)\right)$$

$$X_2(s) = s^{-1}\left[\left(b_2 U(s) - a_2 Y(s)\right) + s^{-1}\left(b_3 U(s) - a_3 Y(s)\right)\right]$$

$$= s^{-1}\left[\left(b_2 U(s) - a_2 Y(s)\right) + X_1(s)\right]$$

$$X_3(s) = s^{-1}\left\{ \left(b_1 U(s) - a_1 Y(s)\right) + s^{-1}\left[\left(b_2 U(s) - a_2 Y(s)\right) + s^{-1}\left(b_3 U(s) - a_3 Y(s)\right)\right]\right\}$$

$$= s^{-1}\left\{ \left(b_1 U(s) - a_1 Y(s)\right) + X_2(s)\right\}$$

$$Y(s) = X_3(s) + b_0 U(s)$$

one obtains

$$sX_1(s) = -a_3 X_3(s) + \left(b_3 - a_3 b_0\right)U(s)$$

$$sX_2(s) = X_1(s) - a_2 X_3(s) + \left(b_2 - a_2 b_0\right)U(s)$$

$$sX_3(s) = X_2(s) - a_1 X_3(s) + \left(b_1 - a_1 b_0\right)U(s)$$

Taking the inverse Laplace transform of the equations leads to the time-domain relations

$$\frac{dx_1(t)}{dt} = -a_3 x_3(t) + (b_3 - a_3 b_0) u(t)$$

$$\frac{dx_2(t)}{dt} = x_1(t) - a_2 x_3(t) + (b_2 - a_2 b_0) u(t)$$

$$\frac{dx_3(t)}{dt} = x_2(t) - a_1 x_3(t) + (b_1 - a_1 b_0) u(t)$$

which can be put in the matrix form (4.109) with

$$A_0 = \begin{bmatrix} 0 & 0 & -a_3 \\ 1 & 0 & -a_2 \\ 0 & 1 & -a_1 \end{bmatrix}; \ B_0 = \begin{bmatrix} b_3 - a_3 b_0 \\ b_2 - a_2 b_0 \\ b_1 - a_1 b_0 \end{bmatrix}; \ H_0^T = \begin{bmatrix} 0 \\ 0 \\ 1 \end{bmatrix}; \ D_0 = b_0$$

If $b_0 = 0$ in eq. (4.106), that is the order of numerator is at least one less than the order of the denominator, an alternative representation of the system can be obtained from the transfer function model. It is called control canonical form, and can be derived from transfer function model by using the relation

$$\frac{Y(s)}{U(s)} = \frac{b_1 s^{n-1} + \cdots + b_n}{s^n + a_1 s^{n-1} + \cdots + a_n} \frac{P(s)}{P(s)}$$

where P is an auxiliary variable, from which one obtains

$$s^n P(s) = -a_1 s^{n-1} P(s) - \cdots - a_n P(s) + U(s)$$
$$Y(s) = b_1 s^{n-1} P(s) + \cdots + b_n P(s)$$

Of course, any coefficient a_i or b_i may be zero. By introducing the state coordinates as

$$X_1(s) = P(s)$$
$$X_2(s) = sP(s)$$
$$X_3(s) = s^2 P(s)$$
$$\vdots$$
$$X_n(s) = s^{n-1} P(s)$$

one obtains

$$sX_n(s) = s^n P(s) = -a_1 s^{n-1} P(s) - \cdots - a_n P(s) + U(s)$$
$$= -a_1 X_n(s) - \cdots - a_n X_1(s) + U(s)$$
$$Y(s) = b_1 X_n(s) + \cdots + b_n X_1(s)$$

Taking the inverse Laplace transform of the equations, further follows

$$\frac{dp(t)}{dt} = \frac{dx_1(t)}{dt} = x_2(t)$$

$$\frac{d^2p(t)}{dt^2} = \frac{d}{dt}\left(\frac{dp(t)}{dt}\right) = \frac{dx_2(t)}{dt} = x_3(t)$$

$$\vdots$$

$$\frac{d^{n-1}p(t)}{dt} = \frac{d}{dt}\left(\frac{d^{n-2}p(t)}{dt}\right) = \frac{dx_{n-1}(t)}{dt} = x_n(t)$$

$$\frac{dx_n(t)}{dt} = -a_1 x_n(t) - \cdots - a_n x_1(t) + u(t)$$

$$y(t) = b_1 x_n(t) + \cdots + b_n x_1(t)$$

By introducing the state vector $x^T(t) = [x_1(t),...,x_n(t)]$ the above equations can be put in the matrix form, named the *control canonical form*,

$$\frac{dx(t)}{dt} = A_c x(t) + B_c u(t) \tag{4.110}$$

$$y(t) = H_c x(t) \tag{4.111}$$

where

$$A_c = \begin{bmatrix} 0 & & \\ \vdots & & I_{n-1} \\ 0 & & \\ \hline -a_n & -a_{n-1} \cdots -a_1 \end{bmatrix} ; \ B_c = \begin{bmatrix} 0 \\ \vdots \\ 0 \\ 1 \end{bmatrix} ; \ H_c^T = \begin{bmatrix} b_n \\ b_{n-1} \\ \vdots \\ b_1 \end{bmatrix} \tag{4.112}$$

The reasons for these particular names of the representations of the transfer function model are from modern control theory and are related to the concepts of observability and controllability.

Similarly as in the discrete-time case, observability of a system means that the initial state of a system can be determined from a suitable measurement of its output. Loosely speaking, observability is the ability to choose measurement variables that enable the construction of the state variables that cannot be measured directly.

A system is said to be completely observable if for any initial condition $x(0)$ in the state space there exists a finite $t_1 > t_0$ such that knowledge of the input $u(t)$ and output $y(t)$ is sufficient to specify $x(0)$. The simplest example of an observable system is one in which the H matrix in eq. (4.104) is $n \times n$ (output equal the number n of states) and invertible. Then from the measurement equation (4.11) we have

$$x(t) = H^{-1} y(t) \tag{4.113}$$

Therefore, given any measurement vector y, we can reconstruct any state x. In general, we only make p measurements $(p < n)$, that is, the matrix H is $p \times n$ and we would like to reconstruct $x(0)$. Recall that

$$y(t) = H\Phi(t,0)x(0) + \int_0^t H\Phi(t,\tau) Bu(\tau) d\tau \tag{4.114}$$

Thus we see that

$$y^*(t) = H\Phi(t,0)x(0) \tag{4.115}$$

where

$$y^*(t) = y(t) - \int_0^t H\Psi(t,\tau)Bu(\tau)d\tau \tag{4.116}$$

From the Cayley-Hamilton theorem (see appendix B), it is possible to show that

$$\Phi(t) = \exp(At) = \sum_{k=0}^{n-1} \alpha(t,k)A^k \tag{4.117}$$

Therefore, eq. (4.115) can be written

$$y^*(t) = H\sum_{k=0}^{n-1} \alpha(t,k)A^k x(0)$$

$$= \left[\alpha(t,0)I \mid \alpha(t,1)I \mid \cdots \mid \alpha(t,n-1)I\right] \begin{bmatrix} H \\ \overline{HA} \\ \vdots \\ \overline{HA^{n-1}} \end{bmatrix} x(0) \tag{4.118}$$

It can be shown that the system is completely observable if the observability matrix O given by

$$O = \begin{bmatrix} H \\ \overline{HA} \\ \vdots \\ \overline{HA^{n-1}} \end{bmatrix} \tag{4.119}$$

is of rank n. In this case the system of linear equations (4.118) can be solved with respect to $x(0)$. So we see that checking that all the measurements for a LTI system contain the necessary information to reconstruct all the states becomes that of checking the rank of the observability matrix in eq. (4.119).

The concept of completely controllability is analogous to observability in that it examines the requirements of a system to be controlled by an input u. A system is said to be completely controllable if every initial state $x(0)$ can be transferred to any final state $x(t)$ in a finite time by an input vector u. Recall that

$$x(t) = \Phi(t,0)x(0) + \int_0^t \Phi(t,\tau)Bu(\tau)d\tau$$
$$= \exp(At)x(0) + \exp(At)\int_0^t \exp(-A\tau)Bu(\tau)d\tau \tag{4.120}$$

from which one obtains, assuming $x(t) = 0$

$$x(0) = -\int_0^t \exp(-A\tau)Bu(\tau)d\tau \tag{4.121}$$

But, from the Cayley-Hamilton theorem, it is possible to show that

$$\exp(-A\tau) = \sum_{k=0}^{n-1} \alpha(\tau,k)A^k \tag{4.122}$$

Therefore, eq. (4.121) can be written as

$$x(0) = -\sum_{k=0}^{n-1} A^k B \int_0^{\cdot} \alpha(\tau,k) u(\tau) d\tau$$

$$= \sum_{k=0}^{n-1} A^k B r_k = \begin{bmatrix} B & | & AB & | & \cdots & | & A^{n-1}B \end{bmatrix} \begin{bmatrix} \dfrac{r_0}{r_1} \\ \dfrac{\vdots}{r_{n-1}} \end{bmatrix} \qquad (4.123)$$

where

$$r_k = -\int_0^{\cdot} \alpha(\tau,k) u(\tau) d\tau$$

It can be shown that the system is completely controllable if the controllability matrix C given by

$$C = \begin{bmatrix} B & | & AB & | & \cdots & | & A^{n-1}B \end{bmatrix} \qquad (4.124)$$

is of rank n (in this case the system of linear equations (4.123) can be solved with respect to $r_i, i = 0,...,n-1$ which depend on the control u).

CHAPTER 5

FUNDAMENTALS OF ESTIMATION

Suppose we have two (scalar or vector) random variables X, Y with a joint density function $f_{X,Y}(\cdot,\cdot)$. Assume that in particular experiment, the random variable Y can be measured and takes the value y. What can be said about the corresponding values, say x, of the unobservable variable X?

Suppose we make an estimate, say $\hat{x}$, of the value of X when $Y = y$, according to the rule

$$\hat{x} = h(y) \qquad (5.1)$$

where x is unobservable true value of X when $Y=y$, and $h(\cdot)$ is one specified function. The domain of this function is the set of values of y, or the random variable Y. As a function of a random variable, it is itself a random variable, which we shall call $\hat{X}$, i.e.

$$\hat{X} = h(Y) \qquad (5.2)$$

A particular value of $\hat{X}$ is $\hat{x}$ and is given by (5.1). Thus, we use the term estimate $\hat{x} = h(y)$ for a particular value y taken by Y, and we shall term $\hat{X}$ an estimator of X in terms of Y. In this way, the estimator is a rule, or a function, for associating particular values of two variables. In contrast, an estimate is a value taken by the estimator, regarded as a function. This distinction is illustrated in Fig. 5.1. In summary, the estimator is a function or a device for assigning a number, given a measurement [1, 17, 23, 42, 44].

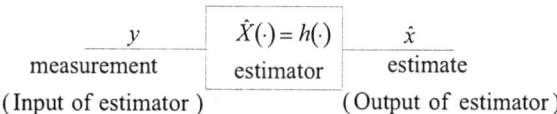

$$
\begin{array}{ccc}
\underline{\quad y \quad} & \boxed{\hat{X}(\cdot) = h(\cdot)} & \underline{\quad \hat{x} \quad} \\
\text{measurement} & \text{estimator} & \text{estimate} \\
(\,\text{Input of estimator}\,) & & (\,\text{Output of estimator}\,)
\end{array}
$$

Fig. 5.1: Block scheme of estimation procedure

One can generate an great number of different estimators depending on the choice of estimator rule (function) $h(\cdot)$. A number of different types of estimators have been developed in the following, and several relationships between the estimators have been studied. We began our study of estimation with Bayes cost method, which was used to derive the mean-square error, maximum a posterior, maximum likelihood and absolute-cost estimators. Since these estimators require a rather complete probabilistic description of the estimation problem, the linear minimum variance and least-square estimators, that need only minimal statistical structure, are considered as a complement to the other methods. Furthermore, if we assume that x is a constant, the problems of this sort are referred to as parameter or point estimation. A brief introduction to state estimation is presented in the next chapters. There the quantity to be estimated is the time-varying state of a dynamical system represented as $x(t)$.

5.1. Bayes cost method

Let us denote

$$\tilde{x} = x - \hat{x} = x - h(y) \tag{5.3}$$

where $\hat{x} = h(y)$ is an estimate of the value x of the random variable X when the random variable Y takes the value y (see Fig. 5.1). We can never hope to make $\tilde{x} = 0$ always (i.e. for every value of Y that may arise) because X and Y are random variables. Therefore, all we can hope is to try to choose $h(\cdot)$ so as to minimize the expected (average) value of some function of the error (5.3). This approach is known as the Bayesian approach. Thus, the Bayesian estimation problem is to find the estimate that minimizes the expected value, or the so-called Bayes risk,

$$B(\hat{X}) = \underset{X,Y}{E} \left\{ C(\hat{X}) \right\} = \int_{-\infty}^{\infty} \int_{-\infty}^{\infty} C(x - \hat{x}) f_{X,Y}(x,y) \, dxdy \tag{5.4}$$

where $C(\cdot)$ is a cost (risk) function which assigns to each value of the estimation error $\tilde{x}$ a unique cost. Here we have written the expected value as a function of $\hat{x}$ to emphasize the dependence on the estimation rule $h(\cdot)$. If we write $f_{X,Y}(x,y)$ as $f_{Y/X}(y/x) f_X(x)$ then eq. (5.4) can be written as

$$B(\hat{X}) = \int_{-\infty}^{\infty} \left[\int_{-\infty}^{\infty} C(x - \hat{x}) f_{Y/X}(y/x) dy \right] f_X(x) \, dx \tag{5.5}$$

The inner integral is the conditional cost given x, which we write as

$$B(\hat{X}/X) = E\left\{ C(\tilde{X})/X \right\} = \int_{-\infty}^{\infty} C(x - \hat{x}) f_{Y/X}(y/x) dy \tag{5.6}$$

This integral represents the expected value of the cost for a given value of $X = x$. In terms of the conditional cost (5.6), the Bayes cost (5.5) can be written as

$$B(\hat{X}) = \int_{-\infty}^{\infty} B(\hat{X}/x) f_X(x) dx = E_X \left\{ B(\hat{X}/X) \right\} \tag{5.7}$$

The *Bayes criterion* can be stated as follows: For a given cost function of the error $C(\tilde{X}) = C(X - \hat{X})$, the estimator rule $\hat{x}_B = h_B(y)$ is optimal if

$$B(\hat{X}_B) \leq B(\hat{X}) \tag{5.8}$$

for any other $\hat{x} = h(y) \neq \hat{x}_B$.

In other words, the Bayes estimator minimizes the Bayes cost.

The choice of an appropriate cost function $C(\cdot)$ is obviously not a clear one and must depend upon the significance of different values of the error for the problem at hand. We shall consider three of the most useful choices for the cost function and develop the associated estimator rule $h(\cdot)$.

5.1.1. Mean Square Error Criterion - Minimum Variance Estimator

Before beginning the development for the mean square error (MSE) criterion, it is convenient to rewrite eq. (5.4). If the density $f_{X,Y}(x,y)$ is written as $f_{X/Y}(x/y) f_Y(y)$ then eq. (5.4) becomes

$$B(\hat{X}) = \int_{-\infty}^{\infty}\left[\int_{-\infty}^{\infty}C(x-\hat{x})f_{X/Y}(x/y)dx\right]f_Y(y)dy = E_Y\left\{E\left\{C(X-\hat{X})/Y\right\}\right\} \qquad (5.9)$$

If we can minimize the inner integral of eq. (5.9) for every possible value of y, then $B(\hat{X})$ must also be minimized since $f_Y(y)$ is nonnegative. Hence we shall select $\hat{X}$ to minimize

$$B(\hat{X}/Y) = \int_{-\infty}^{\infty}C(x-\hat{x})f_{X/Y}(x/y)dx = E\left\{C(X-\hat{X})/Y\right\} \qquad (5.10)$$

The mean square (MS) cost function is

$$C_{MS}(x-\hat{x}) = \|x-\hat{x}\|^2 = \sum_{i=1}^{n}(x_i-\hat{x}_i)^2 = (x-\hat{x})^T(x-\hat{x}) \qquad (5.11)$$

where $\|x\|^2$ is the Euclidean norm of the n-dimensional vector x. If we substitute eq. (5.11) into the expression (5.10), we have

$$B_{MS}(\hat{X}/Y) = \int_{-\infty}^{\infty}\|x-\hat{x}\|^2 f_{X/Y}(x/y)dx = E\left\{\|\tilde{X}\|^2\right\}E\left\{\|\tilde{X}\|^2/Y\right\} \qquad (5.12)$$

In order to minimize $B_{MS}(\hat{X}/Y)$, we take the partial derivative of $B_{MS}(\hat{X}/Y)$ with respect to $\hat{X}$. This derivative should be zero for the optimum choice of $\hat{X}$, which we will call $\hat{X}_{MS}$. Hence $\hat{X}_{MS}$ is given by

$$\left.\frac{\partial B_{MS}(\hat{x}/y)}{\partial \hat{x}}\right|_{\hat{x}=\hat{x}_{MS}} = 0 = -2\int_{-\infty}^{\infty}(x-\hat{x}_{MS})f_{X/Y}(x/y)dx \qquad (5.13)$$

Rearranging this result slightly yields

$$\int_{-\infty}^{\infty}\hat{x}_{MS}f_{X/Y}(x/y)dx = \int_{-\infty}^{\infty}xf_{X/Y}(x/y)dx$$

Now since $\hat{x}_{MS}$ is not function of x, we can remove it from the left integral and obtain

$$\int_{-\infty}^{\infty}\hat{x}_{MS}f_{X/Y}(x/y)dx = \hat{x}_{MS}\int_{-\infty}^{\infty}f_{X/Y}(x/y)dx = \hat{x}_{MS}$$

Therefore, we see that the optimum estimator for the MSE criterion is given by

$$\hat{x}_{MS} = \int_{-\infty}^{\infty}xf_{X/Y}(x/y)dx = E\{X/Y=y\} \qquad (5.14)$$

The estimator is simply the conditional mean (expectation) of X given the observation $Y=y$. As a result, the estimator is also referred to as the conditional mean estimator $\hat{X}_{CM}$. The point is that since Y is a random variable we cannot say which value y it will assume in any particular occurrence, and therefore the optimum estimate is also a random variable, which we shall denote $\hat{X}_{MS}$ or $\hat{X}_{CM}$,

$$\hat{X}_{MS} = \hat{X}_{CM} \quad \hat{x}_{MS} = \hat{x}_{CM} = h_{MS}(Y) = E\{X/Y\} \qquad (5.15)$$

It is easy to show that the expected value of the estimation error is zero. The estimation error is given by

$$\tilde{X} = X - \hat{X} = X - E\{X/Y\}$$

The expected value of $\tilde{x}$ is given by

$$E\{\tilde{X}\} = E_Y\left\{E\{\tilde{X}/Y\}\right\} = E_Y\left\{E\{X - E\{X/Y\}/Y\}\right\}$$
$$= E_Y\left\{E\{X/Y\} - E\{E\{X/Y\}/Y\}\right\}$$

However, since $Y=y$ is known, $\hat{X}_{MS}\,\hat{x}_{MS}(y) = E\{X/Y\}$ is deterministic, therefore $E\{E\{X/Y\}\} = E\{X/Y\}$, so that we have

$$E\{\tilde{X}\} = E_Y\left\{E\{X/Y\} - E\{X/Y\}\right\} = 0 \tag{5.16}$$

Because $\tilde{X}$ is zero-mean, the covariance matrix of the estimation error $\tilde{X}$ can be written as

$$R_{\tilde{X}} = \mathrm{cov}\{\tilde{X}\} = E\{\tilde{X}\tilde{X}^T\} = E_Y\left\{E\{\tilde{X}\tilde{X}^T/Y\}\right\} \tag{5.17}$$

The sum of the main diagonal terms of the matrix (5.17), which is determined as the trace of the matrix is given by

$$\mathrm{tr}\{R_{\tilde{X}}\} = \mathrm{tr}\{E\{\tilde{X}\tilde{X}^T\}\} = E\{\mathrm{tr}\{\tilde{X}\tilde{X}^T\}\} = E\{\tilde{X}^T\tilde{X}\} = E\{\|\tilde{X}\|^2\}$$
$$= E\left\{\sum_{i=1}^{n}\tilde{x}_i^2\right\} = \sum_{i=1}^{n}E\{\tilde{x}_i^2\} = \sum_{i=1}^{n}\sigma^2(\tilde{x}_i) \tag{5.18}$$

where $\tilde{x}_i$ is the i-th component of the n vector x and $\sigma^2(\tilde{x}_i)$ is the corresponding variance. Here is used the fact that

$$\mathrm{tr}\{\tilde{X}\tilde{X}^T\} = \tilde{X}^T\tilde{X} = \|\tilde{X}\|^2 = \sum_{i=1}^{n}\tilde{x}_i^2 \tag{5.19}$$

where $\tilde{x}$ is a n-dimensional column vector. On the other hand, by taking into account (5.11) and (5.12), one concludes

$$B_{MS}(\hat{X}) = E_Y\left\{B_{MS}(\hat{X}/Y)\right\} = E_Y\left\{E\{\|\tilde{X}\|^2/Y\}\right\} = E\{\|\tilde{X}\|^2\} \tag{5.20}$$

By comparing (5.18) and (5.20), we obtain

$$B_{MS}(\hat{X}) = \mathrm{tr}\{R_{\tilde{X}}\} = \sum_{i=1}^{n}\sigma^2(\tilde{x}_i) \tag{5.21}$$

Thus, the criterion $B_{MS}(\hat{X})$ is the sum of the variances of the estimation errors $\tilde{x}_i$, $i = 1,...,n$ ($\tilde{x}_i$ is the i-th term of the error vector $\tilde{x}$), and because of this fact $\tilde{x}_{MS}$ is also called the minimum-variance estimator $\tilde{x}_{MV}$.

The form of eq. (5.14) is not computationally convenient since the density $f_{X/Y}(x/y)$ generally is not easy to compute. If we express $f_{X/Y}(x/y)$ by the use of Bayes rule as

$$f_{X/Y}(x/y) = \frac{f_{Y/X}(y/x)f_X(x)}{f_Y(y)}$$

then eq. (5.14) becomes

$$\hat{x}_{MS} = \frac{\int_{-\infty}^{\infty}xf_{Y/X}(y/x)f_X(x)\,dx}{f_Y(y)}$$

Next $f_Y(y)$ can be written as

$$f_Y(y) = \int_{-\infty}^{\infty} f_{X,Y}(x,y)\,dx = \int_{-\infty}^{\infty} f_{Y/X}(y/x)f_X(x)\,dx$$

so that the final form for $\hat{x}_{MS}$ is

$$\hat{x}_{MS} = \frac{\int_{-\infty}^{\infty} x f_{Y/X}(y/x)f_X(x)\,dx}{\int_{-\infty}^{\infty} f_{Y/X}(y/x)f_X(x)\,dx} \tag{5.22}$$

In this form, we need only $f_X(x)$ and $f_{Y/X}(y/x)$, which are generally given or easy to find.

Example 5.1 (nonlinear MS-estimator for scalar random parameter): Let us find the MS-estimator for the scalar parameter x based on the scalar observation y

$$y = \ln x + n$$

where

$$f_X(x) = \begin{cases} 1; & 0 \le x \le 1 \\ 0; & \text{otherwise} \end{cases}$$

$$f_n(n) = \begin{cases} \exp(-n); & n \ge 0 \\ 0; & \text{otherwise} \end{cases}$$

The conditional density $f_{Y/X}(y/x)$ is easily determined as

$$f_{Y/X}(y/x)$$

Now, let us calculate first the denominator of eq. (5.22)

$$f_Y(y) = \int_{-\infty}^{\infty} f_{Y/X}(y/x)f_X(x)\,dx = \begin{cases} \int_0^1 \exp(-(y-\ln x))\,dx = \dfrac{\exp(-y)}{2}; & y \ge 0 \\[2mm] \int_0^{} \exp(-(y-\ln x))\,dx = \dfrac{\exp(y)}{2}; & y < 0 \end{cases}$$

The numerator is given by

$$\int_{-\infty}^{\infty} x f_{Y/X}(y/x)f_X(x)\,dx = \begin{cases} \int_0^1 x\exp(-(y-\ln x))\,dx = \dfrac{\exp(-y)}{3}; & y \ge 0 \\[2mm] \int_0^{} x\exp(-(y-\ln x))\,dx = \dfrac{\exp(2y)}{3}; & y < 0 \end{cases}$$

and therefore

$$\hat{x}_{MS} = \begin{cases} \dfrac{2}{3}; & y \ge 0 \\[2mm] \dfrac{2\exp(y)}{3}; & y < 0 \end{cases}$$

Note that $\hat{x}_{MS}$ can never become larger than $\dfrac{2}{3}$ even though X can take any value in the unit interval.

Example 5.2 (Gaussian linear MS-estimator for scalar random parameter): Consider the following observations of a scalar parameter x

$$y_i = x + n_i \; ; \; i = 1, 2, ..., N$$

where the n_i's are independent and identically distributed normal random variables with mean zero and variance σ^2. We shall assume that x is also Gaussian with mean m_0 and variance σ_0^2. Find the MS-estimator of the parameter x.

For convenient notation, it is desirable to write the observation sequence in vector form as

$$y = Hx + n$$

where

$$H = [1\,1 \cdots 1]^T \;,\; y = [y_1, y_2, ..., y_{Nn}]^T \;\; \text{and} \;\; n = [n_1, n_2, ..., n_N]^T$$

Instead of carrying out the direct integration involved in eq. (5.22), in this case it is easier to make use of the fact that all the associated density functions are Gaussian and can be completely determined by finding the appropriate mean and covariance matrix. Let us consider first the density $f_Y(y)$. The mean value of y is

$$m_Y = E\{y\} = E\{Hx + n\} = HE\{x\} + E\{n\} = Hm_0$$

while the covariance matrix is

$$\begin{aligned} \text{cov}\{y\} &= E\left\{(y - m_y)(y - m_y)^T\right\} \\ &= E\left\{[H(x - m_0) + n][H(x - m_0) + n]^T\right\} = HE\left\{(x - m_0)(x - m_0)^T\right\}H^T + E\{nn^T\} \\ &= \sigma_0^2 HH^T + \sigma_n^2 I \end{aligned}$$

where I is an identity matrix. Here we have made use of the fact that the n_i sequence is white and have assumed that n_i's and x are uncorrelated. The density function $f_Y(y)$ is therefore

$$f_Y(y) = k_1 \exp\left\{-\frac{1}{2}(y - Hm_0)^T\left(\sigma_0^2 HH^T + \sigma_n^2 I\right)^{-1}(y - Hm_0)\right\}$$

where k_1 is a normalizing constant.

Consider, next, the density $f_{Y/X}(y/x)$. The mean is

$$E\{Y/X\} = E\{Hx + n/x\} = Hx$$

while the covariance matrix is

$$\begin{aligned} R_{Y/X} &= \text{cov}\{y/x\} = E\left\{(y - E\{y/x\})(y - E\{y/x\})^T / x\right\} \\ &= E\left\{(y - Hx)(y - Hx)^T / x\right\} = E\{nn^T\} = \sigma_n^2 I \end{aligned}$$

We combine these two results with the known $f_X(x)$ to obtain

$$f_{X/Y}\left(x/y\right) = \frac{f_{Y/X}\left(y/x\right)f_X\left(x\right)}{f_Y\left(y\right)}$$

$$= k\exp\left(-\frac{1}{2}\left(y-Hx\right)^T\frac{1}{\sigma_n^2}\left(y-Hx\right)-\frac{1}{2\sigma_0^2}\left(x-m_0\right)^2+\frac{1}{2}\left(y-Hm_0\right)^T\left(\sigma_0^2HH^T+\sigma_n^2I\right)^{-1}\left(y-Hm_0\right)\right)$$

Because the density function is Gaussian, it can be written in the form

$$f_{X/Y}\left(x/y\right) = k\exp\left\{-\frac{1}{2\sigma^2}\left(x-E\{X/Y\}\right)^2\right\}$$

$$= k\exp\left\{-\frac{1}{2\sigma^2}\left(x-\hat{x}_{MS}\right)^2\right\} = k\exp\left\{-\frac{1}{2}\left(x^2-2x\hat{x}_{MS}+\hat{x}_{MS}^2\right)\right\}$$

Simply by equating the terms which are first and second order in x, we obtain (see the derivation of eq. (5.62))

$$\frac{1}{\sigma^2} = \frac{H^TH}{\sigma_n^2}+\frac{1}{\sigma_0^2}$$

$$\frac{2}{\sigma^2}\hat{x}_{MS} = 2H^Ty\left(\frac{1}{\sigma_n^2}\right)+2\frac{m_0}{\sigma_0^2}$$

Solving from $\hat{x}_{MS}$, we find that

$$\hat{x}_{MS} = \left(H^TH\sigma_0^2+\sigma_n^2\right)^{-1}\left(\sigma_0^2H^Ty+\sigma_n^2m_0\right)$$

Note that $H^TH = N$ and $H^Ty = \sum_{i=1}^{N}y_i$, so that

$$\hat{x}_{MS} = \frac{\sum_{i=1}^{N}y_i}{N+\frac{\sigma_n^2}{\sigma_0^2}}+\frac{\frac{\sigma_n^2}{\sigma_0^2}}{N+\frac{\sigma_n^2}{\sigma_0^2}}m_0$$

For large N, or large σ_0^2, this becomes

$$\hat{x}_{MS} \sim \frac{1}{N}\sum_{i=1}^{N}y_i$$

This result is called the sample mean. This result says that if enough samples are received, the a priori information about x becomes unimportant.

5.1.2. Mean Square Estimation given a random process

A step in generalization is to consider the MS-estimate of a random variable X given observations of a section of a stochastic process $\{y(\tau), a\leq\tau\leq b\}$. We might reasonably expect, by an extension of the previous arguments, that [23, 29, 36, 48, 49, 50, 57]

$$\hat{x}_{MS} = E\{X/y(\tau), a\leq\tau\leq b\} \tag{5.23}$$

This is correct, but only if the right-hand side is properly defined. The difficulty lies in truing to compute the conditional density $f(x/y(\tau), a \leq \tau \leq b)$. A first guess might be the following: let us assume that the random process $\{y(\tau), a \leq \tau \leq b\}$ is "smooth" enough that it can be represented by a countable number of random variables (samples) $\{y_i, i = 1,...,n\}$. Then it is reasonable to define

$$f(x/y(\tau)) = \lim_{n \to \infty} f(x/y_1,...,y_n)$$

Unfortunately, it often turns out that the limit of the quantity on the right is not generally well behaved; it may not exist or it may be infinite. The above difficulty can be circumvented by using a definition of conditional expectation that does not very on the notion of conditional density.

We shall use the following definition, justified in many books on probability theory.

Definition: The *conditional expectation* of a random variable X given a stochastic process $y(\tau)$ (the cases of a random variable or a random vector are subsumed by this) is the unique random variable that:

1) is a functional of $y(\tau)$, say $h(y(\tau), a \leq \tau \leq b)$;

2) satisfies the orthogonality condition

$$E\{[X - h(y(\tau), a \leq \tau \leq b)]g(y(\tau), a \leq \tau \leq b)\} = 0 \qquad (5.24)$$

for all functionals $g(\cdot)$ (that are random variables for which the above expected vale is meaningful). Such a functional $h(\cdot)$ will be called the conditional expectation of X given $y(\tau)$ and will be written $h(y(\tau)) = E\{X/y(\tau), a \leq \tau \leq b\} = E\{X/y(\cdot)\}$.

It is shown in many textbooks that this descriptive definition coincides with the usual classical definition when ever the latter is meaningful [42, 47]. We shall only provide the following simple example.

Example 5.3 (Equivalence of two definitions of conditional expectations): Let X and Y be random variables with a conditional density $f_{X/Y}(\cdot/\cdot)$ so that we can define the conditional expectation on a classical way

$$E\{X/Y\} = \int x f_{X/Y}(x/y) dx$$

Then we can show that

$$E\{[X - E\{X/Y\}]g(Y)\} = 0 \quad \text{for all } g(Y)$$

or equivalently that

$$E\{E\{X/Y\}g(Y)\} = E\{Xg(Y)\} \quad \text{for all } g(Y)$$

To do this, note that the left hand side (LHS)

$$LHS = \int_{-\infty}^{\infty} \int_{-\infty}^{\infty} \left[\int_{-\infty}^{\infty} u f_{X/Y}(u/y) du\right] g(y) f_{X,Y}(x,y) dxdy$$

$$LHS = \int_{-\infty}^{\infty} \int_{-\infty}^{\infty} \left[\int_{-\infty}^{\infty} u f_{X/Y}(u/y) dy\right] g(y) f_{X,Y}(x,y) dxdy$$

$$= \int_{-\infty}^{\infty} \int_{-\infty}^{\infty} g(y) u \left[\int_{-\infty}^{\infty} f_{X,Y}(x,y) dx\right] f_{X/Y}(u/y) dudy$$

$$= \int_{-\infty}^{\infty} \int_{-\infty}^{\infty} g(y) u f_Y(y) f_{X/Y}(u/y) dudy$$

$$= \int_{-\infty}^{\infty} \int_{-\infty}^{\infty} g(y) u f_{X,Y}(u,y) dudy = RHS$$

where RHS denotes the right hand side.

Conversely, suppose $h(y)$ is a function of y such that

$$E\{Xg(Y)\} = E\{h(Y)g(Y)\} \quad \text{for all } Y$$

Then

$$LHS = \int_{-\infty}^{\infty}\int_{-\infty}^{\infty} xg(y)f_{X,Y}(x,y)\,dxdy$$

$$= \int_{-\infty}^{\infty} g(y)f_{Y}(y)\int_{-\infty}^{\infty} xf_{X/Y}(x/y)\,dxdy$$

Comparing with the right hand side (RHS)

$$RHS = \int_{-\infty}^{\infty}\int_{-\infty}^{\infty} h(y)g(y)f_{X,Y}(x,y)\,dxdy$$

$$= \int_{-\infty}^{\infty} h(y)g(y)\int_{-\infty}^{\infty} f_{X,Y}(x,y)\,dxdy = \int_{-\infty}^{\infty} h(y)g(y)f_{Y}(y)\,dy$$

shows that we must have

$$h(y) = \int_{-\infty}^{\infty} xf_{X/Y}(x/y)\,dx = E\{X/Y\}$$

wherever $f_{Y}(y) > 0$ (and we do not care what values are assumed when $f_{Y}(y) = 0$, since such values make no difference to the integrals).

We shall now show how the new definition of conditional expectation gives a simple approach to the mean square (MS) estimation problem. Given scalar X and an observed scalar process $\{Y(\cdot)\}$, suppose we wish to find a functional $h(Y(\cdot))$ such that the mean-square error criterion

$$E\left\{\left[X - h(Y(\cdot))\right]^2\right\}$$

be minimal. Note that we can write

$$E\left\{\left[X - h(Y(\cdot))\right]^2\right\} = E\left\{\left[X - E\{X/Y(\cdot)\} + E\{X/Y(\cdot)\} - h(Y(\cdot))\right]^2\right\}$$

$$= E\left\{\left[X - E\{X/Y(\cdot)\}\right]^2\right\} + E\left\{\left[E\{X/Y(\cdot)\} - h(\cdot)\right]^2\right\}$$

$$+ 2E\left\{\left[X - E\{X/Y(\cdot)\}\right]\left[E\{X/Y(\cdot)\} - h(Y(\cdot))\right]\right\}$$

But the last term is zero by the orthogonality condition 2 of the conditional expectation definition (here $g(Y(\cdot)) = E\{X/Y(\cdot)\} - h(Y(\cdot))$, and now it is clear that the sum of the two squared terms will be minimal if and only if the second term is equal to zero, i.e.

$$h(Y(\cdot)) = E\{X/Y(\cdot)\}$$

This is certainly a simple derivation and a nice results, but the problem is of course that we have not shown how the conditional expectation may actually be obtained in any practical problem. Unfortunately, there is only a small number of cases for which explicit analytical solutions can be obtained, or even computationally feasible approximation procedures are known. Therefore suboptimal estimates of various kinds are sought, with the most common estimates that are constrained to being linear functionals of the observations. The benefits of such a constraint will be made clear in the next sections.

The orthogonality condition admit of the following geometric interpretation. It suffices to say that we can think of zero-mean random variables $X, Y, Z, \ldots$ as a vectors in some abstract space with an inner product defined by

$$\langle X,Y\rangle = E\{XY\}$$

It can readily be verified that this is a legitimate inner product because it satisfies three requirements:

a)   linearity $\langle k_1 X_1 + k_2 X_2, Y\rangle = k_1 \langle X_1, Y\rangle + k_2 \langle X_2, Y\rangle$

b)   symmetry $\langle X,Y\rangle = \langle Y,X\rangle$

c)   nondegeneracy $\|X\|^2 = \langle X,X\rangle$ is strictly positive unless $X = 0$.

Furthermore, in the case of stochastic process $\{y(\tau); a \leq \tau \leq b\}$, consider an infinite-dimensional space with one coordinate axis for each τ in $[a,b]$. Then any function (random variable) $Y(\tau)$ can be regarded as a vector in this space defined by having component $Y(\tau)$ along the coordinate axis corresponding to τ. The orthogonality principle says that the vector $\tilde{X} = X - h(Y(\tau)) = X - E\{X/Y\}$ is, orthogonal to all other ·vectors $g(Y(\tau))$; that is the inner product $\langle \tilde{X}, g(Y(\tau))\rangle = 0$. The idea behind this can be illustrated as in fig. 5.2, where orthogonal random variables (vectors) are represented as being at right angles.

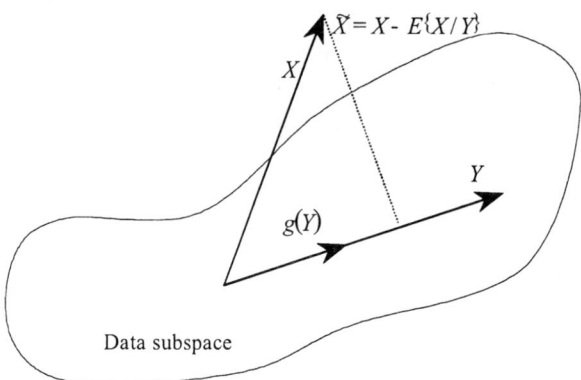

Fig. 5.2: Geometrical interpretation of orthogonality principle

In this figure all functions of Y are represented as being in the direction of Y. It is emphasized the this figure is extremely heuristic, since if nonlinear functions $g(\cdot)$ are allowed, then random variables of the form $g(Y)$ do not form a subspace. However, the figure should only be considered as a mnemonic device. Furthermore, an advantage of the geometric interpretation is that the same minimum mean square error solution can be applied to a large variety of apparently different problems. We shall show in the sequel how to apply this geometric picture to a linear mean-square estimation.

5.2. Uniform cost function: Maximum Aposteriori Estimator

Let us turn now to another cost function which we will refer to as the uniform cost function and is given by

$$C_{UC}(\tilde{X}) = C_{UC}(X - \hat{X}) = \begin{cases} 0 \text{ if } |x_i - \hat{x}_i| < \varepsilon \text{ for } i = 1,...,n \\ 1 \text{ otherwise} \end{cases} \qquad (5.26)$$

where ε is small. This cost function gives zero penalty if all components of the estimation error vector are small and a unit penalty if any component of the estimation error vector becomes larger than ε. For a scalar case $(n=1)$ the uniform cost function is depicted in fig. 5.3.

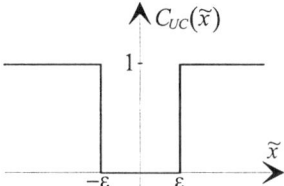

Fig. 5.3.: Uniform cost function

Here we say that an error occurs whenever one or more of the components of the error $\tilde{X} = X - \hat{X}$ exceed ε. If we substitute this cost function into eq. (5.12) we obtain

$$B_{UC}\left(\hat{X}/y\right) = \int_{\substack{|X_i - \hat{X}_i| > \varepsilon \\ i=1,\ldots,n}} f\left(X/y\right)dX$$

$$= \int_{-\infty}^{\infty} f\left(X/y\right)dX - \int_{\hat{X}-E}^{\hat{X}+E} f\left(X/y\right)dX \quad ; \quad E = \begin{bmatrix} \varepsilon \\ \vdots \\ \varepsilon \end{bmatrix}$$

where the sign $\int$ denotes a multiple integral, since $\hat{X}$ and E are n-dimensional vectors. The first integral on the right hand side of eq. (5.27) equals 1. If we apply the mean value theorem for integrals to the second term, then for small ε B_{UC} becomes

$$B_{UC}\left(\hat{X}/y\right) = 1 - \left(2\varepsilon\right)^n f\left(\hat{X}/y\right) \tag{5.28}$$

To minimize B_{UC}, we must make $f\left(\hat{X}/y\right)$ as large as possible, hence the optimum estimator $\hat{X}_{UC}$ is defined by

$$f\left(\hat{X}_{UC}/y\right) \geq f\left(\hat{X}/y\right) \tag{5.29}$$

for all $\hat{X} \neq \hat{X}_{UC}$. In other words, $\hat{X}_{UC}$ is the mode (the mode of a density function $f(X)$ is the value of X for which $f(X)$ is maximum) of the conditional density function $f(X/y)$. Because $\hat{X}_{UC}$ maximized the a posterior density $f(X/y)$, that is the density of X after y is obtained, $\hat{X}_{UC}$ is also called the maximum a posterior (MAP) estimator $\hat{X}_{MAP}$. Since it is possible that $f(X/y)$ may be multimodal, the MAP estimator may not be unique. Using Bayes rule, we can write

$$f_{X/Y}\left(x/y\right) = \frac{f_{Y/X}\left(y/x\right)f_X\left(x\right)}{f_Y\left(y\right)}$$

Since $f_Y\left(y\right)$ does not depend on x, an equivalent definition of the MAP estimator is

$$f_X\left(\hat{x}_{MAP}\right)f_{Y/X}\left(y/\hat{x}_{MAP}\right) \geq f_X\left(\hat{x}\right)f_{Y/X}\left(y/\hat{x}\right) \tag{5.30}$$

for all $\hat{x} \neq \hat{x}_{MAP}$. Quite often $\hat{x}_{MAP}$ occurs at a stationary point of $f_{X/Y}\left(x/y\right)$ which is given by

$$\left. \frac{\partial f_{X/Y}(x/y)}{\partial x} \right|_{x=\hat{x}_{MAP}} = 0 \tag{5.31}$$

or equivalently

$$\left. \frac{\partial f_{Y/X}(y/x)f_X(x)}{\partial x} \right|_{x=\hat{x}_{MAP}} = 0 \tag{5.32}$$

For Gaussian problems, it is usually helpful to find the stationary points of $\ln f_{X/Y}(x/y)$ rather than $f_{X/Y}(x/y)$ so that $\hat{x}_{MAP}$ is given by

$$\left. \frac{\partial \ln f_{X/Y}(x/y)}{\partial x} \right|_{x=\hat{x}_{MAP}} = \frac{\partial \ln f_{Y/X}(y/x)}{\partial x} + \left. \frac{\partial \ln f_X(x)}{\partial x} \right|_{x=\hat{x}_{MAP}} = 0 \tag{5.33}$$

Example 5.4 (MAP estimator of a scalar random variable): Let us consider again the estimation problem of the example 5.1. The density function $f_{Y/X}(y/x)$ was found there

$$f_{Y/X}(y/x) = \begin{cases} \exp(-(y-\ln x)) & \text{if } y > \ln x \\ 0 \text{ ; otherwise} \end{cases}$$

The product

$$f_{Y/X}(y/x)f_X(x) = \begin{cases} \exp(-(y-\ln x)) & \text{if } 0 \le x \le 1 \text{ and } y > \ln(x) \\ 0 \text{ ; otherwise} \end{cases}$$

We can maximize this product by making the quantity $(y-\ln x)$ as positive as possible. Thus

$$\hat{x}_{MAP} = \begin{cases} \exp(y) & \text{if } y \le 0 \\ 1 & \text{if } y > 0 \end{cases}$$

Example 5.5 (Gaussian MAP estimator): As a second example of MAP estimation, let us consider again the Gaussian problem of the example 5.2. There we found that $f_{Y/X}(y/x)$ was given by

$$f_{Y/X}(y/x) = k_2 \exp\left\{ -\frac{1}{2}(y-Hx)^T \frac{1}{\sigma^2}(y-Hx) \right\}$$

Therefore, the product $f_{Y/X}(y/x)f_X(x)$ becomes

$$f_{Y/X}(y/x)f_X(x) = k_3 \exp\left\{ -\frac{1}{2}(y-Hx)^T \frac{1}{\sigma_n^2}(y-Hx) - \frac{1}{2\sigma_0^2}(x-m_0)^2 \right\}$$

To find $\hat{x}_{MAP}$ we shall use eq. (5.32) to write following equation

$$\left. \frac{\partial \ln f_{Y/X}(y/x)f_X(x)}{\partial x} \right|_{x=\hat{x}_{MAP}} = 0 = \frac{1}{\sigma_n^2}H^T(y-H\hat{x}_{MAP}) - \frac{1}{\sigma_0^2}(\hat{x}_{MAP}-m_0)$$

Solving for $\hat{x}_{MAP}$ yields

$$\hat{x}_{MAP} = \left(H^T H \sigma_0^2 + \sigma_n^2 \right)^{-1} \left(\sigma_0^2 H^T y + \sigma_n^2 m_0 \right)$$

We note that this result is identical to $\hat{x}_{MS}$. Of course, this result is not surprising since we know that the conditional mean $\hat{x}_{MS}$ and conditional mode $\hat{x}_{MAP}$ of a Gaussian density are identical.

5.3. Maximum likelihood estimation

The probability density function of the observation Y given the unknown parameter X, that is, $f_{Y/X}(y/x)$ is assumed to be known. No probability density of X is required. The maximum likelihood (ML) estimate is defined by the following criterion. For the observation y, $\hat{x}_{ML}$ is the ML estimate of a parameter x if [17, 23, 29, 31, 37, 44]

$$f_{Y/X}(y/\hat{x}_{ML}) \ge f_{Y/X}(y/\hat{x}) \tag{5.34}$$

for any other $\hat{x} \ne \hat{x}_{ML}$. In other words, $\hat{x}_{ML}$ maximizes the likelihood function $f_{Y/X}(y/x)$ for a given y.

Let us now consider the case where X is uniformly distributed over enough of the n-dimensional Euclidean space. By "enough" we mean that $\hat{x}_{ML}$ is included in this region. In other words, we require that

$$f_X(x) = \begin{cases} Const. \; ; x \in R \\ 0 \quad ; \text{otherwise} \end{cases} \tag{5.35}$$

where $\hat{x}_{ML} \in R$ (we assume that the region R is defined in such a way that the derivative of $f_X(x)$ is well defined). If this is the case, then $\hat{x}_{MAP}$ is clearly in R and, for $x \in R$ eq. (5.33) becomes

$$\left. \frac{\partial \ln f_{X/Y}(x/y)}{\partial x} \right|_{x=\hat{x}_{MAP}} = 0 \tag{5.36}$$

However, from. eq. (5.34) we know that

$$\left. \frac{\partial \ln f_{Y/X}(y/x)}{\partial x} \right|_{\hat{x}=\hat{x}_{ML}} = 0 \tag{5.37}$$

A comparison of eqs. (5.36) and (5.37) indicates that $\hat{x}_{MAP} = \hat{x}_{ML}$ for the conditions (5.35). In other words, if we have no other a priori knowledge concerning x than that it is in a given region, then the ML and MAP estimates are equal. In general, the ML estimate is different from and inferior to the MAP estimate if any other a priori knowledge about x is available.

Example 5.6 (Gaussian ML estimator): Let us consider again the Gaussian problem exposed in the example 5.2. The density function $f_{Y/X}(y/x)$ is given by

$$f_{Y/X}(y/x) = \prod_{i=1}^{N} f_{Y_i/X}(y_i/x) = \left(\frac{1}{\sqrt{2\pi}\sigma_n}\right)^N \prod_{i=1}^{N} \exp\left(-\frac{1}{2}\left(\frac{y_i - x}{\sigma_n}\right)^2\right)$$

$$= \left(\sqrt{2\pi}\sigma_n\right)^{-N} \exp\left(-\frac{1}{2\sigma_n^2}\sum_{i=1}^{N}(y_i - x)^2\right)$$

We can maximize this function by setting the partial derivative of $f_{Y/X}(y/x)$ with respect to x to zero. However, it is easier to work with the $\ln f_{Y/X}(y/x)$ in this case, and so we have

$$\left.\frac{\partial \ln f_{Y/X}(y/x)}{\partial x}\right|_{x=\hat{x}_{ML}} = 0 = \frac{1}{\sigma_n^2}\sum_{i=1}^{N}(y_i - \hat{x}_{ML}) = \frac{1}{\sigma_n^2}\left[\sum_{i=1}^{N} y_i - N\hat{x}_{ML}\right]$$

If we solve for $\hat{x}_{ML}$ we obtain

$$\hat{x}_{ML} = \frac{1}{N}\sum_{i=1}^{N} y_i = \hat{x}_{MS}$$

One of the important properties of the ML estimate is its invariance under invertible transformation. This result is stated in the following theorem [37, 38, 48, 55]:

Theorem: If $g(x)$ is an invertible function defined for all x, then the ML estimator of g given by $\hat{g}_{ML}$ is

$$\hat{g}_{ML} = g(\hat{x}_{ML}) \tag{5.38}$$

In other words, the ML estimate of g is just $g(\hat{x}_{ML})$.

Proof: Let the inverse of g be given by g^{-1} so that

$$g^{-1}(g(x)) = x$$

for all x. The conditional density function of y given g can be written as

$$f_{Y/g}(y/g) = f_{Y/x}(y/g^{-1}(g)) \tag{5.39}$$

By definition of the ML estimate

$$f_{Y/g}(y/\hat{g}_{ML}) \geq f_{Y/g}(y/g)$$

for all $g \neq \hat{g}_{ML}$. Now let $g^* = g(\hat{x}_{ML})$, then from eq. (5.39)

$$f_{Y/g}(y/g^*) = f_{Y/X}(y/\hat{x}_{ML})$$

By definition of $\hat{x}_{ML}$ (see eq. (5.34)) we know that

$$f_{Y/X}(y/\hat{x}_{ML}) \geq f_{Y/X}(y/x)$$

for all $x \neq \hat{x}_{ML}$; therefore

$$f_{Y/g}(y/g^*) \geq f_{Y/g}(y/g)$$

for all $g \neq g^*$. Hence $\hat{g}_{ML} = g^* = g(\hat{x}_{ML})$, which completes the proof.

5.4. Absolute-value cost function

We conclude our discussion of Bayes estimation procedures by examining one more cost function and its associated estimator. The usefulness of this class of estimators is less then the three preceding techniques (MS, MAP and ML) for two reasons. First is limited to scalar parameters x, and second, it is generally computationally more difficult to use.

The absolute value cost function is defined by [37, 48, 49]

$$C_{AB}(\tilde{x}) = |\tilde{x}| = |x - \hat{x}| \tag{5.40}$$

and is depicted in fig. 5.4.

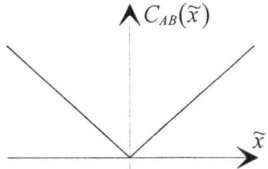

Fig. 5.4.: Absolute value cost function

By substituting this cost function in eq. (5.12), we obtain

$$B_{AB}(\hat{X}/Y) = \int_{-\infty}^{\infty} |x - \hat{x}| f_{X/Y}(x/y) dx$$

$$= -\int_{-\infty}^{\hat{x}} (x - \hat{x}) f_{X/Y}(x/y) dx + \int_{\hat{x}}^{\infty} (x - \hat{x}) f_{X/Y}(x/y) dx$$

Now if we take the partial derivative with respect to $\hat{x}$, we obtain

$$\left. \frac{\partial B_{AB}(\hat{x}/y)}{\partial \hat{x}} \right|_{\hat{x}=\hat{x}_{AB}} = 0 = \int_{-\infty}^{\hat{x}_{AB}} f_{X/Y}(x/y) dx - \int_{\hat{x}_{AB}}^{\infty} f_{X/Y}(x/y) dx \tag{5.41a}$$

or

$$\int_{-\infty}^{\hat{x}_{AB}} f_{X/Y}(x/y) dx = \int_{\hat{x}_{AB}}^{\infty} f_{X/Y}(x/y) dx \tag{5.41b}$$

Hence, we see that $\hat{x}_{AB}$ is the median of the conditional density $f_{X/Y}(x/y)$. In other words, if $F_{X/Y}(x/y)$ is the distribution function of X given y, then $\hat{x}_{AB}$ is defined as

$$F_{X/Y}(\hat{x}_{AB}/y) = \int_{-\infty}^{\hat{x}_{AB}} f_{X/Y}(x/y) dx = \frac{1}{2} \tag{5.42}$$

To illustrate the determination of an absolute-value cost function estimator, let us consider again the simple scalar problem.

Example 5.7 (median estimator for a scalar problem): Using the result obtained in example .5.1, we can write

$$f_{X/Y}(x/y) = \frac{f_{Y/X}(y/x) f_X(x)}{f_Y(y)} = \begin{cases} 2x & \text{if } 0 \le x \le 1 \text{ and } y \ge 0 \\ 2\exp(-2y)x & \text{if } 0 \le x \le \exp(y) \text{ and } y \le 0 \end{cases}$$

For $y \ge 0$, $\hat{x}_{AB}$ is defined by

$$\int_{-\infty}^{\hat{x}_{AB}} 2x dx = \hat{x}_{AB}^2 = \frac{1}{2}$$

or $\hat{x}_{AB} = 1/\sqrt{2}$. For $y \le 0$, $\hat{x}_{AB}$ is defined by

$$\exp(-2y) \int_{0}^{\hat{x}_{AB}} 2x dx = \exp(-2y) \hat{x}_{AB}^2 = \frac{1}{2}$$

or $\hat{x}_{AB} = \exp(y)/\sqrt{2}$.

The nonlinear estimate $\hat{x}_{AB}$ is sketched in fig. 5.5 along with the estimates $\hat{x}_{MS}$ and $\hat{x}_{MAP}$ for the same problem.

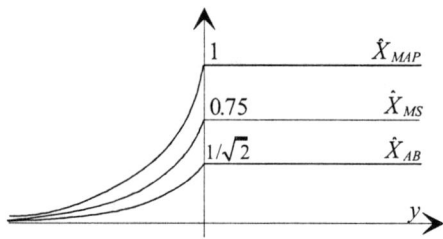

Fig. 5.5. Nonlinear Bayes estimates for different cost functions

5.5. Linear mean-square estimation (Linear minimum variance estimation)

Until now, we have found general solutions for the best possible estimates in various situations. We have not been concerned with the structure of these estimators, but only with their optimality. In this section we will do something a little different. We will find the best estimator of a certain type, even though it may not be the overall optimum. In the section 5.1.1. we derived the mean-square (or minimum variance) estimation method. We found that the estimator was the mean of the conditional density $f_{X/Y}(x/y)$. In general, the conditional mean will not be a linear function of the observation y.

In this section, we shall restrict our attention to linear estimators of the form [11, 25, 48]

$$\hat{X}(y) = Ay + b \qquad (5.43)$$

for some constant matrix A and vector b. Here is assumed that y is a random vector of the corresponding dimension. Since the class of admissible estimates is restricted, the linear mean-square (LMS) estimate (5.43) will not, in general, be as good in the sense of minimizing (5.9),(5.11) as the general MS estimate (5.15).

The constants A and b depend on the measurement process and on a priori information available before the measurement is made. They must be chose to minimize the MS-error (5.9), (5.11). Recall that $\text{tr}(CD) = \text{tr}(DC)$ for compatible matrices C and D and write MS-criterion (5.11) as

$$B_{LMS}\left(\hat{X}\right) = E\left\{\left\|X - \hat{X}\right\|^2\right\} = E\left\{\left(X - \hat{X}\right)^T\left(X - \hat{X}\right)\right\}$$

$$= \text{tr}\left(E\left\{\left(X - \hat{X}\right)\left(X - \hat{X}\right)^T\right\}\right) = \text{tr}\left(E\left\{\left(X - AY - b\right)\left(X - AY - b\right)^T\right\}\right) \tag{5.44}$$

$$= \text{tr}\left(E\left\{\left[\left(X - m_X\right) - \left(AY + b - m_X\right)\right]\left[\left(X - m_X\right) - \left(AY + b - m_X\right)\right]^T\right\}\right)$$

where $m_X = E\{X\}$. After some straightforward work we arrive at

$$B_{LMS}\left(\hat{X}\right) = \text{tr}\left(\left[P_X + A\left(P_Y + m_Y m_Y^T\right)A^T + \left(b - m_X\right)\left(b - m_X\right)^T + 2Am_Y\left(b - m_X\right)^T - 2AP_{YX}\right]\right) \tag{5.45}$$

where $\quad P_X = E\left\{\left(X - m_X\right)\left(X - m_X\right)^T\right\}, \quad m_Y = E\{Y\}, \quad P_Y = E\left\{\left(Y - m_Y\right)\left(Y - m_Y\right)^T\right\} \quad$ and $P_{YX} = E\left\{\left(Y - m_Y\right)\left(X - m_X\right)^T\right\}$.

A necessary condition for the MLS estimator is

$$\left.\frac{\partial B_{LMS}\left(\hat{X}\right)}{\partial \hat{X}}\right|_{\hat{X} = \hat{X}_{LMS}} = 0 \tag{5.46}$$

The following results are required to evaluate an expression such as eq. (5.46).

$$\frac{d}{dA}\text{tr}\left(ABA^T\right) = 2AB \ ; \ \frac{d}{dA}\text{tr}\left(BAC\right) = B^T C^T \tag{5.47}$$

since A is a symmetric matrix. Now if we use eq. (5.44) for $B\left(\hat{X}\right)$ and carry out the operations indicated by eq. (5.47) we obtain

$$\left.\frac{\partial B_{MLS}\left(\hat{X}\right)}{\partial b}\right|_{\hat{X} = \hat{X}_{MLS}} = 2\left(b - \hat{X}_{MLS}\right) + 2Am_Y = 0 \tag{5.48a}$$

and

$$\left.\frac{\partial B_{MLS}\left(\hat{X}\right)}{\partial A}\right|_{\hat{X} = \hat{X}_{MLS}} = 2A\left(P_Y + m_Y m_Y^T\right) - 2P_{XY} + 2\left(b - m_X\right)m_Y^T = 0 \tag{5.48b}$$

Here is used the fact that $P_{YX}^T = P_{XY}$. From (5.48a)

$$b = m_X - Am_Y \tag{5.49}$$

and substituting this into (5.48b)

$$AP_Y - P_{XY} = 0$$

or

$$A = P_{XY}P_Y^{-1} \tag{5.50}$$

Using this A and b in (5.43), the optimal MLS estimate is determined as

$$\hat{X}_{MLS} = m_X + P_{XY}P_Y^{-1}\left(y - m_Y\right) \tag{5.51}$$

This equation is typical for the form of many estimators. We can consider $m_Y = E\{Y\}$ as an estimate for the measurement y. We also define the residual as

$$\tilde{y} = y - m_Y \tag{5.52}$$

The mean m_X is an a priori estimate for X, and the second term on the right-hand side of (5.51) is a correction based on the residual (5.52). If we happen to measure exactly what we expect, so that $y = m_Y$, then the a priori estimate m_X was correct and it does not need to be modified; in this case $\hat{x}_{MLS} = m_X$. The weighting $P_{XY}P_Y^{-1}$ by which the measurement error (residual) is incorporated into the estimate $\hat{x}_{MLS}$ is dependent on the second order joint statistics. Thus, if P_{XY} is smaller then x and y depend on each other to a lesser degree and measurement of y yields less information about x. In fact, if x and y are independent then $P_{XY} = 0$ and y yields no information about x. In this case $\hat{x} = m_X$ for any value of y. Or again, if P_Y is very large, then we have little confidence in y and so $\tilde{y}$ is weighted less heavily into $\hat{x}_{MLS}$. If we have great confidence in y then P_Y is small, and the residual $\tilde{y}$ has a greater role in determining $\hat{x}_{MLS}$.

Let us compare now $\hat{X}_{MLS}$ in eq. (5.51) to the general MS estimate $\hat{X}_{MS}$ given by (5.14). To find $\hat{X}_{MS}$ all the joint statistics of X and Y are required since the density $f_{X/Y}(x/y)$, or equivalently $f_{X,Y}(x,y)$ and $f_Y(y)$, is needed. To find the best linear MS estimate $\hat{X}_{LMS}$, all that is required is the first order statistics, i.e. means m_X and m_Y, and second order statistics, i.e. covariances P_{XY} and P_Y. To find the error covariances associated with $\hat{X}_{LMS}$ first verify that $E\{\hat{X}_{LMS}\} = m_X$ (it is said that $\hat{X}_{LMS}$ is unbiased) by writing

$$
\begin{aligned}
E\{\hat{X}_{LMS}\} &= E\{m_X\} + P_{XY}P_Y^{-1}E\{(Y - m_y)\} \\
&= m_X + P_{XY}P_Y^{-1}\left[E\{Y\} - E\{m_Y\}\right] \\
&= m_X + P_{XY}P_Y^{-1}(m_Y - m_Y) = m_X
\end{aligned} \tag{5.53}
$$

Therefore $E\{\tilde{X}_{LMS}\} = E\{X - \hat{X}_{LMS}\} = E\{X\} - E\{\hat{X}_{LMS}\} = 0$ and the error covariance matrix

$$
\begin{aligned}
P_{\hat{X}_{LMS}} &= E\{\tilde{X}_{LMS}\tilde{X}_{LMS}^T\} = E\left\{\left(X - \hat{X}_{LMS}\right)\left(X - \hat{X}_{LMS}\right)^T\right\} \\
&= E\left\{\left[(X - m_X) - P_{XY}P_Y^{-1}\tilde{Y}\right]\left[(X - m_X) - P_{XY}P_Y^{-1}\tilde{Y}^T\right]\right\} \\
&= P_X - P_{XY}P_Y^{-1}P_{YX} - P_{XY}P_Y^{-1}P_{YX} + P_{XY}P_Y^{-1}P_YP_Y^{-1}P_{YX} \\
&= P_X - P_{XY}P_Y^{-1}P_{YX}
\end{aligned} \tag{5.54}
$$

In eq. (5.54), P_X represents the a priori error covariance and the second term on the right-hand side represents the reduction in uncertainty because of the measurement. We see again that if X and Y are independent then $P_{XY} = 0$ and $P_{\hat{X}_{LMS}} = P_X$, so measuring y provides no increase in our knowledge of x.

A very important point will be now made. In the Gaussian case, the optimal MS estimate $\hat{X}_{MS}$ is linear, since $\hat{X}_{MS} = E\{X/Y\}$ and $\hat{X}_{LMS}$ in eq. (5.51) are the same.

No assumptions are made here on the relation between y and x, i.e. the presented results are valid for any possible nonlinear relations between measurement y and unknown x. Suppose now that x and y are related by the linear measurement model

$$y = Hx + n \tag{5.55}$$

where n is zero-mean measurement noise, with the covariance matrix $R_n = E\{nn^T\}$, and measurement matrix H is known and deterministic (i.e. constant). Assume also that x and n are uncorrelated; that is $R_{Xn} = E\{(X - m_X)n^T\} = 0$. The measurement y has mean $m_Y = E\{Y\}$ and covariance matrix $P_Y = E\{(Y - m_Y)(Y - m_Y)^T\}$, and it is not difficult to determine these quantities in terms of the known quantities H, R_n, m_X and P_X. Thus, since $m_n = E\{n\} = 0$, we have

$$m_Y = E\{HX + n\} = HE\{X\} + E\{n\} = Hm_X + m_n = Hm_X \tag{5.56}$$

Furthermore

$$
\begin{aligned}
P_Y &= E\{(Y - m_Y)(Y - m_Y)^T\} = E\{(HX + n - m_Y)(HX + n - m_Y)^T\} \\
&= E\{[H(X - m_X) + n][H(X - m_X) + n]^T\} \\
&= HE\{(X - m_X)(X - m_X)^T\}H^T + HE\{(X - m_X)n^T\} + E\{n(X - m_X)^T\}H^T + E\{nn^T\} \\
&= HP_X H^T + HR_{Xn} + R_{nX}H^T + R_n
\end{aligned}
$$

or, since $R_{Xn} = 0$ and $R_{nX} = R_{Xn}^T$

$$P_Y = HP_X H^T + R_n \tag{5.57a}$$

To find the cross-covariance of X and Y write

$$
\begin{aligned}
P_{XY} &= E\{(X - m_X)(Y - m_Y)^T\} = E\{(X - m_X)(HX + n - Hm_X)^T\} \\
&= E\{(X - m_X)(X - m_X)^T\}H^T + E\{(X - m_X)n^T\} \\
&= P_X H^T + R_{Xn}
\end{aligned}
$$

or, due to the fact that $R_{Xn} = 0$, $P_X = P_X^T$ and $P_{YX} = P_{XY}^T$

$$P_{XY} = P_X H^T \;\; ; \;\; P_{YX} = HP_X \tag{5.57b}$$

The LMS estimate is found by substituting for P_{XY}, P_Y and m_Y in (5.51)

$$\hat{X}_{LMS} = m_X + P_X H^T \left(HP_X H^T + R_n\right)^{-1}(Y - Hm_X) \tag{5.58}$$

The error covariance is obtained by using the above expressions for P_Y, P_{XY} and P_{YX} in (5.54), which yields

$$P_{\tilde{X}_{LMS}} = P_X - P_X H^T \left(HP_X H^T + R_n\right)^{-1} HP_X \tag{5.59}$$

By using the matrix inversion lemma

$$\left(A^{-1} + BCD\right)^{-1} = A - AB\left(DAB + C^{-1}\right)^{-1} DA \tag{5.60}$$

this can be written as

$$P_{\hat{X}_{LMS}} = \left(P_X^{-1} + H^T R_n^{-1} H\right)^{-1} \tag{5.61}$$

If we make use of the matrix inversion lemma, it is possible to rewrite eq. (5.58) in a form that is often computationally more convenient. Using matrix inversion lemma, we can write

$$\left(HP_X H^T + R_n\right)^{-1} = R_n^{-1} - R_n^{-1} H \left(H^T R_n^{-1} H + P_X^{-1}\right)^{-1} H^T R_n^{-1}$$

If we premultiply this by $P_X H^T$ and then factor $H^T R_n^{-1}$ out, we obtain

$$P_X H^T \left(HP_X H^T + R_n\right)^{-1} = \left\{P_X - P_X H^T R_n^{-1} H \left[H^T R_n^{-1} H + P_X^{-1}\right]^{-1}\right\} H^T R_n^{-1}$$

Next the quantity $\left(H^T R_n^{-1} H + P_X^{-1}\right)^{-1}$ is factored out on the right-hand side, so that we obtain

$$P_X H^T \left(HP_X H^T + R_n\right)^{-1} = \left\{P_X \left[H^T R_n^{-1} H + P_X^{-1}\right] - P_X H^T R_n^{-1} H\right\} \left(H^T R_n^{-1} H + P_X^{-1}\right)^{-1} H^T R_n^{-1}$$

$$= \left(H^T R_n^{-1} H + P_X^{-1}\right)^{-1} H^T R_n^{-1}$$

Therefore $\hat{X}_{LMS}$ can be written as

$$\hat{X}_{LMS} = m_X + \left[H^T R_n^{-1} H + P_X^{-1}\right]^{-1} H^T R_n^{-1} \left(y - Hm_X\right)$$

Combining the two terms involving m_X yields

$$\hat{X}_{LMS} = \left[H^T R_n^{-1} H + P_X^{-1}\right]^{-1} H^T R_n^{-1} y + \left\{I - \left[H^T R_n^{-1} H + P_X^{-1}\right] H^T R_n^{-1} H\right\} m_X$$

If we factor $\left(H^T R_n^{-1} H + P_X^{-1}\right)^{-1}$ out of the second term, we have

$$\hat{X}_{LMS} = \left[H^T R_n^{-1} H + P_X^{-1}\right]^{-1} H^T R_n^{-1} y + \left[H^T R_n^{-1} H + P_X^{-1}\right] \left[H^T R_n^{-1} H + P_X^{-1} - H^T R_n^{-1} H\right] m_X$$

or

$$\hat{X}_{LMS} = \left[H^T R_n^{-1} H + P_X^{-1}\right]^{-1} \left(H^T R_n^{-1} y + P_X^{-1} m_X\right) \tag{5.62}$$

The advantage of the form (5.62) over eq. (5.58) is the size of the matrix to be inverted. In eq. (5.58) the matrix to be inverted has the dimensionality of the observation vector y, while in eq. (5.62) the matrix to be inverted has the dimensionality of the parameter vector X, which is generally much smaller than the first one. Thus, eq. (5.62) generally requires less computational effort. However, we shall see in the following that for sequential (recursive) estimation. the eq. (5.58) is actually the more desirable one.

Example 5.8 (LMS estimator of a scalar random parameter): Let us consider a two-dimensional observation of a scalar parameter X given by

$$Y = \begin{bmatrix} Y_1 \\ Y_2 \end{bmatrix} = \begin{bmatrix} 1 \\ 1 \end{bmatrix} X + \begin{bmatrix} n_1 \\ n_2 \end{bmatrix} = HX + n$$

where the statistical parameters are

$$m_X = E\{X\} = 0 \; ; \; E\left\{(X - m_X)^2\right\} = \sigma_X^2$$

$$E\{n\} = \begin{bmatrix} E\{n_1\} \\ E\{n_2\} \end{bmatrix} = \begin{bmatrix} 0 \\ 0 \end{bmatrix}$$

$$P_n = E\left\{(n - E\{n\})(n - E\{n\})^T\right\} = \sigma_n^2 \begin{bmatrix} 1 & 0 \\ 0 & 1 \end{bmatrix} = \sigma_n^2 I$$

If we substitute into eq. (5.58) we obtain

$$\hat{X}_{LMS} = \sigma_X^2 \begin{bmatrix} 1 & 1 \end{bmatrix} \left\{ \sigma_X^2 \begin{bmatrix} 1 \\ 1 \end{bmatrix} \begin{bmatrix} 1 & 1 \end{bmatrix} + \sigma_n^2 \begin{bmatrix} 1 & 0 \\ 0 & 1 \end{bmatrix} \right\}^{-1} y = \sigma_X^2 \begin{bmatrix} 1 & 1 \end{bmatrix} \begin{bmatrix} \sigma_X^2 + \sigma_n^2 & \sigma_X^2 \\ \sigma_X^2 & \sigma_X^2 + \sigma_n^2 \end{bmatrix}^{-1} y$$

$$= \sigma_X^2 \begin{bmatrix} 1 & 1 \end{bmatrix} \frac{1}{\sigma_n^2(2\sigma_X^2 + \sigma_n^2)} \begin{bmatrix} \sigma_X^2 + \sigma_n^2 & -\sigma_X^2 \\ -\sigma_X^2 & \sigma_X^2 + \sigma_n^2 \end{bmatrix} y = \frac{1}{2 + \sigma_n^2/\sigma_X^2}(y_1 + y_2)$$

Let us repeat example making use of eq. (5.62). The quantity $H^T R_n^{-1} H + P_X^{-1}$ is given by

$$H^T P_n^{-1} H + (\sigma_X^2)^{-1} = \begin{bmatrix} 1 & 1 \end{bmatrix} \frac{1}{\sigma_n^2} \begin{bmatrix} 1 \\ 1 \end{bmatrix} + \sigma_X^{-2} = \frac{2}{\sigma_n^2} + \frac{1}{\sigma_X^2} = \frac{2 + \sigma_n^2/\sigma_X^2}{\sigma_n^2}$$

Therefore $\hat{X}_{LMS}$ is

$$\hat{X}_{LMS} = \frac{\sigma_n^2}{2 + \dfrac{\sigma_n^2}{\sigma_X^2}} \begin{bmatrix} 1 & 1 \end{bmatrix} \begin{bmatrix} y_1 \\ y_2 \end{bmatrix} \frac{1}{\sigma_n^2} = \frac{y_1 + y_2}{2 + \sigma_n^2/\sigma_X^2}$$

as before. Note that there was no matrix inversion in this case because X is a scalar.

Example 5.9 (LMS estimation of AR model parameters): A number of physical phenomena are described by an autoregressive (AR) model of the form

$$y_i = \sum_{k=1}^{n} \Theta_k y_{i-k} + n_i = h_i^T \Theta + n_i \ ; \ i = 1, 2, ..., N$$

where

$$h_i^T = \begin{bmatrix} y_{i-1} & \cdots & y_{i-n} \end{bmatrix} \ ; \ \Theta^T = \begin{bmatrix} \Theta_1 & \cdots & \Theta_n \end{bmatrix}$$

The noise samples n_i are assumed to be uncorrelated and identically distributed so that

$$P_n = E\left\{[n - E\{n\}][n - E\{n\}]^T\right\} = \sigma_n^2 I \ ; \ E\{n\} = 0$$

where

$$n^T = \begin{bmatrix} n_1 & \cdots & n_N \end{bmatrix}$$

and I is an $N \times N$ dimensional identity matrix. A standard approach is to assume no previous knowledge of Θ. No knowledge of Θ would imply an infinite covariance matrix $P_\Theta = E\left\{[\Theta - E\{\Theta\}][\Theta - E\{\Theta\}]^T\right\}$; it is more convenient to represent this as

$$P_\Theta^{-1} = 0 \ \text{ and } \ m_\Theta = E\{\Theta\} = 0$$

The observation model for this problem is given by

$$Y = H\Theta + n$$

where

$$H = \begin{bmatrix} h_1^T \\ h_2^T \\ \vdots \\ h_N^T \end{bmatrix} \; ; \; Y = \begin{bmatrix} y_1 \\ y_2 \\ \vdots \\ y_N \end{bmatrix} \; ; \; \Theta = \begin{bmatrix} \Theta_1 \\ \Theta_2 \\ \vdots \\ \Theta_n \end{bmatrix} \; ; \; n = \begin{bmatrix} n_1 \\ n_2 \\ \vdots \\ n_N \end{bmatrix}$$

The linear MS estimate given by eq. (5.62) is then

$$\hat{\Theta}_{LMS} = \left[\frac{1}{\sigma_n^2} H^T H \right]^{-1} \frac{1}{\sigma_n^2} H^T Y$$

The element of the matrix $H^T H = R$ are given by

$$r_{ij} = \sum_{l=1}^{N} y_{l-i} y_{l-j}$$

while the elements of the vector $H^T Y = \Phi$ are

$$\phi_i = \sum_{l=1}^{N} y_{l-i} y_i$$

In terms of R and Φ

$$\hat{\Theta}_{MLS} = R^{-1}\Phi$$

Example 5.10. (Experiment design): To estimate a deterministic voltage X there are given two alternatives for designing an experiment. Two expensive meters can be used, each of which add Gaussian noise n to X, with zero mean and variance $\sigma_n^2 = 2$, or four inexpensive meters can be used, each of which add Gaussian noise n to X with zero mean and variance $\sigma_n^2 = 3.5$. Which design would result in the more reliable estimate of X.

Design number one is described by

$$Y = \begin{bmatrix} y_1 \\ y_2 \end{bmatrix} = \begin{bmatrix} 1 \\ 1 \end{bmatrix} X + \begin{bmatrix} n_1 \\ n_2 \end{bmatrix} = HX + n$$

where

$$E\{n\} = 0 \; ; \; R_n = \text{cov}\{n\} = E\{nn^T\} = \begin{bmatrix} E\{n_1^2\} & E\{n_1 n_2\} \\ E\{n_1 n_2\} & E\{n_2^2\} \end{bmatrix} = \begin{bmatrix} 2 & 0 \\ 0 & 2 \end{bmatrix}$$

This gives an error covariance of eq. (5.61) (it should be noted that $P_X^{-1} = 0$, since there is no a priori information about X)

$$P_{\hat{X}} = \left(H^T R_n^{-1} H \right)^{-1} = \begin{bmatrix} 1 & 1 \end{bmatrix} \begin{bmatrix} 2 & 0 \\ 0 & 2 \end{bmatrix}^{-1} \begin{bmatrix} 1 \\ 1 \end{bmatrix} = 1$$

Design number two is described by

$$Y = \begin{bmatrix} y_1 \\ y_2 \\ y_3 \\ y_4 \end{bmatrix} = \begin{bmatrix} 1 \\ 1 \\ 1 \\ 1 \end{bmatrix} X + \begin{bmatrix} n_1 \\ n_2 \\ n_3 \\ n_4 \end{bmatrix} = HX + n$$

where

$$E\{n\} = 0 \; ; \; R_n = E\{nn^T\} = \sigma_n^2 I_4 = 3.5 I_4$$

with I_n being the $n \times n$ dimensional identity matrix, so

$$P_{\tilde{X}} = \left(H^T R_n^{-1} H\right)^{-1} = \frac{7}{8}$$

Design number two using four inexpensive meters is therefore a more reliable scheme since it has a lower error covariance.

There are some interesting properties of the LMS estimator. We have shown that this estimate is unbiased; that is the expectation of the estimation error $E\{\tilde{X}_{LMS}\} = 0$, where LMS estimate is given by eq. (5.51). Furthermore, consider the covariance between $\hat{X}_{LMS}$ and $\tilde{X}_{LMS}$. Since $E\{\hat{X}_{LMS}\} = m_X$, and using eq. (5.51) for $\hat{X}_{LMS}$, this becomes

$$\text{cov}\{\hat{X}_{LMS}, \tilde{X}_{LMS}\} = E\left\{\left(\hat{X}_{LMS} - m_X\right)\tilde{X}_{LMS}\right\} = E\left\{P_{XY}P_Y^{-1}(Y - m_Y)\left[X - m_X - P_{XY}P_Y^{-1}(Y - m_Y)\right]^T\right\}$$

$$= P_{XY}P_Y^{-1}E\left\{(Y - m_Y)(X - m_X)^T\right\} - P_{XY}P_Y^{-1}E\left\{(Y - m_Y)(Y - m_Y)^T\right\}P_Y^{-1}P_{XY}^T \qquad (5.63)$$

$$= P_{XY}P_Y^{-1}P_{YX} - P_{XY}P_Y^{-1}P_Y P_Y^{-1}P_{YX} = 0$$

This result says that the estimate $\hat{X}_{LMS}$ and estimation error $\tilde{X}_{LMS}$ are orthogonal. Equation (5.63) is a statement of a general result known as the *orthogonal projection principle* which says that the linear MS is the orthogonal projection of X onto the space spanned by observations y. By beginning with this fact, one can derive the linear MS estimator of eq. (5.51), as it will be shown in the next section (see Fig. 5.2).

An alternative method for deriving the LMS estimator is to assume the $f(X/y)$ is Gaussian with the appropriate first- and second-order moments. One then derives the conditional mean estimator which is linear, and this is the best linear mean square or minimum variance estimator (see example 5.2).

A very important points will now be made

- In the Gaussian case, the optimal MS is linear, since $\hat{X}_{LMS}$ and $\hat{X}_{MS}$ are the same;

- The linear MS (5.51) requires only knowledge of the means and covariances of the random variables involved and does not require knowledge of the joint density function as would be needed for the true MS, or CM, estimate in (5.15);

- The expressions in eq. (5.58) and (5.59) are the optimal LMS estimate and error covariance for the case of linear measurements even if the random variables involved are not Gaussian.

5.6. Least-squares method

The linear MS method, discussed in sec. 5.5, requires only the first- and second-order moments of the random variables. In this section, we consider the least-squares (LS) approach, which uses no stochastic information but instead treats the parameter-estimation task as a deterministic optimization problem [32, 33, 36, 55].

We assume an additive noise model for the measurement; that is

$$Y = HX + n \tag{5.64}$$

where n is an unknown disturbance or noise. As usual, the measurement Y is N-dimensional while X is p-dimensional with p normally much smaller than N. The problem is then to select an estimate $\hat{X}$ of X such that the mean-square error criterion

$$J(\hat{X}) = \frac{1}{2}(Y - H\hat{X})^T W (Y - H\hat{X}) \tag{5.65}$$

is minimized. The matrix W is assumed to be positive definite $(W > 0)$ and symmetric $(W^T = W)$ and is called the weighting matrix. The weighting matrix may be used to assign different costs to each of the errors $\tilde{y}_i = (y - H\hat{X})_i$, $i = 1,..., N$. The estimator which minimizes the quadratic performance index (5.65) is known as the least-squares (LS) estimator $\hat{X}_{LS}$. This method is sometimes referred to as weighted LS (WLS) because of the use of the weighting matrix W.

A necessary condition for the LS estimator is

$$\left. \frac{\partial J(\hat{X})}{\partial \hat{X}} \right|_{\hat{X} = \hat{X}_{LS}} = 0 \tag{5.66}$$

Note that $J(\hat{X})$ is a quadratic form in the variable $\hat{X}$, that is, it contains terms where the degree of $\hat{X}$ is no higher than 2. The derivative of a quadratic form appears with some regularity in estimation and control problems, and it is useful to examine it in some depth. The following results are required to evaluate an expression such as eq. (5.66).

For any square matrix A and vectors X and Y the following derivatives are true:

$$\frac{\partial}{\partial X} X^T AY = AY$$

$$\frac{\partial}{\partial X} Y^T AX = A^T Y$$

$$\frac{\partial}{\partial X} X^T AX = (A + A^T) X$$

Now if we use eq. (5.65) for $J(\hat{X})$ and carry out the operations indicated by eq. (5.66) we obtain

$$H^T W (Y - H\hat{X}_{LS}) = 0$$

so that the LS estimator is

$$\hat{X}_{LS} = (H^T W H)^{-1} H^T W Y \tag{5.67}$$

Note that estimator depends only on H and the weighting matrix W.

Let us consider some of the properties of the estimator (5.66). First, let us find the expected value of the estimation error. The estimation error is given by

$$\tilde{X}_{LS} = X - \hat{X}_{LS} = X - (H^T W H)^{-1} H^T W Y$$

If we use eq. (5.64) for Y, then

$$\tilde{X}_{LS} = X - \left(H^T WH\right)^{-1} H^T W \left(HX + n\right)$$

$$= X - \left(H^T WH\right)^{-1} \left(H^T WH\right) X - \left(H^T WH\right)^{-1} H^T Wn$$

or

$$\tilde{X}_{LS} = -\left(H^T WH\right)^{-1} H^T Wn \qquad (5.68)$$

Therefore the expected value of $\tilde{X}$ is

$$E\left\{\tilde{X}_{LS}\right\} = -\left(H^T WH\right)^{-1} H^T WE\{n\}$$

Hence we see that if $E\{n\} = 0$, then $E\left\{\tilde{X}_{LS}\right\} = 0$, that is $\hat{X}_{LS}$ is unbiased estimate. The covariance matrix of the LS estimation error is given by

$$\text{cov}\left\{\tilde{X}_{LS}\right\} = \left(H^T WH\right)^{-1} H^T WR_n WH \left(H^T WH\right)^{-1} \qquad (5.69)$$

where $R_n = \text{cov}\{n\} = E\left\{nn^T\right\}$.

For a linear MS (MV) estimator with no a priori information about X, so that $P_X^{-1} = 0$ and $m_X = 0$, the LMS estimator as given by eq. (5.62) is

$$\hat{X}_{LMS} = \left(H^T R_n^{-1} H\right)^{-1} H^T R_n^{-1} y \qquad (5.70)$$

and the associated error covariance is given by eq. (5.61)

$$P_{\hat{X}_{LMS}} = \text{cov}\left\{\tilde{X}_{LMS}\right\} = \left(H^T R_n^{-1} H\right)^{-1} \qquad (5.71)$$

A comparison of eqs. (5.67) and (5.70) indicates that if $W = R_n^{-1}$ then the LS estimator will be a LMS estimator with no a priori parameter information. And we see that the error covariance matrices for these two cases are also identical. Since the LS estimator is linear and the linear MS or MV estimator has minimum trace of the covariance matrix (see section 5.5), it is clear that

$$\text{tr}\left\{\text{cov}\left(\tilde{X}_{LMS}\right)\right\} \le \text{tr}\left\{\text{cov}\left(\tilde{X}_{LS}\right)\right\} \qquad (5.72)$$

for all W, where 'tr' denotes the trace. The equality in eq. (5.72) will hold for $W = R_n^{-1}$.

Example 5.11 (LS estimator of a scalar parameters): Let us consider again a two-dimensional observation y of a scalar parameter X given by

$$y = \begin{bmatrix} 1 \\ 1 \end{bmatrix} X + n = Hx + n$$

Let us use the following weighting matrix W and the noise covariance matrix, respectively

$$W = \begin{bmatrix} 2 & 1 \\ 1 & 2 \end{bmatrix} \; ; \quad R_n = \text{cov}\{n\} = \sigma_n^2 I$$

Then $\hat{X}_{LS}$ is given by eq. (5.67)

$$\hat{X}_{LS} = \left[\begin{bmatrix} 1 & 1 \end{bmatrix}\begin{bmatrix} 2 & 1 \\ 1 & 2 \end{bmatrix}\begin{bmatrix} 1 \\ 1 \end{bmatrix}\right]^{-1} \begin{bmatrix} 1 & 1 \end{bmatrix}\begin{bmatrix} 2 & 1 \\ 1 & 2 \end{bmatrix} y = \frac{y_1 + y_2}{2}$$

For the LS estimator the estimation error variance is given by eq. (5.69)

$$\text{var}\{\hat{X}_{LS}\} = \left[\begin{bmatrix}1 & 1\end{bmatrix}\begin{bmatrix}2 & 1 \\ 1 & 2\end{bmatrix}\begin{bmatrix}1 \\ 1\end{bmatrix}\right]^{-1}\begin{bmatrix}1 & 1\end{bmatrix}\begin{bmatrix}2 & 1 \\ 1 & 2\end{bmatrix}\sigma_u^2\begin{bmatrix}2 & 1 \\ 1 & 2\end{bmatrix}\left[\begin{bmatrix}1 & 1\end{bmatrix}\begin{bmatrix}2 & 1 \\ 1 & 2\end{bmatrix}\begin{bmatrix}1 \\ 1\end{bmatrix}\right]^{-1} = \frac{\sigma_n^2}{2}$$

For the linear MS or MV estimator, the estimation error variance is given by eq. (61)

$$\text{var}\{\tilde{X}_{LMS}\} = \left\{\begin{bmatrix}1 & 1\end{bmatrix}\frac{1}{\sigma_n^2}\begin{bmatrix}1 \\ 1\end{bmatrix} + \frac{1}{\sigma_X^2}\right\}^{-1} = \left[\frac{2}{\sigma_n^2} + \frac{1}{\sigma_X^2}\right]^{-1} = \frac{\sigma_n^2}{2 + \frac{\sigma_n^2}{\sigma_X^2}}$$

Hence we see that

$$\text{var}\{\tilde{X}_{LMS}\} < \text{var}\{\tilde{X}_{LS}\}$$

Note, however, that the two error variances will be equal if $1/\sigma_X^2 = 0$ even though $W \neq R_n^{-1}$. Of course, both estimators will also be equal when $W = R_n^{-1}$.

Example 5.12 (Linear regression analysis) : Consider the case where the observation y_i is related to a known parameter ξ by a $(p-1)$ st-degree polynomial of the form [4, 32, 51]

$$y_i = \Theta_1 + \Theta_2\xi_i + \Theta_3\xi_i^2 + \cdots + \Theta_p\xi_i^{p-1} + n_i$$

Now we take N observations for different values of ξ_i to obtain

$$Y = H\Theta + n$$

where

$$H = \begin{bmatrix} 1 & \xi_1 & \xi_1^2 & \cdots & \xi_1^{p-1} \\ 1 & \xi_2 & \xi_2^2 & \cdots & \xi_2^{p-1} \\ \vdots & \vdots & \vdots & \ddots & \vdots \\ 1 & \xi_N & \xi_N^2 & \cdots & \xi_N^{p-1} \end{bmatrix} ; Y = \begin{bmatrix} y_1 \\ y_2 \\ \vdots \\ y_N \end{bmatrix} ; n = \begin{bmatrix} n_1 \\ n_2 \\ \vdots \\ n_N \end{bmatrix}$$

Now by the use of eq. (5.67) we can estimate the polynomial coefficients $\Theta_1, \ldots, \Theta_p$.

A special case of this problem occurs when $p=2$. This case is called linear regression analysis and is an attempt to find the best, in the least-square sense, straight line to fit a given set of data. We consider here only the case when $W = I$. For this case, the matrix product $H^T H$ becomes

$$H^T H = \begin{bmatrix} 1 & 1 & \cdots & 1 \\ \varphi_1 & \varphi_2 & \cdots & \varphi_N \end{bmatrix}\begin{bmatrix} 1 & \varphi_1 \\ 1 & \varphi_2 \\ \vdots & \vdots \\ 1 & \varphi_N \end{bmatrix} = \begin{bmatrix} N & \sum_{i=1}^{N}\varphi_i \\ \sum_{i=1}^{N}\varphi_i & \sum_{i=1}^{N}\varphi_i^2 \end{bmatrix}$$

where we adopted $\xi_i = \varphi_i$, so that $\hat{X}_{LS}$ is given by eq. (5.67)

$$\hat{X}_{LS} = \frac{1}{N\sum_{i=1}^{N}\varphi_i^2 - \left[\sum_{i=1}^{N}\varphi_i\right]^2}\begin{bmatrix} \sum_{i=1}^{N}\varphi_i^2 & -\sum_{i=1}^{N}\varphi_i \\ -\sum_{i=1}^{N}\varphi_i & N \end{bmatrix}\begin{bmatrix} 1 & 1 & \cdots & 1 \\ \varphi_1 & \varphi_2 & \cdots & \varphi_N \end{bmatrix}Y$$

If we carry out the indicated matrix manipulations, we find that

$$\hat{X}_1 = \frac{\sum_{i=1}^{N}\varphi_i^2 \sum_{i=1}^{N} y_i - \sum_{i=1}^{N}\varphi_i \sum_{i=1}^{N}\varphi_i y_i}{N\sum_{i=1}^{N}\varphi_i^2 - \left[\sum_{i=1}^{N}\varphi_i\right]^2}$$

$$\hat{X}_2 = \frac{N\sum_{i=1}^{N}\varphi_i y_i - \sum_{i=1}^{N}\varphi_i \sum_{i=1}^{N} y_i}{N\sum_{i=1}^{N}\varphi_i^2 - \left[\sum_{i=1}^{N}\varphi_i\right]^2}$$

Now given a set of parameter values φ_i and observations y_i, we can easily calculate $\hat{X}$. Moreover, this example is a special case of the general problem of curve fitting by generalized functions. In this case, we write observation as

$$y_i = \sum_{k=1}^{p}\Theta_k f_k(\varphi_i) + n_i$$

where the f_k are known functions of the parameter φ_i. As usual, the problem is to determine the value of the coefficients Θ_k. By using functions which are orthogonal over the given data set, it is possible to cause the matrix $H^T H$ to be diagonal, thereby significantly reducing the computational requirements. Typical examples of such functions are Legendre polynomials.

5.6.1. Sequential (recursive) least-squares method

The algorithm in eq. (5.67) does not take the earlier estimate into account; that is to say $\hat{X}_{LS}$ based on $(N+1)$-dimensional measurement vector $Y = \{y_1,...,y_{N+1}\}^T$ must be computed from the entire collection of $N+1$ measurements y_i, $i=1,...,N+1$. No use of $\hat{X}_{LS}$ based on N measurements y_i, $i=1,...,N$ is made during the calculation of $\hat{X}_{LS}$ based on $N+1$ measurements. In other words, the entire measurement time history is available ahead of time. This situation is often known as an off-line or batch-processing procedure. However, this seems quite wasteful. We intuitively expect that it should be possible to compute the estimate based on $N+1$ measurements from the estimate based on N measurements, and a modification of this earlier estimate to account for the new y_{N+1} measurement. This situation, in which the expanding memory concept is important and very useful, is known as on-line or recursive estimation procedure. Let us proceed to justify our intuition [32, 36].

The "batch" LS-estimator can be calculated recursively, or sequentially, for a scalar measurement y_{N+1} by defining in eq. (5.64)

$$Y = Y(N) = [y_1 \; y_2 \; \cdots \; y_N]^T \; ; \; H = H(N) = [h(1) \; h(2) \; \cdots \; h(N)]^T$$

$$X = X(N) = [X_1(N) \; X_2(N) \; \cdots \; X_p(N)]^T \quad \text{for } X \text{ and } h \in R^p, Y \in R^N, H \in R^{N \times p}$$

The batch LS estimate of X based on N pieces of data is given by eq. (5.67) (here $W = I$)

$$\hat{X}_{LS}(N) = P(N)H^T(N)Y(N) \qquad (5.73)$$

where $P(N) = [H^T(N)H(N)]^{-1}$. Now suppose we want to add a new piece of data y_{N+1}; then we have

$$\hat{X}_{LS}(N+1) = P(N+1)H^T(N+1)Y(N+1)$$

$$= P(N+1)\left[H^T(N)\ h(N+1)\right]^T \begin{bmatrix} Y(N) \\ y_{N+1} \end{bmatrix} \tag{5.74}$$

$$= P(N+1)\left[H^T(N)Y(N) + h(N+1)y_{N+1}\right]$$

The new P matrix is given by

$$P^{-1}(N+1) = H^T(N+1)H(N+1) = \left[H^T(N)\ h(N+1)\right]\begin{bmatrix} H(N) \\ h^T(N+1) \end{bmatrix} \tag{5.75}$$

$$= H^T(N)H(N) + h(N+1)h^T(N+1) = P^{-1}(N) + h(N+1)h^T(N+1)$$

Applying the matrix inversion lemma in eq. (5.60) with $A = P^{-1}(N), B = h(N+1), C = 1$ and $D = h^T(N+1)$, we have

$$P(N+1) = \left[I - K(N+1)h^T(N+1)\right]P(N) \tag{5.76}$$

where

$$K(N+1) = P(N)h(N+1)\left[h^T(N+1)P(N)h(N+1) + 1\right]^{-1} \tag{5.77}$$

Thus, the LS estimate of eq. (5.74) can be written using eqs. (5.75) and (5.76) as

$$\hat{X}_{LS}(N+1) = \left[I - K(N+1)h^T(N+1)\right]\hat{X}_{LS}(N) + \left[I - K(N+1)h^T(N+1)\right]P(N)h(N+1)y_{N+1}$$

or

$$\hat{X}_{LS}(N+1) = \hat{X}_{LS}(N) + K(N+1)\left[y_{N+1} - h^T(N+1)\hat{X}_{LS}(N)\right] \tag{5.78}$$

We summarize the results for the generalized case of weighted LS $(W \neq I)$ and vector measurement (y_{N+1} is a vector) with $N \to t-1$ in the following table:

$$\hat{X}_{LS}(t) = \hat{X}_{LS}(t-1) + K(t)\left[y(t) - H(t)\hat{X}_{LS}(t-1)\right] \qquad (X \text{ correction})$$

$$\uparrow \qquad\quad \uparrow \qquad\quad \uparrow \qquad\quad \uparrow$$

$$\text{new} \qquad \text{old} \qquad \text{weight} \quad \text{error or residual}$$

$$\left(\hat{X} \text{ correction}\right)$$

$$K(t) = P(t-1)H^T(t)\left[H(t)P(t-1)H^T(t) + W(t)\right]^{-1} \qquad (\text{weight})$$

$$P(t) = \left[I - K(t)H(t)\right]P(t-1) \qquad (P \text{ correction})$$

Initial conditions: $\hat{X}(0) = X_0$, $P(0) = P_0$, $X \in R^p$, $y \in R^m$, H^T and $K \in R^{p \times m}$, $P \in R^{p \times p}$, $W \in R^{m \times m}$. It should be noted that the value 1 in the gain matrix (5.77) for the LS estimator in the scalar measurement case is replaced by the weighting matrix W for the general case of weighted LS estimator with vector measurements.

5.7. Properties of estimators

One of the major tasks in the estimation problem is to describe how "good" a given estimation algorithm is. The only complete description of the effectiveness of the estimation algorithm is the conditional probability density $f(\hat{X}/X)$. Since this function can differ for every value of X, it is quite a cumbersome description. In an effort to simplify the problem of describing how good a given estimation procedure is, a number of classes of estimation algorithms have been defined. In this chapter we shall review these definitions and show how they can be used to describe estimation algorithms.

5.7.1. Unbiased estimators

An estimation algorithm can be characterized as a mapping from the true quantity X to the estimate $\hat{X}$. This is a probabilistic mapping and therefore must be described by the probability density function $f(\hat{X}/X)$. One typical example of such a density function for the one-dimensional case is given in fig. 5.6, which shows the actual value of the parameter X [44].

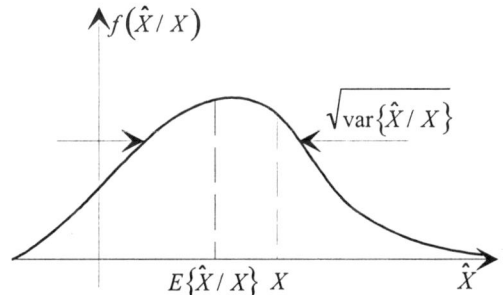

Fig. 5.6: Typical conditional probability density $f(\hat{X}/X)$

The "goodness" of the estimation procedure is characterized by how likely it is that the estimate $\hat{X}$ is close to the actual value of X. Therefore, if the density function is closely clustered about X, it is probably a good estimation procedure; if the density function is not clustered or if it is clustered about some other point, it is a less good estimation scheme.

The variance of a density function of a scalar is a good measure of how clustered it is, and the mean of the density function tells where it is clustered. In the instrumentation theory, the effectiveness of a measuring instrument is specified by its accuracy and precision. Accuracy is a measure of how well the conditional mean $E\{\hat{X}/X\}$ matches X, and precision measures the conditional variance $\text{var}\{\hat{X}/X\}$. Clearly, then a good estimation algorithm is one for which the conditional mean $E\{\hat{X}/X\}$ is close to X and for which the conditional variance $\text{var}\{\hat{X}/X\}$ is small.

It is sometimes possible to design estimators for which the mean of $\hat{X}$, when conditioned on X, is always exactly equal to X, which is generally a desirable property. When this is true for all values of X, such an estimator is referred to as *conditionally unbiased*. In other words, an estimator is conditionally unbiased when

$$E\left\{\hat{X}/X\right\}=X \tag{5.79}$$

One example of such an estimator is as follows.

Example 5.13 (conditionally unbiased ML estimator): Consider forming the ML-estimator of a scalar parameter X from the set of observations

$$y_i = X + n_i \ ; \ i=1,...,N$$

where the noise terms n_i are independent zero-mean, Gaussian random variables of variances σ^2. The joint conditional density function is

$$f\left(y_1,...,y_N / X\right)=\prod_{i=1}^{N} f\left(y_i / X\right)=\prod_{i=1}^{N}\frac{1}{\sqrt{2\pi}\sigma}\exp\left(-\frac{\left(y_i-X\right)^2}{2\sigma^2}\right)$$

$$=\left(2\pi\sigma\right)^{-N/2}\exp\left(-\frac{1}{2\sigma^2}\sum_{i=1}^{N}\left(y_i-X\right)^2\right)$$

The derivative of the log of this function is

$$\frac{\partial}{\partial X}\ln f\left(y_1,...,y_n / X\right)=\frac{1}{\sigma^2}\sum_{i=1}^{N}\left(y_i-X\right)$$

Therefore, the ML-estimate is given by

$$\frac{\partial}{\partial X}\ln f\left(y_1,...,y_N / X\right)\Big|_{X=\hat{X}_{ML}}=0$$

from which one obtains

$$\hat{X}_{ML}=\frac{1}{N}\sum_{i=1}^{N}y_i$$

so that

$$E\left\{\hat{X}_{ML}/X\right\}=\frac{1}{N}\sum_{i=1}^{N}E\left\{y_i/X\right\}=\frac{1}{N}\sum_{i=1}^{N}E\left\{X+n_i/X\right\}=X$$

since $E\left\{n_i/X\right\}=E\left\{n_i\right\}=0$. Thus, since $E\left\{\hat{X}_{ML}/X\right\}=X$, the estimate $\hat{X}_{ML}$ is unbiased. To determine the precision of the estimate $\hat{X}_{ML}$, we must compute the conditional variance

$$\operatorname{var}\left\{\hat{X}_{ML}/X\right\}=E\left\{\left(\hat{X}_{ML}-E\left\{\hat{X}_{ML}/X\right\}\right)^2/X\right\}=E\left\{\left(\frac{1}{N}\sum_{i=1}^{N}y_i-X\right)^2/X\right\}$$

$$=E\left\{\left[\frac{1}{N}\sum_{i=1}^{N}\left(y_i-X\right)\right]^2/X\right\}=E\left\{\left(\frac{1}{N}\sum_{i=1}^{N}n_i^2\right)/X\right\}=\frac{1}{N^2}\sum_{i=1}^{N}E\left\{n_i^2/X\right\}$$

$$=\frac{1}{N^2}\sum_{i=1}^{N}\sigma^2=\frac{\sigma^2}{N}$$

Clearly, the precision improves as N increases. Note that for an unbiased estimate of a scalar, the conditional variance of $\hat{X}$ is the same as the conditional mean square error $E\left\{\left(\hat{X}-X\right)^2/X\right\}$. This is not true in general, and minimizing the conditional variance may not minimize the conditional mean square error.

For those estimation situations in which the density function of X is known, it is possible to define the unconditional expected value, $\hat{X}$ as well as the conditional expected value. If this quantity is equal to the expected value of X, the estimator is called *unconditionally unbiased*. In other words, an estimator in unconditionally unbiased if

$$E_X\left\{E\left\{\hat{X}/X\right\}\right\} = E\left\{\hat{X}\right\} = E\left\{X\right\} \tag{5.80}$$

It is clear that any conditionally unbiased estimator is also unconditionally unbiased. The converse, however, is not necessarily true. It is possible to have an estimate that is unconditionally unbiased but that is not conditionally unbiased. The following example demonstrates the above fact.

Example 5.14 (conditionally and unconditionally unbiased estimators): An observation y is the product of a parameter X and a noise term n ($y = nx$). The probability density function of the noise is

$$f(n) = \begin{cases} 2n \; ; \; 0 \le n \le 1 \\ 0 \; ; \; \text{otherwise} \end{cases}$$

The density function is shown in fig. 5.7a. The conditional density of y given X is just the scaled version of $f(n)$, as shown in fig. 5.7b.

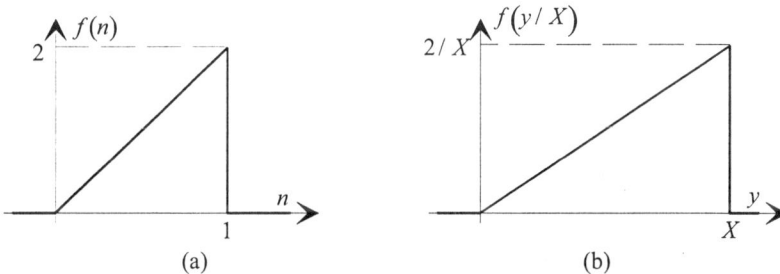

(a)   (b)

Fig. 5.7: Density functions: a) Density of n ; b) Density of y given X

The value of X which maximize $f(y/X)$ is just y and

$$\hat{X}_{ML} = y = nX$$

Clearly this is conditionally biased estimate since

$$E\left\{\hat{X}_{ML}/X\right\} = E\{y/X\} = E\{nX/X\} = XE\{n\} = X\int_0^1 2n\,dn = \frac{2X}{3}$$

However, if X has a zero-mean; that is $E\{X\} = 0$, the estimate would be unconditionally unbiased, since

$$E\left\{\hat{X}_{ML}\right\} = E_X\left\{E\left\{\hat{X}_{ML}/X\right\}\right\} = \frac{2}{3}E\{X\} = 0$$

In fact, the estimator $\hat{X} = 0$ would be conditionally biased but unconditionally unbiased if $E\{X\} = 0$. However, it would not be very good.

Occasionally, the familiar ML estimate will not be unbiased, but there may be another estimate which is. This is demonstrated by the following example.

Example 5.15 (ML estimator for mean and variance): Assume we have a set of observations

$$y_i = \sigma n_i + m; \quad i = 1, ..., N$$

where n_i are independent zero-mean, unit-variance Gaussian random variables. We wish to estimate the parameter vector $X = \begin{bmatrix} m & \sigma^2 \end{bmatrix}^T$. The joint conditional density function of observations is

$$f(y_1, ..., y_N / m, \sigma^2) = \prod_{i=1}^{N} f(y_i / m, \sigma^2)$$

where $f(y_i / m, \sigma^2)$ is a normal density function. Since

$$E\{y_i / m, \sigma^2\} = \sigma E\{n_i\} + m = m$$

$$\text{var}\{y_i / m, \sigma^2\} = E\left\{\left(y_i - E\{y_i / m, \sigma^2\}\right)^2 / m, \sigma^2\right\}$$

$$= E\left\{(y_i - m)^2 / m, \sigma^2\right\} = \sigma^2 E\{n_i^2\} = \sigma^2$$

one concludes

$$f(y_i / m, \sigma^2) = \frac{1}{\sqrt{2\pi}\sigma} \exp\left(-\frac{(y_i - m)^2}{2\sigma^2}\right)$$

so that

$$f(y_1, ..., y_N / m, \sigma^2) = \left(2\pi\sigma^2\right)^{-N/2} \exp\left\{-\frac{1}{2\sigma^2} \sum_{i=1}^{N} (y_i - m)^2\right\}$$

Therefore, the ML-estimate is given by

$$\frac{\partial \ln f(y_1, ..., y_N / m, \sigma^2)}{\partial m} = \frac{1}{\sigma^2} \sum_{i=1}^{N} (y_i - m)\Bigg|_{m = \hat{m}_{ML}} = 0$$

$$\frac{\partial \ln f(y_1, ..., y_N / m, \sigma^2)}{\partial \sigma^2} = -\frac{N}{2\sigma^2} + \frac{1}{2\sigma^4} \sum_{i=1}^{N} (y_i - m)^2\Bigg|_{\sigma^2 = \hat{\sigma}_{ML}^2} = 0$$

from which one obtains

$$\hat{m}_{ML} = \frac{1}{N} \sum_{i=1}^{N} y_i$$

$$\hat{\sigma}_{ML}^2 = \frac{1}{N} \sum_{i=1}^{N} (y_i - \hat{m}_{ML})^2$$

which are known as sample mean and sample variance, respectively. Obviously, the estimate of m is conditionally unbiased, since

$$E\{\hat{m}_{ML} / m, \sigma^2\} = \frac{1}{N} \sum_{i=1}^{N} E\{y_i / m, \sigma^2\} = \frac{1}{N} \sum_{i=1}^{N} E\{\sigma n_i + m / m, \sigma^2\}$$

$$= \frac{1}{N} \sum_{i=1}^{N} \left[\sigma E\{n_i\} + m\right] = \frac{1}{N} \sum_{i=1}^{N} m = m$$

However, the conditional expected value of $\hat{\sigma}^2_{ML}$ is

$$E\left\{\hat{\sigma}^2_{ML}/m,\sigma^2\right\}=E\left\{\frac{1}{N}\sum_{i=1}^{N}\left[\sigma n_i+m-\frac{1}{N}\sum_{j=1}^{N}(\sigma n_j+m)\right]^2\right\}=E\left\{\frac{1}{N}\sum_{i=1}^{N}\left[\sigma n_i+m-\frac{1}{N}\sum_{j=1}^{N}\sigma n_j-\frac{1}{N}\sum_{j=1}^{N}m\right]^2\right\}$$

$$=\frac{\sigma^2}{N}E\left\{\sum_{i=1}^{N}\left(n_i-\frac{1}{N}\sum_{j=1}^{N}n_j\right)^2\right\}=\frac{\sigma^2}{N}E\left\{\sum_{i=1}^{N}\left[n_i^2-\frac{2}{N}n_i\sum_{j=1}^{N}n_j+\frac{1}{N^2}\left(\sum_{j=1}^{N}n_j\right)^2\right]\right\}$$

$$=\frac{\sigma^2}{N}\left\{\sum_{i=1}^{N}E\left\{n_i^2\right\}-\frac{2}{N}\sum_{i=1}^{N}E\left\{\sum_{j=1}^{N}n_i n_j\right\}+\frac{1}{N^2}\sum_{i=1}^{N}\sum_{j=1}^{N}\sum_{j=1}^{N}E\left\{n_j n_k\right\}\right\}$$

$$=\frac{\sigma^2}{N}\left\{\sum_{i=1}^{N}1-\frac{2}{N}\sum_{i=1}^{N}\sum_{j=1}^{N}E\left\{n_i n_j\right\}+\frac{1}{N^2}\sum_{i=1}^{M}\sum_{j=1}^{N}1\right\}=\frac{\sigma^2}{N}\left\{N-\frac{2}{N}\sum_{i=1}^{N}1+\frac{1}{N^2}\sum_{i=1}^{N}N\right\}$$

$$=\frac{\sigma^2}{N}\left\{N-2+1\right\}=\frac{N-1}{N}\sigma^2$$

and $\hat{\sigma}^2_{ML}$ is not an unbiased estimate (let us denote an unbiased estimate as 'ub'). Obviously, it is possible to form an unbiased estimate by simple scaling $\hat{\sigma}^2_{ML}$ as follows

$$\hat{\sigma}^2_{ub}=\frac{N}{N-1}\hat{\sigma}^2_{ML}$$

since

$$E\left\{\hat{\sigma}^2_{ub}\middle|m,\sigma\right\}=\frac{N}{N-1}E\left\{\hat{\sigma}^2_{ML}\middle|m,\sigma\right\}=\frac{N}{N-1}\frac{N-1}{N}\sigma^2=\sigma^2$$

In this way, one obtains

$$\hat{\sigma}^2_{ub}=\frac{N}{N-1}\hat{\sigma}^2_{ML}=\frac{N}{N-1}\frac{1}{N}\sum_{i=1}^{N}(y_i-\hat{m}_{ML})^2$$

However, although this estimate is unbiased, it is not necessarily better than the ML estimate. Note that the conditional variance of the unbiased estimate is

$$\text{var}\left\{\hat{\sigma}^2_{ub}\middle|m,\sigma^2\right\}=E\left\{\left(\hat{\sigma}^2_{ub}-E\left\{\hat{\sigma}^2_{ub}\right\}\right)^2\middle|m,\sigma^2\right\}$$

$$=E\left\{\left[\frac{1}{N-1}\sum_{i=1}^{N}(y_i-\hat{m}_{ML})^2-\sigma^2\right]^2\middle|m,\sigma^2\right\}$$

$$=\frac{1}{(N-1)^2}E\left\{\left[\sum_{i=1}^{N}(y_i-\hat{m}_{ML})^2-(N-1)\sigma^2\right]^2\middle|m,\sigma^2\right\}$$

Here is used the fact that $E\left\{\hat{\sigma}^2_{ub}\right\}=\sigma^2$. On the other hand,

$$\mathrm{var}\left\{\hat{\sigma}_{ML}^2 \,\middle|\, m,\sigma^2\right\} = E\left\{\left(\hat{\sigma}_{ML}^2 - E\left\{\hat{\sigma}_{ML}^2\right\}\right)^2 \,\middle|\, m,\sigma^2\right\}$$

$$= E\left\{\left[\frac{1}{N}\sum_{i=1}^{N}(y_i - \hat{m}_{ML})^2 - \frac{N-1}{N}\sigma^2\right]^2 \,\middle|\, m,\sigma^2\right\}$$

$$= \frac{1}{N^2}E\left\{\left[\sum_{i=1}^{N}(y_i - \hat{m}_{ML}) - (N-1)\sigma^2\right]^2 \,\middle|\, m,\sigma^2\right\}$$

Thus, the variance of the ML estimate is lower then that of the unbiased estimate.

5.7.2. Efficient Estimators

We have seen that when estimating a scalar quantity, the conditional mean and conditional variance let us determine whether we have a good estimator. If we restrict our attention to only those estimators that are unbiased, we shall reduce our measure of goodness to the conditional variance. In other words, it will be possible to define the "best" estimator in the class of unbiased estimators as the one that minimizes the conditional variance. Such estimators are formally referred to as *minimum variance (MV) conditionally unbiased estimators*. At first, such a concept does not seem particularly useful since it appears that we would have to look at all unbiased estimators to determine which one is the best. Fortunately, this is not always the case, since we can establish a lower bound on the conditional variance of any unbiased estimator (the bound was first stated by Fisher, but is generally attributed by Cramer and Rao). Clearly, then, any estimator whose variance equals this bound must be the best in the above sense. As we shall see, if any estimator meets the bound, then the ML estimator does. The *Cramer-Rao* bound is stated as follows [17, 30, 31, 44, 48]:

For any unbiased estimate $\hat{X}$ of a scalar X, the conditional variance is bounded by

$$\mathrm{var}\left\{\hat{X}/X\right\} \ge E\left\{\left[\frac{\partial}{\partial X}\ln f(y/X)\right]^2\right\}^{-1} \tag{5.81a}$$

or, alternatively,

$$\mathrm{var}\left\{\hat{X}/X\right\} \ge \left[-E\left\{\frac{\partial^2}{\partial X^2}\ln f(y/X)\right\}\right]^{-1} \tag{5.81b}$$

where the quantity

$$I(f) = E\left\{\left[\frac{\partial}{\partial X}\ln f(y/X)\right]^2\right\} = \int_{-\infty}^{\infty}\left[\frac{\partial}{\partial X}\ln f(y/X)\right]^2 f(y/X)\,dy$$

$$= \int_{-\infty}^{\infty}\frac{\left[\frac{\partial}{\partial X}f(y/X)\right]^2}{f(y/X)}\,dy \tag{5.82}$$

is known as the Fisher information. The critical feature of these bounds is that they depend only on the conditional probability density function $f(y/X)$. The two forms are given since one may be easier to evaluate than the other. To prove the theorem, note that since $\hat{X}$ is required to be unbiased, we can write

$$E\{\hat{X}(y)/X\} = X \tag{5.83}$$

or, equivalently,

$$\int_{-\infty}^{\infty} \hat{X}(y) f(y/X) dy - X = \int_{-\infty}^{\infty} \hat{X}(y) f(y/X) dy - X \int_{-\infty}^{\infty} f(y/X) dX$$

$$= \int_{-\infty}^{\infty} \left[\hat{X}(y) - X \right] f(y/X) dy = 0 \tag{5.84}$$

where we have indicated explicitly that the estimate $\hat{X}$ is a function of the observation y. Now, if we differentiate both sides of eq.(5.84) with respect to X, we can interchange the differentiation with the integration to obtain

$$\int_{-\infty}^{\infty} \left[-f(y/X) + \left[\hat{X}(y) - X \right] \frac{\partial f(y/X)}{\partial X} \right] dy = 0 \tag{5.85}$$

Using the fact that the integral of any probability density is 1 ($\int_{-\infty}^{\infty} f(y/X) dy = 1$) and that

$$\frac{\partial f}{\partial X} = f \frac{\partial \ln f}{\partial X} \tag{5.86}$$

the eq. (5.85) can be written

$$\int_{-\infty}^{\infty} \left[\hat{X}(y) - X \right] f(y/X) \frac{\partial \ln f(y/X)}{\partial X} dy = 1 \tag{5.87}$$

Now we shall need the Schwartz inequality. It states that for any two functions $f(x)$ and $g(x)$

$$\left[\int_a^b f(x) g(x) dx \right]^2 \leq \int_a^b f^2(x) dx \int_a^b g^2(x) dx \tag{5.88}$$

with equality if and only if

$$g(x) = cf(x) \tag{5.89}$$

If eq. (5.87) is squared, we can use eqs. (5.88) and (5.89) to show that

$$1 = \left[\int_{-\infty}^{\infty} \left\{ \left[\hat{X}(y) - X \right] \sqrt{f(y/X)} \right\} \left\{ \sqrt{f(y/X)} \frac{\partial \ln f(y/X)}{\partial X} \right\} dy \right]^2$$

$$\leq \int_{-\infty}^{\infty} \left[\hat{X}(y) - X \right]^2 f(y/X) dy \int_{-\infty}^{\infty} \left\{ \frac{\partial \ln f(y/X)}{\partial X} \right\}^2 f(y/X) dy$$

or equivalently

$$1 \leq \text{var}\{\hat{X}(y)/X\} E\left\{ \left[\frac{\partial \ln f(y/X)}{\partial X} \right]^2 | X \right\}$$

which is the same as eq. (5.81a). It holds with equality if and only if

$$\frac{\partial}{\partial X} \ln f(y/X) = k \left[\hat{X}(y) - X \right] \tag{5.90}$$

for some constant k. Note that k can depend on X, i.e. $k = k(X)$, but it cannot be zero for any X. If it was zero, the left side of eq. (5.87) would be zero and would not be consistent with the right side.

To establish the alternative form of the bound shown in eq. (5.81b) we need first to differentiate the equality

$$\int_{-\infty}^{\infty} f(y/X)\,dy = 1$$

with respect to X and use eq. (5.86) to give

$$\int_{-\infty}^{\infty} \frac{\partial \ln f(y/X)}{\partial X} f(y/X)\,dy = 0$$

A second round of differentiation and substitution with eq. (5.86) yields

$$\int_{-\infty}^{\infty} \frac{\partial^2 \ln f(y/X)}{\partial X^2} f(y/X)\,dy + \int_{-\infty}^{\infty} \left[\frac{\partial \ln f(y/X)}{\partial X}\right]^2 f(y/X)\,dy = 0$$

But this is just

$$E\left\{\frac{\partial^2}{\partial X^2} \ln f(y/X)\right\} = -E\left\{\left[\frac{\partial}{\partial X}\ln f(y/X)\right]^2\right\}$$

which, together with eq. (5.81a), establishes eq. (5.81b).

The true beauty of the Cramer-Rao bound is that it lets us define and find an efficient estimate. The definition of an efficient estimate is the following:

An efficient estimate is an unbiased estimate whose conditional variance satisfies the Cramer-Rao bound with equality.

There is no guarantee that an efficient estimate exists for a given problem; it may not. However, we can show that if one does exist, the ML estimate will be efficient. In other words, if an efficient estimate exists, the ML estimate is efficient. To prove this, recall that the ML estimate was the one which satisfied

$$\left.\frac{\partial}{\partial X}\ln f(y/X)\right|_{X=\hat{X}_{ML}} = 0$$

However, if an efficient estimate $\hat{X}_{eff}$ exists, eq. (5.90) says that for $\hat{X}_{eff}$

$$\frac{\partial}{\partial X}\ln f(y/X) = k\left(\hat{X}_{eff} - X\right)$$

Evaluating this for $X = \hat{X}_{ML}$ the left side is zero, so the right side must be zero. But, since k cannot be zero, $\hat{X}_{eff}$ must be the same as $\hat{X}_{ML}$.

As an example of an efficient estimate, consider the following example.

Example 5.16 (revisited example 5.13): From example 5.13, the ML estimate was unbiased and the conditional variance was

$$\mathrm{var}\left\{\hat{X}_{ML}|X\right\} = \frac{\sigma^2}{N}$$

To find the Cramer Rao bound, we must compute the expected value of the square of

$$\frac{\partial}{\partial X}\ln f(y/X) = \frac{1}{\sigma^2}\sum_{i=1}^{N}(y_i - X)$$

or the expected value of

$$\frac{\partial^2}{\partial X^2} \ln f(y/X) = -\frac{N}{\sigma^2}$$

Clearly, the second option is easier, and it is obvious that

$$\left[-E\left\{ \frac{\partial^2}{\partial X^2} \ln f(y/X) \right\} \right]^{-1} = \text{var}\left\{ \hat{X}_{ML} | X \right\}$$

Thus, the ML estimate is efficient. Note that since the ML-estimate is the sample mean

$$\hat{X}_{ML} = \frac{1}{N} \sum_{i=1}^{N} y_i$$

the equation

$$\frac{\partial}{\partial X} \ln f(y/X) = \frac{\partial}{\partial X} \ln f(y_1,...,y_N / X) = \frac{1}{\sigma^2} \sum_{i=1}^{N} (y_i - X)$$

can be put into the form

$$\frac{\partial}{\partial X} \ln f(y/X) = \frac{1}{\sigma^2} \left[\sum_{i=1}^{N} y_i - NX \right] = \frac{N}{\sigma^2} \left[\frac{1}{N} \sum_{i=1}^{N} y_i - X \right] = k\left[\hat{X}_{ML} - X \right]$$

where $k = \dfrac{N}{\sigma^2} > 0$, and we could have known at that point that the estimate $\hat{X}_{ML}$ was efficient. In fact, if we had noticed this, we could have used the equation

$$\frac{\partial^2 \ln f(y/X)}{\partial X^2} = -k = -\frac{N}{\sigma^2}$$

to determine the variance σ^2.

As we have stated, the Cramer-Rao bound applies only to estimates of scalars. There is a form of the bound for estimating vectors, but it is more complex and not as useful. We shall state the Cramer-Rao bound for vector case without the proof. In the general case, the conditional mean $\overline{X}$ and conditional covariance matrix $R_{\hat{X}}$ of the estimate $\hat{X}$ are given by

$$\overline{X} = E\left\{ \hat{X}(y) | X \right\} = \int_{-\infty}^{\infty} \hat{X}(y) f(y/X) dy \tag{5.91}$$

$$R_{\hat{X}} = E\left\{ \left(\hat{X} - \overline{X} \right)\left(\hat{X} - \overline{X} \right)^T | X \right\} = \int_{-\infty}^{\infty} \left[\hat{X}(y) - \overline{X} \right]\left[\hat{X}(y) - \overline{X} \right]^T f(y/X) dy \tag{5.92}$$

where the sign $\int$ denotes a multiple integral. We see from eq. (5.91) that $\overline{X}$ is a function of the true value of the parameter vector X. If this function is differentiable, then we can form the matrix

$$P = \frac{\partial \overline{X}}{\partial X} = \int_{-\infty}^{\infty} \hat{X}(y) \left[\frac{\partial f(y/X)}{\partial X} \right]^T dy \tag{5.93}$$

We also define a matrix J by

$$J = E\left\{\left[\frac{\partial \ln f(y/X)}{\partial X}\right]\left[\frac{\partial \ln f(y/X)}{\partial X}\right]^{T}\right\}$$

$$= \int_{-\infty}^{\infty}\left[\frac{\partial \ln f(y/X)}{\partial X}\right]\left[\frac{\partial \ln f(y/X)}{\partial X}\right]^{T} f(y/X)dy$$

(5.94)

The Cramer-Rao theorem asserts that the matrix $R_{\hat{X}} - PJ^{-1}P^{T}$ is positive semidefinite, i.e.

$$R_{\hat{X}} - PJ^{-1}P^{T} \geq 0 \tag{5.95}$$

and that it is null if and only if there exists a matrix A (whose elements may be functions of X) such that

$$\hat{X} - X = A(X)\frac{\partial \ln f(y/X)}{\partial X} \tag{5.96}$$

The matrix $PJ^{-1}P^{T}$ is called the *minimum variance bound* (MVB).

Since the diagonal elements of a positive semidefinite matrix must be nonnegative, we have for the variance of each element $\hat{X}_{\alpha}$ ($\alpha = 1,...,n$) of the $n \times 1$ vector $\hat{X}$

$$\text{var}\{\hat{X}_{\alpha}\} = \sigma_{\hat{X}_{\alpha}}^{2} \geq \sum_{\beta}\sum_{\gamma} P_{\alpha\beta}J_{\beta\gamma}^{-1}P_{\alpha\gamma} \tag{5.97}$$

where $A_{\alpha\beta}$ denotes an element in the α-th row and β-th column of a matrix A, with equality holding only when eq. (5.96) is satisfied. In other words, the equality for the α-th element holds if and only if

$$\hat{X}_{\alpha} - X_{\alpha} = a_{\alpha}^{T}\frac{\partial}{\partial X}\ln f(y/X) \; ; \; \alpha = 1,...,n \tag{5.97}$$

where a_{α}^{T} is the α-th row of the matrix $A(X)$ in eq. (5.96).

An estimate $\hat{X}(y)$ is called efficient if its variance is the lowest theoretically attainable, i.e. when

$$R_{\hat{X}} = PJ^{-1}P^{T} \tag{5.98}$$

When the estimate is unbiased, we have $\overline{\hat{X}} = X$, and hence $P = I$. The Cramer-Rao theorem reduces to the statement that $R_{\hat{X}} - J^{-1}$ is positive semidefinite, i.e.

$$R_{\hat{X}} - J^{-1} \geq 0 \tag{5.99}$$

and an efficient unbiased estimate has the conditional covariance

$$R_{\hat{X}} = J^{-1} \tag{5.100}$$

As with unbiasedness, efficiency can be attained only in a small class of relatively simple models. It should be pointed out that in some cases, although no efficient estimate exists, among those estimates that do exist, or among estimates of a certain class, there may be one whose variance is least. Moreover, the ML estimate for the general Gaussian case satisfies eq. (5.97), and therefore it is efficient.

Example 5.17. (general Gaussian problem): Let the observation model

$$y = HX + n$$

where n is a zero-mean Gaussian noise. The conditional mean and covariance matrix of y are defined by

$$E\{y/X\} = HX$$

$$\text{cov}\{y/X\} = E\left\{\left[y - E\{y/X\}\right]\left[y - E\{y/X\}\right]^T \mid X\right\} = E\{nn^T\} = R_n$$

where R_n is the covariance matrix of n, so that the conditional density function of y is

$$f(y/X) = \frac{1}{(2\pi)^{N/2}|R_n|^{1/2}} \exp\left\{-\frac{1}{2}(y - HX)^T R_n^{-1}(y - HX)\right\}$$

where $|\cdot|$ denotes the determinant, and the derivative of the natural log of this is

$$\frac{\partial}{\partial X}\ln f(y/X) = H^T R_n^{-1}(y - HX)$$

ML estimate is defined by

$$\left.\frac{\partial}{\partial X}\ln f(y/X)\right|_{X = \hat{X}_{ML}} = 0$$

from which one obtains

$$\hat{X}_{ML} = \left(H^T R_n^{-1} H\right)^{-1} H^T R_n^{-1} y$$

Thus, one concludes that

$$\hat{X}_{ML} - X = \left(H^T R_n^{-1} H\right)^{-1} H^T R_n^{-1} y - X$$

$$= \left(H^T R_n^{-1} H\right)^{-1}\left[\frac{\partial}{\partial X}\ln f(y/X) + H^T R_n^{-1} HX\right] - X$$

$$= \left(H^T R_n^{-1} H\right)^{-1}\frac{\partial \ln f(y/X)}{\partial X}$$

Since the derivative of the log of the conditional density has the form required by eq. (5.96), with $A = \left(H^T R_n^{-1} H\right)^{-1}$, for equality in (5.99) (it should be noted that ML is an unbiased estimate, i.e. $E\{\hat{X}_{ML}|X\} = \left(H^T R_n^{-1} H\right)^{-1} H^T R_n^{-1} E\{y|X\} = X$) the Cramer-Rao bound (5.100) also must be satisfied. Therefore, we can compute the conditional covariance matrix from eqs. (5.94) and (5.100)

$$\text{cov}\{\hat{X}_{ML}|X\} = R_{\hat{X}_{ML}} = J^{-1} \tag{5.101}$$

where

$$J^{-1} = E\left\{ \left[\frac{\partial \log f(y/X)}{\partial X}\right]\left[\frac{\partial \log f(y/X)}{\partial X}\right]^{T} |X\right\}$$

$$= H^{T}R_{n}^{-1}E\left\{(y-HX)(y-HX)^{T}|X\right\}R_{n}^{-1}H$$

$$= H^{T}R_{n}^{-1}E\left\{nn^{T}\right\}R_{n}^{-1}H = H^{T}R_{n}^{-1}R_{n}R_{n}^{-1}H = H^{T}R_{n}^{-1}H$$

In addition to the bound on conditional variance for conditionally unbiased estimators, there is a similar bound on the mean-square error for estimating a random variable. It is stated as the following theorem:

The mean-square error in estimating a random variable X can be bounded by

$$E\left\{\left(\hat{X}-X\right)^{2}\right\} \geq \left(E\left\{\left[\frac{\partial \log f(y,X)}{\partial X}\right]^{2}\right\}\right)^{-1} = \left[-E\left\{\frac{\partial^{2}\log f(y,X)}{\partial X^{2}}\right\}\right]^{-1}$$

if the following conditions are met:

(a) $\dfrac{\partial \log f(y,X)}{\partial X}$ and $\dfrac{\partial^{2}\log f(y,X)}{\partial X^{2}}$ are both absolutely integrable with respect to y and X;

(b) $\lim\limits_{X\to\infty} f(X)E\left\{\hat{X}-X|X\right\} = 0$;

(c) $\lim\limits_{X\to-\infty} f(X)E\left\{\hat{X}-X|X\right\} = 0$.

The proof is quit similar to that for the Cramer-Rao bound. Consider differentiating $f(X)E\left\{\hat{X}-X|X\right\}$ with respect to X

$$\frac{d}{dX}\left(f(X)E\left\{\hat{X}-X|X\right\}\right) = \frac{d}{dX}\left(f(X)\int_{-\infty}^{\infty}\left(\hat{X}-X\right)f(y/X)dy\right)$$

$$= \frac{d}{dX}\int_{-\infty}^{\infty}\left(\hat{X}-X\right)f(y/X)f(X)dy = \frac{d}{dX}\int_{-\infty}^{\infty}\left(\hat{X}-X\right)f(X,y)dy$$

$$= \int_{-\infty}^{\infty}\left[\left[\frac{d}{dX}\left(\hat{X}-X\right)\right]f(X,y)+\left(\hat{X}-X\right)\frac{\partial f(X,y)}{\partial X}\right]dy$$

$$= \int_{-\infty}^{\infty}\left[-f(X,y)+\left(\hat{X}(y)-X\right)\frac{\partial f(X,y)}{\partial X}\right]dy$$

If this is integrated with respect to X from $-\infty$ to $+\infty$, we have

$$f(X)E\left\{\left(\hat{X}-X\right)|X\right\}\Big|_{-\infty}^{\infty} = \int_{-\infty}^{\infty}\int_{-\infty}^{\infty}\left[-f(X,y)+\left(\hat{X}(y)-X\right)\frac{\partial f(X,y)}{\partial X}\right]dydX \qquad (5.102)$$

Conditions (b) and (c) ensure that the left side is zero; therefore, eq. (5.102) is the same as eq. (5.85). From this point, the derivations is essentially unchanged. The major difference is that when the Schwartz inequality is used, the variable of integration includes X, as well as y. Therefore, the conditions for equality is

$$\frac{\partial}{\partial X}\log f(y,X) = k\left[\hat{X}(y)-X\right] \qquad (5.103)$$

and k cannot depend on X and cannot be zero. An estimator which satisfies the bound with equality is again called an *efficient estimator*. Therefore, since the definition of the MAP estimate requires that

$$\frac{\partial}{\partial X}\log f(y,X)\Big|_{X=\hat{X}_{MAP}}=0$$

then if an efficient estimate exists, the MAP estimate is efficient. In addition, since $\hat{X}_{MAP}$ has minimum mean-square error, it must be equal to conditional mean estimate $\hat{X}_{CM}$. Additionally, replacing $f(X,y)$ by $f(X/y)f(y)$ in eq. (5.103) and taking another derivatives, gives

$$\frac{\partial^2}{\partial X^2}\log f(X/y)=-k \tag{5.104}$$

Therefore, $f(X/y)$ must be of the form

$$f(X/y)=K\exp\left\{-kX^2+k_1X+k_2\right\} \tag{5.105}$$

which is just a Gaussian density, with K, k, k_1 and k_2 being appropriate constants. Thus, the density function $f(X/y)$ must be Gaussian for an efficient estimate to exists.

For a stochastic unknown vector X, a result similar to the Cramer-Rao bound in eq. (5.99) can be stated as follows.

If $\hat{X}$ is any estimate of a stochastic vector X based on measurement vector Y, then the covariance of the estimation error $\tilde{X}=X-\hat{X}$ is bounded by

$$P_{\hat{X}}=E\left\{\tilde{X}\tilde{X}^T\right\}\geq L^{-1} \tag{5.106}$$

where the information matrix L is given by

$$L=E\left\{\left[\frac{\partial\log f(X,Y)}{\partial X}\right]\left[\frac{\partial\log f(X,Y)}{\partial X}\right]^T\right\}=-E\left\{\frac{\partial^2\log f(X,Y)}{\partial X^2}\right\} \tag{5.107}$$

Equality holds in (5.106) if and only if

$$\frac{\partial\log f(X,Y)}{\partial X}=k\left(X-\hat{X}\right) \tag{5.108}$$

where k is a constant. It is assumed that $\dfrac{\partial f(X,Y)}{\partial X}$ and $\dfrac{\partial^2 f(X,Y)}{\partial X^2}$ exist and are absolutely integrable with respect to both variables, and that

$$\lim_{X\to\pm\infty}B(X)f(X)=0 \tag{5.109}$$

$$B(X)=E\left\{X-\hat{X}|X\right\}=\int_{-\infty}^{\infty}\left(X-\hat{X}\right)f(Y/X)dY \tag{5.110}$$

An estimate $\hat{X}$ for which the bound is satisfied with equality, i.e. the condition (5.108) on the form of $f(X,Y)$ holds, is said to be efficient.

Several points should be noted.

(1) The bound in (5.106), (5.107) depends on the joint density $f(X,Y)$ while the Cramer-Rao bound (for deterministic unknown vector X) in (5.94) and (5.100) depends on the likelihood function $f(Y/X)$.

(2) The Cramer-Rao bound (5.100) deals with unbiased estimate $\hat{X}$, while the bound in (5.106) substitutes for unbiasedness the requirements (5.109).

(3) The efficiency condition (5.108) in terms of the constant k should be contrasted with the efficiency condition (5.96) of the Cramer-Rao bound in terms of the function $A(X)$ of X.

(4) Similarly as in the scalar case, differentiating (5.108) yields the equivalent condition for efficiency

$$\frac{\partial^2 \log f(X,Y)}{\partial X^2} = k \tag{5.111}$$

or, since $f(X,Y) = f(X/Y)f(Y)$ and $f(Y)$ is independent of X,

$$\frac{\partial^2 \log f(X/Y)}{\partial X^2} = k \tag{5.112}$$

Integrating twice and taking antilog yields

$$f(X/Y) = \exp\left(-\frac{k}{2}X^T X + bX + c\right)$$

so that the a posterior density $f(X/Y)$ must be Gaussian for an efficient estimate to exist.

(5) Since $f(X,Y) = f(Y/X)f(X)$, the information matrix L in (5.107) and the Fisher information matrix J in (5.94) are related through the equation

$$L = J - E\left\{\frac{\partial^2 \log f(X)}{\partial X^2}\right\} \tag{5.113}$$

so that the stochastic description of X enters independently of the noise statistics.

We can gain a little more insight by considering the general Gaussian problem.

Example 5.18. (General Gaussian problem; example 5.17. revisited): Let us assume that the $p \times 1$ dimensional observation Y depends linearly on the $n \times 1$ dimensional X according to

$$Y = HX + V$$

Assume that the measurement noise V is zero-mean normal so that

$$f_V(v) = \frac{1}{(2\pi)^{p/2}(\det(R_V))^{1/2}} \exp\left(-\frac{1}{2}v^T R_V^{-1} v\right)$$

Then by the example 5.17 we know that the Fisher information matrix

$$J = H^T R_V^{-1} H$$

To find L, note that

$$f(X) = \frac{1}{(2\pi)^{n/2}(\det(P_X))^{1/2}} \exp\left(-\frac{1}{2}(X - \bar{X})^T P_X^{-1}(X - \bar{X})\right)$$

where $\bar{X} = E\{X\}$, so that

$$\frac{\partial \log f(X)}{\partial X} = -P_X^{-1}(X - \bar{X})$$

Then, by (5.113)

$$L = H^T R_V^{-1} H + P_X^{-1}$$

Thus, the total information available has been increased by a priori information on X, so that the bound is decreased, i.e. $L^{-1} \leq J^{-1}$.

5.7.3. Asymptotic properties

In many estimation problems, the observation actually consists of a sequence of observations. In such cases, the number of observations in the sequence may be a parameter of the estimator design. It is important, therefore, to examine the behavior of an estimator as a function of the number of observations. In general, we shall expect the estimate to get better as the number of observations increases. We shall use $\hat{X}_N$ to denote the estimate of X based on the first N observations $Y_1, Y_2, ..., Y_N$.

As an illustration, recall the example 5.15, where we computed the variance of a scalar Gaussian process from a set of samples.

Example 5.19. (example 5.15 revisited): ML estimate of the variance of a Gaussian process from a set of samples $y_1, y_2, ..., y_N$ is given by

$$\hat{\sigma}_{ML}^2(N) = \frac{1}{N} \sum_{i=1}^{N} \left(y_i - \frac{1}{N} \sum_{j=1}^{N} y_j \right)^2$$

The conditional mean of this estimate is

$$E\{\hat{\sigma}_{ML}^2(N)|\sigma^2\} = \frac{N-1}{N}\sigma^2$$

and the estimate is biased. However, in the limit of large N, $E\{\hat{\sigma}_{ML}^2|\sigma^2\}$ becomes σ^2, and the estimate becomes unbiased. Such an estimator is called *asymptotically unbiased* and is defined as follows [17, 37, 48].

An estimate $\hat{X}_N$ is asymptotically unbiased if

$$\lim_{N \to \infty} E\{\hat{X}_N|X\} = X \tag{5.114}$$

Generally, one would expect a good estimate to get better as the number of observations N increases. In fact, $\hat{X}_N$ should be essentially the same as X when the number of observations increases beyond bound ($N \to \infty$). There are two simple ways to state this formally: an estimate $\hat{X}_N$ can be *simple consistent*, or to converge in probability to X, and mean-square consistent. To discuss these two approaches we need the concept of stochastic convergence. As mentioned in the section 4.1, the sequence $X_1, X_2, ...$ of random variables converges in probability to random variable X if for all $\varepsilon > 0$

$$P\left(\left\|\hat{X}_N - X\right\| > \varepsilon\right) \to 0 \quad \text{for } N \to \infty \tag{5.115a}$$

where $\|\cdot\|$ denotes the Euclidean norm, or equivalently

$$\lim_{N \to \infty} P\left(\left\|\hat{X}_N - X\right\| < \varepsilon\right) = 1 \tag{5.115b}$$

It is said to be mean-square convergent if

$$\lim_{N \to \infty} E\left\{\left(\hat{X}_N - X\right)^T \left(\hat{X}_N - X\right)\right\} = 0 \tag{5.116}$$

As with unbiased estimators, a consistent estimator can be conditionally consistent or, if $f(X)$ is defined, unconditionally consistent, which leads to four definitions.

An estimate is *conditionally simple consistent*, or *conditionally convergent with probability* 1, if for all $\varepsilon > 0$

$$\lim_{N \to \infty} P\left\{\left\|\hat{X}_N - X\right\| < \varepsilon \big| X\right\} = 1 \tag{5.117}$$

An estimate of a random vector X is *unconditionally simple consistent*, or *unconditionally convergent with probability* 1, if for any $\varepsilon > 0$

$$\lim_{N \to \infty} P\left\{\left\|\hat{X}_N - X\right\| < \varepsilon\right\} = 1 \tag{5.118}$$

An estimate is *conditionally mean-square consistent*, or *mean-square convergent*, if

$$\lim_{N \to \infty} E\left\{\left(\hat{X}_N - X\right)\left(\hat{X}_N - X\right)^T \big| X\right\} = 0 \tag{5.119}$$

An estimate of a random variable X is *unconditionally mean-square consistent* if

$$\lim_{N \to \infty} E\left\{\left(\hat{X}_N - X\right)\left(\hat{X}_N - X\right)^T\right\} = 0 \tag{5.120}$$

Implicit in the mean-square consistent definitions is the concept of an inequality between estimation error covariance matrices $P_N = E\left\{\left(\hat{X}_N - X\right)\left(\hat{X}_N - X\right)^T \big| X\right\}$ or $P_N = E\left\{\left(\hat{X}_N - X\right)\left(\hat{X}_N - X\right)^T\right\}$. When we say that $P_1 \leq P_2$, we mean that $P_2 - P_1$ is nonnegative definite, i.e. $P_2 - P_1 \geq 0$. This concept lets us compare two mean-square-consistent estimators and determine which is better. Therefore, we can define an *asymptotically efficient estimate* as follows.

An unconditionally mean-square-consistent estimator $\hat{X}_N$ is *unconditionally asymptotically efficient* if there is some N_0 such that for any other unconditionally mean-square-consistent estimator $\hat{X}'_N$, the unconditional estimation error covariance matrices satisfy the matrix inequality

$$E\left\{\left(\hat{X}_N - X\right)^T \left(\hat{X}_N - X\right)\right\} \leq E\left\{\left(\hat{X}'_N - X\right)^T \left(\hat{X}'_N - X\right)\right\} \text{ for all } N \geq N_0 \tag{5.121}$$

An conditionally mean-square-consistent estimator $\hat{X}_N$ is *conditionally asymptotically efficient* if there is some N_0 such that for any other conditionally mean-square-consistent estimator $\hat{X}'_N$ the conditional estimation error covariance matrices satisfy

$$E\left\{\left(\hat{X}_N - X\right)^T \left(\hat{X}_N - X\right) / X\right\} \leq E\left\{\left(\hat{X}'_N - X\right)^T \left(\hat{X}'_N - X\right) / X\right\} \text{ for all } N \geq N_0 \tag{5.122}$$

Finally, we can define an *asymptotically normal estimate* as one for which the distribution of the random variable $\sqrt{N}\left(\hat{X}_N - X\right)$ approaches the Gaussian distribution in the limit ($N \to \infty$). The best, or optimal, asymptotically normal estimator would be the one for which the mean of $\sqrt{N}\left(\hat{X}_N - X\right)$ would be zero, and its variance would be a minimum over all asymptotically normal estimators, i.e. $\lim_{N \to \infty} E\left\{N\left(\hat{X}_N - X\right)\left(\hat{X}_N - X\right)^T\right\}$ would be a minimum over all asymptotically normal estimators. Here the concept of the inequality between covariance matrices has to be used.

Several points should be also used:

(1) The variances of most of the estimates approach zero as $1/N$ when N increases. Hence, if the estimator is consistent, or asymptotically unbiased, for $\hat{X}$, it is consistent for any well-behaved function of $\hat{X}$. Thus consistency is a more significant concept then unbiasedness.

(2) Both $R_{\hat{x}}$ and J^{-1} in (5.92) and (5.94) tend to zero as $1/N$ for most relevant estimators. Hence, we call a consistent estimator asymptotically efficient if with probability one

$$\lim_{N \to \infty} N\left(R_{\hat{x}} - J^{-1}\right) = 0 \qquad (5.123)$$

(3) Under fairly general conditions, the ML estimate has the following limiting properties: it is consistent, asymptotically efficient and asymptotically normal.

(4) Our discussion of estimation criteria would not be completed without the mention of *sufficient statistics*. A statistics ρ of a sample $Y = \{Y_1, Y_2, ...\}$ is any function computed from the values of the sample for the purpose of extracting relevant information $\rho = r(Y)$. In particular, any estimate $\hat{X}$ is a statistics defined by $\hat{X} = h(Y)$. A statistic ρ is deemed sufficient for the parameter X if the value of ρ conveys as much information concerning the value of X as did the original sample Y. In other words, we may compute the value of ρ from the sample Y, and then discard the data Y without loosing any information relevant to the estimation of X. It can be proved that if an efficient estimate exists, it must also be sufficient. Conversely, if a sufficient estimate exists, some function of it is an efficient estimate of some function of the parameters.

(5) The value of an estimate usually depends on the form of the probability distribution that we assume. We rarely possess exact knowledge of the distribution, and must usually confer ourselves with a rough approximation. We desire, therefore, that our estimate be robust, i.e. that it be only slightly affected by seemingly unimportant changes in the form of the assumed distribution.

(6) The choice of parameters appearing in a model is often arbitrary. We may replace the original parameters X with a different set of parameters X' which are single valued functions of X, e.g. $X' = s(X)$, where s is a vector of functions. It is desirable that our estimators be invariant under reparametrization. That is, if $\hat{X}$ and $\hat{X}'$ are the estimates obtained when the model is represented respectively in terms of X and X', then we expect to find that $\hat{X}' = s(\hat{X})$.

(7) An estimation procedure which cannot be implemented on available computing machine is of little use. An estimate which is readily computable is, from a practical point of view, more valuable than a statistically more efficient estimate which is computable only with an excessive amount of labor.

(8) Among the properties that we have defined, the most important in practice are small bias, small variance, robustness and computability.

(9) Useful linear estimates valid over a wide range of data values can be found only when the model itself is linear in the parameters.

Example 5.20 (comparison of different estimators): Let's consider the measurement series described by the model of linear regression:

$$y_i = \Theta + \xi_i \; ; i = 1,...,n$$

Fig. 5.8 presents the MATLAB program for generating the samples of such measurements. The actual value of the parameter Θ is defined by the user, while the observation noise realizations ξ_i can be chosen from one of three possible distribution laws. First of them is Gaussian and the second one is Laplacian. Both of them are with zero mean and the variance that can be chosen arbitrary. The third one is generated as a Gaussian mixture: $0.4N(0.6,1) + 0.6N(-0.4,1)$. The samples from Gaussian probability density function (pdf) with mean value a and variance b, say $N(\cdot|a,b)$, are generated using the MATLAB function 'randn', which gives a sample ξ is a sample from $N(\xi|0,1)$, the sample ξ^* from $N(\xi^*|a,b)$ can be generated as $\xi^* = a + \sqrt{b}\xi$. The samples from Laplacian pdf are obtained using the methodology exposed in Chapter 1. Namely, these samples can be generated by posing the sample r from the uniform pdf on the interval $(0,1)$, say $U(0,1)$, through the nonlinearity $g(\cdot)$ which is equal to the inverse of the required distribution $F_\xi(\cdot)$, that is $\xi = g(r) = F_\xi^{-1}(r)$. Since the Laplacian pdf is given by $f_\xi(z) = \lambda \exp\{-\lambda|z|\}/2$, the corresponding distribution

$$F_\xi(y) = \int_{-\infty}^y f_\xi(z)dz = \begin{cases} 0.5\exp\{\lambda y\} \; ; y \le 0 \\ 1 - 0.5\exp(-\lambda y) \; ; y > 0 \end{cases}$$

so that the equation

$$r = g^{-1}(\xi) = F_\xi(\xi)$$

has the solution

$$\xi = \begin{cases} \ln(2r)/\lambda \; ; r \le 0.5 \\ -\ln(2(1-r))/; r > 0.5 \end{cases}$$

Finally, the samples from the chosen Gaussian mixture pdf can be generated by first taking a uniform deviate $r \sim U(0,1)$. If $r > 0.4$, then ξ is generated by the pdf $N(\xi|0.6,1)$, otherwise, ξ is a sample from $N(\xi|-0.4,1)$.

In addition, the program offers the possibility of parameter estimation based on three different estimation procedures: minimizing the mean square error, minimizing the mean absolute error and maximum likelihood approach. At the end of the program, it is also possible to estimate the variance of estimation error, using by Monte Carlo simulation.

function ex520

```
clear all;
global YYY type sig2
randn('seed',0);rand('seed',0);
teta=input('type the real value of the
parameter:  ');
N=input('number of measurements:  ');
disp('the following pdfs of the measurement...
     noisese are available');
disp('case 1: Gaussian pdf with arbitrary
variance');
disp('case 2: Laplace pdf with arbitrary
variance');
disp('case 3: Gaussian mixture');
type=input('your choice is :  ');
if type<3;sig2=input('variance =');end
if type==2
  lam=sqrt(2/sig2);
end
for i=1:N
  if type==1
    YYY(i)=teta+sqrt(sig2)*randn;
  end
  if type==2
    r=rand;
    if r<0.5
      YYY(i)=teta+log(2*r)/lam;
    else
      YYY(i)=teta-log(2-2*r)/lam;
    end
  end
  if type==3
    if rand<0.4
      YYY(i)=teta+randn-0.6;
    else
      YYY(i)=teta+randn+0.4;
    end
  end
end
y=YYY;
if N<15
disp('*****************************
*******');
  disp('Obtained measurements are:');
  str1=num2str(y(1));for i=2:N; str1=[str1 ' '
num2str(y(i))];end;
  disp(str1);
disp('*****************************
*******');
end
```

```
disp('*****************************
*******');
disp('The following estimatation procedures
can be applied: ');
disp('method 1: mean square algorithm');
disp('method 2: mean absolute values
algorithm');
disp('method 3: maximum likelihood
approach');
method=input('your choice is: ');
disp('*****************************
*******');
if method==1; est=mean(y); end
if method==2; est=median(y);end
if method==3
    est=fmin('jpdf',-2*abs(teta),2*abs(teta));
end
disp('*****************************
*******');
disp('                              ');
disp([' ESTIMATED VALUE IS :         '
num2str(est)]);
disp('                              ');
disp('*****************************
*******');
reply=input('Would you like to estimate the
variance of estimation error:  ','s');

if (reply=='y')|(reply=='Y')
  M=10000;
  for j=1:M
    for i=1:N
      if type==1
        YYY(i)=teta+sqrt(sig2)*randn;
      end
      if type==2
        r=rand;
        if r<0.5
          YYY(i)=teta+log(2*r)/lam;
        else
          YYY(i)=teta-log(2-2*r)/lam;
        end
      end
      if type==3
        if rand<0.4
          YYY(i)=teta+randn-0.6;
        else
          YYY(i)=teta+randn+0.4;
        end
      end
    end
```

```
YYY–YYY(1:N);
y=YYY;
if method==1
  est=mean(y);
end
if method==2
  est=median(y);
end
if method==3
  est=fmin('jpdf',-
2*abs(teta),2*abs(teta));
end
err(j)=teta-est;
end
disp('*******************************
*******');
  disp('The variance of estimation error for ...
    proposed procedure');
  disp(['and given pdf is equal to: '...
    num2str(std(err)^2)]);
disp('*******************************
*******');
end
```

```
function j=jpdf(x);

global YYY type sig2;
y=YYY;
N=length(y);
if type==2;  lam=sqrt(2/sig2);end;
j=1;
for i=1:N
  if type==1
    j=j*exp(-0.5*(y(i)-
x)^2/sig2)/sqrt(2*pi*sig2);
  end
  if type==2
    j=j*exp(-lam*abs(y(i)-x))*lam/2;
  end
  if type==3
    j=j*(0.4*exp(-0.5*(y(i)-
x+0.6)^2)+0.6*exp(-0.5*(y(i)-x-
0.4)^2))/sqrt(2*pi);
  end
end
j=-j;
end
```

Fig. 5.8: MATLAB program code for ex. 5.20

If we decide to execute the program, choosing the true parameter value $\Theta = 5$ and number of measurements $n = 5$, with the pdf of the measurement noise ξ to be Gaussian with unit variance, the program will generate 5 measurements:

$$y_1 = 6.165 \; ; \; y_2 = 5.6268; y_3 = 5.0751 \; ; \; y_4 = 5.3516 \; ; \; y_5 = 4.3035$$

Minimizing the mean square error, the corresponding MS parameter estimate would be the arithmetic mean of the measurement $\hat{\Theta}_1 = 5.3044$, minimization of the mean absolute error would generate the median of the sample, as the estimate $\hat{\Theta}_2 = 5.3516$, while the third approach, maximum likelihood (ML), would calculate the parameter estimate $\hat{\Theta}_3 = 5.3044$ by maximizing the joint probability density of the measurements. It is interesting to note that the first and the third estimates are equal, but such result was expected since for the Gaussian measurement noise these two estimation procedures are equivalent. If we now execute the program with choosing Laplace pdf for the measurement noise, the program will generate the following measurements:

$$y_1 = 4.4161 \; ; \; y_2 = 3.3287 \; ; \; y_3 = 5.3131 \; ; \; y_4 = 5.314 \; ; \; y_5 = 6.4393$$

The corresponding estimates are $\hat{\Theta}_1 = 4.9623 \; ; \; \hat{\Theta}_2 = 5.3131 \; ; \; \hat{\Theta}_3 = 5.3131$. The similar conclusion can be derived now: the estimations based on minimal mean absolute value and the maximum likelihood approach are equivalent, because the actual measurement noise pdf is Laplacian.

It is interesting to see what's the matter with the estimation error variance if we change the probability density function of the measurement noise and estimation procedure. Table 5.1 shows the corresponding variances for 9 different cases, while Monte Carlo simulation has been done with 10000 repetitions and the set of 5 measurements for each of them.

$\sigma_{\hat{\Theta}}^2$	Minimum mean square error approach	Minimum mean absolute error approach	Maximum likelihood approach
Gaussian pdf	0.20088	0.2831	0.20088
Laplacean pdf	0.20223	0.17198	0.17198
Gaussian mixture	0.24477	0.35308	0.23542

Tab. 5.1: Estimated variances of estimation error for different measurement noise pdfs and different estimation approaches

It can be seen, again, that for Gaussian pdf minimum mean square error approach and maximum likelihood algorithm generate the estimation with the same variance (theoretical minimum obtained by Cramer-Rao inequality is equal to 0.2), while for the Laplacean pdf the minimum mean absolute error algorithm and the maximum likelihood approach generate the same estimates. For a pdf that is neither Gaussian nor Laplacian the best results can be obtained by using the maximum likelihood approach.

CHAPTER 6

OPTIMUM NONRECURSIVE LINEAR ESTIMATION:
WIENER FILTERING

This part of the book deals with extraction of signal from noisy measured data. We have seen in chapter 5 that the mean-square error is a useful criterion showing how good an estimation process is. Therefore, the mean-square error is taken as the fundamental criterion. The estimates that minimize the mean-square error are taken as the best, or optimum estimates, and are also referred to as the least mean-square (l.m.s.) estimates. The case we considered firstly is a scalar (one-dimensional) parameter with a random distribution of its values. The optimum nonrecursive estimator derived in section 6.1.1 is the scalar Wiener filter whose coefficients are solutions of the Wiener-Hopf equation. This is followed in section 6.1.2 by the extension of scalar results to vector case. Additionally, in section 6.1.3 the case of a stationary random time-varying signal is considered. It will be shown that the assumption of stationarity gives that the optimum Wiener filter is in fact time invariant linear dynamic system with the corresponding pulse transfer function. A case of nonstationary multivariate random time varying signal is discussed in the chapter 6.1.4. Finally, this is followed in section 6.2 by the extension of discrete-time results to the continuous-time stationary signals.

6. 1. Optimum nonrecursive estimator from discrete-time noisy measurement data

6.1.1. Optimum nonrecursive estimator for scalar random variable

We adopt the notation x for a random parameter, representing for example, an unknown amplitude of a random signal, and $x(k)$ for the time-varying signal. Measurement of this signal, denoted by $y(k)$, is linearly related to the signal x but has an additive noise component $v(k)$, introduced by random errors in measurements or any other causes. Therefore, we have [11, 37, 49]

$$y(k) = x + v(k) \tag{6.1}$$

Thus, the signal parameter considered here is a random variable with some mean m_x and variance σ_x^2. The noise samples are assumed to be of zero-mean $(m_v = 0)$ with identical variances σ_v^2, and also to be uncorrelated. This condition is not necessary, but it is introduced to simplify analysis.

It is assumed that n data samples, as specified by eq. (6.1), are to be processed using the nonrecursive filter structure shown in fig. 6.1., where the input sampled-data signals are $y(i)$; $i = 1,...,n$, and the output is taken as an estimate $\hat{x}$ of unknown parameter x in eq. (6.1).

Thus, in general, the nonrecursive filter processor with different weights can be written as

$$\hat{x} = \sum_{i=1}^{n} h(i) y(i) + g \tag{6.2}$$

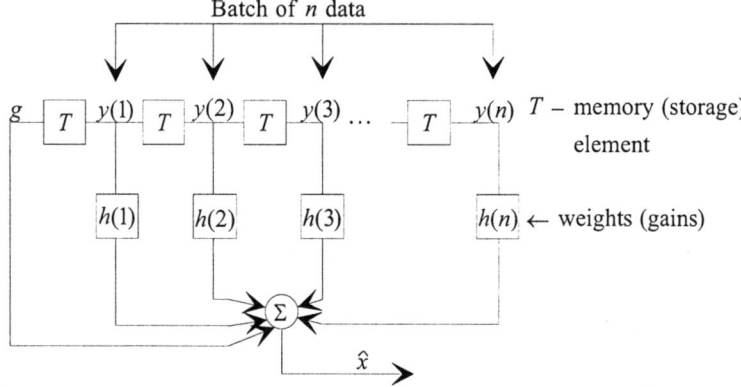

Fig. 6.1. Nonrecursive filter structure

This is a modified convolution equation for a finite length filter (the filter length is n) and observation time $k > n$ (see chapter 3). The time index k does not appear since the time of data recording is of no importance in the processing of a batch $\{y(1),...,y(n)\}$. Particularly, for $h(i) = 1/n$, $i = 1,...,n$, the nonrecursive processor of equation (6.2) is the familiar sample mean estimator (see chapter 5), which we often use as a first approximation to estimate a quantity x from n data. Now, we want to choose these coefficients $h(i)$, $i = 1,...,n$ in such a way that the mean square error

$$J_e = E\{e^2\} = E\{(x - \hat{x})^2\}$$ (6.3)

is minimized, where $E\{\cdot\}$ represents the expected (mean or average) operation. Note that x is desired signal and $\hat{x}$ is its estimate, in this case given by eq. (6.2). The solution to the posed problem is the least mean-squares or minimum variance estimator discussed in chapter 5. However, it is instructive to derive the general solution to this problem for random x, as well as to extend the obtained result to time-varying stochastic sequence $\{x(k)\}$. Substituting for $\hat{x}$, we have

$$J_e = E\left\{\left[x - \sum_{i=1}^{n} h(i) y(i) - g\right]^2\right\}$$ (6.4)

The least (or minimum) mean-square error is obtained by differentiating of the above equation with respect to g and each of the n parameters $h(i)$. This can be written in the following concise form

$$\frac{\partial J_e}{\partial g} = -2E\left\{\left[x - \sum_{i=1}^{n} h(i) y(i) - g\right]\right\} = 0$$ (6.5)

or

$$g = E\{x\} - \sum_{i=1}^{n} h(i) E\{y(i)\} = m_x - \sum_{i=1}^{n} h(i) m_y$$ (6.6)

Substituting eq. (6.6) into (6.4), we obtain

$$J_e = E\left\{\left[\tilde{x} - \sum_{i=1}^{n} h(i) \tilde{y}(i)\right]^2\right\}$$ (6.7)

where $\tilde{x} = x - m_x$ and $\tilde{y}(i) = y(i) - m_y$. Differentiating eq. (6.7) with respect to parameters $h(i); i = 1, \ldots, n$, we have

$$\frac{\partial J_e}{\partial h(i)} = -2E\left\{\left[\tilde{x} - \sum_{i=1}^{n} h(i)\tilde{y}(i)\right]\tilde{y}(i)\right\} = 0 \tag{6.8}$$

or

$$\sum_{i=1}^{n} h(i) E\{\tilde{y}(i)\tilde{y}(j)\} = E\{\tilde{x}\tilde{y}(j)\} \; ; \; j = 1, \ldots, n \tag{6.9}$$

It is worthwhile to note that from eq. (6.8) we can also write

$$E\{e\tilde{y}(j)\} = 0 \; ; \; j = 1, \ldots, n \tag{6.10}$$

where $e = x - \hat{x}$ is the estimation error. This is the well-known *orthogonality principle* in discussions on estimation (see chapter 5.). It means that the product of the estimation error, $e = x - \hat{x}$, with each of the centered measured sample, $\tilde{y}(j) = y(j) - m_y$, is equal to zero in an expectation (or average) sense.

Returning to eq. (6.9), we introduce

$$E\{\tilde{y}(i)\tilde{y}(j)\} = R_y(i,j) \tag{6.11}$$

which is, in fact, the data autocovariance function. It should be also noted that $R_y(i,j)$ refers to the non-stationary case, while for the stationary cases it would be $R_y(i-j)$. Similarly, we introduce

$$E\{\tilde{x}\tilde{y}(j)\} = R_{xy}(j) \tag{6.12}$$

which is the cross-covariance between the random variables x and $y(j)$. Using equations (6.11) and (6.12), we rewrite eq. (6.9) as

$$\sum_{i=1}^{n} h(i) R_y(i,j) = R_{xy}(j) \; ; \; j = 1, \ldots, n \tag{6.13}$$

Expanding over $i = 1, \ldots, n$ we have

$$R_y(1,j)h(1) + R_y(2,j)h(2) + \cdots + R_y(n,j)h(n) = R_{xy}(j)$$

and then expanding over $j = 1, \ldots, n$, we obtain

$$R_y(1,1)h(1) + R_y(2,1)h(2) + \cdots + R_y(n,1)h(n) = R_{xy}(1)$$
$$R_y(1,2)h(1) + R_y(2,2)h(2) + \cdots + R_y(n,2)h(n) = R_{xy}(2)$$
$$\vdots$$
$$R_y(1,n)h(1) + R_y(2,n)h(2) + \cdots + R_y(n,n)h(n) = R_{xy}(n)$$

$$\tag{6.14}$$

where $R_y(i,j) = R_y(j,i)$, since function R_y is symmetrical. The known quantities are $R_y(i,j)$; that is the autocovariance coefficients of the input data, and $R_{xy}(j)$, representing the cross-covariance coefficients between the desired output x and the input data $y(j)$. The unknown quantities $h(i); i = 1, \ldots, n$, are the optimum filter coefficients.

The least mean-square error corresponding to the above optimum solution is obtained from

$$J_e = E\left\{e^2\right\} = E\left\{e\left[\tilde{x} - \sum_{i=1}^{n} h(i)\tilde{y}(i)\right]\right\} = E\left\{e\tilde{x}\right\}$$

on the basis of eq. (6.10). Therefore, we have

$$J_e = E\left\{\tilde{x}^2\right\} - \sum_{i=1}^{n} h(i)E\left\{\tilde{x}\tilde{y}(i)\right\} = \sigma_{\tilde{x}}^2 - \sum_{i=1}^{n} h(i)R_{xy}(i) \tag{6.15}$$

where we have used eq. (6.12) to establish the result (6.15).

The compete solution to the posed estimation problem (we wish to estimate the value of x as a linear combination (6.2) of the values of n other random variables $y(1),...,y(n)$ so that the mean-square errors (6.3) is minimized) is given by the set of eqs. (6.14), the estimate eq. (6.2), and the corresponding least mean-square error given by eq. (6.15). Matrix form of these equations are

$$R_y h = R_{yx} \tag{6.16}$$

where $R_y = E\left\{\tilde{y}\tilde{y}^T\right\}$ is the $n\times n$ autocovariance matrix, $\tilde{y}$ and h are $(n\times1)$ column vectors, and $R_{yx} = E\left\{\tilde{y}\tilde{x}\right\}$ is $(n\times1)$ cross-covariance column vector. The formal solution of eq. (6.16) is

$$h = R_y^{-1} R_{yx} \tag{6.17}$$

However, taking into account eq. (6.6), the estimate eq. (6.2) can be written as

$$\hat{x} = m_x + h^T \tilde{y} \tag{6.18}$$

where h and $\tilde{y}$ are $n\times1$ column vectors, and h^T is a row vector. Substituting eq. (6.17) into (6.18) we obtain for the estimate

$$\hat{x} = m_x + R_{yx}^T R_y^{-1} \tilde{y} = m_x + R_{xy} R_y^{-1}\left[y - m_y\right] \tag{6.19}$$

and similarly for the least mean-square error (6.15)

$$J_e = E\left\{x^2\right\} - h^T R_{yx} = E\left\{x^2\right\} - R_{yx}^T R_y^{-1} R_{yx}$$
$$= R_x - R_{xy} R_y^{-1} R_{yx} \tag{6.20}$$

since the matrix R_y is symmetrical and $R_{xy}^T = R_{yx}$. Here σ_x^2 is denoted as R_x. Equations with form similar to (6.17) also arise in control theory where the least-square criterion is   used for system identification.

A filter of the type described above is often called a scalar discrete Wiener filter, and eq. (6.13) is known as the scalar discrete Wiener-Hopf equation. The term scalar has been used here to distinguish this single signal case from the vector (multidimensional) case considered later (chapter 6.1.2), while the term discrete denotes that the estimate of the signal parameter x is based on the discrete-time random samples $y(i)$, $i = 1,...,n$.

It should be noted that the solution (6.19) only requires knowledge of first and second order statistics; that is the means and covariances, and that it coincides with the true mean-square estimate when the $\{x, y(1),...,y(n)\}$ are jointly Gaussian (see chapter 5). Finally, let us observe that the observation model $y(k) = x + v(k)$; eq. (6.1), has not been used in the above derivations. Therefore, the result is more general than it appears. It states that if the data samples $\{y(i); i = 1,...n\}$ "somehow" contain the unknown random variable x; that is the signal, the best linear-filter operation, in the mean-square sense, carried out on the samples in order to estimate x is given by the Wiener filter.

Example 6.1 (estimation of constant signal amplitude): Let measured data and signal be related linearly by eq. (6.1); that is $y(k) = x + v(k)$, where $v(k)$ is additive noise and x is a random signal amplitude in stationary situation. We assume, as before, that the noise samples are of zero mean, with variance σ_v^2, uncorrelated with each other and with the signal x. This means that we assume white noise, so

$$E\{v(j)v(k)\} = \sigma_v^2 \delta(j,k) \; ; \; E\{xv(j)\} = 0$$

where $\delta(j,k)$ is the Kronecker delta, i.e.

$$\delta(j,k) = \begin{cases} 0 \; ; \; j \neq k \\ 1 \; ; \; j = k \end{cases}$$

To solve the problem, we calculate first

$$\begin{aligned} R_{xy}(i,j) &= E\{y(i)y(j)\} = E\{[x+v(i)][x+v(j)]\} \\ &= \sigma_x^2 + \sigma_v^2 \delta(i,j) \end{aligned} \tag{I}$$

Next we have

$$R_{xj}(j) = E\{xy(j)\} = E\{x[x+v(j)]\} = E\{x^2\} = \sigma_x^2 \tag{II}$$

Substituting eqs. (I) and (II) into eqs. (6.14), we have

$$(\sigma_x^2 + \sigma_v^2)h(1) + \sigma_x^2 h(2) + \cdots + \sigma_x^2 h(n) = \sigma_x^2$$
$$\sigma_x^2 h(1) + (\sigma_x^2 + \sigma_v^2)h(2) + \cdots + \sigma_x^2 h(n) = \sigma_x^2$$
$$\vdots$$
$$\sigma_x^2 h(1) + \sigma_x^2 h(2) + \cdots + (\sigma_x^2 + \sigma_v^2)h(n) = \sigma_x^2$$

or

$$\sigma_v^2 h(1) + \sigma_x^2 \sum_{i=1}^{n} h(i) = \sigma_x^2$$
$$\sigma_v^2 h(2) + \sigma_x^2 \sum_{i=1}^{n} h(i) = \sigma_x^2 \tag{III}$$
$$\vdots$$
$$\sigma_v^2 h(n) + \sigma_x^2 \sum_{i=1}^{n} h(i) = \sigma_x^2$$

Summing both sides

$$(\sigma_v^2 + n\sigma_x^2) \sum_{i=1}^{n} h(i) = n\sigma_x^2$$

for which we have

$$\sum_{i=1}^{n} h(i) = \frac{n\sigma_x^2}{n\sigma_x^2 + \sigma_v^2}$$

Substituting this into each of eqs. (III), we obtain

$$h(1) = h(2) = \cdots = h(n) = \frac{\sigma_x^2}{n\sigma_x^2 + \sigma_v^2} \tag{IV}$$

Substituting the solution, eq. (IV) , in eq. (6.2) with $g = 0$ (due to eq. (6.6)), we have for the signal amplitude estimate

$$\hat{x} = \frac{1}{n+\gamma} \sum_{i=1}^{n} y(i) \; ; \; \gamma = \frac{\sigma_v^2}{\sigma_x^2} \tag{V}$$

The corresponding least mean-square error value, from eqs. (6.15), (II) and (IV), is given by

$$J_e = \frac{\sigma_v^2}{n+\gamma} \tag{VI}$$

Note that for large signal to noise ration $\left(\sigma_x^2 \gg \sigma_v^2\right)$ we have $\gamma \ll n$, and the least mean-square estimator reduces to the sample mean estimator discussed in chapter 5.

Example 6.2 (estimation of unknown amplitude of given signal waveform): Estimate the random amplitude x of a sinusoidal waveform of known frequency ω, in the presence of additive noise $v(t)$. The measurement is represented as

$$y(t) = x\cos(\omega t) + v(t) \tag{I}$$

where the additive noise $v(t)$ accounts for both receiver noise and inaccuracies of the instruments used. Let $y(t)$ be sampled at $\omega t = 0$ and $\omega t = \pi/4$, giving us measured data $y(1) = y(\omega t = 0)$ and $y(2) = y(\omega t = \pi/4)$. Assume that $m_x = E\{x\} = 0$, $E\{x^2\} = \sigma_x^2$, $m_v = E\{v(t)\} = 0$, $E\{v(1)v(2)\} = 0$, $E\{v^2(1)\} = E\{v^2(2)\} = \sigma_v^2$, and also that x is independent of $v(t)$.

The linear estimator for this case is given by eq. (6.2) with optimum weights expressed by eq. (6.14). It should be noted that $g = 0$ in eq. (6.2), due to the eq. (6.6), since $m_x = 0$ and $m_y = m_x \cos(\omega t) + m_v = 0$. We have two measurement samples $y(1) = x + v(1)$ and $y(2) = x/\sqrt{2} + v(2)$. Therefore, using the eq.(6.14), we have

$$\begin{aligned} h(1)R_y(1,1) + h(2)R_y(2,1) &= R_{xy}(1) \\ h(2)R_y(1,2) + h(2)R_y(2,2) &= R_{xy}(2) \end{aligned} \tag{II}$$

where $R_y(i,j)$ and $R_{xy}(j)$ are given by eqs. (6.11) and (6.12), respectively. In this way, since $m_y = 0$, we have

$$\begin{aligned} R_y(i,j) &= E\{y(i)y(j)\} \; ; \; i,j = 1,2 \\ R_{xy}(j) &= E\{xy(j)\} \; ; \; j = 1,2 \end{aligned} \tag{III}$$

or

$$\begin{aligned} R_y(i,j) &= E\{[x+v(i)][x+v(j)]\} \\ &= E\{x^2\} + E\{xv(i)\} + E\{xv(j)\} + E\{v(i)v(j)\} \\ &= \sigma_x^2 + \sigma_v^2\delta(i,j) \end{aligned} \tag{IV}$$

and

$$R_{xy}(j) = E\{x[x+v(j)]\} = E\{x^2\} + E\{xv(j)\} = \sigma_x^2 \qquad \text{(V)}$$

where $\delta(i,j)$ is the Kronecker's delta. Solving eqs. (II) for $h(1)$ and $h(2)$, and substituting the values for $R_y(i,j)$ and $R_{xy}(j)$ from eqs (IV) and (V), respectively, we have

$$h(1) = \frac{\sigma_x^2}{\frac{3}{2}\sigma_x^2 + \sigma_v^2} \quad ; \quad h(2) = \frac{h(1)}{\sqrt{2}} \qquad \text{(VI)}$$

Using these solutions in eq.(6.2), we obtain the estimate x as

$$\hat{x} = \frac{\sigma_x^2}{\frac{3}{2}\sigma_x^2 + \sigma_v^2} y(1) + \frac{1}{\sqrt{2}} \frac{\sigma_x^2}{\frac{3}{2}\sigma_x^2 + \sigma_v^2} y(2) \qquad \text{(VII)}$$

where $y(1)$ and $y(2)$ are the measurement samples.

Geometric interpretations and orthogonality conditions

The optimum Wiener solution (6.2), (6.14) can be also derived by using the geometric interpretations and the orthogonality principle, which are exposed in chapter 5. Let us suppose we have vectors in some abstract space and that we have a notion of inner product $\langle \cdot, \cdot \rangle$ and of length $\|\cdot\|^2 = \langle \cdot, \cdot \rangle$ in this space.

Now suppose we wish to find $\{h(1), h(2), ..., h(n)\}$ so that

$$\hat{X}_0 = \sum_{i=1}^{n} h(i) Y(i) \qquad (6.21)$$

forms a best approximation of a given vector X in the sense that

$$\|X - \hat{X}_0\|^2 = \text{minimum}$$

Extending our geometric intuition from 3-dimensional space (see chapter 5), we would say that the optimal $\hat{X}_0$ would be the vector determined by the projection X on the hyper-plane determined by the $\{Y_i\}$, as it is denoted in Fig. 6.2. In other words, we would say that the optimum $\{h(i)\}$ would be uniquely determined by the *orthogonality conditions*

$$X - \sum_{i=1}^{n} h(i) Y(i) \perp Y_j \quad ; \quad j=1,...,n \qquad (6.23)$$

where $\perp$ stands for "is orthogonal to".

These orthogonality conditions, which are also known as the *projection theorem*, give us the equations

$$\left\langle X - \sum_{i=1}^{n} h(i) Y(i), Y(j) \right\rangle = 0 \; ; \; j=1,...,n \qquad (6.24a)$$

or

$$\left\langle X, Y(j) \right\rangle = \sum_{i=1}^{n} h(i) \left\langle Y(i), Y(j) \right\rangle \; ; \; j = 1, \ldots, n \tag{6.24b}$$

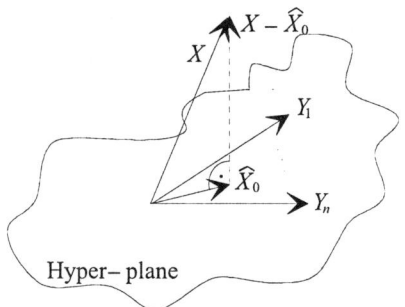

Fig. 6.2. Orthogonality principle or projection theorem

Expanding over $i = 1, 2, \ldots, n$ and then over $j = 1, 2, \ldots, n$, we obtain the system of equations

$$\left\langle Y(1), Y(1) \right\rangle h(1) + \left\langle Y(2), Y(1) \right\rangle h(2) + \cdots + \left\langle Y(n), Y(1) \right\rangle = \left\langle X, Y(1) \right\rangle$$

$$\left\langle Y(1), Y(2) \right\rangle h(1) + \left\langle Y(2), Y(2) \right\rangle h(2) + \cdots + \left\langle Y(n), Y(2) \right\rangle = \left\langle X, Y(2) \right\rangle \tag{6.25a}$$

$$\vdots$$

$$\left\langle Y(1), Y(n) \right\rangle h(1) + \left\langle Y(2), Y(n) \right\rangle h(2) + \cdots + \left\langle Y(n), Y(n) \right\rangle = \left\langle X, Y(n) \right\rangle$$

which can be expressed in the form of matrix equation (in an obvious notation)

$$\left[\left\langle Y(i), Y(j) \right\rangle \right]_{i,j=1}^{n} h = \left[\left\langle X, Y(i) \right\rangle \right]_{i=1}^{n} \tag{6.25b}$$

where $h = \left[h(1), \ldots, h(n) \right]^{T}$.

This seems very reasonable and reduces nicely to the 3-dimensional space. But one may be bothered by the fact that there are nonintuitive phenomena in multidimensional space. Therefore, how can one be sure that the geometrical arguments just used to find $\hat{X}_0$ are as valid in multidimensional space as they obviously are in 3-dimensional space, especially as we have abstract definitions of length and inner-product. The answer is that our geometrical arguments will be valid if the definition used for the inner product $\left\langle \cdot, \cdot \right\rangle$ satisfies three requirements: linearity, symmetry and nondegeneracy (see chapter 5.).

We shall show now how to apply this geometric picture to the statistical estimation problem concerned. It suffices to say that we can think of zero-mean random variables X, Y ... as vectors in some abstract space with an inner products defined as (see chapter 5)

$$\left\langle X, Y \right\rangle = E\{XY\} \tag{6.26}$$

It can readily be verified that this is a legitimate inner product, because it has the mentioned three properties. Therefore, to find $\{h(i)\}$ so that (see eq.(6.2) with $g = 0$)

$$E\left\{ \left[x - \sum_{i=1}^{n} h(i) y(i) \right]^2 \right\} = \left\| x - \sum_{i=1}^{n} h(i) y(i) \right\|^2 = \text{minimum} \tag{6.27}$$

it suffices to note that we must have

$$x - \sum_{i=1}^{n} h(i) y(i) \perp y(j) \; ; \; j = 1, ..., n \tag{6.28a}$$

or equivalently

$$\left\langle x - \sum_{i=1}^{n} h(i) y(i), y(j) \right\rangle = E\left\{ \left[x - \sum_{i=1}^{n} h(i) y(i) \right] y(j) \right\} = 0 \; ; \; j = 1, ..., n \tag{6.28b}$$

It will be seen that this, of course, leads to the equation (6.13), or after expanding over $i, j = 1, ..., n$ to the system of equations (6.14). The solution of eqs. (6.28b); that is of eqs. (6.14) is given by eq. (6.17), and the optimal estimate is defined by eq. (6.2) with $g = 0$, or equivalently by eq. (6.19) with $m_x = 0$ and $m_y = 0$. Moreover, we see that the above result will be valid even in the general nonzero means case, if we first assume that the random variables x and $y(i)$ are centered, namely x is replaced by $\tilde{x} = x - m_x$ and $y(i)$ by $\tilde{y}(i) = y(i) - m_y$. Therefore, we can assume that all random variables arising in the estimation problem have zero-means.

6.1.2 Optimum nonrecursive estimator for multivariate random variables

So far the most complicated problem we have discussed involved the estimation of a single random variable X based on observations $\{y(i); i = 1, ..., n\}$ of a single (scalar) random process $\{y(\cdot)\}$. It is easy to handle the case of a vector process $\{Y(\cdot)\}$. Clearly, we shall have for the scalar parameter estimate [11, 48, 49]

$$\hat{x} = \sum_{i=1}^{n} \sum_{\beta=1}^{r} h_\beta(i) y_\beta(i) = \sum_{i=1}^{n} h^T(i) Y(i) \tag{6.29}$$

where $Y(i) = [y_1(i) \cdots y_r(i)]^T$ is a $r \times 1$ random vector and $h^T(i) = [h_1(i) \cdots h_r(i)]^T$, with n being the number of multivariate measurements $Y(\cdot)$. The best approximation $\hat{x}$ of parameter x in the sense that

$$\|x - \hat{x}\|^2 = \left\| x - \sum_{i=1}^{n} h^T(i) Y(i) \right\|^2 = E\left\{ \left[x - \sum_{i=1}^{n} h^T(i) Y(i) \right]^2 \right\} = \text{minimum minimum}$$

is uniquely determined by the orthogonality conditions

$$\left\langle x - \sum_{i=1}^{n} h^T(i) Y(i), Y(j) \right\rangle = 0 \; ; \; j = 1, ..., n \tag{6.30a}$$

or

$$\sum_{i=1}^{n} h^T(i) \langle Y(i), Y(j) \rangle = \langle x, Y(j) \rangle \tag{6.30b}$$

where

$$\langle Y(i), Y(j) \rangle = E\{Y(i) Y^T(j)\} = R_Y(i, j) \tag{6.31}$$

$$\langle x, Y(j) \rangle = E\{x Y^T(j)\} = R_{xY}(j) \tag{6.32}$$

Using eqs. (6.30b), (6.31) and (6.32), and expanding over $i, j = 1, ..., n$ we have

$$h^T(1)R_Y(1,1)+h^T(2)R_Y(2,1)+\cdots+h^T(n)R_Y(n,1)=R_{xY}(1)$$
$$h^T(1)R_Y(1,2)+h^T(2)R_Y(2,2)+\cdots+h^T(n)R_Y(n,2)=R_{xY}(2)$$
$$\vdots \qquad\qquad (6.33)$$
$$h^T(1)R_Y(1,n)+h^T(2)R_Y(2,n)+\cdots+h^T(n)R_Y(n,n)=R_{xY}(n)$$

The known quantities are $n\times n$ autocovariance matrices $R_Y(\cdot,\cdot)$ and $1\times r$ cross-covariance vectors $R_{xY}(\cdot)$, while the unknown quantities are the coefficients of the $1\times r$ vectors $h^T(\cdot)$. After transposing the eqs. (6.33), we can write in a matrix form

$$R_{Yx}=R_{YY}H \qquad\qquad (6.34)$$

where R_{Yx} is $nr\times 1$ vector, R_{YY} is $nr\times nr$ matrix and H is $nr\times 1$ vector. The solution of the matrix eq. (6.34) is

$$H=R_{YY}^{-1}R_{Yx} \qquad\qquad (6.35)$$

The optimum parameter estimate is given by eq. (6.29); that is

$$\hat{x}=H^T Y \qquad\qquad (6.36)$$

where $Y=\begin{bmatrix} Y^T(1) & Y^T(n)\end{bmatrix}^T$ is $nr\times 1$ vector. Finally, we obtain for the least mean-square error

$$J_e=E\{e^2\}=E\{e(x-\hat{x})\}=E\{ex\}$$

since $E\{e\hat{x}\}=0$ due to the projection theorem. In this way, we have

$$J_e=E\{(x-\hat{x})x\}=E\left\{\left[x-\sum_{i=1}^n h^T(i)Y(i)\right]x\right\}$$
$$=E\{x^2\}-\sum_{i=1}^n h^T(i)E\{xY(i)\}=\sigma_x^2-\sum_{i=1}^n h^T(i)R_{Yx}(i)=\sigma_x^2-H^T R_{Yx} \qquad (6.37)$$

The question of $q\times 1$ vector parameter X is just a bit more delicate. We can compute the least mean-square estimate $\hat{x}_\alpha$ of each component x_α of X based on $r\times 1$ dimensional observations of $Y(\cdot)$, using eq.(6.29) or (6.36); that is

$$\hat{x}_\alpha=\sum_{i=1}^n\sum_{\beta=1}^r h_{\alpha,\beta}(i)y_\beta(i)=\sum_{i=1}^n h_\alpha^T(i)Y(i)=H_\alpha^T Y \qquad (6.38)$$

and then we can define

$$\hat{X}=\begin{bmatrix}\hat{x}_1\\ \vdots\\ \hat{x}_q\end{bmatrix}=\begin{bmatrix}H_1^T\\ \vdots\\ H_q^T\end{bmatrix}Y=HY \qquad\qquad (6.39a)$$

where H is the $q\times nr$ matrix and Y is the $nr\times 1$ vector. The eq. (6.39a) can be also rewritten as

$$\hat{X}=HY=\sum_{i=1}^n H(i)Y(i) \qquad\qquad (6.39b)$$

where $H(i)$ is $q\times r$ matrix. The covariance matrix of the estimation error is given by

$$P_e=E\{EE^T\}=E\{(X-\hat{X})E^T\}=E\{XE^T\}$$

since $E\{X\mathrm{E}^T\} = 0$ due to the orthogonality principle. In this way, one obtains

$$P_e = E\left\{X\left(X - \hat{X}\right)^T\right\} = E\{XX^T\} - E\{X\hat{X}^T\}$$

$$= E\{XX^T\} - E\left\{X\left[\sum_{i=1}^{n} H(i)Y(i)\right]^T\right\} \qquad (6.40)$$

$$= E\{XX^T\} - \sum_{i=1}^{n} E\{XY^T(i)\} H^T(i) = P_X - \sum_{i=1}^{n} R_{XY}(i) H^T(i)$$

We could think of obtaining the $q \times nr$ matrix H in (6.39a) by using the eq. (6.35) for determining its rows H_α^T ; $\alpha = 1,...,q$. The same result can be obtained by using the orthogonality conditions

$$\left\langle X - \sum_{i=1}^{n} H(i)Y(i), Y(j) \right\rangle = 0 \; ; \; j = 1,...,n \qquad (6.41)$$

with

$$\langle X, Y \rangle = E\{XY^T\} \qquad (6.42)$$

It is easy to check that this leads to the correct equation but it must be noted that $\langle X, Y \rangle$ as defined in (6.42) is not an inner product, because it does not satisfy the symmetry property; that is

$$E\{XY^T\} \neq E\{YX^T\}$$

However, if we ignore this deficiency, the projection theorem, i.e. the orthogonality conditions, can be used as a quick mnemonic way of obtaining the appropriate optimality conditions, and this is often done. Of course, there is a rigorous theory in which inner products can be matrices, but this is not within the scope of this book.

6.1.3 Optimum nonrecursive estimator for time-varying discrete-time signals

In the previous sections we developed Wiener filter for a constant signal x. Namely, in eq. (6.1) signal x is a constant, and we operate on n data $y(i)$, $i = 1,...,n$ to obtain the estimate $\hat{x}$ using eq. (6.2). We want now to extend this result to a time varying signal denoted as $x(k)$. This means that now data samples $y(k)$ will be time-varying not only due to the additive noise $v(k)$ in eq. (6.1), but due to $x(k)$ time changes. Thus suppose that, instead of estimating just a single random variable x, we wish to estimate the values of a stochastic process $\{d(k) = x(k + \lambda)\}$ given $\{y(i), -\infty < i \leq k\}$, where λ is a fixed integer constant. Now as k increases, λ can be specified in different ways: if $\lambda > 0$ we say that we are finding λ-step predicted estimate $\hat{d}(k) = \hat{x}(k + \lambda | y(i), -\infty < i \leq k)$; if $\lambda = 0$ then $\hat{d}(k) = \hat{x}(k | y(i), -\infty < i \leq k)$ is known as filtered estimate; if $\lambda < 0$ we say that $\hat{d}(k) = \hat{x}(k + \lambda | y(i), -\infty < i \leq k)$ is fixed-lag smoothed estimated with lag λ. It seems logical then to assume for eq. (6.2) to hold in this case, but the filter coefficients $h(i)$ should become variable with k. Introducing k via $x(k)$ on the right-hand-side of eq. (6.2) and via $h(k,i)$ on the left-hand-side, we have the Wiener estimator for time-varying discrete-time signal [11, 25, 37, 48]

$$\hat{d}(k) = \sum_{i=-\infty}^{k} h(k,i) y(i) \tag{6.43}$$

The function $h(k,i)$ can be regarded as the impulse response of a causal (i.e. $h(k,i)=0$ for $k<i$) linear system whose response at time k to the input signal $\{y(i), -\infty < i < k\}$ is the desired estimate $\hat{d}(k) = \hat{x}(k + \lambda | y(i), -\infty < i < k)$, as it is denoted in Fig. 6.3.

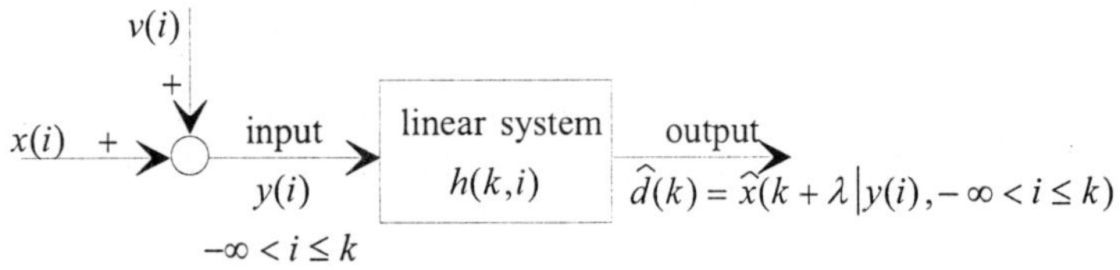

Fig. 6.3. Wiener estimator for time-varying signal

To determine the optimum system; that is the optimum coefficients $h(k,i)$, we use the orthogonality property (projection theorem for linear mean-square estimates)

$$d(k) - \hat{d}(k) \perp y(j) ; -\infty < j \le k \tag{6.44}$$

which yields that the estimation error $\{d(k) - \hat{d}(k)\}$ is normal to each of the measurements $y(j), -\infty < j \le k$; that is the inner product

$$\langle d(k) - \hat{d}(k), y(j) \rangle = 0 ; -\infty < j \le i \tag{6.45}$$

or

$$\langle d(k), y(j) \rangle = \langle \hat{d}(k), y(j) \rangle = \sum_{i=-\infty}^{k} h(k,i) \langle y(i), y(j) \rangle ; -\infty < j \le k \tag{6.46}$$

where

$$\langle d(k), y(j) \rangle = E\{d(k) y(j)\} = R_{dy}(k,j) \tag{6.47}$$

$$\langle y(i), y(j) \rangle = E\{y(i) y(j)\} = R_{y}(i,j) \tag{6.48}$$

with R_{dy} being the cross-covariance function of $d(\cdot)$ and $y(\cdot)$, and R_{y} the auto-covariance function of $y(\cdot)$, respectively. In this way, eq. (6.46) can be rewritten as

$$R_{dy}(k,j) = \sum_{i=-\infty}^{k} h(k,i) R_{y}(i,j) \tag{6.49}$$

Additionally, the assumption of jointly stationary $x(\cdot)$ and $y(\cdot)$ further yields

$$R_{dy}(k-j) = \sum_{i=-\infty}^{k} h(k,i) R_{y}(i-j) = \sum_{i=0}^{\infty} h(k,k-i) R_{y}(k-i-j) \tag{6.50}$$

If we change the variable k-j to k, the eq. (6.50) becomes

$$R_{dy}(k) = \sum_{i=0}^{\infty} h(k+j,k+j-i) R_{y}(k-i) ; k > 0 \tag{6.51}$$

The left hand side of eq. (6.51) does not depend upon the value of j, not does $R_y(k-i)$. Therefore $h(k+j, k+j-i)$ must not depend on j; that is, must be a function only of the difference of its arguments, which by an abuse of notation we shall still write as $h(\cdot)$,

$$h(k+j, k+j-i) = h[(k+j)-(k+j-i)] = h(i) \tag{6.52}$$

Therefore, the assumption of jointly stationary $x(\cdot)$ and $y(\cdot)$ gives that the optimum system is in fact time-invariant, i.e. $h(k,i)$ is of the form $h(k-i)$. The equation for $h(\cdot)$ is

$$R_{dy}(k) = \sum_{i=0}^{\infty} h(i) R_y(k-i) \; ; \; k > 0 \tag{6.53a}$$

Since

$$h(i) = 0 \quad \text{for} \quad i < 0$$

we can also write

$$R_{dy}(k) = \sum_{i=-\infty}^{\infty} h(i) R_y(k-i) \; ; \; k > 0 \tag{6.53b}$$

In this form it might appear that eq. (6.53) can be readily solved by taking Z-transform, but the presence of the constraint $k > 0$ prevents this. The easiest way of seeing why taking Z-transform would not work is to try it. Thus if

$$H(z) = Z\{h(i)\} = \sum_{i=-\infty}^{\infty} h(i) z^{-i}$$

$$S_y(z) = Z\{R_y(i)\} = \sum_{i=-\infty}^{\infty} R_y(i) z^{-i}$$

then we can write

$$\sum_{k=0}^{\infty} R_{dy}(k) z^{-k} = \sum_{i=-\infty}^{\infty} h(i) z^{-i} \sum_{k=0}^{\infty} R_y(k-i) z^{-(k-i)}$$

$$= \sum_{i=-\infty}^{\infty} h(i) z^{-i} \sum_{j=-i}^{\infty} R_y(j) z^{-j}$$

The presence of $(-i)$ in the summation over j prevents any easy solution. The $(-i)$ appears because the range of summation is restricted by the constraint $k > 0$; if there were no constraints, these would be no difficulty. There will also be no difficulty if $\{y(i)\}$ is white stochastic sequence, or if we restrict our search to the classical finite impulse response (FIR) solution for the Wiener estimation problem (6.43). The last case is known in signal-processing literature as the *finite impulse response* (FIR) or *all-zero* Wiener estimator. Let us consider firstly the FIR Wiener estimator.

6.1.3.1 Finite impulse response Wiener estimator (all-zero Wiener estimator or FIR Wiener filter)

In this section we discuss the classical finite impulse response (FIR) solution to the Wiener estimation problem (6.43). As before, we require the estimator that minimizes the error variance

$$J(k) = E\left\{ \left[\hat{d}(k) - d(k) \right]^2 \right\}$$

We restrict our search for the optimum causal linear system characterized by the impulse response $h(i)$ ($h(i)=0$ for $i<0$) such that the linear estimator is given by the finite impulse response (FIR), or all-zero time invariant system ($h(i)=0$ for $i>n$) [41, 43, 45].

$$\hat{d}(k) = \sum_{i=-\infty}^{k} h(k-i)y(i) = \sum_{i=0}^{\infty} h(i) y(k-i) = \sum_{i=0}^{n} h(i) y(k-i) \qquad (6.54)$$

The eq. (6.54) follows from the general eq. (6.43) where $h(k,i) = h(k-i)$, and taking into account that the impulse response $h(i)$ is finite; that is $h(i)=0$ for $i>n$. Therefore, the discrete Wiener-Hopf equation (6.53) becomes

$$R_{dy}(k) = \sum_{i=0}^{n} h(i) R_y(k-i) \qquad (6.55)$$

Expanding it over $k = 0,1,...,n$ we obtain the vector-matrix equation

$$R_{dy}(n) = \begin{bmatrix} R_{dy}(0) \\ R_{dy}(1) \\ \vdots \\ R_{dy}(n) \end{bmatrix} = \begin{bmatrix} R_y(0) & \cdots & R_y(-n) \\ R_y(1) & \cdots & R_y(1-n) \\ \vdots & \cdots & \vdots \\ R_y(n) & \cdots & R_y(0) \end{bmatrix} \begin{bmatrix} h(0) \\ h(1) \\ \vdots \\ h(n) \end{bmatrix} = R_y(n) H(n) \qquad (6.56)$$

where $R_{dy}(n), R_y(n)$ and $H(n)$ are $(n+1)$ dimensional column vector, $(n+1) \times (n+1)$ symmetric matrix and $(n+1)$ dimensional column vector, respectively. Thus, the optimum FIR estimator (6.54) is defined by the optimal impulse response samples vector

$$H(n) = R_y^{-1}(n) R_{dy}(n) \qquad (6.57)$$

Taking into account eqs. (6.54) and (6.57), one obtains for the optimal FIR estimates

$$\hat{d}(k) = H^T(n) Y(k) = R_{dy}^T(n) R_y^{-1}(n) Y(k) \; ; k = 0,1,...,n \qquad (6.58)$$

where $(n+1)$-column vector $Y(k) = \left[y(k) \; y(k-1) \cdots y(k-n) \right]^T$.

The associated minimum error variances can be calculated as before in eq. (6.15) or (6.37) by substituting x with $d(k)$, $\hat{x}$ with $\hat{d}(k)$ and $y(i)$ with $y(k-i)$, that is

$$J_e(k) = \sigma_d^2 - \sum_{i=0}^{n} h(i) E\{d(k) y(k-i)\}$$

$$= \sigma_d^2(k) - \sum_{i=0}^{n} h(i) R_{dy}(i) \qquad (6.59a)$$

or in vector-matrix notation

$$J_e(k) = R_d(0) - H^T(n) R_{dy}(n)$$

$$= R_d(0) - R_{dy}^T(n) R_y^{-1}(n) R_{dy}(n) \qquad (6.59b)$$

$$= R_d(0) - H^T(n) R_y(n) H(n)$$

where $R_d(0) = \text{var}\{d(k)\} = \sigma_d^2(k)$. We note that since $R_y(n)$ is Toeplitz matrix, then efficient recursive algorithms exist to perform the required inversion. Moreover, the FIR Wiener solution is useful when signal characteristics can be measured directly in order to obtain the cross-correlation matrix R_{dy}. However, usually we can only obtain a noisy measurement of $x(k)$; that is

$$y(k) = x(k) + v(k) \; ; \; k = 0, 1, ..., n$$

and, as before, $d(k)$ is either the signal $x(k)$ ($\lambda = 0$) or its shifted version; that is $d(k) = x(k + \lambda)$, where λ is a fixed integer constant.

Consider the following examples of an all-zero or FIR Wiener estimator.

Example 6.3 (FIR or all-zero Wiener estimators) : Suppose we are to obtain the Wiener FIR estimate of a stationary random signal $\{x(k)\}$ with covariance function $R_x(\tau) = 4 \cdot 0.5^{|\tau|}$ from noisy measurement data $y(k) = x(k) + v(k)$ in additive zero-mean white noise $v(k)$ with variance $\sigma_v^2 = 2$. Let us further suppose that the FIR system is of the order two, and that the output of the FIR system has to approximate the following signals: a) $d(k) = x(k)$-optimal Wiener filtering; b) $d(k) = x(k+1)$- optimal Wiener one-step prediction; c) $d(k) = x(k-1)$ - optimal Wiener one-lag smoothed estimate.

The covariance function of the measurement signal is given by

$$R_y(\tau) = E\{y(k+\tau)y(k)\} = E\{[x(k+\tau)+v(k+\tau)][x(k)+v(k)]\}$$
$$= E\{x(k+\tau)x(k)\} + E\{x(k+\tau)v(k)\} + E\{v(k+\tau)x(k)\} + E\{v(k+\tau)v(k)\}$$
$$= R_x(\tau) + R_{xv}(\tau) + R_{vx}(\tau) + R_v(\tau) = R_x(\tau) + R_v(\tau)$$

Here is used the fact that $\{v(k)\}$ is white noise uncorrelated with the signal $\{x(k)\}$, so that the cross-covariance functions R_{xv} and R_{vx} are equal to zero. Moreover, $R_v(\tau) = \sigma_v^2 \delta(\tau)$ where $\delta(\tau)$ is the Kronecker's delta symbol, i.e. $\delta(\tau) = 1$ for $\tau = 0$ and $\delta(\tau) = 0$ for $\tau \neq 0$. Therefore,

$$R_y(\tau) = 4 \cdot 0.5^{|\tau|} + 2\delta(\tau)$$

Furthermore, since the optimum system is of the order 2, this means that $n = 2$ in eq. (6.55) and the optimum estimator is defined by the vector-matrix form of the Wiener-Hopf equation (6.56)

$$\begin{bmatrix} R_{dy}(0) \\ R_{dy}(1) \\ R_{dy}(2) \end{bmatrix} = \begin{bmatrix} R_y(0) & R_y(-1) & R_y(-2) \\ R_y(1) & R_y(0) & R_y(-1) \\ R_y(2) & R_y(1) & R_y(0) \end{bmatrix} \begin{bmatrix} h(0) \\ h(1) \\ h(2) \end{bmatrix}$$

which can be rewritten as the system of linear equations

$$6h(0) + 2h(1) + h(2) = R_{dy}(0)$$
$$2h(0) + 6h(1) + 2h(2) = R_{dy}(1)$$
$$h(0) + 2h(1) + 6h(2) = R_{dy}(2)$$

Here is used the fact that $h(0) = 6$, $R_y(-1) = R_y(1) = 2$ and $R_y(-2) = R_y(2) = 1$.

a) In the case of optimal filtering, we have

$$R_{dy}(\tau) = E\{d(k+\tau)y(k)\} = E\{x(k+\tau)[x(k)+v(k)]\}$$
$$= E\{x(k+\tau)x(k)\} + E\{x(k=\tau)v(k)\} = R_x(\tau)$$

so that $R_{dy}(0) = R_x(0) = 4$, $R_{dy}(1) = R_x(1) = 2$ and $R_{dy}(2) = R_x(2) = 1$. In this way, the solution of the above system of linear equations is given by $h(0) = 0.6235$, $h(1) = 0.1176$, $h(2) = 0.0235$.

b) In the case of one-step optimal prediction $(\lambda = 1)$, the cross-covariance function R_{dy} is given by

$$R_{dy}(\tau) = E\{d(k+\tau)y(k)\} = E\{x(k+1+\tau)[x(k)+v(k)]\}$$
$$= E\{x(k+1+\tau)x(k)\} + E\{x(k+1+\tau)v(k)\}$$
$$= E\{x(k+1+\tau)x(k)\} = R_x(\tau+1)$$

so that, we have

$$R_{dy}(0) = R_x(1) = 2 \ ; \ R_{dy}(1) = R_x(2) = 1 \ ; \ R_{dy}(2) = R_x(3) = 0.5$$

The corresponding solution of the above system of linear equations is given by

$$h(0) = 0.3118 \ ; \ h(1) = 0.0588 \ ; \ h(2) = 0.0188$$

c) For the fixed-lag smoothed estimator $(\lambda = -1)$, we have

$$R_{dy}(\tau) = E\{d(k+\tau)y(k)\} = E\{x(k-1+\tau)[x(k)+v(k)]\}$$
$$= E\{x(k-1+\tau)x(k)\} = R_x(\tau-1)$$

Therefore,

$$R_{dy}(0) = R_x(-1) = 2 \ ; \ R_{dy}(1) = R_x(0) = 4 \ ; \ R_{dy}(2) = R_x(1) = 2$$

so that the solution of the system of linear equations is

$$h(0) = 0.1176 \ , \ h(1) = 0.5882 \ , \ h(2) = 0.1176$$

The pulse transfer function of the FIR Wiener estimate is given by

$$H(z) = \sum_{i=0}^{2} h(i) z^{-i}$$

6.1.3.2 Infinite impulse response Wiener estimator (Pole-zero or IIR Wiener estimator)

As mentioned earlier, a more sophisticated technique than Fourier or bilateral Z-transform is needed to solve the Wiener-Hopf equation (6.53), rewritten now as

$$R_{xy}(k+\lambda) = \sum_{i=0}^{\infty} h(i) R_y(k-i) \ ; \ k > 0 \tag{6.60}$$

The equation (6.60) is obtained from (6.53) by using

$$R_{dy}(k) = E\{d(t+k)y(t)\} = E\{x(t+\lambda+k)y(t)\}$$
$$= R_{xy}\{(t+\lambda+k)-t\} = R_{xy}(\lambda+k) \tag{6.61}$$

Although there are several methods of solving eq. (6.60), we shall present here a somewhat special form of such a method as applied to stationary process with rational power spectral density $S(z)$.

Such process are also known as lumped process (a lumped process will be one that can be obtained by passing white noise through a lumped or finite-dimensional linear system). A simple property of the Z-transform, combined with the spectral factorization of the power spectral density $S_y(z)$, is the key to the proposed technique [41, 42, 43, 45].

Let

$$g(k) = R_{xy}(k+\lambda) - \sum_{i=0}^{\infty} h(i) R_y(k-i) \tag{6.62}$$

Then we have from eq. (6.60)

$$g(k) = \begin{cases} 0 & \text{for } k \geq 0 \\ \text{unknown} & \text{for } k < 0 \end{cases} \tag{6.63}$$

However, we know, from the property of Z-transform, that for such a function $G(z) = Z\{g(k)\}$ represents a function with no poles inside the unit circle within the z-plane. This can be seen from the closed complex integral for the inverse Z-transform

$$g(k) = Z^{-1}\{G(z)\} = \oint_C G(z) z^k \frac{dz}{z} \tag{6.64}$$

where the contour C is a circle in the region of convergence of $G(z)$. The above integral can be evaluated, under certain conditions that include the case of rational function $G(z)$, by Cauchy residue theorem (see appendix A)

$$\oint_C G(z) z^k \frac{dz}{z} = \sum_i \operatorname*{Res}_{z=p_i} \{G(z) z^{k-1}\} \tag{6.65}$$

where p_i denotes a pole within the contour C. Thus, if $G(z)$ converges inside the unit circle in the z-plane, then for $k \geq 0$ there are no poles of the integrand $z^{k-1} G(z)$ in eq. (64) within the contour C, so that $g(k) = 0$ for $k \geq 0$.

Furthermore, we observe from eq. (6.62) that

$$G(z) = z^{\lambda} S_{xy}(z) - H(z) S_y(z) \tag{6.66}$$

where $H(z) = Z\{h(i)\}$ and $S_{xy}(z), S_y(z)$ represent the cross-power spectral density and power spectral density, respectively.

Moreover, using the spectral factorization of $S_y(z)$, we have

$$S_y(z) = S_y^+(z) S_y^-(z) = S_y^+(z) S_y^+(z^{-1}) \tag{6.67}$$

where $S_y^+(z)$ is the causal part of $S_y(z)$ containing only poles that lie inside the unit circle. Additionally, if all zeros of $S_y^+(z)$ lies inside the unit circle, such a rational function $S_y^+(z)$ is known as *causal-minimum-phase transfer function* (the point is that if we used a zero outside the unit circle in place of a zero inside the unit circle, the magnitude of the transfer function for $z = \exp(j\omega)$; $\omega \geq 0$ would not be changed, but its phase angle would be increased). Therefore, $S_y^-(z)$ contains only poles and zeros that lie outside the unit circle.

We can now write

$$\frac{G(z)}{S_y^-(z)} = \frac{S_{xy}(z)z^\lambda}{S_y^-(z)} - H(z)S_y^+(z) \tag{6.68}$$

where $\dfrac{G(z)}{S_y^-(z)}$ has no poles inside the unit circle (because $G(z)$ has none, and neither has $S_y^-(z)$);

$\dfrac{S_{xy}(z)z^\lambda}{S_y^-(z)}$ has poles inside and outside the unit circle; $H(z)S_y^+(z)$ has no poles outside the unit

circle (because $h(k)$ is a causal impulse response $(h(k)=0$ for $k<0)$ and, owing to (6.64) and
(6.65), $H(z)=Z\{h(k)\}$ has no poles outside the unit circle, and neither has $S_y^+(z)$).

Since we restrict the optimum estimator to be causal, i.e. physically realizable (this means that the response of a system does not exist before the excitation), we must extract the causal parts from eq. (6.68). By applying the operator $\{\cdot\}_+$ to both sides of this equation, we have

$$\left\{\frac{G(z)}{S_y^-(z)}\right\}^+ = \left\{\frac{S_{xy}(z)z^\lambda}{S_y^-(z)}\right\}^+ - H(z)S_y^+(z) = 0$$

since $G(z)/S_y^-(z)$ has no poles inside the unit circle, that is $\{G(z)/S_y^-(z)\}^+ = 0$, from which one concludes

$$H(z) = \frac{1}{S_y^+(z)}\left\{\frac{S_{xy}(z)z^\lambda}{S_y^-(z)}\right\}^+ \tag{6.69}$$

Thus, the causal IIR Wiener estimator is given by eq. (6.69), where the $(+)$ notation signifies the causal part of $\{\cdot\}$. Furthermore, since the Wiener solution (6.69) can be rewritten as

$$H(z) = \left\{\frac{z^\lambda S_{xy}(z)}{S_y(z)}\right\}^+ \tag{6.70}$$

we observe that the Wiener solution can be obtained by taking the causal part of the noncausal transfer function which follows directly from the Wiener-Hopf equation (6.60), after applying the Z-transform. As mentioned earlier, the obtained noncausal transfer function is not the solution of the Wiener-Hopf equation (6.60), due to the constraint $k \geq 0$. Particularly, if we take $\lambda = 0$, the Wiener solution (6.69) or (6.70) is known as the infinite-impulse (IIR) Wiener filter. This means that the impulse response of the causal transfer function $H(z)$ given by eq. (6.69) or (6.70)

$$h(i) = Z^{-1}\{H(z)\} \; ; \; i \geq 0 \tag{6.71}$$

lasts infinite time, while the filtered or estimated signal is given by

$$\hat{x}(k) = \sum_{i=0}^{\infty} h(i)y(k-i) = h \otimes y \tag{6.72}$$

where the $\otimes$ notation signifies the convolution. The associated mean square error corresponding to the above optimal solution is given by eq. (6.15), which can be now rewritten as

$$P_e(k) = R_x(0) - \sum_{i=0}^{\infty} h(i)R_{xy}(i) \tag{6.73}$$

where $R_x(0) = E\{x^2(k)\}$ and $R_{xy}(i) = E\{x(k)y(k-i)\}$.

A slightly different approach for deriving the IIR Wiener estimator is given in the sequel.

6.1.3.3 The innovations approach

Alternatively, we can obtain the result (6.69) using the whitening filter or innovations approach. This method is based on two observations [1, 17, 23, 25, 30]:

1) The Wiener-Hopf equation (6.60) is trivial to solve if the observed process $y(\cdot)$ is white noise.

2) For stationary lumped processes it is easy to transform, without loss of information, an arbitrary observed process into a white noise process.

Let us consider these points separately.

1) Estimation based on white noise observation

Suppose that we are to estimate $x(k+\lambda)$, given observations of a white noise process $\{v(i); -\infty < i < k\}$ with

$$R_{v(\tau)}(\tau) = \delta(\tau) \; ; \; S_v(z)\big|_{z=\exp(j\omega)} = 1 \tag{6.74}$$

where $\delta(\tau)$ is the Kronecker's delta, i.e. $\delta(\tau) = 1$ for $\tau = 0$ and $\delta(\tau) = 0$ for $\tau \neq 0$.

The corresponding Wiener-Hopf equation for the optimum estimator is given by eq. (6.60) which can be now rewritten as

$$R_{xv}(k+\lambda) = \sum_{i=0}^{\infty} h(i) R_v(k-i) \; ; \; k \geq 0 \tag{6.75}$$

But since $v(\cdot)$ is white

$$\sum_{i=0}^{\infty} h(i) R_v(k-i) = \sum_{i=0}^{\infty} h(i) \delta(k-i) = h(k) \tag{6.76}$$

and we have, from eq. (6.75), that the optimum estimator for $x(k+\lambda)$ given $\{v(i); -\infty < i < k\}$ is

$$h(k) = R_{xv}(k+\lambda) \; ; \; k \geq 0 \tag{6.77}$$

Thus, the optimum causal estimator is immediately determined. Note that its Z-transfer function is

$$H(z) = Z\{h(k)\} = Z\{R_{xv}(k+\lambda)\} = \{z^\lambda S_{xv}(z)\}^+ \tag{6.78}$$

where $S_{xv}(z) = Z\{R_{xv}(k)\}$.

Conversion to white noise

A process $\{y(i)\}$ with power-spectral density $S_y(z)$ can be converted to a white noise process $\{v(i)\}$ by passing it through a linear system with pulse-transfer function $W(z)$ (Fig. 6.4) such that

$$W(z)W(z^{-1}) = \frac{1}{S_y(z)} \tag{6.79}$$

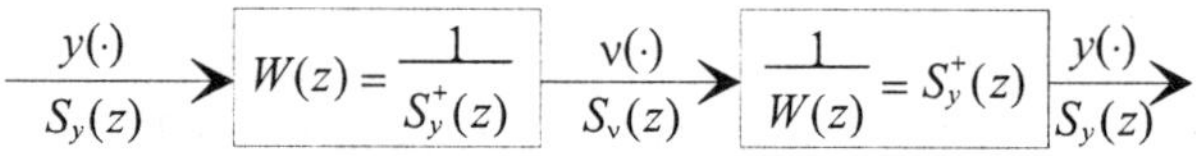

Fig. 6.4. Frequency domain whitening filter and its inverse

Really, by definition, the power spectral density of the output $\{v(i)\}$ will be

$$S_v(z) = E\{v(z)v(z^{-1})\}$$

or, since, from Fig. 6.4, $v(z) = W(z)Y(z)$, we have

$$S_v(z) = E\{W(z)W(z^{-1})Y(z)Y(z^{-1})\}$$
$$= W(z)W(z^{-1})E\{Y(z)Y(z^{-1})\}$$
$$= W(z)W(z^{-1})S_y(z) = S_y^{-1}(z)S_y(z) = 1$$

Here is used eq. (6.79) and the fact that, by definition, $S_y(z) = E\{Y(z)Y(z^{-1})\}$. Clearly, there are many filters $W(z)$ that will "whiten" a given process. However, one requirement on $W(z)$ in our problem is that it be causal, for otherwise an estimate of the white noise output up to time k would actually depend on observations of the input signal $\{y(i)\}$ for times i greater than k. The requirement of causality on $W(z)$ can be met by requiring that all poles of $W(z)$ be inside the unit circle. But this still does not specify $W(z)$ uniquely.

However, note that in addition to causality we should also require that the transformation to white noise be done without loss of information. There are various definitions of information, but the property of interest here is that all linear combinations of the input random variables up to time k be in a one-to-one relationship with some linear combinations of output random variables up to time k. This feature will be ensured if and only if, in addition to causality of $W(z)$, we also have causality of the inverse transfer function $1/W(z)$. This means that both the poles and zeros of $W(z)$ must be inside the unit circle, i.e. $W(z)$ has to be a minimum-phase transfer function.

Bearing in mind the spectral factorization theorem, we may now recall that this requirement can be met if and only if we choose

$$W(z) = 1/S_y^+(z) \tag{6.80}$$

where $S_y^+(z)$ is the causal minimum-phase, or canonical, spectral factor of $S_y(z)$ (see Fig. 6.4); that is

$$S_y(z) = S_y^+(z)S_y^-(z) \text{ and } S_y^-(z) = S_y^+(z^{-1}).$$

We shall denote the response to $\{y(i)\}$ of the canonical whitening linear system, or filter, (6.80) by $\{v(i)\}$, which will be called the innovations, or new information, process of $\{y(i)\}$. The name arises from the fact that the values of $v(\cdot)$ at different time instants are uncorrelated unlike the values of $y(\cdot)$. Therefore, the observation $y(k)$ of $\{y(\cdot)\}$ at instant k does not bring us completely new information, because some information of $y(k)$ can be obtained from other values of $y(\cdot)$; on the other hand, the value $v(k)$ cannot be predicted from other values of $\{v(\cdot)\}$; or more precisely the predicted value of $v(k)$ given, for example, $v(s)$ is zero because $E\{v(k)v(s)\}=0$. Therefore, each $v(k)$ brings us new information, and $\{v(\cdot)\}$ may be called the new information or innovations process of $\{y(\cdot)\}$.

We can now combine the previous results to obtain a solution of the original problem of determining $\hat{x}(k+\lambda)$ given $\{y(i)\,;\,-\infty<i\leq k\}$. The optimum estimator, or filter, can be repeated (see Fig. 6.5.) as a cascade of the canonical whitening filter (Fig. 6.4) and the optimum filter for white noise observations (6.78):

$$H(z)=\frac{1}{S_y^+(z)}\left\{S_{xv}(z)z^\lambda\right\}^+ \tag{6.81}$$

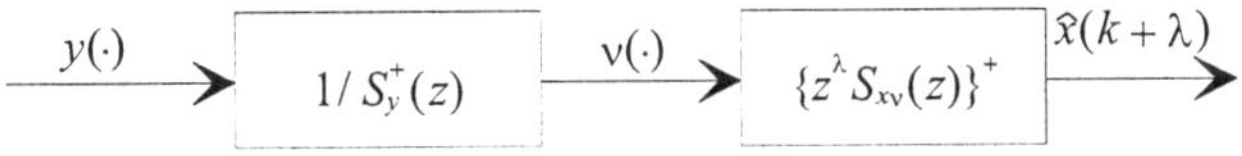

$$y(\cdot) \longrightarrow \boxed{1/S_y^+(z)} \xrightarrow{\ v(\cdot)\ } \boxed{\{z^\lambda S_{xv}(z)\}^+} \xrightarrow{\hat{x}(k+\lambda)}$$

Fig. 6.5: Optimum filter: cascade of canonical whitening filter and optimum filter for white noise

Now by linear-system relationships we know that, since $v(\cdot)$ is obtained by passing $y(\cdot)$ through the linear system, or filter, with the pulse-transfer function $1/S_y^+(z)$ (Fig. 6.5),

$$v(z)=\frac{1}{S_y^+(z)}Y(z)$$

while the cross-spectral power density is defined by

$$S_{xv}(z)=E\left\{X(z)v(z^{-1})\right\}$$

By combining the last two relations, we have

$$S_{xv}(z)=\frac{1}{S_y^+(z^{-1})}E\left\{X(z)Y(z^{-1})\right\}=\frac{1}{S_y^+(z^{-1})}S_{xy}(z)=\frac{1}{S_y^-(z)}S_{xy}(z) \tag{6.82}$$

Therefore,

$$H(z)=\frac{1}{S_y^+(z)}\left\{\frac{z^\lambda S_{xy}(z)}{S_y^-(z)}\right\}^+$$

that is the same result as the formula (6.69), which is obtained previously by a different method. The causal, or realizable, part operator $\{\cdot\}^+$ means that we must do a partial fraction expansion of the term in the square brackets of eq. (6.69) or (6.81) and retain the realizable portion, which has no poles outside the unit circle, i.e.

$$\left\{z^\lambda S_{xv}(z)\right\}=\left\{z^\lambda S_{xv}(z)\right\}^+ +\left\{z^\lambda S_{xv}(z)\right\}^- \tag{6.83}$$

where $S_{xv}(z)$ is given by (6.82). The second term on the right-hand-side of eq. (6.83) has poles outside the unit circle and corresponds to the noncausal (i.e. $i < 0$). portion, denoted as $\{\cdot\}^-$, of $z^\lambda S_{xv}(z)$. Any constant terms in $z^\lambda S_{xv}(z)$ are includes in the realizable portion, denoted as $\{\cdot\}^+$ (the first term on the right-hand-side of eq. (6.83)).

The IIR Wiener filter solution (6.69) can be generalized to multivariable processes. In fact, if the unknown $\{x(\cdot)\}$ and the observations $\{y(\cdot)\}$ are $n\times 1$ and $r\times 1$ discrete-time vector random processes, respectively, then eq. (6.79) can be rewritten as

$$S_y(z) = S_y^+(z)\left[S_y^-(z)\right]^T = S_y^+(z)\left[S_y^+(z^{-1})\right]^T$$

$$S_y^{-1}(z) = \left\{\left[S_y^+(z^{-1})\right]^T\right\}^{-1}\left\{S_y^+(z)\right\}^{-1} = W^T(z^{-1})W(z) \tag{6.84}$$

where $W(z) = \left\{S_y^+(z)\right\}^{-1}$. In this case, the causal part, denoted with the sign $(+)$, means to retain the portion with poles inside or on the unit circle. If $\det S_y(z) \neq 0$ on the unit circle $|z| = 1$, where 'det' denotes the determinant of the matrix, then $W(z)$ is a minimum phase and the realizable poles and zeros are strictly within the unit circle. Thus, for the multivariable case the transfer function matrix of the canonical whitening filter represents a multidimensional version of the scalar pulse-transfer function in eq. (6.80), i.e.

$$W(z) = \left[S_y^+(z)\right]^{-1} \tag{6.85}$$

In this way, the innovations vector

$$v(z) = W(z)Y(z)$$

so that the cross-spectral power density matrix is defined by

$$S_{xv}(z) = E\left\{X(z)v^T(z^{-1})\right\}$$

or, after replacing the expression for $v(z)$,

$$S_{xv}(z) = E\left\{X(z)Y^T(z^{-1})\right\}W^T(z^{-1})$$
$$= S_{xy}(z)W^T(z^{-1}) \tag{6.86}$$

It should be noted that the spectral power density matrix of the innovations vector v is given by

$$S_v(z) = E\left\{v(z)v^T(z^{-1})\right\} = W(z)E\left\{Y(z)Y^T(z^{-1})\right\}W^T(z^{-1})$$
$$= W(z)S_y(z)W^T(z^{-1}) = W(z)\left\{W^{-1}(z)\left[W^{-1}(z^{-1})\right]^T\right\}W^T(z^{-1})$$
$$= W(z)W^{-1}(z)\left[W^T(z^{-1})\right]^{-1}W^T(z^{-1}) = I$$

where I is an identity matrix. Finally, the pulse transfer function matrix $H(z)$ of the multivariable Wiener filter is the product of the pulse transfer function matrix of the whitening filter in eq. (6.85) and the pulse transfer function matrix of the optimum filter for the white innovations process in eq. (6.86); that is

$$H(z) = \left\{S_{xy}(z)W^T(z^{-1})\right\}^T W(z) \tag{6.87}$$

It should be noted that here $\lambda = 0$ (filtering problem). Before we close this chapter, consider the following example.

Example 6.4 (IIR or pole-zero Wiener filter) : Suppose we are to obtain the Wiener estimate of a random signal with covariance $R_x(\tau) = 0.5^{|\tau|}$ from noisy measurement data in zero-mean additive white noise $v(k)$ with unit variance, that is $R_v(\tau) = \delta(\tau)$, where $\delta(\tau) = 1$ for $\tau = 0$ and $\delta(\tau) = 0$ for $\tau \neq 0$. The noncausal Wiener solution is found as (see eq. (6.70) where $\lambda = 0$)

$$H(z) = \frac{S_{xy}(z)}{S_y(z)}$$

Therefore,

$$S_{xy}(z) = E\left\{X(z)Y(z^{-1})\right\} = E\left\{X(z)\left[X(z^{-1}) + V(z^{-1})\right]\right\}$$
$$= E\left\{X(z)X(z^{-1})\right\} + E\left\{X(z)V(z^{-1})\right\} = S_x(z) + S_{xv}(z) = S_x(z)$$

and

$$S_y(z) = E\left\{Y(z)Y(z^{-1})\right\} = E\left\{\left[X(z) + V(z)\right]\left[X(z^{-1}) + V(z^{-1})\right]\right\}$$
$$= E\left\{X(z)X(z^{-1})\right\} + E\left\{X(z)V(z^{-1})\right\} + E\left\{V(z)X(z^{-1})\right\} + E\left\{V(z)V(z^{-1})\right\}$$
$$= S_x(z) + S_{xv}(z) + S_{vx}(z) + S_v(z) = S_x(z) + S_v(z)$$

since $v(k)$ is white noise so that $S_{xv}(z) = S_{vx}(z) = 0$. The corresponding spectra are found from the sum decomposition; that is

$$S_x(z) = S_x^+(z) + S_x^-(z) - R_x(0)$$

where

$$S_x^+(z) = \sum_{\tau=0}^{\infty} R_x(\tau)z^{-\tau} = \sum_{\tau=0}^{\infty} 0.5^\tau z^{-\tau}$$
$$= \sum_{\tau=0}^{\infty}(0.5z^{-1})^\tau = \frac{1}{1-0.5z^{-1}}$$

and

$$S_x^-(z) = S_x^+(z^{-1}) = \frac{1}{1-0.5z}$$

In this way, one obtains

$$S_x(z) = \frac{1}{1-0.5z^{-1}} + \frac{1}{1-0.5z} - 1 = \frac{\frac{3}{4}}{\left(1-\frac{z^{-1}}{2}\right)\left(1-\frac{z}{2}\right)}$$

Thus, since

$$S_v(z) = Z\{\delta(\tau)\} = \sum_{\tau=-\infty}^{\infty} \delta(\tau)z^{-\tau} = \delta(0) = 1$$

the output spectrum is given by

$$S_y(z) = S_x(z) + S_v(z) = \frac{2 - \frac{1}{2}(z + z^{-1})}{\left(1 - \frac{z^{-1}}{2}\right)\left(1 - \frac{z}{2}\right)}$$

The causal Wiener solution is obtained immediately from eq. (6.70), where $\lambda = 0$, as

$$H(z) = \left\{\frac{S_{xv}(z)}{S_y(z)}\right\}^+ = \left\{\frac{\frac{3}{2}}{4 - (z + z^{-1})}\right\}^+$$

To find the causal part of the above pulse-transfer function denoted as $\{\cdot\}^+$, we must factor the denominator as

$$a_0 + a_1(z + z^{-1}) = (b_0 - b_1 z^{-1})(b_0 - b_1 z)$$

where $a_0 = 4$ and $a_1 = -1$. Equating the coefficients, we have

$$b_0^2 + b_1^2 = a_0 = 4 \; ; \; b_0 b_1 = -a_1 = 1$$

Solving for b_0 and substituting, we obtain

$$b_1^4 - 4b_1^2 + 1 = 0 \text{ or } b_1 = 0.52 \text{ and } b_0 = 1.93,$$

which gives the causal filter of eq. (6.70) as

$$H(z) = \left\{\frac{\frac{3}{2}}{(1.93 - 0.52z^{-1})(1.93 - 0.52z)}\right\}^+ = \frac{\frac{3}{2}}{1.93 - 0.52z^{-1}} = \frac{0.403}{1 - 0.268z^{-1}}$$

or

$$h(k) = Z^{-1}\{H(z)\} = 0.403(0.268)^k \; ; \; k \geq 0$$

The same result can be obtained by using whitening filter or innovations approach. Here we first perform a spectral factorization of $S_y(z)$ to obtain the whitening filter in eq. (6.80). To find the causal part $S_y^+(z)$ we must also factor the nominator of $S_y(z)$ as

$$\alpha_0 + \alpha_1(z + z^{-1}) = (\beta_0 - \beta_1 z^{-1})(\beta_0 - \beta_1 z)$$

Equating the coefficients, we have

$$\beta_0^2 + \beta_1^2 = \alpha_0 = 2 \; ; \; \beta_0 \beta_1 = -\alpha_1 = 0.5$$

Solving for β_0 and substituting, we obtain

$$4\beta_1^4 - 8\beta_1^2 + 1 = 0 \text{ or } \beta_1 = 0.366 \text{ and } \beta_0 = 1.37$$

which gives

$$S_y(z) = \frac{1.37(1 - 0.268z^{-1})}{1 - 0.5z^{-1}} \frac{1.37(1 - 0.268z)}{1 - 0.5z} = S_y^+(z)S_y^-(z)$$

Then we extract the causal part of $S_{xy}(z)/S_y^-(z)$; that is

$$\left\{ \frac{S_{xy}(z)}{S_y^-(z)} \right\}^+ = \left\{ \frac{\frac{3}{4}}{(1-0.5z^{-1})(1-0.5z)} \left[\frac{1.37(1-0.268z)}{1-0.5z} \right]^{-1} \right\}^+$$

$$= \left\{ \frac{\frac{3}{4}}{1.37(1-0.268z)(1-0.5z^{-1})} \right\}^+$$

$$= \left\{ \frac{-1.0275z}{(z-0.5)(z-1/0.268)} \right\}^+$$

Taking partial fraction expansion, further follows

$$\left\{ \frac{S_{xy}(z)}{S_y^-(z)} \right\}^+ = \left\{ \left[\frac{A}{z-0.5} + \frac{B}{z-1/0.268} \right] z \right\}^+$$

$$= \left\{ \frac{(A+B)z+(-A/0.268-0.5B)}{(z-0.5)(z-1/0.268)} z \right\}^+$$

Equating the coefficients, we have

$$A+B=0 ; -A/0.268-0.5B = -1.0275$$

from which one obtains

$$A = -B = 0.55$$

In this way, we conclude

$$\left\{ \frac{S_{xy}(z)}{S_y^-(z)} \right\}^+ = \left\{ \frac{0.55}{1-0.5z^{-1}} - \frac{0.55z}{z-1/0.268} \right\}^+ = \frac{0.55}{1-0.5z^{-1}}$$

The causal IIR Wiener filter of eq. (6.69) is given by $(\lambda=0)$

$$H(z) = \left\{ \frac{S_{xy}(z)}{S_y^-(z)} \right\}^+ \left[S_y^+(z) \right]^{-1} = \frac{0.55}{1-0.5z^{-1}} \left[\frac{1.37(1-0.268z^{-1})}{(1-0.5z^{-1})} \right]^{-1} = \frac{0.403}{1-0.268z^{-1}}$$

what is the same result as before.

Example 6.5: Let us design the Wiener filter for the case when signal and measurement noise are correlated. Let the measurement noise $n(k)$ be a white stochastic process with zero mean and unit variance, while the signal $x(k)$ is described by a difference equation:

$$x(k+2)-x(k+1)+0.24x(k)=n(k)$$

The measurement sequence $y(k)$ is defined in a standard form:

$$y(k)=x(k)+n(k)$$

In order to design the Wiener filter, it is necessary to compute the power spectral density $S_y(z)$ and cross-spectral density $S_{xy}(z)$, but before that , we need some intermediate results. Firstly, concerning that $n(k)$ is said to be a white stochastic process with unit variance, we can say that the corresponding power spectral density is:

$$S_n(z) = 1$$

On the other hand, using the spectral factorization theorem, we can calculate the power spectral density of the signal x as:

$$z^2 X(z) - z X(z) + 0.24 X(z) = N(z)$$

from which it follows

$$G(z) = \frac{X(z)}{N(z)} = \frac{1}{z^2 - z + 0.24}$$

and

$$S_x(z) = G(z) G(z^{-1}) S_n(z) = \frac{1}{(z^2 - z + 0.24)(z^{-2} - z^{-1} + 0.24)}$$

Also, it is necessary to calculate the cross-correlation function between the signals x and n as:

$$R_{xn}(k) = E\{x(i+k)n(i)\} = E\{x(i+k)[x(i+2) - x(i+1) + 0.24x(i)]\}$$
$$= R_x(k-2) - R_x(k-1) + 0.24 R_x(k)$$

and

$$R_{nx}(k) = E\{n(i+k)x(i)\} = E\{[x(i+k+2) - x(i+k+1) + 0.24x(i+k)]x(i)\}$$
$$= R_x(k+2) - R_x(k+1) + 0.24 R_x(k)$$

Finally, the cross-spectral power density $S_{xy}(z)$ and power spectral density $S_y(z)$ can be calculated as:

$$S_{xy}(z) = Z_{II}\{R_{xy}(k)\} = Z_{II}\{E\{x(i+k)y(i)\}\} = Z_{II}\{E\{x(i+k)(x(i) + n(i))\}\}$$
$$= Z_{II}\{R_x(k) + R_{xn}(k)\} = S_x(z) + S_{xn}(z) = S_x(z)[1 + z^{-2} - z^{-1} + 0.24] = (1.24 - z^{-1} + z^{-2}) S_x(z)$$

and

$$S_y(z) = Z_{II}\{R_y(k)\} = Z_{II}\{E\{y(i+k)y(i)\}\}$$
$$= Z_{II}\{E\{(x(i+k) + n(i+k))(x(i) + n(i))\}\}$$
$$= Z_{II}\{R_x(k) + R_{xn}(k) + R_{nx}(k) + R_n(k)\}$$
$$= S_x(z) + S_{xn}(z) + S_{nx}(z) + S_n(z) = S_x(1 + z^{-2} - z^{-1} + 0.24 + z^2 - z^1 + 0.24) + S_n(z)$$
$$= S_x(z)(1.48 + z^2 - z - z^{-1} + z^{-2}) + S_n(z)$$

where $Z_{II}\{\}$ denotes the bilateral (two-side) Z-transform. If we factorize the function $S_y(z)$:

$$S_y(z) = \frac{1.24\left(z^{-2} - 0.8065z^{-1} + 0.8065\right)\left(z^2 - 0.8065z + 0.8065\right)}{\left(z^{-2} - z^{-1} + 0.24\right)\left(z^2 - z + 0.24\right)}$$

it can be written very easily in the form

$$S_y(z) = A^+(z)A^-(z)$$

where

$$A^+(z) = \sqrt{1.24}\,\frac{\left(z^2 - 0.8065z + 0.8065\right)}{\left(z^2 - z + 0.24\right)}\;;\; A^-(z) = \sqrt{1.24}\,\frac{\left(z^{-2} - 0.8065z^{-1} + 0.8065\right)}{\left(z^{-2} - z^{-1} + 0.24\right)}$$

Now, the ratio $S_{xy}(z)/A^-(z)$ has to be written in the form of partial fractions:

$$\frac{S_{xy}(z)}{A^-(z)} = \frac{1.3807z^2 - 1.1135z + 1.1135}{\left(z^2 - z + 0.24\right)\left(z^2 - z + 1.24\right)} = \frac{0.2672z + 0.7821}{z^2 - z + 0.24} + \frac{-0.2672z + 0.5986}{z^2 - z + 1.24} = B^+(z) + B^-(z)$$

Finally, the transfer function of the Wiener filter becomes:

$$H(z) = \frac{B^+(z)}{A^+(z)} = \frac{0.24z + 0.7023}{z^2 - 0.8065z + 0.8065}.$$

6.1.4 Wiener estimator design for nonstationary time-varying discrete-time multivariate signals

We can obtain the solution to the discrete Wiener estimation problem using a slightly different approach. We start with the assumption that we operate on n data $y(i); i = 1,...,n$ to obtain the estimate $\hat{d}(k) = \hat{x}(k)$ of a time-varying discrete signal $x(k)$; $k=1,...,n$ using the Wiener estimator given by eq. (6.43), rewritten now in a smoothing form

$$\hat{x}(k) = \sum_{i=0}^{n} h(k,i)y(i)\;;\; k = 1,...,n \tag{6.88}$$

The optimal filter coefficients $h(\cdot,\cdot)$ are defined by the Wiener-Hopf equation given by eq. (6.13), which can be rewritten now as

$$\sum_{i=1}^{n} h(k,i)E\{y(i)y(j)\} = E\{x(k)y(j)\}\;;\; j,k = 1,...,n \tag{6.89}$$

To derive the vector equation we first consider the right-hand side. Expanding it over k and j, and neglecting the operator $E\{\cdot\}$, we obtain the following array

$$
\begin{array}{c}
\qquad\qquad \rightarrow j \\
\downarrow \atop k
\begin{bmatrix}
x(1)y(1) & \cdots & x(1)y(j) & \cdots & x(1)y(n) \\
\vdots & & & & \\
x(k)y(1) & \cdots & x(k)y(j) & \cdots & x(k)y(n) \\
\vdots & & & & \\
x(n)y(1) & \cdots & x(n)y(j) & \cdots & x(n)y(n)
\end{bmatrix}
\end{array}
$$

This can be expressed as vector product xy^T, where the vectors are defined as

$$x = \begin{bmatrix} x(1) \\ \vdots \\ x(k) \\ \vdots \\ x(n) \end{bmatrix} \; ; \; y^T = \left[y(1) \cdots y(j) \cdots y(n) \right]$$

(6.90)

where x is an $(n \times 1)$ column vector and y^T is a $(1 \times n)$ row vector. Therefore, the right-hand side of eq. (6.89) can be written as

$$E\{xy^T\}$$

(6.91)

which is the cross-covariance $n \times n$ matrix of the signal and data samples.

Consider now a (k,j) element on the left-hand side of eq. (6.89), which for fixed (k,j) can be written as

$$\sum_{i=1}^{n} h(k,i) E\{y(i) y(j)\} = h(k,1) E\{y(1) y(j)\} + \cdots + h(k,n) E\{y(n) y(j)\}$$

In vector notation we can express the above as $h^T(k) R_y(j)$, where

$$h^T(k) = \left[h(k,1) \; h(k,2) \cdots h(k,n) \right]$$

(6.92)

and

$$R_y(j) = \begin{bmatrix} R_y(1) \\ R_y(2) \\ \vdots \\ R_y(n) \end{bmatrix} = \begin{bmatrix} E\{y(1) y(j)\} \\ E\{y(2) y(j)\} \\ \vdots \\ E\{y(n) y(j)\} \end{bmatrix}$$

(6.93)

Expanding $h^T(k) R_y(j)$ over k and j we obtain the following array

$$\begin{bmatrix} h^T(1) R_y(1) & \cdots & h^T(1) R_y(j) & \cdots & h^T(1) R_y(n) \\ \vdots & & & & \\ h^T(k) R_y(1) & \cdots & h^T(k) R_y(j) & \cdots & h^T(k) R_y(n) \\ \vdots & & & & \\ h^T(n) R_y(1) & \cdots & h^T(n) R_y(j) & \cdots & h^T(n) R_y(n) \end{bmatrix}$$

This can be expressed as the following matrix product

$$HE\{yy^T\}$$

(6.94)

where

$$H = \begin{bmatrix} h^T(1) \\ h^T(2) \\ \vdots \\ h^T(n) \end{bmatrix} \; ; \; E\{yy^T\} = \left[R_y(1) \; R_y(2) \cdots R_y(n) \right]$$

(6.95)

with $h^T(k)$ and $R_y(j)$ being defined by eqs. (6.92) and (6.93), respectively. Therefore, the Wiener estimator for a time-varying signal, given by eq. (6.88), can be written in matrix form as

$$HE\{yy^T\} = E\{xy^T\} \tag{6.96}$$

where $E\{xy^T\}$, $E\{yy^T\}$ are $(n \times n)$ covariance matrices, and H is the $(n \times n)$ matrix of filter coefficients

$$H = \begin{bmatrix} h(1,1) & \cdots & h(1,j) & \cdots & h(1,n) \\ \vdots & & & & \\ h(k,1) & \cdots & h(k,j) & \cdots & h(k,n) \\ \vdots & & & & \\ h(n,1) & \cdots & h(n,j) & \cdots & h(n,n) \end{bmatrix} \tag{6.97}$$

In a multidimensional system we have, at time k, perhaps q signals $x_1(k), x_2(k), \cdots, x_q(k)$ and r observation values $y_1(k), y_2(k), \cdots, y_r(k)$. The estimator equation for the multidimensional case, analogous to eq. (6.88), is given by

$$\hat{x}(k) = \sum_{i=1}^{n} H(k,i)y(i) \; ; \; k = 1, ..., n \tag{6.98}$$

where

$$\hat{x}(k) = \begin{bmatrix} \hat{x}_1(k) \\ \hat{x}_2(k) \\ \vdots \\ \hat{x}_q(k) \end{bmatrix} \; ; \; y(i) = \begin{bmatrix} y_1(k) \\ y_2(k) \\ \vdots \\ y_r(k) \end{bmatrix} \tag{6.99}$$

and $H(k,i)$ is the following $q \times r$ matrix

$$H(k,i) = \begin{bmatrix} h_{1,1}(k,i) & \cdots & h_{1,r}(k,i) \\ h_{2,1}(k,i) & \cdots & h_{2,r}(k,i) \\ \vdots & & \\ h_{q,1}(k,i) & \cdots & h_{q,r}(k,i) \end{bmatrix} \tag{6.100}$$

The Wiener-Hopf equation, analogous to eq. (6.89), is given by

$$\sum_{i=1}^{n} H(k,i)R_y(i,j) = R_{xy}(k,j) \; ; \; j,k = 1, ..., n \tag{6.101}$$

where $R_y(i,j)$ is the $(r \times r)$ auto-correlation matrix of $y(i)$ and $y(j)$, and $R_{xy}(k,j)$ is the cross-correlation of $x(k)$ and $y(j)$. Expanding eq. (6.101) over $i,j,k = 1, ..., n$ we have

$$R_y = \begin{bmatrix} R_y(1,1) & \cdots & R_y(1,n) \\ \vdots & & \\ R_y(n,1) & & R_y(n,n) \end{bmatrix} \; ; \; R_{xy} = \begin{bmatrix} R_{xy}(1,1) & \cdots & R_{xy}(1,n) \\ \vdots & & \\ R_{xy}(n,1) & & R_{xy}(n,n) \end{bmatrix} \tag{6.102}$$

Similarly, $H(k,i)$ produces the $(qn \times rn)$ matrix

$$H = \begin{bmatrix} H(1,1) & \cdots & H(1,n) \\ \vdots & & \\ H(n,1) & \cdots & H(n,n) \end{bmatrix} \tag{6.103}$$

Therefore, eq. (6.101) can be written in the compact form as

$$HR_y = R_{xy}$$

with the solution

$$H = R_{xy} R_y^{-1} \tag{6.104}$$

The error covariance matrix, corresponding to eq. (6.40) for a constant signal case, can be now expanded as

$$P_e(k) = R_x(k,k) - \sum_{i=1}^{n} H(k,i) R_{yx}(i,k) \tag{6.105}$$

where $R_x(k,k) = E\{x(k)x^T(k)\}$ and $R_{yx}(i,k) = E\{y(i)x^T(k)\}$. The vector or multidimensional Wiener estimator is then given by eqs. (6.98), (6.101) and (6.105). To solve it, we require the correlation matrices R_{xy} and R_y.

For the case of a linear measurement model

$$y(k) = Cx(k) + v(k) \tag{6.106}$$

where $y(k)$ and $v(k)$ are $(r \times 1)$ column vectors, $x(k)$ is a $(q \times 1)$ row vector, and C is a $(r \times q)$ observation matrix (usually, it is assumed $r < q$), we have

$$
\begin{aligned}
R_{xy}(k,j) &= E\{x(k)[Cx(j) + v(j)]^T\} \\
&= E\{x(k)x^T(j)\}C^T + E\{x(k)v^T(j)\} \\
&= R_x(k,j)C^T
\end{aligned} \tag{6.107}
$$

and

$$
\begin{aligned}
R_{yy}(k,j) &= E\{[Cx(i) + v(i)][Cx(j) + v(j)]^T\} \\
&= CE\{x(i)x^T(j)\}C^T + CE\{x(i)v^T(j)\} + E\{v(i)x^T(j)\}C^T + E\{v(i)v^T(j)\} \\
&= CE\{x(i)x^T(j)\}C^T + E\{v(i)v^T(j)\} \\
&= CR_x(i,j)C^T + R_v(i)\delta(i,j)
\end{aligned} \tag{6.108}
$$

assuming signal $\{x(k)\}$ and measurement zero-mean white noise $\{v(k)\}$ to be uncorrelated. In the above, $R_v(i)$ is the observation noise covariance matrix and $\delta(i,j)$ is the Kronecker's delta discussed earlier. From the above two equations, we see that to solve the filtering problem it is sufficient to know the signal auto-correlation function $R_x(i,j)$ and the noise covariance matrix $R_v(i)$. If the signal generation equation is also known, further modifications are possible, as illustrated in the following example.

Example 6.6 (Wiener filter application to falling body): Consider a noise free second-order system representing a falling body in a constant field [11, 49]

$$\ddot{z} = -g \; ; \; t \geq 0 \tag{I}$$

Let the position be $z = x_1$ and velocity $\dot{z} = x_2$. Then, defining the two-state vector

$$x(t) = \begin{bmatrix} x_1(t) \\ x_2(t) \end{bmatrix}$$

eq. (I) can be written in the state space form

$$\begin{aligned}
\dot{x}_1 &= \dot{z} = x_2 \\
\dot{x}_2 &= \ddot{z} = -g
\end{aligned} \tag{IIa}$$

or in the matrix notation

$$\dot{x}(t) = \begin{bmatrix} 0 & 1 \\ 0 & 0 \end{bmatrix} x(t) + \begin{bmatrix} 0 \\ -g \end{bmatrix} = Ax(t) + Bu(t) \; ; \; t \geq 0 \tag{IIb}$$

where $u(t) = 1$ for $t \geq 0$. The state transition matrix of the system $\Phi(t, \tau)$ is the solution of the autonomous matrix equation, that is (see chapter 4)

$$\dot{\Phi}(t, \tau) = A\Phi(t, \tau) \; ; \; \Phi(t, t) = I \tag{III}$$

where

$$\Phi(t, \tau) = \begin{bmatrix} \Phi_{1,1}(t, \tau) & \Phi_{1,2}(t, \tau) \\ \Phi_{2,1}(t, \tau) & \Phi_{2,2}(t, \tau) \end{bmatrix} \tag{IV}$$

Taking into account eqs. (IIb), (III) and (IV), one obtains the system of scalar differential equations

$$\begin{aligned}
\dot{\Phi}_{1,1}(t, \tau) &= \Phi_{2,1}(t, \tau) \; ; \; \Phi_{1,1}(\tau, \tau) = 1 \\
\dot{\Phi}_{1,2}(t, \tau) &= \Phi_{2,2}(t, \tau) \; ; \; \Phi_{1,2}(\tau, \tau) = 0 \\
\dot{\Phi}_{2,1}(t, \tau) &= 0 \; ; \; \Phi_{2,1}(\tau, \tau) = 0 \\
\dot{\Phi}_{2,2}(t, \tau) &= 0 \; ; \; \Phi_{2,2}(\tau, \tau) = 1
\end{aligned}$$

The solution of the above system is

$$\begin{aligned}
\Phi_{1,1}(t, \tau) &= 1 \\
\Phi_{2,1}(t, \tau) &= 0 \\
\Phi_{1,2}(t, \tau) &= t - \tau \\
\Phi_{2,2}(t, \tau) &= 1
\end{aligned}$$

Therefore, the state transition matrix of the system is

$$\Phi(t, \tau) = \begin{bmatrix} 1 & t - \tau \\ 0 & 1 \end{bmatrix} \tag{V}$$

so that the solution for the state $x(t)$ is given by (see chapter 4)

$$x(t) = \Phi(t,\tau)x(\tau) + \int_\tau^t \Phi(t,s)Bu(s)\,ds$$

$$= \Phi(t,\tau)x(\tau) + \int_\tau^t \begin{bmatrix} 1 & t-s \\ 0 & 1 \end{bmatrix}\begin{bmatrix} 0 \\ -g \end{bmatrix}ds \qquad \text{(VI)}$$

$$= \Phi(t,\tau)x(\tau) - g\begin{bmatrix} (t-\tau)^2/2 \\ (t-\tau) \end{bmatrix}$$

Substituting for $\Phi(t,\tau)$ from eq. (V), we can write the state solution (VI) as

$$x_1(t) = x_1(\tau) + (t-\tau)x_2(\tau) - \frac{g}{2}(t-\tau)^2$$

$$x_2(t) = x_2(\tau) - g(t-\tau) \qquad \text{(VII)}$$

where the first equation in (VII) is the body position, and the second one its velocity. These are exact solutions which we have used to calculate the true values for $x_1(t)$ and $x_2(t)$ at times $t = 1,2,...,6$. It has been assumed that the true initial conditions are $x_1(0) = z(0) = 100$; $x_2(0) = \dot{z}(0) = 0$ and $g=1$. The results are shown in the first two columns of Tab. 6.1, and a falling body experiment is depicted in Fig. 6.6.

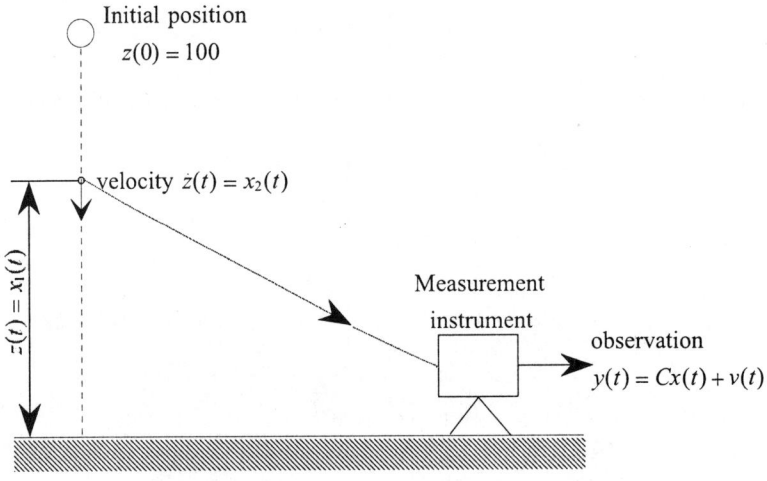

Fig. 6.6. Measurements of falling body

The system has been observed by measuring the falling body position $x_1(t)$ in discrete-times $t=1,2,...,6$ (Fig. 6.6), and the results of measurements are shown in the third column of Tab 6.1. They have been affected by some independent zero-mean random disturbance v, so that the observation equation, in this case, can be written as

$$y(k) = Cx(k) + v(k) = [1 \ 0]x(k) + v(k) \qquad \text{(VIII)}$$

with the random disturbance, or noise, variance given by $R_v(k) = \sigma_v^2 = 1$. Therefore, the general dynamic behavior of the system is described in continuous-time. We can easily obtain its discrete-time form by setting $t=kT$, $\tau=(k-1)T$, and taking $T=1$. Applying this to eqs. (V) and (VI), we obtain (see chapter 4)

$$x(k)=\begin{bmatrix}1 & 1\\ 0 & 1\end{bmatrix}x(k-1)+\begin{bmatrix}0.5\\ 1\end{bmatrix}(-g)=A_d x(k-1)+B_d u(k) \tag{IX}$$

which is the difference equation representing our multidimensional system; i.e. falling body in a constant field.

The states of the discrete-time system (IV) is to be determined from a set of measurement (VIII), as indicated in fig. 6.6. This is a situation where we can use vector Wiener filter to find the system states. We start with the vector Wiener-Hopf equation (6.101). This means applying this equation to the falling body; that is

$$\sum_{i=1}^{6}H(k,i)E\{y(i)y^T(j)\}=E\{x(k)y^T(j)\} \tag{X}$$

Since the system has two states (x is 2×1 column vector $x^T=[\,x_1\ x_2\,]$) and we measure only one state (position x_1) y is a scalar and this equation reduces to

$$\sum_{i=1}^{6}H(k,i)E\{y(i)y^T(j)\}=E\left\{\begin{bmatrix}x_1(k)y(j)\\ x_2(k)y(j)\end{bmatrix}\right\} \tag{XI}$$

Introducing

$$R_y(i,j)=E\{y(i)y(j)\}\ ;\ R_{x_1y}(k,j)=E\{x_1(k)y(j)\}\ ;\ R_{x_2y}(k,j)=E\{x_2(k)y(j)\}$$

and splitting $H(k,i)$ into two parts to correspond to the right-hand side of matrix eq. (XI), we have

$$\sum_{i=1}^{6}\begin{bmatrix}H_1(k,i)\\ H_2(k,i)\end{bmatrix}R_y(i,j)=\begin{bmatrix}R_{x_1y}(k,j)\\ R_{x_2y}(k,j)\end{bmatrix} \tag{XII}$$

From the above we obtain two scalar equations

$$\sum_{i=1}^{6}H_1(k,i)R_y(i,j)=R_{x_1y}(k,j) \tag{XIIIa}$$

$$\sum_{i=1}^{6}H_2(k,i)R_y(i,j)=R_{x_2y}(k,j) \tag{XIIIb}$$

These equations must be solved for $H_1(k,i)$ and $H_2(k,i)$, which are required for calculations of the state estimates

$$\hat{x}_1(k)=\sum_{i=1}^{6}H_1(k,i)y(i) \tag{XIVa}$$

$$\hat{x}_2(k)=\sum_{i=1}^{6}H_2(k,i)y(i) \tag{XIVb}$$

The mean-square errors are given by the diagonal terms in the error covariance matrix (6.105), which in this case are

$$P_{11}(k)=E\{x_1^2(k)\}-\sum_{i=1}^{6}H_1(k,i)E\{y(i)x_1(k)\} \tag{XVa}$$

$$P_{22}(k) = E\{x_2^2(k)\} - \sum_{i=1}^{6} H_2(k,i) E\{y(i)x_2(k)\}$$ (XVb)

As before, we introduce the notation

$$R_{x_1}(k,k) = E\{x_1^2(k)\} \; ; \; R_{x_2}(k,k) = E\{x_2^2(k)\} \; ;$$

$$R_{yx_1}(i,k) = E\{y(i)x_1(k)\} \; ; \; R_{yx_2}(i,k) = E\{y(i)x_2(k)\}$$

so that the above equations become

$$P_{11}(k) = R_{x_1}(k,k) - \sum_{i=1}^{6} H_1(k,i) R_{yx_1}(i,k)$$ (XVIa)

$$P_{22}(k) = R_{x_2}(k,k) - \sum_{i=1}^{6} H_2(k,i) R_{yx_2}(i,k)$$ (XVIb)

We proceed now by using the measurement equation

$$y(k) = x_1(k) + v(k)$$

to calculate the correlation quantities in eqs. (XIII)

$$R_{x_1 y}(k,j) = E\{x_1(k)y(j)\} = E\{x_1(k)[x_1(j) + v(j)]\}$$
$$= E\{x_1(k)x_1(j)\} = R_{x_1}(k,j)$$ (XVIIa)

$$R_{x_2 y}(k,j) = E\{x_2(k)y(j)\} = E\{x_2(k)[x_1(j) + v(j)]\}$$
$$= E\{x_2(k)x_1(j)\} = R_{x_2 x_1}(k,j)$$ (XVIIb)

$$R_y(i,j) = E\{y(i)y(j)\} = E\{[x_1(i) + v(i)][x_1(j) + v(j)]\}$$
$$= E\{x_1(i)x_1(j)\} + E\{v(i)v(j)\} = R_{x_1}(i,j) + \sigma_v^2 \delta(i,j)$$ (XVIIc)

where $\delta(i,j)$ is the Kronecker's delta and $\sigma_v^2 = 1$. In the error eqs. (XVI) $R_{x_1}(k,k)$ is a special case of $R_{x_1}(k,j)$ for $k=j$, while $R_{x_2}(k,k)$ will be calculated separately. Furthermore, we have

$$R_{yx_1}(i,k) = E\{y(i)x_1(k)\} = E\{[x_1(i) + v(i)]x_1(k)\}$$
$$= E\{x_1(i)x_1(k)\} = R_{x_1}(i,k)$$

$$R_{yx_2}(i,k) = E\{y(i)x_2(k)\} = E\{[x_1(i) + v(i)]x_2(k)\}$$
$$= E\{[x_1(i) + v(i)]x_2(k)\} = E\{x_1(i)x_2(k)\} = R_{x_1 x_2}(i,k)$$

These two quantities can be expressed in terms of the previous quantities, i.e. $R_{x_1}(i,k) = R_{x_1}(k,i)$ and also $R_{x_1 x_2}(i,k) = R_{x_2 x_1}(k,i)$. The last equation means, for example, $E\{x_1(i)x_2(k)\} = E\{x_2(k)x_1(i)\}$ which is correct since the quantities within the brackets are scalars. From eqs. (XVII) we see that all of the required quantities are expressed in terms of R_{x_1} and $R_{x_2 x_1}(k,j)$. To calculate these correlation terms, we use the system model (IX). The solution of the first order vector difference equation (IX) is given by (see chapter 3)

$$x(k) = A_d^k x(0) + \sum_{i=0}^{k-1} A_d^i B_d u(k-i-1) \qquad \text{(XVIII)}$$

Taking $u(k) = -g = -1$, and substituting A_d and B_d from (IX), we can rewrite (XVIII) as

$$x(k) = \begin{bmatrix} 1 & 1 \\ 0 & 1 \end{bmatrix}^k x(0) - \sum_{i=0}^{k-1} \begin{bmatrix} 1 & 1 \\ 0 & 1 \end{bmatrix}^i \begin{bmatrix} 0.5 \\ 1 \end{bmatrix}$$

$$= \begin{bmatrix} 1 & k \\ 0 & 1 \end{bmatrix} x(0) - \begin{bmatrix} \sum_{i=0}^{k-1}(0.5+i) \\ \sum_{i=0}^{k-1} 1 \end{bmatrix} = \begin{bmatrix} 1 & k \\ 0 & 1 \end{bmatrix} x(0) + \begin{bmatrix} k^2/2 \\ k \end{bmatrix} \qquad \text{(XIX)}$$

To obtain the second term, a summation of the arithmetic progression has been performed. The individual state vector components, from the above result, are given by

$$x_1(k) = x_1(0) + kx_2(0) - k^2/2 \qquad \text{(XXa)}$$

$$x_2(k) = x_2(0) - k \qquad \text{(XXb)}$$

where $x_1(0)$ and $x_2(0)$ are the initial position and velocity, respectively. These values are unknown, and will be considered as random variables with mean values $E\{x_1(0)\} = p_0$ and $E\{x_2(0)\} = v_0$, and variances p_1 and p_2, respectively. We also assume they are not correlated, i.e. $E\{x_1(0)x_2(0)\} = p_0 v_0$. Now we can calculate $R_{x_1}(i,k)$ and $R_{x_1 x_2}(k,j)$ as follows:

$$R_{x_1}(k,j) = E\{x_1(k)x_1(j)\}$$
$$= E\{[x_1(0) + kx_2(0) - k^2/2][x_1(0) + jx_2(0) - j^2/2]\} \qquad \text{(XXIa)}$$
$$= (p_0 + kv_0 - k^2/2)(p_0 + jv_0 - j^2/2) + p_1 + kjp_2$$

$$R_{x_1 x_2}(k,j) = E\{x_1(k)x_2(j)\}$$
$$= E\{[x_1(0) + kx_2(0) - k^2/2][x_2(0) - j]\} \qquad \text{(XXIb)}$$
$$= (p_0 + kv_0 - k^2/2)(v_0 - j) + kp_2$$

since $E\{x_i^2(0)\} = [E\{x_i(0)\}]^2 + p_i$. It should be noted that the auto-correlation function $R_{x_1}(k,j)$ is not a function of the difference $(k-j)$; that is $\{x_1(k)\}$ is not a wide sense stationary signal.

The first terms in eqs. (XVI), $R_{x_1}(k,k)$ and $R_{x_2}(k,k)$ are obtained from eq. (XXIa), by setting $j=k$, and using $R_{x_2}(k,k)$ from

$$R_{x_2}(k,k) = E\{x_2^2(k)\} = [E\{x_2(k)\}]^2 + p_2$$
$$= [E\{(x_2(0) - k)\}]^2 + p_2 = (v_0 - k)^2 + p_2 \qquad \text{(XXII)}$$

We develop the computer program, using MATLAB package, to calculate the position and velocity estimates and their mean-square errors. Results for the initial conditions $p_0 = 95$, $v_0 = 0$ and their variances $p_1 = 10$, $p_2 = 1$ are shown in tab. 6.1.

time	True values		position observations	Estimates		Estimation errors	
	position	velocity		position	velocity	position	velocity
$t=kT$	$x_1(t)$	$x_2(t)$	$y(t)$	$\hat{x}_1(t)$	$\hat{x}_2(t)$	$p_{11}(t)$	$p_{22}(t)$
0	100.0	0	-	95.0	0.00	10	1
1	99.5	-1.0	100.0	99.31	0.07	0.49	0.99
2	98	-2.0	97.9	99.45	-0.91	45.94	1.15
3	95.5	-3.0	94.4	91.64	-1.89	11.66	1.31
4	92.0	-4	92.7	89.89	-2.88	7.80	1.49
5	87.5	-5.0	87.3	86.43	-3.86	2.89	1.68
6	82.0	-6.0	82.1	82.09	-4.84	1.00	1.88

Table 6.1. Falling body in a constant field: Wiener filter estimates

The derivation and discussion of Wiener filter in the case of a constant signal estimation give the impression that the Wiener filter does not require a system equation. However, the application of the Wiener filter to the problem of a falling body, which is a dynamic system, shows quite stronger that to solve the estimation problem we must use the system equations. This means that the Wiener filter requires a knowledge of the system dynamics, unless the auto-correlation and cross-correlation matrices R_y and R_{xy} can be obtained by same other means.

6.2. Optimum nonrecursive estimator from continuous-time noisy measurements

6.2.1 Optimum nonrecursive estimator for scalar random variable

Suppose we wish to find the linear mean-square (LMS) estimate of a scalar random variable X given observations of a random process $\{Y(\tau), a \leq \tau \leq b\}$. We shall assume that the most general linear functional of Y will have the form [25, 48, 49]

$$\int_a^b h(\tau)Y(\tau)d\tau$$

Then the problem of choosing $h(\cdot)$ so as to minimize the mean-square error (MSE)

$$E\left\{\left(X - \int_a^b h(\tau)Y(\tau)d\tau\right)^2\right\} = E\left\{E\left\{\left(X - \int_a^b h(\tau)Y(\tau)d\tau\right)\middle| Y(\tau)\right\}\right\}$$

will reduce to an "infinite-dimensional" minimization problem

$$\min_{h(\cdot)} E\left\{\left(X - \int_a^b h(\tau)Y(\tau)\right)^2 \middle| Y(\tau)\right\} = \min_{h(\cdot)} \int_{-\infty}^{\infty}\left(X - \int_a^b h(\tau)Y(\tau)\right)^2 f(X|Y(\tau))dX$$

As mentioned in section 2, if the random process $\{Y(\tau), a \leq \tau \leq b\}$ is smooth enough it can be represented by a countable number of random variables $\{Y_i, i = 1,...,n\}$ and then it is reasonable to define $f(X|Y(\tau)) = \lim_{n \to \infty} f(X|Y_1,...,Y_n)$. From a such discussion, it is clearly seen that the problem concerned can be solved by a simple route, based on first approximating the infinite-dimensional problem by a finite-dimensional one and then taking the limit (we rejected this approach in section 2, but here is the difference in the fact that because of the constraint on linearity we shall not need to consider infinite-dimensional conditional density functions). An alternative approach can be based on calculus of variations (see chapter 6.1.3).

To carry through the first approach, let us first note that a smooth random process $\{Y(\tau), a \le \tau \le b\}$ can be approximately represented as in Fig. 6.7.; that is

$$Y(\tau) = \sum_{i=0}^{n-1} Y(\tau_i)\sqrt{\Delta}\Pi(\tau - i\Delta) = \sum_{i=0}^{n-1} Y_i\Pi(\tau - i\Delta) \qquad (6.109)$$

where

$$\Pi(\tau) = \begin{cases} \dfrac{1}{\sqrt{\Delta}} \; ; \; 0 \le \tau \le \Delta \\[2mm] 0 \; ; \; \text{elsewhere} \end{cases}$$

Then the LMS estimate of a random variable X given $\{Y(\tau), a \le \tau \le b\}$ can be approximately obtained by virtue of the result of the previous finite-dimensional problem (see eq. 6.2) as

$$\hat{X} = h^T Y = \sum_{i=1}^{n} h_i Y_i \; ; \; Y^T = \{Y_1,...,Y_n\} \; ; \; h^T = \{h_1,...,h_n\} \qquad (6.110a)$$

where h is defined by the discrete Wiener-Hopf equation (6.16), i.e.

$$h^T R_Y = R_{XY} \qquad (6.110b)$$

with R_Y and R_{XY} being the auto-correlation and cross-correlation matrices, respectively; that is $R_Y = E\{YY^T\}$ and $R_{XY} = E\{XY^T\}$.

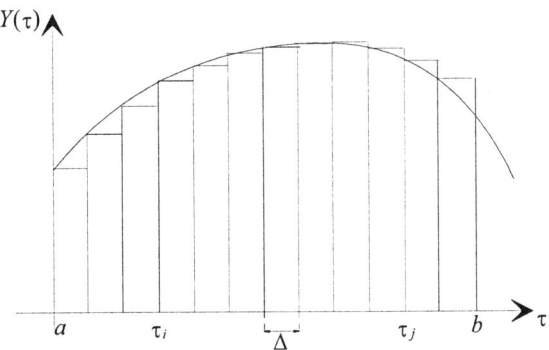

Fig. 6.7. Approximation of a smooth random process

More explicitly

$$\sum_{j=1}^{n} h_j E\{Y_j Y_i\} = E\{XY_i\} \; ; \; i = 1,...,n \qquad (6.111a)$$

or

$$\sum_{j=1}^{n} h_j R_Y(\tau_j, \tau_i)\Delta = R_{XY}(\tau_i)\sqrt{\Delta} \; ; \; \tau_i = i\Delta \; ; \; i = 1,...,n \qquad (6.111b)$$

Now let $h(\cdot)$ be a function approximately represented as

$$h(\tau) = \sum_{i=1}^{n} h(\tau_i)\sqrt{\Delta}\Pi(\tau - i\Delta) = \sum_{i=1}^{n} h_i \Pi(\tau - i\Delta); \ a \leq \tau \leq b \tag{6.112}$$

Then the equation (6.111) becomes

$$\sum_{j=1}^{n} h(\tau_j) R_Y(\tau_j, \tau_i)\Delta = R_{XY}(\tau_i) \ ; \ i=1,...,n \tag{6.113}$$

since $h_j = h(\tau_j)\sqrt{\Delta}$. Finally let τ_i be a fixed point t in $[a,b]$ and consider the limit as $\Delta \to 0$. In the limit we get the following equation for $h(\cdot)$

$$\int_a^b h(\tau) R_Y(\tau, t)d\tau = R_{XY}(t) \ ; \ a \leq \tau \leq b \tag{6.114}$$

This is known as the Wiener-Hopf integral equation because the unknown function $h(\cdot)$ appears under an integral sign. Note also that the equation (6.110a) reduces to

$$\hat{X} = \sum_{i=1}^{n} h_i Y_i = \sum_{i=1}^{n} h(\tau_i) Y(\tau_i)\Delta$$

and in the limit as $\Delta \to 0$, $n \to \infty$ we have

$$\hat{X} = \int_a^b h(\tau) Y(\tau)d\tau \tag{6.115}$$

Finally, similarly as in (6.15), the mean-square error (MSE) associated with the estimate (6.115) is

$$MSE = R_X - \int_a^b \int_a^b h(\tau) h(\sigma) R_Y(\tau, \sigma)d\tau d\sigma \tag{6.116}$$

We can verify that the result (6.114), (6.115) can be obtained via the geometric viewpoint arguments and the orthogonality conditions. Namely, the real rewards of the geometric interpretation arise in the problem of estimating X from observations of a random process $Y(\cdot)$. If we think of the random process as an collection of random variables $\{Y(\tau), a \leq \tau \leq b\}$, then the optimum estimate

$$\hat{X} = \int_a^b h(\tau) Y(\tau)d\tau$$

will be uniquely defined by the orthogonality conditions

$$X - \int_a^b h(\tau) Y(\tau)d\tau \perp Y(t) \ ; \ a \leq t \leq b \tag{6.117}$$

which are equivalent to

$$\left\langle X - \int_a^b h(\tau) Y(\tau)d\tau, Y(t) \right\rangle = 0 \tag{6.118}$$

where $\langle \cdot, \cdot \rangle$ denotes the inner product defined by $\langle X, Y \rangle = E\{XY\}$. In this way, one obtains

$$E\{XY(t)\} = \int_a^b h(\tau) E\{Y(\tau) Y(t)\}d\tau$$

or

$$R_{XY}(t) = \int_a^b h(\tau) R_Y(t, \tau)d\tau$$

which represents the equation (6.114) that was obtained much less directly by a limiting argument. Of course, no matter how his equation is obtained, we are still left with the problem of solving the integral equation. Explicit solutions can only be obtained in special, but still widely useful cases, and these will be the main object of attention in later chapters.

6.2.2 Optimum nonrecursive estimator for multivariate random variable

So far we have considered the estimation of a single random variable X based on observations of a single random process $Y(\cdot)$. It is easy to handle the case of a $m \times 1$ vector random process $Y(\cdot)$. Clearly we shall have [25]

$$\hat{X} = \int_a^b h^T(\tau) Y(\tau) d\tau \tag{6.119}$$

where $1 \times n$ vector $h^T(\cdot)$ obeys the vector integral equation

$$\int_a^b h^T(\tau) R_Y(t,\tau) d\tau = R_{XY}(t) \; ; \; a \le t \le b \tag{6.120}$$

The question of vector X is just a bit more delicate. We can compute the MS estimate $\hat{X}_i$ of each component of X based on observation of $Y(\cdot)$, and then we can define $\hat{X} = \left[\hat{X}_1, ..., \hat{X}_n \right]^T$. The interesting thing is that the joint statistics of $\{X_i\}$ do not enter the problem. Therefore, to summarize, to find the LMS estimate of a n vector X based on observations of an m-vector $Y(\cdot)$ we shall have

$$\hat{X} = \begin{bmatrix} \hat{X}_1 \\ \vdots \\ \hat{X}_n \end{bmatrix} = \begin{bmatrix} \int_a^b h_1^T(\tau) Y(\tau) d\tau \\ \vdots \\ \int_a^b h_n^T(\tau) Y(\tau) d\tau \end{bmatrix} = \int_a^b H(\tau) Y(\tau) d\tau \tag{6.121}$$

where $H(\cdot)$ is a $n \times m$ matrix

$$H(\tau) = \begin{bmatrix} h_1^T(\tau)_{1 \times m} \\ \vdots \\ h_n^T(\tau)_{1 \times m} \end{bmatrix}_{n \times m} \tag{6.122}$$

satisfying the matrix integral equation

$$\begin{bmatrix} \int_a^b h_1^T(\tau) R_Y(t,\tau) d\tau \\ \vdots \\ \int_a^b h_n^T(\tau) R_Y(t,\tau) d\tau \end{bmatrix} = \begin{bmatrix} R_{X_1 Y}(t) \\ \vdots \\ R_{X_n Y}(t) \end{bmatrix} \; ; \; a \le t \le b \tag{6.123a}$$

or equivalently

$$\int_a^b H(\tau) R_Y(t,\tau) d\tau = R_{XY}(t) \; ; \; a \le t \le b \tag{6.123b}$$

We could think of obtaining this integral equation by using the orthogonality condition

$$\left(X - \hat{X} \right) \perp Y(t) \; ; \; a \le t \le b$$

or equivalently

$$\left\langle X - \hat{X}, Y(t) \right\rangle = 0 \; ; \; a \le t \le b$$

with $\hat{X}$ being given by (6.121), and

$$\left\langle X, Y \right\rangle = E\left\{ XY^T \right\}.$$

In this way, one obtains

$$\left\langle X - \int_a^b H(\tau) Y(\tau) d\tau, Y(t) \right\rangle = 0$$

or equivalently

$$\left\langle X, Y(t) \right\rangle = \int_a^b H(\tau) \left\langle Y(\tau), Y(t) \right\rangle d\tau$$

which is exactly the equation (6.123). Thus, it is easy to check that this leads to the correct matrix integral Wiener-Hopf equation. But, as mentioned in chapter 5, it must be noted that $\left\langle X, Y \right\rangle$ as defined above is not an inner product because for vector X and Y $E\left\{ XY^T \right\} \ne E\left\{ YX^T \right\}$.

6.2.3 Optimum nonrecursive estimator for time-varying stationary continuous-time signals

Given two zero-mean jointly stationary random processes $x(\cdot)$ and $y(\cdot)$ with known auto and cross-covariance functions, the basic problem concerned is the following: given observations $\left\{ y(\tau) ; -\infty < \tau < t \right\}$ find the LMS estimate of $x(t+\lambda)$, with λ being a fixed constant [25, 48, 49].

If we denote the estimate by $\hat{x}(t+\lambda|t)$, the task is to find $h(t,\tau)$ such that

$$\hat{x}(t+\lambda|t) = \int_{-\infty}^t h(t,\tau) y(\tau) d\tau \tag{6.124}$$

and

$$E\left\{ \left[x(t+\lambda) - \hat{x}(t+\lambda|t) \right]^2 \right\} = \text{minimum} \; ; \; -\infty < t < \infty \tag{6.125}$$

The function $h(t,\tau)$ can be regarded as the impulse response of a causal (i.e. $h(t,\tau) = 0$ for $t < \tau$) linear system whose response at time t to the input waveform $\left\{ y(\tau), -\infty < \tau < t \right\}$ is the desired estimate $\hat{x}(t+\lambda|t)$. To determine the optimum system, we use the orthogonality property; that is the projection theorem for LMS estimates

$$\left(x(t+\lambda) - \hat{x}(t+\lambda|t) \right) \perp y(\sigma) \; ; \; -\infty < \sigma < t$$

which yields

$$\left\langle x(t+\lambda) - \hat{x}(t+\lambda|t), y(\sigma) \right\rangle = 0$$

with

$$\left\langle x, y \right\rangle = E\left\{ xy \right\}$$

or equivalently

$$E\left\{ x(t+\lambda) y(\sigma) \right\} = \int_{-\infty}^t h(t,\tau) E\left\{ y(\tau) y(\sigma) \right\} d\tau$$

The expectation on the left-hand side is the cross-covariance function $R_{xy}(t+\lambda,\sigma)$, while the expectation on the right-hand side is the auto-covariance function $R_y(\tau,\sigma)$. However, since $x(\cdot)$

and $y(\cdot)$ are stationary processes, we have $R_{xy}(t+\lambda,\sigma) = R_{xy}(t+\lambda-\sigma)$ and $R_y(\tau,\sigma) = R_y(\tau-\sigma)$, so that the last equation reduces to

$$R_{xy}(t+\lambda-\sigma) = \int_{-\infty}^{t} h(t,\tau)R_y(\tau-\sigma)d\tau \; ; \; -\infty < \sigma < t$$

or

$$R_{xy}(t+\lambda-\sigma) = \int_{0}^{\infty} h(t,t-\tau)R_y(t-\sigma-\tau)d\tau \; ; \; -\infty < \sigma < t$$

If we change the variable $t-\sigma$ to t, the equation becomes

$$R_{xy}(t+\lambda) = \int_{0}^{\infty} h(t+\sigma,t+\sigma-\tau)R_y(t-\tau)d\tau \; ; \; t > 0$$

The left hand side of this equation does not depend upon σ, nor does $R_y(t-\tau)$. Thus $h(t+\sigma,t+\sigma-\tau)$ must be a function only of the difference of its arguments; that is $h(t+\sigma,t+\sigma-\tau) = h(\tau)$. In this way, the equation for $h(\cdot)$ is

$$R_{xy}(t+\lambda) = \int_{0}^{\infty} h(\tau)R_y(t-\tau)d\tau \; ; \; t > 0$$

or, since $h(\tau) = 0$ for $\tau < 0$,

$$R_{xy}(t+\lambda) = \int_{-\infty}^{\infty} h(\tau)R_y(t-\tau)d\tau \; ; \; t > 0 \tag{6.126}$$

The equation (6.126) is known as continuous-time Wiener-Hopf equation, and it might appear that this equation can be readily solved by taking Fourier or bilateral Laplace transforms. However, the presence of the constraint $t > 0$ prevents a such approach. Analogous to discrete-time case in section 6.1.3, the easiest way of seeing why these transforms would not work is to try them. Thus if the system transfer function is

$$H(s) = \int_{-\infty}^{\infty} h(t)e^{-st}dt$$

and the spectral density

$$S_y(s) = \int_{-\infty}^{\infty} R_y(t)e^{-st}dt$$

then we can write

$$\int_{0}^{\infty} R_{xy}(t+\lambda)e^{-st}dt = \int_{-\infty}^{\infty} h(\tau)e^{-st}d\tau \int_{0}^{\infty} R_y(t-\tau)e^{-s(t-\tau)}dt$$

$$= \int_{-\infty}^{\infty} h(\tau)e^{-st}d\tau \int_{-\tau}^{\infty} R_y(\sigma)e^{-s\sigma}d\sigma$$

The presence of $-\tau$ in the limits of the integral over σ prevents any easy solution of the equation (6.126). Therefore, a more sophisticated technique is needed to solve this equation and such techniques were developed in chapter 6.1.3 for discrete-time random sequences. We can develop the same techniques for continuous-time stochastic processes using Laplace transforms instead of Z-transforms. However, we shall use here a slightly different approach based on frequency-domain concepts, such as transfer functions and spectral densities. We can represent the problem concerned as in Fig. 6.8.

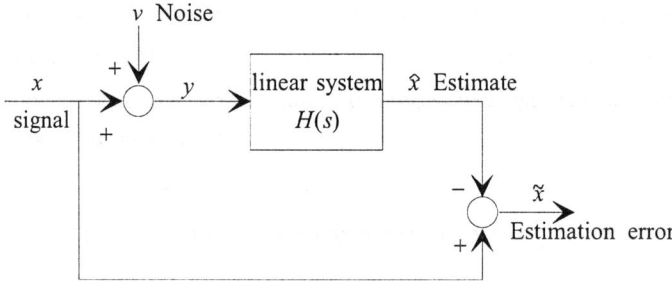

Fig. 6.8. Representation of the Wiener- estimation problem

The message or signal $x(t)$ is contaminated by an additive noise $v(t)$, where $x(t)$ and $v(t)$ are zero-mean stationary random process with spectral and cross-spectral densities $S_y(s)$, $S_v(s)$ and $S_{yv}(s)$, respectively, and are uncorrelated; that is the cross-spectral density $S_{yv}(s) = 0$. The observation $y(t) = x(t) + v(t)$ is passed through a liner time-invariant system, with transfer function $H(s)$, the output of which is denoted by $\hat{x}(t)$; that is $\lambda = 0$ in eq. (6.124) and this is known as *filtering problem* (the case $\lambda > 0$ is denoted as *prediction problem*, while $\lambda < 0$ is known as *smoothing problem*). We wish to select the transfer function, i.e. filter, $H(s)$ such that the output $\hat{x}(t)$ is the optimal estimate of $x(t)$ in the mean square error (MSE) sense (6.125). In summary, we wish to determine the filter transfer function $H(s)$ in such a way to minimize the MSE in eq. (6.125). The MSE can be rewritten as

$$MSE = E\left\{\tilde{x}^T(t)\tilde{x}(t)\right\} = \text{trace}\left[E\left\{\tilde{x}(t)\tilde{x}^T(t)\right\}\right] \qquad (6.127)$$

where the estimation (filtering) error $\tilde{x}(t) = x(t) - \hat{x}(t)$. Here is assumed that x, y and v are vector signals; that is we consider a multivariable case, with $\text{trace}[\cdot]$ being the trace of a matrix. Furthermore, since the auto-covariance function $R_{\tilde{x}}(\tau) = E\left\{\tilde{x}(t+\tau)\tilde{x}^T(t)\right\}$ one concludes $R_{\tilde{x}}(0) = E\left\{\tilde{x}(t)\tilde{x}^T(t)\right\}$; that is

$$MSE = \text{trace}\left[R_{\tilde{x}}(0)\right] \qquad (6.128)$$

Moreover, by the use of Wiener-Hintchin theorem, the auto-covariance function $R_{\tilde{x}}(\tau)$ and the spectral density $S_{\tilde{x}}(s)$ represent a Fourier transform pair, i.e.

$$R_{\tilde{x}}(\tau) = \frac{1}{2\pi j}\int_{-j\infty}^{j\infty} S_{\tilde{x}}(s)e^{s\tau}ds \qquad (6.129)$$

so that from (6.128) follows

$$MSE = \frac{1}{2\pi j}\text{trace}\left[\int_{-j\infty}^{j\infty} S_{\tilde{x}}(s)e^{s\tau}ds\right]_{\tau=0} = \frac{1}{2\pi j}\text{trace}\left[\int_{-j\infty}^{j\infty} S_{\tilde{x}}(s)ds\right] \qquad (6.130)$$

The spectral density for the error is defined by

$$S_{\tilde{x}}(s) = E\left\{\tilde{x}(s)\tilde{x}^T(-s)\right\} \qquad (6.131)$$

Bearing in mind the figure .6.8., one obtains

$$\tilde{x}(s) = x(s) - \hat{x}(s)$$
$$= x(s) - H(s)y(s)$$
$$= x(s) - H(s)\big[x(s) + v(s)\big] \qquad (6.132)$$
$$= \big[I - H(s)\big]x(s) - H(s)v(s)$$

so that

$$S_{\tilde{x}}(s) = \big[I - H(s)\big]E\big\{x(s)x^T(-s)\big\}\big[I - H(s)\big]^T$$
$$-H(s)E\big\{v(s)x^T(-s)\big\} - \big[I - H(s)\big]E\big\{x(s)v^T(-s)\big\}H^T(-s) + H(s)E\big\{v(s)v^T(-s)\big\}H^T(-s)$$
$$= \big[I - H(s)\big]S_x(s)\big[I - H(s)\big]^T + H(s)S_v(s)H^T(-s)$$

$$(6.133)$$

Here is used the fact that $S_{xv}(s) = E\big\{x(s)v^T(-s)\big\} = 0$, since x and v are uncorrelated signals. Substituting this expression into eq. (6.130), we have

$$MSE = \frac{1}{2\pi j}\,\text{trace}\left[\int_{-j\infty}^{j\infty}\Big\{\big[I - H(s)\big]S_x(s)\big[I - H(-s)\big]^T + H(s)S_v(s)H^T(-s)\Big\}ds\right] \qquad (6.134)$$

Thus, the problem is to select a matrix transfer function $H(s)$ such that MSE in (6.134) is minimized; this is a standard problem in the calculus of variations. To solve this problem, let $H(s)$ be expressed as

$$H(s) = H_0(s) + \varepsilon\Sigma(s) \qquad (6.135)$$

where H_0 is the optimum, as yet unknown, matrix and $\varepsilon\Sigma$ is called the variation. Note that MSE must have its minimum value where $\varepsilon = 0$ independent of Σ. Hence we can find H_0 by forcing the following equation to be true

$$\left.\frac{\partial\,MSE}{\partial\,\varepsilon}\right|_{\varepsilon=0} = 0 \qquad (6.136)$$

The logic of this requirement is seen by examining Fig.6.9, where we see that the derivative of MSE with respect to ε becomes zero independent of Σ.

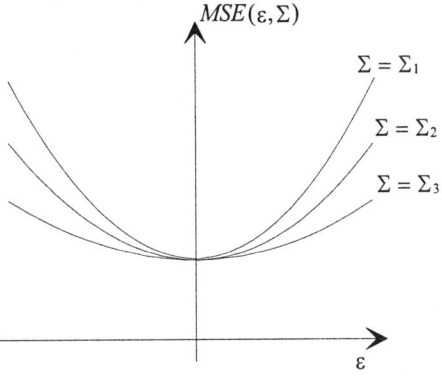

Fig. 6.9. MSE plots for different variations

If we substitute eq. (6.135) into eq. (6.134), we obtain

$$MSE(\varepsilon, \Sigma) = \frac{1}{2\pi j} \text{trace}\left[\int_{-j\infty}^{j\infty} F(s) ds\right]$$

$$F(s) = \left\{\left[I - H_0(s) - \varepsilon \Sigma(s)\right]S_x(s)\left[I - H_0(-s) - \varepsilon \Sigma(-s)\right]^T + \right.$$

$$\left. + \left[H_0(s) + \varepsilon \Sigma(s)\right]S_v(s)\left[H_0(-s) + \varepsilon \Sigma(-s)\right]^T\right\} \tag{6.137}$$

Here we have added the arguments ε and Σ to MSE to emphasize that the value of MSE is dependent on ε and Σ. Equation (6.136) is now

$$0 = \frac{1}{2\pi j} \text{trace}\left[\int_{-j\infty}^{j\infty} \left\{H_0(s)\left[S_x(s) + S_v(s)\right] - S_x(s)\right\} \Sigma^T(-s) ds\right]$$

$$+ \frac{1}{2\pi j} \text{trace}\left[\int_{-j\infty}^{j\infty} \Sigma(s)\left\{\left[S_x(s) + S_v(s)\right]H_0^T(-s) - S_x(s)\right\} ds\right] \tag{6.138}$$

If we make use of the symmetry of the spectral density matrices and the fact that $\text{trace}\left[xy^T\right] = \text{trace}\left[yx^T\right]$, then eq. (6.138) becomes (the two integrals in eq. (6.138) are equal),

$$\text{trace}\left[\int_{-\infty}^{\infty} \left\{H_0(s)\left[S_x(s) + S_v(s)\right] - S_x(s)\right\} \Sigma^T(-s) ds\right] = 0 \tag{6.139}$$

We must select $H_0(s)$ such that it satisfies eq. (6.139). Clearly, eq. (6.139) will be satisfied for any $\Sigma(s)$ if

$$H_0(s) = S_x(s)\left[S_x(s) + S_v(s)\right]^{-1} = S_x(s)S_y^{-1}(s) \tag{6.140}$$

The solution is called the unrealizable Wiener filter, because $H_0(s)$ generally possesses poles in the right half of the s-plane. We remember that since we are using Fourier transforms (e.g. $s = j\omega$) poles in the right half-plane do not indicate an unstable system, but rather a noncausal system, requiring response before excitation, which is physically unrealizable.

The difficulty is that we have not examined the structure of the problem closely enough. Although it is surely true that a $H_0(s)$ which satisfies (6.140) will satisfy (6.139), it is also possible to satisfy (6.139) without satisfying (6.140), since $\Sigma(s)$, and hence $\Sigma(-s)$, are not completely arbitrary. For $H(s)$ in (6.135) to be an admissible solution, $H(s)$, $H_0(s)$ and $\Sigma(s)$ must be physically realizable, or in other words, must have all their poles in the left half of the s-plane. By taking advantage of this requirement on $\Sigma(s)$, it is possible to select $H_0(s)$ to satisfy eq. (6.139) and to be physically realizable.

Let the spectral density matrix $S_y(s) = S_x(s) + S_v(s)$ be spectrum factored into the form

$$S_y(s) = S_x(s) + S_v(s) = W(s)W^T(-s) \tag{6.141}$$

where the matrix $W(s)$ is such that rational function $\det(W(s))$, with $\det(\cdot)$ denoting the determinant, has all its poles and zeros in the left half of the s-plane. The actual process of finding $W(s)$ is, in general, a difficult computational task, and can normally be accomplished numerically only by means of lengthy algorithms.

Eq. (6.139) may now be written as

$$\text{trace}\left[\int_{-j\infty}^{j\infty}\left[H_0(s)W(s)-S_x(s)W^{-T}(-s)\right]W^T(-s)\Sigma^T(-s)ds\right]=0 \qquad (6.142)$$

The next step is to make a partial-fraction expansion of the quantity $S_x(s)W^{-T}(-s)$ as

$$S_x(s)W^{-T}(-s)=A(s)+B(-s) \qquad (6.143)$$

where $A(s)$ contains all terms with poles in the left half-plane, and $B(-s)$ contains all terms with poles in the right half-plane. The matrix $A(s)$, which can also be defined as the Laplace transform of the positive time portion of the time response of the transformed quantity on the left hand side of (6.143), will be defined as the physically realizable portion and noted as

$$A(s)=\left[S_x(s)W^{-T}(-s)\right]^+ \qquad (6.144)$$

If eq. (6.143) is substituted into eq. (6.142), the required necessary condition is now

$$\text{trace}\left[\int_{-j\infty}^{j\infty}\left\{H_0(s)W(s)-A(s)\right\}W^T(-s)\Sigma^T(-s)ds+\int_{-j\infty}^{j\infty}B(-s)W^T(-s)\Sigma^T(-s)ds\right]=0 \quad (6.145)$$

However, the second integral is equal zero, owing to the Cauchy theorem, since all the poles of the quantity $B(-s)W^T(-s)\Sigma^T(-s)$ lie in the right half plane. If the contour of integration is closed to the left, no poles are encircled, and therefore the sum of residuals is equal to zero; that is the value of the integral is zero. In this way, eq. (6.145) becomes

$$\text{trace}\left[\int_{-j\infty}^{j\infty}\left\{H_0(s)W(s)-A(s)\right\}W^T(-s)\Sigma^T(-s)ds\right]=0 \qquad (6.146)$$

The optimal physically realizable filter is given by

$$H_0(s)=A(s)W^{-1}(s) \qquad (6.147a)$$

or equivalently

$$H_0(s)=\left[S_x(s)W^{-T}(-s)\right]^+W^{-1}(s) \qquad (6.147b)$$

This is the final form of the matrix Wiener filter for multivariable stationary estimation problem. Eq. (6.147b) can be also rewritten as

$$H_0(s)=\left[S_{yx}(s)W^{-T}(-s)\right]^+W^{-1}(s) \qquad (6.148)$$

since the cross-covariance function

$$R_{yx}(\tau)=E\left\{y(t+\tau)x^T(t)\right\}=E\left\{\left[x(t+\tau)+v(t+\tau)\right]x^T(t)\right\}$$
$$=E\left\{x(t+\tau)x^T(t)\right\}+E\left\{v(t+\tau)x^T(t)\right\}=R_x(\tau)+R_{vx}(t)$$

so that the spectral density matrices satisfy

$$S_{yx}(s)=S_x(s)+S_{vx}(s) \qquad (6.149)$$

However, for the case when the signal $x(t)$ and the noise $v(t)$ are uncorrelated $R_{vx}(\tau)=0$, i.e. $S_{vx}(s)=0$, from which follows $S_{yx}(s)=S_x(s)$. If this result is substituted into eq. (147b), we obtain eq. (6.148). It should be noted that eq. (6.148) is valid for the general case when $x(\tau)$ and $v(\tau)$ are correlated, but in this case eq. (6.141) reduces to

$$S_y(s) = S_x(s) + S_v(s) + S_{xv}(s) + S_{vx}(s) = W(s)W^T(-s) \tag{6.150}$$

Eq. (6.150) follows from the fact that $S_y(s)$ is Fourier transform of the cross-covariance function

$$R_y(\tau) = E\{y(t+\tau)y^T(t)\} = E\{[x(t+\tau)+v(t+\tau)]^T[x(t)+v(t)]\}$$
$$= E\{x(t+\tau)x^T(t)\} + E\{x(t+\tau)v^T(t)\} + E\{v(t+\tau)x^T(t)\} + E\{v(t+\tau)v^T(t)\}$$
$$= R_x(\tau) + R_{xv}(\tau) + R_{vx}(\tau) + R_v(\tau)$$

Let us illustrate the use of the Wiener filter by considering a simple scalar example. However, the matrix Wiener filter for the multivariable stationary estimation problem never gained wide acceptance in engineering practice , because of the sever computational problems associated with the required matrix spectrum factorization. Moreover, the utilization of the Kalman filter algorithms has essentially superseded many uses of the matrix Wiener filter.

Example 6.7 (scalar continuous Wiener filter): The signal spectral density is given by

$$S_x(s) = \frac{3600}{-s^2(169 - s^2)}$$

and the noise is white, so that

$$S_v(s) = 1.$$

The signal and noise are uncorrelated. For this problem, $S_y(s)$ is

$$S_y(s) = S_x(s) + S_v(s) = \frac{3600 - 169s^2 + s^4}{-s^2(169 - s^2)}$$

The required spectral factorization in eq. (6.141) is easily accomplished to give

$$W(s) = \frac{60 + 17s + s^2}{s(13 + s)} \;;\; W^T(-s) = \frac{60 - 17s + s^2}{-s(13 - s)}$$

Note that two poles at the origin have been divided, so that one is treated as being in the right half plane and the other as being in the left half plane. Substituting the results into eq. (6.147b), we obtain

$$H_0(s) = \frac{\left\{ \dfrac{3600/\left[-s^2(13+s)(13-s)\right]}{(60-17s+s^2)/\left[-s(13-s)\right]} \right\}}{(60+17s+s^2)/\left[s(13+s)\right]} = \frac{\left\{ \dfrac{3600}{s(60-17s+s^2)(s+13)} \right\}^+}{(60+17s+s^2)/\left[s(13+s)\right]}$$

$$H_0(s) = \frac{\left\{ \dfrac{3600/\left[-s^2(13+s)(13-s)\right]}{(60-17s+s^2)/\left[-s(13-s)\right]} \right\}}{(60+17s+s^2)/\left[s(13+s)\right]} = \frac{\left\{ \dfrac{3600}{s(60-17s+s^2)(s+13)} \right\}^+}{(60+17s+s^2)/\left[s(13+s)\right]}$$

Let us consider the numerator of this expression. The partial fraction expansion takes the form

$$\frac{3600}{s(s+13)(s^2 - 17s + 60)} = \frac{\frac{60}{13}}{s} - \frac{\frac{8}{13}}{s+13} + \frac{k_1 s + k_2}{s^2 - 17s + 60}$$

The last term is unimportant, since we need the portion $A(s)$ of the partial fraction expansion (see eq. (6.143)) which has poles in the left half plane. Since $s^2 - 17s + 60$ is in the numerator polynomial of $W^T(-s)$, it must have roots in the right half s-plane only. Hence we have

$$\left\{ \frac{3600}{s(s+13)(s^2-17s+60)} \right\}^+ = \frac{60}{13}{s} - \frac{\frac{8}{13}}{s+13} = \frac{4s+60}{s(s+13)}$$

and the optimal Wiener filter is given by eqs. (6.147); that is

$$H_0(s) = \frac{(4s+60)/[s(s+13)]}{(s^2+17s+60)[s(s+13)]} = \frac{4s+60}{s^2+17s+60}$$

The minimum value of the MSE criterion is calculated by substituting the expression for $H_0(s)$ into eq. (6.130) and carrying out the indicated contour integration. This work is considerably simplified because integrals of the form

$$I_n = \frac{1}{2\pi j} \int_{-j\infty}^{j\infty} A_n(s) A_n(-s) \, ds \tag{6.151}$$

where

$$A_n(s) = \frac{\displaystyle\sum_{i=0}^{n-1} c_i s^i}{\displaystyle\sum_{i=0}^{n} d_i s^i} \tag{6.152}$$

have been tabulated for values of n as high as 10. Particularly, the values of I_n for $n=1,2,3$ are

$$I_1 = \frac{c_0^2}{2d_0 d_1} \; ; I_2 = \frac{c_0^2 d_2 + c_1^2 d_0}{2 d_0 d_1 d_2} \; ; I_3 = \frac{c_2^2 d_0 d_1 + \left(c_1^2 - 2c_0 c_2\right) d_0 d_3 + c_0^2 d_2 d_3}{2 d_0 d_3 \left(-d_0 d_3 + d_1 d_2\right)} \tag{6.153}$$

In the case of uncorrelated signal and noise, $S_{\tilde{x}}(s)$ may be written as in eq. (6.133); that is

$$S_{\tilde{x}}(s) = [I - H_0(s)] S_x(s) [I - H_0(-s)]^T + H_0(s) S_v(s) H_0^T(s) \tag{6.154}$$

By treating each of the terms in the expression separately, the factorization necessary to apply the integral forms is easily accomplished by just factoring $S_x(s)$ and $S_v(s)$, which are often given in factored form. A desirable feature of this approach is that the contribution to the MSE criterion due to signal and noise errors is specified. If we define MSE_x and MSE_v signal and noise MSE, respectively, then we have from eq. (6.134)

$$\text{MSE} = \text{MSE}_x + \text{MSE}_v \tag{6.155}$$

where

$$\text{MSE}_x = \frac{1}{2\pi j} \text{trace} \left[\int_{-j\infty}^{j\infty} [I - H_0(s)] S_x(s) [I - H_0(-s)]^T \, ds \right] \tag{6.156}$$

and

$$\text{MSE}_v = \frac{1}{2\pi j} \text{trace} \left[\int_{-j\infty}^{j\infty} H_0(s) S_v(s) H_0^T(s) \, ds \right] \tag{6.157}$$

Let us use the foregoing procedure to find the minimum value of the MSE criterion for the optimal filter developed in the preceding example.

Example 6.8 (MSE criterion for the optimal filter; revisited example 6.7) : In the preceding example the signal and noise were uncorrelated, so that the simplified eqs. (6.155)-(6.157) are applicable. The MSE is given by eq. (6.156), where $S_x(s)$, $S_v(s)$ and $H_0(s)$ are given in the preceding example, i.e.

$$MSE_x = \frac{1}{2\pi j} \int_{-j\infty}^{j\infty} \left(1 - \frac{4s+60}{s^2+17s+60}\right)\left(1 - \frac{-4s+60}{s^2-17s+60}\right)\frac{3600}{-s^2(169-s^2)}\,ds$$

$$= \frac{1}{2\pi j}\int_{-j\infty}^{j\infty} \frac{60}{s^2+17s+60}\frac{60}{s^2-17s+60}\,ds = \frac{1}{2\pi j}\int_{-j\infty}^{j\infty} A_2(s)A_2(-s)\,ds$$

This expression is now in the form to apply the I_2 integral expression of eq. (6.151), with $c_1 = 0$; $c_0 = d_0 = 60$; $d_1 = 17$ and $d_2 = 1$, so that from (6.153) follows

$$MSE_x = I_2 = \frac{60^2 \cdot 1}{2 \cdot 60 \cdot 17 \cdot 1} = \frac{60}{34} = 1.765$$

From eq. (6.157) the MSE_v is given by

$$MSE_v = \frac{1}{2\pi j}\int_{-j\infty}^{j\infty} \frac{4s+60}{s^2+17s+60}\cdot\frac{-4s+60}{s^2-17s+60}\,ds$$

Once again we use the I_2 expression in eq. (6.153), but this time we have $c_0 = 60, c_1 = 4, d_0 = 60, d_1 = 17, d_2 = 1$, so that

$$MSE_v = \frac{60^2 \cdot 1 + 4^2 \cdot 60}{2 \cdot 60 \cdot 17 \cdot 1} = \frac{76}{34} = 2.235$$

Therefore, the total MSE is

$$MSE = 1.765 + 2.235 = 4.00$$

CHAPTER 7

OPTIMUM RECURSIVE LINEAR ESTIMATION:
KALMAN FILTERING

In this chapter we develop model-based processor for the dynamic estimation problem; that is, the estimation of processes that vary with time. Here the state space representation is used as the basic model. first, the fitting of noisy discrete-time observation to a continuous-time state space model is approached by block processing using minimum variance linear estimator. Then we develop the linear sequential continuous-discrete estimator, the so-called continuous-discrete Kalman filter, for estimating continuous-time states from noisy discrete observations. We discuss the operation of the filter as a predictor-corrector algorithm. This is followed by the derivation of the discrete Kalman filter for estimating the discrete-time states from discrete observations, as well as the development of the continuous Kalman filter for estimating the continuous-time states from the continuous time observations. The innovations sequence is also discussed in detail to point out various properties useful for tuning the Kalman filter, and some of the geometric concepts are developed. Finally, the stochastic controllability, stochastic observability, the stability of the Kalman filter and the statistical steady state is discussed. The corresponding steady-state Kalman filter is given; and the link between the Kalman filter and the Wiener filter is established.

7.1 Continuous estimation with discrete observations

Estimation is an attempt to determine a set of variables such as Cartesian components of position, velocity, acceleration, etc., from observations related to those variables. The observations may be:
- *incomplete*: related to some, but not all of the variables to be estimated;
- *indirect*: expressed as some known function of the variables to be estimated;
- *intermittent*: taken at irregularly space instants of time;
- *inexact*: corrupted by biases and by random errors.

In many applications the variable to be estimated are dynamic, i.e. they are components of the solution x to a vector differential equation in the general form:

$$\frac{dx}{dt} = f(x,t) \tag{7.1}$$

which for the case of a linear functional relationship reduces to

$$\frac{dx}{dt} = Ax + w \tag{7.2}$$

where A is a matrix of coefficients describing the system dynamics (A may be function of time but not of x) and w a forcing function with specified characteristics.

For dynamic systems estimation subdivides into these operations:
1) *Extrapolation*: the solution $x(m)$ of Eq. (7.1) or (7.2) at a time t_m is estimated from observations taken prior to t_m.

2) *Filtering*: $x(m)$ is estimated from observations taken at and before t_m.

3) *Smoothing*: $x(m)$ is estimated from observations included those taken after t_m.

These operations are also categorized as to:

1) whether the dynamic process characteristics are known, partially known or largely unknown;

2) whether the applicable equations are linear or nonlinear .

The former case is characterized by linearity in the dynamics (e.g. eq. (7.2)) and in the equations relating x to observations; that is, a scalar measurement at the discrete instant t_m could have a value denoted by

$$z(m) = H(m)x(m) + v(m) \tag{7.3}$$

where H expresses the proportionality between the measurement z and each component of x, while v denotes measurement error, or noise.

For estimators using knowledge of the dynamic process that generates x, the dynamic equations (7.1) or (7.2) in combination with observation equation (7.3) constitute a system or signal model.

Furthermore, many practical applications of estimation theory involve fitting data to an imperfect model, based on equations in nearly linear form. This text specifies guidelines for these applications.

7.1.1 Basic concepts in estimation

The fitting of imperfect data to a model is approached by a method that has gained wide acceptance through a host of successful applications. As a first step in this approach, any linear dynamical system can be described by some number n of linearly independent first-order differential equations, reexpressible in the form (7.2) for a linear n-dimensional vector differential equation (that is, none of the n equations can be obtained from a linear transformation of the others; the equations can still be coupled, however, since A in eq. (7.2) can have nonzero off-diagonal elements). Its solution $x(m)$ at any time t_m is called the state vector, or simply, the states. Thus $x(m)$ can represent the "initial conditions" applicable to the dynamic differential equation (7.2) at time t_m, from which it follows that [1, 2, 17, 19, 23, 25, 26, 27, 30, 31, 48]:

1) The state is unambiguous: a complete set of values for the state variables at time t_m would enable a complete and unique determination of x for all time in eq. (7.2), if A and w were fully specified.

2) The state is nonredudant: no fewer than n variables could enable the full determination of dynamic behavior just described.

Continuous dynamic equations (7.2) and discrete updates (7.3) involve zero-mean random vectors w and v, respectively, defined such that each component is a zero-mean random variable, i.e.

$$E\{w(t)\} = 0 \; ; \; E\{v(t_m)\} = 0 \tag{7.4}$$

with t and t_m being a continuous-time and discrete-time instant, respectively. In addition, a sequentially uncorrelated property for w and v is supposed. This property is expressed for v in terms of a observable covariance matrices $R(m)$, specified at discrete instants t_m such that

$$E\{v(m)v^T(n)\} = R(m)\delta(m,n) \tag{7.5}$$

where $\delta(m,n)$ is the Kronecker's delta, i.e.

$$\delta(m,n) = \begin{cases} 0 \; ; \; m \neq n \\ 1 \; ; \; m = n \end{cases} \tag{7.6}$$

and for w in terms of a process covariance matrix Q with the Dirac delta function δ, such that

$$E\{w(t)w^T(\tau)\} = Q(t)\delta(t-\tau) \tag{7.7}$$

in which the elements of Q are white noise spectral densities. While allowed to change with time, R and Q are subject to restrictions of positive definiteness and positive semidefiniteness, respectively, i.e. for any nonzero vector α of conformable dimension

$$\begin{aligned} \alpha^T R \alpha > 0 \; ; \; \alpha \neq 0 \\ \alpha^T Q \alpha \geq 0 \; ; \; \alpha \neq 0 \end{aligned} \tag{7.8}$$

The random vector $w(t)$ influences the dynamics of $x(t)$ in accordance with the state transition equation

$$x(t) = \Phi(t,t_{m-1})x(t_{m-1}) + \int_{t_{m-1}}^{t} \Phi(t,\tau)w(\tau)d\tau \tag{7.9}$$

where Φ is the state transition matrix defined by

$$\frac{d\Phi(t,\tau)}{dt} = A\Phi(t,\tau) \; ; \; \Phi(t,t) = I \tag{7.10}$$

The random vector $v(m)$ degrades the observations as follows: a data vector, consisting of r_m simultaneous scalar-measurements, has an error-free value of

$$y(m) = H(m)x(m) \tag{7.11}$$

and an observed value given by the r_m-dimensional generalization of (7.3)

$$z(m) = H(m)x(m) + v(m) \tag{7.12}$$

where $H(m)$ is the $r_m \times n$ matrix expressing the proportionality between each observable, i.e. each measured quantity, and each state variable. A *continuous-discrete* estimation problem may now be defined as follows:

Let a sequence of noise-corrupted data vectors, characterized by (7.12), be used to construct estimates $\hat{x}(m)$ for the state of a dynamic process defined by (7.2) and (7.9), under conditions of known $A(t), Q(t), H(m)$ and $R(m)$. An estimate that minimizes the variance $E\{\tilde{x}_j^2\}$ of every component $\tilde{x}_j$ of the state uncertainty, or estimation error, $\tilde{x}$ defined as

$$\tilde{x} = x - \hat{x} \tag{7.13}$$

is known as the *minimum variance unbiased estimate* (see chapter 5).

Expressions will now be derived relating state estimates to observations under conditions just given. The covariance matrix of estimation error (7.13) is also presented. Processing of all observations in a block, the so-called batch or off-line processing, is first considered, followed by description of repeated (sequential) updating as observations appear in succession, the so-called recursive or on-line processing.

7.1.2 Linear minimum variance estimator by block processing

A special case of the linear estimation problem just stated follows from the condition $w(t) = 0$ throughout the data fitting duration. In this case, eq. (7.12) in combination with a dynamic transformation (7.9) yields

$$z(m) = H(m)\Phi(t_m, t_0)x(0) + v(m) \tag{7.14}$$

With each $r_m \times n$ product $H\Phi$ regarded as a partition of a larger matrix J, the measurements obtained at observation time $t_1, t_2, ..., t_m, ..., t_M$ can be collected in the form

$$z = Jx(0) + V \tag{7.15}$$

in which

$$z = \begin{bmatrix} z(1) \\ z(2) \\ \vdots \\ z(M) \end{bmatrix} ; J = \begin{bmatrix} H(1)\Phi(t_1, t_0) \\ H(2)\Phi(t_2, t_0) \\ \vdots \\ H(M)\Phi(t_M, t_0) \end{bmatrix} ; V = \begin{bmatrix} v(1) \\ v(2) \\ \vdots \\ v(M) \end{bmatrix} \tag{7.16}$$

that is, the number of elements in z and V, and the number of rows of J, is the total number of scalar measurements

$$r = \sum_{m=1}^{M} r_m \tag{7.17}$$

There are then r equations that can be used to estimate the n components of $x(0)$, but in general, the unknown V prevents an exact solution of (7.15). One possible procedure is to seek an estimate $\hat{x}(0)$ that minimizes the quantity

$$\left\| x - J\hat{x}(0) \right\|^2 = \left[z - J\hat{x}(0) \right]^T \left[z - J\hat{x}(0) \right] \tag{7.18}$$

where $\|\cdot\|$ is the Euclidean norm. With unequal accuracy for different measurements, however, a better procedure is to allow the more precise observations a greater influence on the estimate; suppose that

$$U = \left(E\{VV^T\} \right)^{-1/2} \tag{7.19}$$

where $E\{VV^T\}$ is the covariance matrix of the zero-mean random vector V, and the sign -1/2 denotes the square root of the inverse matrix. Namely, $R = \left(E\{VV^T\} \right)^{-1}$ is a symmetric $r \times r$ noise covariance matrix. Furthermore, suppose D and E are, respectively, diagonal and nonsingular $r \times r$ matrices, satisfying

$$R = EDE^T$$

Then EDE^T is referred to as a *spectral decomposition* of R. If R is positive definite, it is possible to obtain spectral decomposition in which $D=I$, the identity matrix, i.e. $R = EE^T$. Of particular interest is the decomposition in which E is a symmetric matrix U; that is $R = UU^T = U^2$. The matrix U is called the *square root* of R.

A weighted sum of squares of difference elements can then be defined in the form

$$J = \left\| U\left(Z - J\hat{x}(0)\right) \right\|^2 = \left[U\left(Z - J\hat{x}(0)\right) \right]^T \left[U\left(Z - J\hat{x}(0)\right) \right]$$
$$= \left(Z - J\hat{x}(0)\right)^T U^T U \left(Z - J\hat{x}(0)\right) = \left(Z - J\hat{x}(0)\right)^T R \left(Z - J\hat{x}(0)\right) \tag{7.20}$$

since $U^T U = U^2 = R$.

As a positive definite quadratic function, J is minimized when its partial derivative with respect to each component of $\hat{x}(0)$ is set to zero

$$\frac{\partial J}{\partial \hat{x}(0)} = 0 \tag{7.21}$$

However, since from (7.20)

$$J = Z^T U^T U Z - \hat{x}^T(0)(UJ)^T UZ - Z^T U^T U J\hat{x}(0) + \hat{x}^T(0)(UJ)^T (UJ)\hat{x}(0) \tag{7.22}$$

and using the following rules for differentiation of a scalar function with respect to a vector a

$$\frac{\partial \left(b^T a\right)}{\partial a} = b \; ; \; \frac{\partial \left(a^T b\right)}{\partial a} = b \; ; \; \frac{\partial \left(a^T B a\right)}{\partial a} = Ba + B^T a \tag{7.23}$$

where a and b are the column vectors (i.e. b^T is a row vector) and B is a square matrix of the corresponding order, one obtains

$$\frac{\partial J}{\partial \hat{x}(0)} = -\frac{\partial}{\partial \hat{x}(0)} \left[\hat{x}^T(0)(UJ)^T UZ \right] - \frac{\partial}{\partial \hat{x}(0)} \left[\left((UJ)^T UZ\right)^T \hat{x}(0) \right] + \frac{\partial}{\partial \hat{x}(0)} \left[\hat{x}^T(0)(UJ)^T (UJ)\hat{x}(0) \right]$$
$$= -(UJ)^T UZ - (UJ)^T UZ - 2(UJ)^T (UJ)\hat{x}(0) = 0$$

or equivalently

$$(UJ)^T \left[U\left(Z - J\hat{x}(0)\right) \right] = 0 \tag{7.24}$$

An estimate that conforms to this minimization will be written $\hat{x}(0) = \hat{x}_0$. The elements of the difference vector

$$\varepsilon = Z - J\hat{x}_0 \tag{7.25}$$

are then called *true residuals* and, with $S = UJ$, the expression (7.24) can be rewritten as

$$S^T UZ - S^T S\hat{x}_0 = 0$$

from which one obtains

$$\hat{x}_0 = \left(S^T S\right)^{-1} S^T UZ \; ; \; \left| S^T S \right| \neq 0 \tag{7.26}$$

where $|\cdot|$ denotes the determinant. The solution (7.26) is called the *weighted least squares estimate*, since it minimizes the weighted sum of squares of the true residuals; that is the criterion J in (7.20) can be rewritten as

$$J = \varepsilon^T R \varepsilon \tag{7.27}$$

The corresponding smoothed estimate for the state at another time, e.g. t_m, is (see eq. (7.9) in which $w = 0$)

$$\hat{x}_m = \Phi(t_m, t_0)\hat{x}_0 \qquad (7.28)$$

When all components of V in (7.19) have equal variance U is the product of a scalar, e.g. $1/\sigma$ with σ being the standard deviation of the component term and an orthogonal matrix (the square matrix O is said to be orthogonal if $O^T O = OO^T = I$); eq. (7.26) then reduces to

$$\hat{x}_0 = \left((UJ)^T UJ\right)^{-1} (UJ)^T UZ = \left(J^T U^T UJ\right)^{-1} J^T U^T UZ$$
$$= \left(J^T \frac{1}{\sigma^2} O^T OJ\right)^{-1} J^T \frac{1}{\sigma^2} O^T OZ = \left(J^T J\right)^{-1} J^T Z \ ; \ \left|J^T J\right| \neq 0 \qquad (7.29)$$

since $U^T U = \frac{1}{\sigma^2} O^T O = \frac{1}{\sigma^2} I$. The solution (7.29) is called the (unweighted) *least squares* solution.

Another special case follows when J is nonsingular, which can occur only when $r=n$. In this case, estimation on the basis of (7.15) and (7.29) reduces to

$$\hat{x}_0 = J^{-1}\left(J^T\right)^{-1} J^T Z = J^{-1} Z \ ; \ \left|J\right| \neq 0 \qquad (7.30)$$

A solution from any of the equations (7.26), (7.29) or (7.30) is contingent upon a nonvanishing determinant. If the necessary inversion can be performed, we say that the state is observable. A classical definition, which formulates an observability matrix and tests its rank, produces a sharp division into two classes: observable and unobservable. However, for many applications a sharp division between observable and nonobservable cases is insufficient. The information matrix $S^T S$ in (7.26) offers a general quantitative assessment. Namely, a quantitative measure of accuracy is needed for these cases in which inversion is possible. From (7.15) and (7.26), the estimation error $\tilde{x}_0$ is

$$\tilde{x}_0 = x(0) - \hat{x}_0 = x(0) - \left(S^T S\right)^{-1} S^T UZ = x(0) - \left(S^T S\right)^{-1} S^T U\left(Jx(0)+V\right)$$
$$= x(0) - \left(S^T S\right)^{-1} S^T Sx(0) - \left(S^T S\right)^{-1} S^T UV = -\left(S^T S\right)^{-1} S^T UV \ ; \ \left|S^T S\right| \neq 0 \qquad (7.31)$$

since $S = UJ$ and $\left(S^T S\right)^{-1} S^T S = I$, with a covariance matrix

$$E\left\{\tilde{x}_0 \tilde{x}_0^T\right\} = \left(S^T S\right)^{-1} S^T UE\left\{VV^T\right\} U^T S\left(S^T S\right)^{-1}$$
$$= \left(S^T S\right)^{-1} S^T UU^{-2} U^T S\left(S^T S\right)^{-1} = \left(S^T S\right)^{-1} S^T S\left(S^T S\right)^{-1}$$
$$= \left(S^T S\right)^{-1} \ ; \ \left|S^T S\right| \neq 0 \qquad (7.32)$$

Here is used the fact that from (7.19) $E\left\{VV^T\right\} = U^{-2}$, as well as that U is a symmetric matrix, i.e. $U^T = U$. The error covariance matrix (7.32) provides an indication of observability, not only by the variance of uncertainty (diagonal elements of the matrix) corresponding to each state, but also by the ease or difficulty of obtaining an accurate inversion from inexact computation. Two situations that suggest attention to possible numerical degradation, especially with large r, are the following:

1) *High precision measurement*: extremely accurate observation can introduce wide spreads between elements of (7.19). In the limit, a measurement known to be exact would reduce a row, let us say $z(m)$, of (7.15) to the form

$$z(m) = J(m)x(0) \qquad (7.33)$$

producing $v(m) = z(m) - J(m)x(0) = 0$, which will result into the zero-row in the matrix (7.19). In this way, the matrix $S^T S = (UJ)^T UJ = J^T U^T UJ = J^T U^2 J$ will not be regular; that is, the condition $|S^T S| \neq 0$ will not be fulfilled.

2) *Highly correlated measurement errors*: close correlation between different observation errors can produce ill-conditioning, in the sense of the small determinant $|S^T S|$. In the limit, full correlation between two measurement errors,, e.g. $v(m) = v(n)$, can be expressed in a form

$$z(m) = J(m)x(0) + v(m) \; ; \; z(n) = J(n)x(0) + v(n)$$
$$z(m) - z(n) = \left[J(m) - J(n) \right] x(0) \tag{7.34}$$

This means that two rows in (7.19) will be identical, and by linear operations equivalent to substitution and elimination, they will result in a zero-row in the matrix (7.19), so that, similarly as in the preceding example, the condition $|S^T S| \neq 0$ will not be satisfied.

Namely, as constraining relations between the state variables eqs. (7.33) and (7.34) reduces the number of independent quantities to be estimated. As mentioned above, by linear operations equivalent to substitution and elimination, these relations can be used to advantage in a proper reformulation. This opportunity is known as *dimensionality reduction*, and, if unrecognized, it appears as a computational disadvantage through violation of independence between states.

As instance of *marginal observability* that differs from the two items just mentioned is a *partial deficiency* in observational data. Situations arise when some state variables are accurately monitored, while others are associated with only the lowest precision measurements. In a case of marginal observability, numerical inversion in (7.26) presents inescapable computational difficulties. An attempt to use (7.26) in that case can often produce large estimation errors, even in those states that could be adequately determined from the data. A computational efficient procedure, particularly for large data blocks, is replacement of (7.26) by

$$\hat{x}_0 = S^\# UZ \tag{7.35}$$

where the operator # denotes the *Penrose generalized inverse*, or *pseudoinverse* defined as

$$S^\# = \lim_{\delta \to 0} \left(S^T S + \delta I \right)^{-1} S^T \tag{7.36}$$

if the matrix S has more rows than columns. Moreover, for nonsingular $S^T S$, the pseudoinverse is defined by

$$S^\# = \left(S^T S \right)^{-1} S^T \tag{7.37}$$

In the solution of linear equations (7.15) this is the so-called *overdetermined case*, where there are more equations (measurements) than unknown states. The resulting solution (7.35) is the best in a weighted least squares (minimum variance) sense (7.27). If S has more columns than rows, the pseudoinverse is defined as

$$S^\# = S^T \lim_{\delta \to 0} \left(S^T S + \delta I \right)^{-1} \tag{7.38}$$

or

$$S^\# = S^T \left(SS^T \right)^{-1} \tag{7.39}$$

for nonsingular SS^T. This corresponds to the *underdetermined case*; there are fewer equations (measurements) than unknowns (states). Typically, such a situation leads to an infinite number of weighted least-squares solutions (consider the least-squares fit of a straight line to a point). The solution resulting from the pseudoinverse is also best in a least-squares (minimum variance) sense (7.27), and the vector $\hat{x}_0$ in (7.35) is the solution of minimum length.

The limits (7.36) and (7.38) always exist, are unique, and can be computed from the four conditions, which for real matrices are given by

$$SS^{\#}S = S \; ; \; S^{\#}SS^{\#} = S^{\#} \; ; \; SS^{\#} = \left(SS^{\#}\right)^{T} \; ; \; S^{\#}S = \left(S^{\#}S\right)^{T} \tag{7.40}$$

A powerful algorithms for computing $S^{\#}$ appears in the numerous literature.

The formulation (7.35) is valid even when $\left(J^T J\right)$ vanishes, so long as U is nonsingular. The case of a vanishing determinant $\left|J^T J\right|$ is exemplified by a column of zeros in S (due to a lack of measured data to one of the states); there will then be a row of zeros in $S^T S$ and (7.26) will be inapplicable. A weighted least squares estimates for the observable states is obtainable in any case from (7.35), which would also null the estimate of the unobservable states (due to the row of zeros in $S^{\#}$). More generally, in the presence of unobservable states, the solution (7.35) simultaneously minimizes J of (7.27) and the length $\|\hat{x}_0\|$ itself. In summary, *weighted least squares smoothing* provides an opportunity to extract maximum information from all measurements within a data block. Its presentation has also produced much insight for the development of linear sequential estimator that follows.

7.1.3 Linear sequential continuous estimation from discrete observations (continuous-discrete Kalman filtering)

Suppose that an estimate for the state at t_{m-1} and a subsequent observation vector $z(m)$ are available for processing, but the earlier data vectors $z(m-1), z(m-2),...$ are no longer accessible. Instead of repeating (7.26) with each added measurement an update can be performed as follows:

- An a priori state estimate $\hat{x}_m^-$ just before the latest observation $z(m)$ is obtained by extrapolating, using eq. (7.28), the state $\hat{x}_{m-1}^+$ estimated just after the preceding observation $z(m-1)$ (see Fig. 7.1):

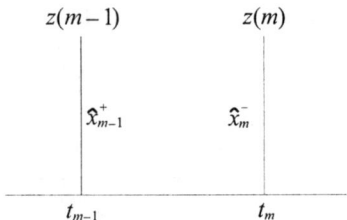

Fig. 7.1: Illustration for computational cycle of sequential estimation

$$\hat{x}_m^- = \Phi\left(t_m, t_{m-1}\right)\hat{x}_{m-1}^+ \tag{7.41}$$

so that an anticipated value $\hat{z}_m^-$, on the basis of eqs. (7.3) and (7.41), for the data vector

$$\hat{z}_m^- = H(m)\hat{x}_m^-$$
(7.42)

can be subtracted from $z(m)$ to form the *predicted residual*

$$\varepsilon(m) = z(m) - \hat{z}_m^- = z(m) - H(m)\hat{x}_m^-$$
(7.43)

which produces the desired adjustment to the estimate when linearly weighted by a gain matrix K_m:

$$\hat{x}_m^+ = \hat{x}_m^- + K_m\varepsilon(m) = \hat{x}_m^- + K_m\left(z(m) - H(m)\hat{x}_m^-\right)$$
(7.44)

This updating procedure is recursive, in the sense that only the most recent values are needed for all variables. Thus, in contrast to repeated application of eq. (7.26), both computation and data storage requirements are eased.

To determine the weighting matrix K_m that produces a minimum variance estimate, eq. (7.44) is combined with (7.12) and (7.13) to yield

$$\begin{aligned}\varepsilon(m) &= z(m) - \hat{z}_m^- = H(m)x(m) + v(m) - H(m)\hat{x}_m^- \\ &= H(m)\left[x(m) - \hat{x}_m^-\right] + v(m) = H(m)\tilde{x}_m^- + v(m)\end{aligned}$$
(7.45)

where

$$\tilde{x}_m^- = x(m) - \hat{x}_m^-$$
(7.46)

Both sides of eq. (7.44) are then subtracted from $x(m)$ and the result is combined with eq. (7.45); that is

$$\begin{aligned}\tilde{x}_m^+ &= x(m) - \hat{x}_m^+ = x(m) - \hat{x}_m^- - K_m z(m) + K_m H(m)\hat{x}_m^- \\ &= x(m) - \hat{x}_m^- - K_m\left[H(m)x(m) + v(m)\right] + K_m H(m)\hat{x}_m^- \\ &= \left[I - K_m H(m)\right]\left[x(m) - \hat{x}_m^-\right] - K_m v(m) \\ &= \left[I - K_m H(m)\right]\tilde{x}_m^- - K_m v(m)\end{aligned}$$
(7.47)

Of immediate interest are the mean values and the variances of the estimation errors, as propagate through the transformation (7.47). In particular, when the initial uncertainty $\tilde{x}_0^-$, as well as the measurement error $v(n)$, has zero mean, i.e.

$$E\{\tilde{x}_0^-\} = 0, \ E\{v(n)\} = 0$$
(7.48)

then every component of $\tilde{x}_0^+$ is also a zero mean random variable, that is from eqs. (7.41) and (7.46)-(7.48) one concludes

$$E\{\tilde{x}_0^+\} = \left[I - K_0 H(0)\right]E\{\tilde{x}_0^-\} - K_0 E\{v(0)\} = 0$$
$$\tilde{x}_1^- = x(1) - \hat{x}_1^- = \Phi(t_1,t_0)x(0) - \Phi(t_1,t_0)\hat{x}_0^+$$
$$= \Phi(t_1,t_0)\left[x(0) - \hat{x}_0^+\right] = \Phi(t_1,t_0)\tilde{x}_0^+$$

so that

$$E\{\tilde{x}_1^-\} = \Phi(t_1,t_0)E\{\tilde{x}_0^+\} = 0$$

and

$$E\{\tilde{x}_1^+\}=\left[I-K_1H(1)\right]E\{\tilde{x}_1^-\}-K_1E\{v(1)\}=0$$

We may now repeat the process for $\tilde{x}_2^-$ and $\tilde{x}_2^+$, i.e.

$$\tilde{x}_2^- = x(2)-\hat{x}_2^- = \Phi(t_2,t_1)x(1)-\Phi(t_2,t_1)\hat{x}_1^+$$
$$= \Phi(t_2,t_1)\left[x(1)-\hat{x}_1^+\right]=\Phi(t_2,t_1)\tilde{x}_1^+$$

and

$$\tilde{x}_2^+ = \left[I-K_2H(2)\right]\tilde{x}_2^- - K_2v(2)$$

from which one concludes

$$E\{\tilde{x}_2^-\}=\Phi(t_2,t_1)E\{\tilde{x}_1^+\}=0$$
$$E\{\tilde{x}_2^+\}=\left[I-K_2H(2)\right]E\{\tilde{x}_2^-\}-K_2E\{v(2)\}=0$$

Thus, by induction, we verify that

$$E\{\tilde{x}_m^+\}=E\{x(m)-\hat{x}_m^+\}=0 \tag{7.49}$$

that is the estimator (7.44) is *unbiased* (see chapter 5). Namely, the goodness of an estimation procedure is characterized by how likely it is that the estimate $\hat{x}^+$ is close to the actual value of x. Therefore, if $E\{\tilde{x}^+\}=0$; i.e. the estimator is unbiased in the sense that $E\{\hat{x}\}=E\{x\}$, it is possible a good estimation procedure, since the estimate $\hat{x}^+$ should be closely clustered about the estimated quantity x. On the other hand, the variance of the every component of the estimation error $\tilde{x}^+$ is a good measure of how clustered it is around the zero-value. As mentioned before, in the instrumentation theory the effectiveness of a measuring instrument is specified by its *accuracy* and *precision*. Accuracy is a measure of how well the mean value $E\{\hat{x}^+\}$ matches $E\{x\}$, or $E\{\tilde{x}^+\}$ matches zero-vector, and the precision measures the trace (sum of diagonal elements) of a covariance matrix for $\tilde{x}^+$

$$P = E\left\{\left[\tilde{x}^+ - E\{\tilde{x}^+\}\right]\left[\tilde{x}^+ - E\{\tilde{x}^+\}\right]^T\right\} \tag{7.50}$$

which reduces to

$$P = E\{\tilde{x}^+\tilde{x}^{+T}\} \tag{7.51}$$

in the case of unbiased estimation procedure $\hat{x}^+$, i.e. $E\{\tilde{x}^+\}=0$.

By substitution of eqs. (7.5) and (7.47) into (7.51), under conditions of no correlation between $\tilde{x}_m^-$ and $v(m)$, i.e.

$$E\{\tilde{x}_m^- v^T(m)\}=0 \tag{7.52}$$

we obtain that the estimation error at instant t_m

$$P_m^+ = E\{\tilde{x}_m^+\tilde{x}_m^{+T}\} \tag{7.53}$$

is given by

$$P_m^+ = \left[I-K_mH(m)\right]P_m^-\left[I-K_mH(m)\right]^T + K_mR_mK_m^T \tag{7.54}$$

where

$$P_m^- = E\left\{\tilde{x}_m^- \tilde{x}_m^{-T}\right\} \tag{7.55}$$

with the noise covariance matrix R_m being defined by eq. (7.5). The criterion for choosing the gain matrix K_m is to minimize a scalar sum of the diagonal elements of the estimation error covariance matrix (7.54); that is the trace of this matrix. Thus, we choose for the cost function the minimum variance criterion

$$J_m = \text{trace}\left\{P_m^+\right\} = E\left\{\tilde{x}_m^{+T} \tilde{x}_m^+\right\} \tag{7.56}$$

This is equivalent to minimizing the length of the estimation error vector $\left\|\tilde{x}_m^+\right\|^2$. To find the value of K_m which provides a minimum, it is necessary to take the partial derivative of J_m with respect to K_m and equate it to zero. Use is made of the relation for the partial derivative of the trace of the product of two matrices A and B, with B symmetric

$$\frac{\partial}{\partial A} \text{trace}\left\{A B A^T\right\} = 2AB \tag{7.57}$$

as well as the relation for the partial derivative of the trace of the product of three matrices A, B and C

$$\frac{\partial}{\partial A} \text{trace}\left\{BAC\right\} = B^T C^T \tag{7.58}$$

From eqs. (7.54)-(7.58), the result is

$$\frac{\partial J_m}{\partial K_m} = \frac{\partial}{\partial K_m} \text{trace}\left\{P_m^- - K_m H(m) P_m^- - P_m^- H^T(m) K_m^T + K_m H(m) P_m^- H^T(m) K_m + K_m R_m K_m^t\right\} = 0$$

or equivalently

$$\frac{\partial}{\partial K_m} = -2\frac{\partial}{\partial K_m} \text{trace}\left\{\left(K_m H(m) P_m^-\right)\right\} +$$

$$\frac{\partial}{\partial K_m} \text{trace}\left\{K_m H(m) P_m^- H^T(m) K_m\right\} + \frac{\partial}{\partial K_m} \text{trace}\left\{K_m R_m K_m^T\right\} = 0$$

from which one obtains, using the rules for differentiation (7.57) and (7.58)

$$-2P_m^- H^T(m) + 2K_m H(m) P_m^- H^T(m) + 2K_m R_m = 0$$

or

$$-2\left(I - K_m H(m)\right) P_m^- H^T(m) + 2K_m R_m = 0$$

Here is used the fact that for a symmetric matrix A $\text{trace}(A) = \text{trace}(A^T)$, that is

$$\text{trace}\left(P_m^- H^T(m) K_m^T\right) = \text{trace}\left(K_m H(m) P_m^-\right)$$

since P_m^- is a symmetric matrix. Solving for K_m

$$K_m = P_m^- H^T(m)\left[H(m) P_m^- H^T(m) + R_m\right]^{-1} \tag{7.59}$$

which is referred to as the Kalman gain matrix. Examination of the Hessian of J_m reveals that this value of K_m does indeed minimize J_m, i.e. using the eq. (7.58)

$$\frac{\partial^2 J_m}{\partial^2 K_m^2} = 2\frac{\partial}{\partial K_m} K_m H(m) P_m^- H^T(m) + 2\frac{\partial}{\partial K_m} K_m R_m$$
$$= H(m) P_m^- H^T(m) + R_m$$

The matrix

$$Z_m = H(m) P_m^- H^T(m) + R_m \tag{7.60}$$

is the covariance matrix of predicted residuals from (7.45). Since this symmetric matrix is positive definite, owing to (7.8a), the Hessian of J_m is positive definite, and thus eq. (7.59) does indeed define a minimum. In other words, every component of $\tilde{x}_m^+$ has minimum variance when the weighting matrix of (7.59) is used in (7.44).

Substitution of eq. (7.59) into eq. (7.54) gives

$$P_m^+ = P_m^- - K_m H(m) P_m^- - P_m^- H^T(m) K_m^T + K_m Z_m K_m^T$$
$$= P_m^- - P_m^- H^T(m) Z_m^{-1} H(m) P_m^- - P_m^- H^T(m) Z_m^{-1} H(m) P_m^- +$$
$$P_m^- H^T(m) Z_m^{-1} Z_m Z_m^{-1} H(m) P_m^- \tag{7.61a}$$
$$= P_m^- - P_m^- H^T(m) Z_m^{-1} H(m) P_m^-$$

or equivalently

$$P_m^+ = \left[I - P_m^- H^T(m) Z_m^{-1} H(m) \right] P_m^- = \left[I - K_m H(m) \right] P_m^- \tag{7.61b}$$

since $Z_m^{-1} Z_m = I$. This relation is the optimized value of the updated estimation error covariance matrix.

There is a matrix inversion relationship which states that, for P_m^+ as given in eq. (7.61), $\left[P_m^+ \right]^{-1}$ is expressible as

$$\left[P_m^+ \right]^{-1} = \left[P_m^- \right]^{-1} + H^T(m) R_m^{-1} H(m) \tag{7.62}$$

This relationship can easily be verified by showing that $P_m^+ \left[P_m^+ \right]^{-1} = I$. Namely, using eqs. (7.60), (7.61a) and (7.62), one obtains

$$P_m^+ \left[P_m^+ \right]^{-1} = \left[P_m^- - P_m^- H^T(m) Z_m^{-1} H(m) P_m^- \right] \left[\left(P_m^- \right)^{-1} + H^T(m) R_m^{-1} H(m) \right]$$
$$= P_m^- \left(P_m^- \right)^{-1} - P_m^- H^T(m) Z_m^{-1} H(m) P_m^- \left(P_m^- \right)^{-1} +$$
$$P_m^- H^T(m) R_m^{-1} H(m) - P_m^- H^T(m) Z_m^{-1} H(m) P_m^- H^T(m) R_m^{-1} H(m)$$
$$= I - P_m^- H^T Z_m^{-1} H(m) + P_m^- H^T(m) \left[I - Z_m^{-1} H(m) P_m^- H^T(m) \right] R_m^{-1} H(m)$$
$$= I - P_m^- H^T Z_m^{-1} H(m) + P_m^- H^T(m) Z_m^{-1} \left[Z_m - H(m) P_m^- H^T(m) \right] R_m^{-1} H(m)$$
$$= I - P_m^- H^T Z_m^{-1} H(m) + P_m^- H^T(m) Z_m^{-1} R_m R_m^{-1} H(m) = I$$

We use this result to manipulate K_m as follows,

$$K_m = P_m^+ \left(P_m^+\right)^{-1} P_m^- H^T(m) Z_m^{-1}$$

$$= P_m^+ \left[\left(P_m^-\right)^{-1} + H^T(m) R_m^{-1} H(m)\right] P_m^- H^T(m) \left[H(m) P_m^- H^T(m) + R_m\right]^{-1}$$

Expanding and collecting terms yields

$$K_m = P_m^+ H^T(m) \left[I + R_m^{-1} H(m) P_m^- H^T(m)\right] \left[H(m) P_m^- H^T(m) + R_m\right]^{-1}$$

$$= P_m^+ H^T(m) R_m^{-1} \left[R_m + H(m) P_m^- H^T(m)\right] \left[H(m) P_m^- H^T(m) + R_m\right]^{-1} \qquad (7.63)$$

$$= P_m^+ H^T(m) R_m^{-1}$$

which is the simpler form sought.

Dependence of the weighting matrix K_m on P_m^- calls for a specification of:
(1) an initial covariance matrix of uncertainty at some reference time, e.g. P_0 at time t_0;
(2) the dynamic behavior of P between observations;
(3) changes in P at discrete measurement instants (see fig. 7.1).

The initial matrix P_0 can start a cycle of dynamic adjustments in the estimator uncertainty, characterized as follows: subtraction of (7.41) from (7.9), evaluated at t_m, yields

$$\tilde{x}_m^- = x(t_m) - \hat{x}_m^-$$

$$= \Phi(t_m, t_{m-1}) x(t_{m-1}) + \int_{t_{m-1}}^{t_m} \Phi(t_m, \tau) w(\tau) d\tau - \Phi(t_m, t_{m-1}) \hat{x}_{m-1}^+$$

$$= \Phi(t_m, t_{m-1}) \left[x(t_{m-1}) - \hat{x}_{m-1}^+\right] + \int_{t_{m-1}}^{t_m} \Phi(t_m, \tau) w(\tau) d\tau \qquad (7.64)$$

$$= \Phi(t_m, t_{m-1}) \tilde{x}_{m-1}^+ + \int_{t_{m-1}}^{t_m} \Phi(t_m, \tau) w(\tau) d\tau$$

and, by substitution into (7.55), under the assumption that $\tilde{x}_{m-1}^+$ and $w(t_{m-1})$ are uncorrelated, further follows:

$$P_m^- = E\left\{\tilde{x}_m^- \left(\tilde{x}_m^-\right)^T\right\} = \Phi(t_m, t_{m-1}) E\left\{\tilde{x}_{m-1}^+ \left(\tilde{x}_{m-1}^+\right)^T\right\} \Phi(t_m, t_{m-1})$$

$$+ \int_{t_{m-1}}^{t_m} \int_{t_{m-1}}^{t_m} \Phi(t_m, \tau) E\left\{w(\tau) w^T(\sigma)\right\} \Phi^T(t_m, \sigma) d\tau d\sigma$$

$$= \Phi(t_m, t_{m-1}) P_{m-1}^+ \Phi^T(t_m, t_{m-1}) + \int_{t_{m-1}}^{t_m} \int_{t_{m-1}}^{t_m} \Phi(t_m, \tau) Q(\tau) \delta(\tau - \sigma) \Phi^T(t_m, \sigma) d\tau d\sigma \qquad (7.65)$$

$$= \Phi(t_m, t_{m-1}) P_{m-1}^+ \Phi^T(t_m, t_{m-1}) + \int_{t_{m-1}}^{t_m} \Phi(t_m, \tau) Q(\tau) \Phi^T(t_m, \tau) d\tau$$

Either this relation, or its derivative

$$\dot{P} = AP + PA^T + Q \qquad (7.66)$$

can provide P_m^-, given the uncertainty covariance immediately after the preceding update (see fig. 7.1), or in the case of $m=1$ given P_0.

The uncertainty adjustment at a discrete update has already been expressed by eq. (61), which corresponds to a change in P by an amount

$$\Delta P_m = P_m^+ - P_m^- = -P_m^- H^T(m) Z_m^{-1} H(m) P_m^- \qquad (7.67)$$

The negative quadratic nature of this symmetric matrix indicates a reduction in uncertainty, as each properly weighted observation is used to adjust $\hat{x}$. The equations of the continuous-discrete Kalman filter are summarized in Table 7.1.

Table 7.1: Summary of continuous-discrete Kalman filter equations

System model	$\dot{x} = Ax + w$; $E\{w\} = 0$; $E\{w(t)w^T(\tau)\} = Q(t)\delta(t-\tau)$
Measurement model	$z(m) = H(m)x(m) + v(m)$; $E\{v\} = 0$ $E\{v(t_m)v^T(\tau_n)\} = R_m\delta(m-n)$
Initial conditions	$E\{x(0)\} = \hat{x}_0$; $E\{[x(0)-\hat{x}_0][x(0)-\hat{x}_0]^T\} = P_0$
Other assumptions	$E\{w(t)v^T(t_m)\} = 0$ for all t, t_m
State estimate extrapolation (Time update)	$\hat{x}_m^- = \Phi(t_m,t_{m-1})\hat{x}_{m-1}^+$; Φ state transition matrix $\dot{\Phi}(t,\tau) = A\Phi(t,\tau)$; $\Phi(t,t) = I$ or the solution of $\dot{\hat{x}}(t) = A\hat{x}(t)$; $\hat{x}_0$ given
Error covariance extrapolation	$P_m^- = \Phi(t_m,t_{m-1})P_{m-1}^+\Phi^T(t_m,t_{m-1}) + \int_{t_{m-1}}^{t_m}\Phi(t_m,\tau)Q(\tau)\Phi^T(t_m,\tau)d\tau$ or the solution of $\dot{P} = AP + PA^T + Q$; P_0 given
State estimate update (measurement update)	$\hat{x}_m^+ = \hat{x}_m^- + K_m\left[z(m) - H(m)\hat{x}_m^-\right]$
Error covariance update	$P_m^+ = \left[I - K_mH(m)\right]P_m^-$
Kalman gain matrix	$K_m = P_m^-H^T(m)\left[H(m)P_m^-H^T(m) + R_m\right]^{-1}$

Figure 7.2 illustrates these equations in block diagram form.

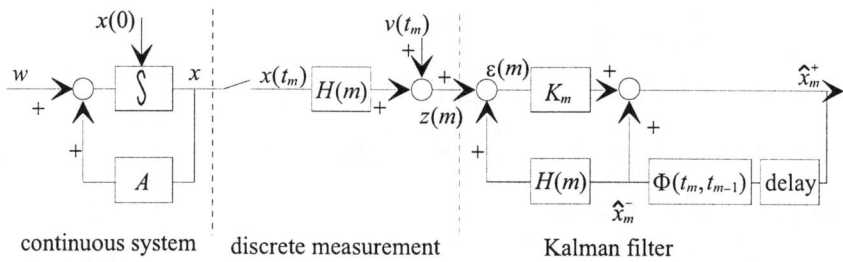

Fig.7.2. Continuous-discrete system model and Kalman filter

The Kalman filter we implement is based upon the mathematical model; that is a model of what we think the system and measurement process are. Later, we shall dwell on the practical consequence of this fact. It is often said that the Kalman filter generates its own error analysis [1, 25, 29, 35, 36, 48]. Clearly, this refers to the computation of P_m, which provides an indication of the accuracy of the estimate.

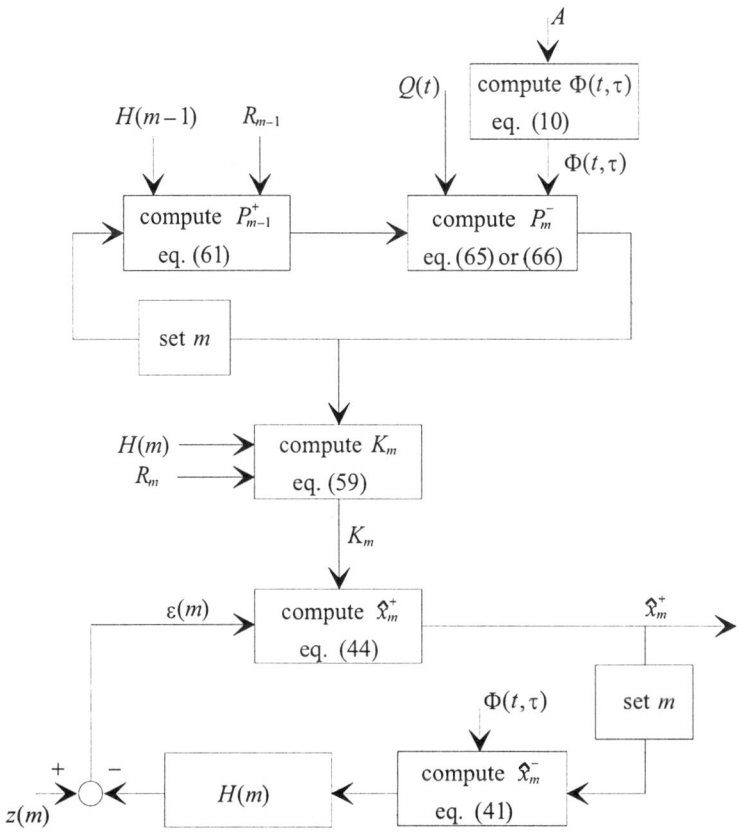

Fig. 7.3. A simplified computer flow diagram of the continuous-discrete Kalman filter

In the linear Kalman filter, calculations at the covariance level ultimately serve to provide K_m, which is then used in the calculation of conditional mean values, i.e. the estimate $\hat{x}_m$ (it should be noted that the minimum variance estimate $\hat{x}_m$ is given by the conditional expectation of the state $x(t_m)$ given the observations at and before t_m). There is no feedback from the state equations to the covariance equations. This is illustrated in fig. 7.3, which is essentially a simplified computer flow diagram of the continuous-discrete Kalman filter.

7.1.4 Geometric concept: Orthogonality principle

We may verify by direct calculation that

$$E\left\{\hat{x}_m^+\left(\tilde{x}_m^+\right)^T\right\} = 0 \tag{7.69}$$

That is, the optimal estimate $\hat{x}_m^+$ and its error $\tilde{x}_m^+$ are orthogonal (uncorrelated), and this is well known *orthogonality principle*. Taking into account eqs. (7.41) and (7.44), one can characterize the optimal estimate $\hat{x}_m^+$ as the linear combination of observations $z(0)$, $z(1),...,z(m)$ such that the estimation error $\tilde{x}_m^+$ is orthogonal into the data space spanned by the observations $z(0),...,z(m)$ (see Fig. 7.4).

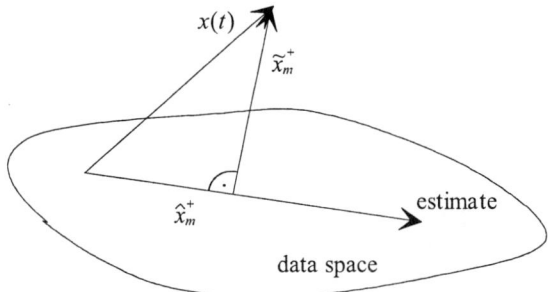

Fig.7.4 Geometric interpretation of orthogonality principle

In other words, the optimal estimate $\hat{x}_m^+$ can be considered geometrically as an orthogonal projector in the sense that the estimate $\hat{x}_m^+$ is the orthogonal projection of the state vector $x(m)$ onto the data space (plane).

The result (7.69) can be also confirmed analytically. The state and its estimate at time t_1 are, in accordance to eqs. (7.9), (7.41) and (7.44),

$$x_1 = x(t_1) = \Phi_0 x_0 + \int_{t_0}^{t_1} \Phi(t_1, \tau) w(\tau) d\tau \tag{7.70}$$

where $\dot{\Phi}_0 = \Phi(t_1, t_0)$ and $x_0 = x(t_0)$, and

$$\begin{aligned}\hat{x}_1^+ &= \Phi_0 \hat{x}_0^+ + K_1 \left[z_1 - H_1 \Phi_0 \hat{x}_0^+ \right] \\ &= \Phi_0 \hat{x}_0^+ + K_1 \left[H_1 \left(\Phi_0 x_0 + \int_{t_0}^{t_1} \Phi(t_1, \tau) w(\tau) d\tau \right) + v_1 - H_1 \Phi_0 \hat{x}_0^+ \right]\end{aligned} \tag{7.71}$$

where the measurement equation (7.12) has been employed. Subtracting eq. (7.71) from eq. (7.70) yields an equation in $\tilde{x}_1^+$, i.e.

$$\tilde{x}_1^+ = \left(\Phi_0 - K_1 H_1 \Phi_0 \right) \tilde{x}_0^+ + \left[I - K_1 H_1 \right] \int_{t_0}^{t_1} \Phi(t_1, \tau) w(\tau) d\tau - K_1 v_1 \tag{7.72}$$

Since $\hat{x}_0^+ = E\{x_0\}$ and $\tilde{x}_0^+ = x_0 - \hat{x}_0^+$, one obtains $E\left\{ \hat{x}_0^+ \left(\tilde{x}_0^+ \right)^T \right\} = \hat{x}_0^+ E\left\{ \left(x_0 \right)^T - \left(\hat{x}_0^+ \right)^T \right\} = 0$, so that we directly calculate the quantity of interest as

$$\begin{aligned}E\left\{ \hat{x}_1^+ \left(\tilde{x}_1^+ \right)^T \right\} &= E\Bigg\{ \left[\Phi_0 \hat{x}_0^+ + K_1 \left(H_1 \Phi_0 \tilde{x}_0^+ + H_1 \int_{t_0}^{t_1} \Phi(t_1, \tau) w(\tau) d\tau + K_1 v_1 \right) \right] \times \\ &\quad \times \left[\left(\tilde{x}_0^+ \right)^T \left(\Phi_0 - K_1 H_1 \Phi_0 \right)^T + \int_{t_0}^{t_1} w^T(s) \Phi^T(t_1, s) ds \left(I - K_1 H_1 \right)^T - v_1^T K_1^T \right] \Bigg\}\end{aligned}$$

$$\begin{aligned}&= K_1 H_1 \Phi_0 P_0^+ \left(\Phi_0^T - \Phi_0^T H_1^T K_1^T \right) + K_1 H_1 \int_{t_0}^{t_1} \int_{t_0}^{t_1} \Phi(t_1, \tau) w(\tau) w^T(s) \Phi^T(t_1, s) d\tau ds \left(I - K_1 H_1 \right)^T - K_1 R_1 K_1^T \\ &= K_1 H_1 \Phi_0 P_0^+ \left(\Phi_0^T - \Phi_0^T H_1^T K_1^T \right) + K_1 H_1 \left(P_1^- - \Phi_0 P_0^+ \Phi_0^T \right) \left(I - K_1 H_1 \right)^T - K_1 R_1 K_1^T \\ &= K_1 H_1 \left(\Phi_0 P_0^+ \Phi_0^T - \Phi_0 P_0^+ \Phi_0^T \right) + K_1 H_1 \left(-\Phi_0 P_0^+ \Phi_0^T + \Phi_0 P_0^+ \Phi_0^T \right) H_1^T K_1^T + K_1 H_1 P_1^- \left(I - H_1^T K_1^T \right) - K_1 R_1 K_1^T \\ &= K_1 H_1 P_1^+ - K_1 R_1 \left(P_1^+ H_1^T R_1^{-1} \right) = K_1 H_1 P_1^+ - K_1 R_1 R_1^{-1} H_1 P_1^+ = 0\end{aligned}$$

where use has been made of eqs. (7.65) and (7.68); that is

$$\int_{t_0}^{t_1}\int_{t_0}^{t_1}\Phi(t_1,\tau)w(\tau)w^T(s)\Phi^T(t_1,s)\,d\tau ds = \int_{t_0}^{t_1}\Phi(t_1,\tau)Q(\tau)\Phi^T(t_1,\tau)\,d\tau = P_1^- - \Phi_0 P_0^+ \Phi_0^T$$

and

$$K_1 = P_1^+ H_1^T R_1^{-1}$$

We may now repeat the process for $E\left\{\hat{x}_2^+\left(\tilde{x}_2^+\right)^T\right\}$ and, by induction, verify eq. (7.69).

7.1.5 Optimal prediction

Optimal prediction can be thought of, quite simply, in terms of optimal filtering in the absence of measurements. This, in turn, is equivalent to optimal filtering with arbitrarily large measurement errors (thus $R^{-1} \to 0$ and hence $K \to 0$). Therefore, if measurements are unavailable beyond some time, say t_{m-1}, the optimal prediction of $x(t)$ for $t_m \geq t_{m-1}$ given $\hat{x}(t_{m-1})$ must be obtained from eq. (7.44) when $K_m = 0$; that is

$$\hat{x}_m^+\big|_{K_m=0} = \hat{x}_m^- \tag{7.73}$$

or, after using eqs. (7.41) and (7.73)

$$\hat{x}_m^- = \Phi(t_m,t_{m-1})\hat{x}_{m-1}^+\big|_{K_{m-1}=0} = \Phi(t_m,t_{m-1})\hat{x}_{m-1}^- \tag{7.74}$$

Generally, if measurements are unavailable beyond some t_0, the optimal prediction of $x(t)$ for $t \geq t_0$ given $\hat{x}(t_0)$ must be

$$\hat{x}(t) = \Phi(t,t_0)\hat{x}(t_0) \tag{7.75}$$

The corresponding equation for uncertainty in the optimal prediction, given $P(t_0) = E\left\{\left[x(t_0)-\hat{x}(t_0)\right]\left[x(t_0)-\hat{x}(t_0)\right]^T\right\}$ is eq. (7.66), i.e.

$$\dot{P} = AP + PA^T + Q \; ; \; P(t_0) \text{ given} \tag{7.76}$$

It is also possible to utilize an alternative form of the prediction equation (7.75). We can expand Φ in a Taylor series

$$\begin{aligned}
\Phi(t,t_0) &= \Phi(t_0,t_0) + \frac{d}{dt}\Phi(t_0,t_0)(t-t_0) + O(t-t_0)^2 \\
&= I + A(t_0)\Phi(t_0,t_0)(t-t_0) + O(t-t_0)^2 \\
&= I + A(t_0)(t-t_0) + O(t-t_0)^2
\end{aligned} \tag{7.77}$$

where the state transition matrix equation (7.10) has been employed. Here $O(t^2)$ is a matrix containing terms of order t^2 and higher.

Substituting eq. (7.77) into eq. (7.75), one obtains

$$\hat{x}(t) = \left[I + A(t_0)(t-t_0) + O(t-t_0)^2\right]\hat{x}(t_0)$$

or equivalently

$$\frac{\hat{x}(t) - \hat{x}(t_0)}{t - t_0} = A(t_0)\hat{x}(t_0) + \frac{O(t - t_0)^2}{t - t_0}$$

Therefore, letting $O(t - t_0)^2 / (t - t_0) \rightarrow 0$ as $t \rightarrow t_0$, into the limit gives

$$\dot{\hat{x}} = \frac{d\hat{x}}{dt} = A(t)\hat{x}(t) \; ; \; \hat{x}(t_0) \text{ given} \tag{7.78}$$

which is the desired form of the prediction equation.

Additionally, when the system under observations is excited by a deterministic time-varying input u, whether due to a control being intentionally applied or a deterministic disturbance which occurs, these known inputs must be accounted for by the optimal predictor. For the case of deterministic input u, a linear functional relationship (7.2) takes the form

$$\dot{x} = \frac{dx}{dt} = Ax + w + Bu \tag{7.79}$$

It is easy to see that the modification

$$\dot{\hat{x}} = A\hat{x} + Bu \tag{7.80}$$

must be made in order for the predictor to remain unbiased. Namely, the estimation error is given by the equation

$$\dot{\tilde{x}} = \dot{x} - \dot{\hat{x}} = Ax + w + Bu - A\hat{x} - Bu = A\tilde{x} + w$$

so that

$$\tilde{x}(t) = \Phi(t, t_0)\tilde{x}(t_0) + \int_{t_0} \Phi(t, \tau)w(\tau)d\tau \tag{7.81}$$

for which one concludes $E\{\tilde{x}\} = 0$ if $E\{\tilde{x}(t_0)\} = 0$ and $E\{w\} = 0$. On the other hand, if we use (7.78) and (7.79)

$$\dot{\tilde{x}} = A\tilde{x} + w + Bu$$

from which one concludes

$$\tilde{x}(t) = \Phi(t, t_0)\tilde{x}(t_0) + \int_{t_0} \Phi(t, \tau)w(\tau)d\tau + \int_{t_0} \Phi(t, \tau)Bu(\tau)d\tau \tag{7.82}$$

yielding

$$E\{\tilde{x}\} = \int_{t_0} \Phi(t, \tau)Bu(\tau)d\tau$$

That is, the estimate is biased. However, if we use eq. (7.80), it is observed that the resultant equation (7.81) for the prediction error $\tilde{x}$ is precisely the equation (7.64) obtained before. Hence, the procedure (7.76) for computing P remains unchanged. Note that in the Kalman filter equations in Tab. 7.1 the state estimate extrapolation has to be replaced by eq. (7.80), while the procedures for computing P, K, etc., remain unchanged. It should be noted also that Kalman filter contains an exact model of the system in its formulation (i.e. the A, B matrices in eq. (7.79)). This provides the mechanism by which past information is extrapolated into the future for the purpose of prediction.

To gain some insight into the design of continuous-discrete Kalman filter, let us consider an example.

Example 7.1 (continuous-time radar tracking with discrete measurements): Consider the two dimensional radar tracking problem depicted in Fig. 7.5 [5, 6, 10].

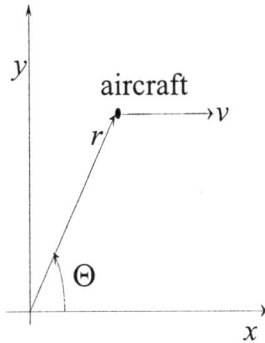

Fig. 7.5 Radar tracking geometry

For simplicity the velociti v is given to be in the x direction. To derive the state model write

$$r^2 = x^2 + y^2$$

or, after differentiation,

$$2r\dot{r} = 2x\dot{x} + 2y\dot{y}$$

from which it follows

$$\dot{r} = \frac{x}{r}\dot{x} + \frac{y}{r}\dot{y}$$

However, since

$$x = r\cos\Theta \; ; \; \dot{x} = v \; ; \; \dot{y} = 0$$

we have

$$\dot{r} = v\cos\Theta \tag{I}$$

or, after differentiation,

$$\ddot{r} = \dot{v}\cos\Theta - v\dot{\Theta}\sin\Theta \tag{II}$$

For Θ, we can write

$$\tan\Theta = \frac{y}{x}$$

from which it follows

$$\dot{\Theta}\frac{1}{\cos^2\Theta} = \frac{x\dot{y} - \dot{x}y}{x^2}$$

or equivalently

$$\dot{\Theta} = \frac{x\dot{y} - \dot{x}y}{x^2}\cos^2\Theta = \frac{x\dot{y} - \dot{x}y}{r^2}$$

But, since

$$y = r\sin\Theta \; ; \; \dot{x} = v \; ; \; \dot{y} = 0$$

we have

$$\dot{\Theta} = -\frac{v}{r}\sin\Theta \tag{III}$$

or, after differentiation,

$$\ddot{\Theta} = -\frac{\dot{v}r - v\dot{r}}{r^2}\sin\Theta - \frac{v}{r}\dot{\Theta}\cos\Theta \qquad (IV)$$

State equations (III) and (IV) are nonlinear. We shall give in chapter 8 how to apply the Kalman filter to nonlinear systems. Here we will apply the Kalman filter in Table 7.1 by converting to a linear system as follows.

Assume that the aircraft is not accelerating so that $\dot{v} = 0$. Then, if r and Θ do not change much during each scan of the radar (about 1-10sec), we have $\dot{r} = 0$ and $\dot{\Theta} = 0$. To account for the small changes in $\dot{r}$ and $\dot{\Theta}$ during the scan time T, introduce random disturbances ω_1 and ω_2 so that

$$\frac{d}{dt}\dot{r} = w_1(t) \qquad (V)$$

$$\frac{d}{dt}\dot{\Theta} = w_2(t) \qquad (VI)$$

It should be noted that the process noise has been added to compensate for the unmodeled dydnamics that we introduced by iggnoring the nonlinear terms. However, the robustness properties of the Kalman filter in the presence of process noise $w = \begin{bmatrix} w_1 & w_2 \end{bmatrix}^T$ will allow tracking of $\dot{r}$ and $\dot{\Theta}$. The state equations are now linear.

$$\frac{d}{dt}\begin{bmatrix} r \\ \dot{r} \\ \Theta \\ \dot{\Theta} \end{bmatrix} = \begin{bmatrix} 0 & 1 & 0 & 0 \\ 0 & 0 & 0 & 0 \\ 0 & 0 & 0 & 1 \\ 0 & 0 & 0 & 0 \end{bmatrix}\begin{bmatrix} r \\ \dot{r} \\ \Theta \\ \dot{\Theta} \end{bmatrix} + \begin{bmatrix} 0 \\ w_1 \\ 0 \\ w_2 \end{bmatrix} \qquad (VII)$$

or equivalently

$$\dot{x} = Ax + w \qquad (VIII)$$

where the process w is known as maneuver noise (do not confuse the state vector $x = \begin{bmatrix} x_1 & x_2 & x_3 & x_4 \end{bmatrix}^T = \begin{bmatrix} r & \dot{r} & \Theta & \dot{\Theta} \end{bmatrix}^T$ with the x coordinate in Fig. 7.5). Let us take that the radar measures range r and azimuth Θ at every T sec, so that the discrete measurement model is given by

$$y_k = \begin{bmatrix} r_k \\ \Theta_k \end{bmatrix} = \begin{bmatrix} 1 & 0 & 0 & 0 \\ 0 & 0 & 1 & 0 \end{bmatrix}x_k + \begin{bmatrix} v_{1k} \\ v_{2k} \end{bmatrix} = Hx_k + v_k \qquad (IX)$$

where we use the notation $t_k = kT$ and $x_k = x(t_k) = x(kT)$. Here v_k is the measurement noise vector. The state space model (VII), (IX) shows that we can simplify computations by running, not one Kalman filter on the fourth-order model (VII), but two parallel Kalman filters on two second order subsystems (see Fig. 7.5)

$$\text{range subsystem:} \quad \frac{d}{dt}\begin{bmatrix} r \\ \dot{r} \end{bmatrix} = \begin{bmatrix} 0 & 1 \\ 0 & 0 \end{bmatrix}\begin{bmatrix} r \\ \dot{r} \end{bmatrix} + \begin{bmatrix} 0 \\ w_1 \end{bmatrix} \qquad (Xa)$$

$$r_k = \begin{bmatrix} 1 & 0 \end{bmatrix}\begin{bmatrix} r_k \\ \dot{r}_k \end{bmatrix} + v_{1,k} \qquad (Xa)$$

$$\text{azimuth subsystem:} \quad \frac{d}{dt}\begin{bmatrix} \Theta \\ \dot{\Theta} \end{bmatrix} = \begin{bmatrix} 0 & 1 \\ 0 & 0 \end{bmatrix}\begin{bmatrix} \Theta \\ \dot{\Theta} \end{bmatrix} + \begin{bmatrix} 0 \\ w_2 \end{bmatrix} \qquad (XIa)$$

$$\Theta_k = \begin{bmatrix} 1 & 0 \end{bmatrix}\begin{bmatrix} \Theta_k \\ \dot{\Theta}_k \end{bmatrix} + v_{2,k} \qquad (XIb)$$

Furthermore, we assume that the continuous-time maneuver noises w_1 and w_2, as well as the discrete-time measurement noises v_{1k} and v_{2k} are zero-mean white Gaussian processes; that is

$$E\{w_1(t)w_1(t+\tau)\} = q_1\delta(t-\tau) \ ; \ E\{w_2(t)w_2(t+\tau)\} = q_2\delta(t-\tau) \qquad \text{(XII)}$$

$$E\{v_{1k}v_{1j}\} = r_1\delta_{kj} \ ; \ E\{v_{2k}v_{2j}\} = r_2\delta_{kj} \qquad \text{(XIII)}$$

where $\delta(\cdot)$ is the impulse (Dirac) function and δ_{kj} is the Kronecker delta symbol. To find the maneuver noise spectral densities q_1 and q_2, note that w_1 is a radial acceleration and w_2 is an angular acceleration. Let us assume that the disturbance accelerations are independent and uniformly distributed between the limits $\pm a$. Then their variances are $(2a)^2/12 = a^2/3$. Thus, w_1 has the spectral density $q_1 = a^2/3$. However, since $w_2 r$ is distributed according to the maneuver noise pdf, then w_2 has the spectral density $q_2 = a^2/3r^2$. Now all we need to do is enter the system matrices and noise statistics into our computer MATLAB program (Fig. 7.6), which implements the continuous-discrete Kalman fitler in Table 7.1 for second-order subsystem.

At each time $t_k = kT$ we give the parallel filters the radar reading $r_k = r(kT)$ and $\Theta_k = \Theta(kT)$ and they will give the best estimates of $r(t), \dot{r}(t), \Theta(t)$ and $\dot{\Theta}(t)$. Thus, up to date, estimates for range, range rate, azimuth and azimuth rate will be available at all times, including times between the appearance of the target on the radar screen. An alternative approach can be based on discretization of the state space models (Xa) and (XIa) and application of the discrete-time Kalman filter from Tab. 7.2. However, the discretization requires to find $\exp(AT)$ and, in some applications, it can be quite complicated to find it. Here we avoided the discretization by using the continuous-discrete filter formulation. Finally, to initialize the filters we could suppose that the a priori information is $x_0 = \{\bar{x}_0, P_0\}$ and then take one initializing measurements r_0 and Θ_0. In this way, for the range subsystem, the initial estimate and the covariance matrix are given by

$$\bar{x}_0 = \begin{bmatrix} r_0 \\ 0 \end{bmatrix} \ ; \ P_0 = \begin{bmatrix} p_1 & 0 \\ 0 & p_1 \end{bmatrix}, \qquad \text{(XIVa)}$$

while the same quantities for the azimuth subsystems are

$$\bar{x}_0 = \begin{bmatrix} \Theta_0 \\ 0 \end{bmatrix} \ ; \ P_0 = \begin{bmatrix} p_2 & 0 \\ 0 & p_2 \end{bmatrix} \qquad \text{(XIVb)}$$

However, it is often more convenient to select $\bar{x}_0 = 0$ and $P_0 = I$, with I being an identity matrix, since the filter converges to the same stochastic steady state for any $\bar{x}_0$ and P_0, if it is properly designed (see section 7.10).

In summary, the continuous discrete Kalman filter for the range subsystem is defined by Tab. 7.1, where the corresponding matrices are given by

$$A = \begin{bmatrix} 0 & 1 \\ 0 & 0 \end{bmatrix} \ ; \ H = \begin{bmatrix} 1 & 0 \end{bmatrix} \ ; \ Q = \begin{bmatrix} 0 & 0 \\ 0 & q_1 \end{bmatrix} \ ; \ R = r_1 \ ; \ \hat{x}_0 = \begin{bmatrix} 0 \\ 0 \end{bmatrix} \ ; \ P_0 = \begin{bmatrix} 1 & 0 \\ 0 & 1 \end{bmatrix}$$

MATLAB programme implementations of the time update between measurements and the measurements update are given in Fig. 7.6. Using this software the true and estimated states are shown in Fig. 7.7 and the error variances for the range and range rate are given in Fig. 7.8. Measurements were taken every $T = 0.5s$, and each time a measurement update was made the error variance decreased.

Let's mention that the equation

$$\dot{P} = AP + PA^T + Q \; ; \; P_0 \text{ given} \tag{XV}$$

from the Tab. 7.1, can be rewritten in the form of linear vector differential equation:

$$\begin{bmatrix} \dot{p}_1 \\ \dot{p}_2 \\ \dot{p}_3 \end{bmatrix} = \begin{bmatrix} 0 & 2 & 0 \\ 0 & 0 & 1 \\ 0 & 0 & 0 \end{bmatrix} \begin{bmatrix} p_1 \\ p_2 \\ p_3 \end{bmatrix} + \begin{bmatrix} 0 \\ 0 \\ 1 \end{bmatrix} Q$$

where it is adopted for the matrix P to be in the form:

$$P = \begin{bmatrix} p_1 & p_2 \\ p_2 & p_3 \end{bmatrix}$$

In this way, in order to compute the solution of (XV), it is possible to use the system MATLAB functions for linear-time systems simulation.

```
function ex71
% design of aircraft trajectory
tfin=50; %total simulation time

x(1)=0;y(1)=0;r(1)=0;theta(1)=0;
vx(1)=10;vy(1)=5;i=1;
for t=0.05:0.05:tfin
  i=i+1;
  if ((t>20)& (t<25))
    vx(i)=vx(i-1)+0.5;
  else
    if ((t>30) & (t<35))
      vx(i)=vx(i-1)-0.5;
    else
      vx(i)=vx(i-1);
    end
  end
  if ((t>15)& (t<20))
    vy(i)=vy(i-1)+0.05;
  else
    if ((t>20) & (t<25))
      vy(i)=vy(i-1)-0.1;
    else
      vy(i)=vy(i-1);
    end
  end
  x(i)=x(i-1)+vx(i-1)*0.05;
  y(i)=y(i-1)+vy(i-1)*0.05;
  r(i)=sqrt(x(i)^2+y(i)^2);
  theta(i)=atan2(y(i),x(i));
  vr(i)=(x(i)*vx(i)+y(i)*vy(i))/r(i);
end

for i=1:10:1000
  robs((i-1)/10+1)=r(i)+10*randn;
  thetaobs((i-1)/10+1)=theta(i)+0.01*randn;
end

% design of Kalman filter for range
rp(1:2,1)=[robs(1);0];pp=[1 0;0 1];A=[0 1;0
0];
Ap=[0 2 0;0 0 1; 0 0 0];T=0.5;
h=[1 0];R=10;Q=.1;
for i=2:100
  rm(1:2,i)=expm(A*T)*rp(1:2,i-1);
  [temp,temp1]=lsim(ss(Ap,[0;0;1],eye(3),[0;0;
0]),...
  Q*ones(11,1),[0:0.05:0.5],[pp(1);pp(2);pp(3)]
);
  pm=[temp(11,1) temp(11,2); temp(11,2)
temp(11,3)];
  K=pm*h'*inv(h*pm*h'+R);
  rp(1:2,i)=rm(1:2,i)+K*(robs(i)-
h*rm(1:2,i));
  pp=(eye(2)-K*h)*pm;
end
figure(1);plot(1:100,robs,1:100,rp(1,:));
title('the observed and estimated range');
figure(2);plot(1:100,vr(1:10:1000),1:100,rp(2,
:));
title('the actual and estimated range rate');
keyboard;
```

Fig. 7.6: MATLAB programme for continuous system-discrete measurements Kalman filter

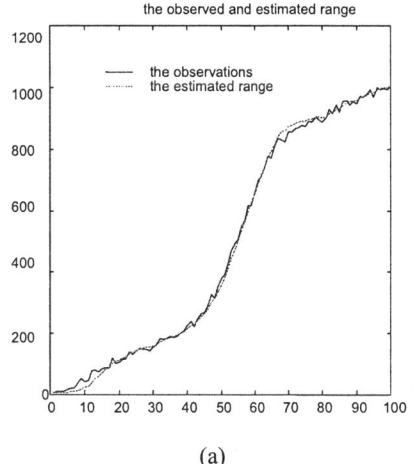
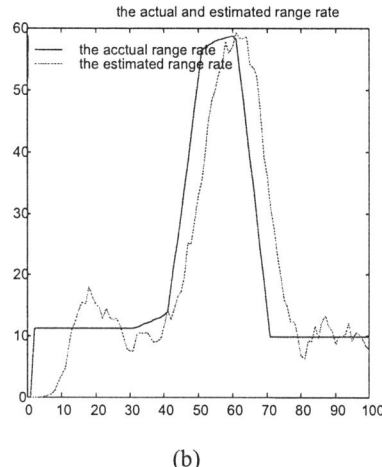

(a)   (b)

Fig. 7.7: a) The sequence of measurements and estimated range; b) The acctual and estimated range rate

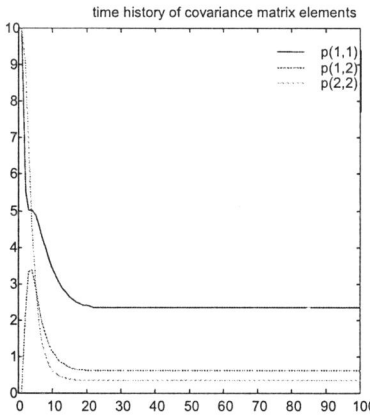

Fig. 7.8: Time history of the covariance matrix elements (Example 7.1)

7.2 Discrete estimation with discrete observations (discrete Kalman filter)

The transition from the continuous-discrete to the discrete formulation of the Kalman filter is readily accomplished. First, in order to go from the continuous system in eq. (7.2); that is

$$\frac{dx(t)}{dt} = Ax(t) + Gw(t) \tag{7.83}$$

where A and G are given matrices (in contrast to eq. (7.2)), now the noise term is Gw instead to w, to the discrete form

$$x(k) = \Phi(k-1)x(k-1) + w(k-1) \tag{7.84}$$

where $x(k)$ is the state at discrete time $t_k = kT$, with T being the uniform sampling interval, and $\{w(k)\}$ is a zero-mean white noise sequence with covariance matrix

$$Q(k) = E\{w(k)w^T(k)\} \tag{7.85}$$

it is necessary to remember the following equivalencies, valid in the limit as $t_k - t_{k-1} = T \to 0$

$$\Phi(k) \to I + AT \; ; \; Q(k) \to GQG^T T \tag{7.86}$$

The relations (7.86) were derived in section 4.

The measurement equation relating the states $x(k)$ to the discrete observations $z(k)$ is given by eq. (7.12); i.e.

$$z(k) = H(k)x(k) + v(k) \tag{7.87}$$

where $\{v(k)\}$ is a zero-mean white noise sequence with covariance matrix $R(k)$.

The discrete form of the Kalman filter may be written as in eq. (7.44); that is [13, 17]

$$\hat{x}_k^+ = \hat{x}_k^- + K_k \varepsilon(k) = \hat{x}_k^- + K_k\left[z(k) - H(k)\hat{x}_k^-\right] \tag{7.88}$$

where the signs (-) and (+) are used to denote the times immediately before and immediately after the discrete measurement $z(k) = z(t_k)$; $t_k = kT$ (Fig. 7.1). If the stages in the continuos-discrete Kalman filter derivation are repeated, one obtains that the Kalman gain matrix K_k is given by eq. (7.59) or eq. (7.63), while the estimation error covariance matrix across a measurement, P_k^+, is given by eqs. (7.61b); that is

$$K_k = P_k^- H^T(k)\left[H(k)P_k^- H^T(k) + R(k)\right]^{-1} \tag{7.89}$$

or

$$K_k = P_k^+ H^T(k) R^{-1}(k) \tag{7.90}$$

and

$$P_k^+ = \left[I - P_k^- H^T(k) Z_k^{-1} H(k)\right]P_k^- ; \; Z_k = H(k)P_k^- H^T(k) + R(k) \tag{7.91}$$

or

$$P_k^+ = \left[I - K_k H(k)\right]P_k(-) \tag{7.92}$$

where

$$P_k^+ = \{\tilde{x}_k^+ \tilde{x}_k^{+T}\} \; ; \; P_k^+ = E\{\tilde{x}_k^+ \tilde{x}_k^{+T}\} \; ; \; \tilde{x}_k^+ = x(k) - \hat{x}_k^+ \tag{7.93}$$

$$P_k^- = \{\tilde{x}_k^- \tilde{x}_k^{-T}\} \; ; \; P_k^- = E\{\tilde{x}_k^- \tilde{x}_k^{-T}\} \; ; \; \tilde{x}_k^- = x(k) - \hat{x}_k^- \tag{7.94}$$

The equation (7.91) is referred to as the *matrix discrete Ricatti equation*. The extrapolation of the state estimate between measurements is in accordance to the state equation (7.84) and the fact that $E\{w(k)\} = 0$

$$\hat{x}_k^- = \Phi(k-1)\hat{x}_{k-1}^+ \tag{7.95}$$

where $\hat{x}_{k-1}^+$ denotes the state estimate immediately after the discrete measurement $z(k-1)$. Moreover, since

$$\tilde{x}_k^- = x(k) - \hat{x}_k^- = \Phi(k-1)x(k-1) + w(k-1) - \Phi(k-1)\hat{x}_{k-1}^-$$
$$= \Phi(k-1)\left[x(k-1) - \hat{x}_{k-1}^+\right] + w(k-1) = \Phi(k-1)\tilde{x}_{k-1}^+ + w(k-1) \quad (7.96)$$

one obtains for the extrapolation of the error covariance matrix between measurements, under conditions of no correlation between $\tilde{x}_{k-1}^+$ and $w(k-1)$

$$P_k^- = E\left\{\tilde{x}_k^-\left(\tilde{x}_k^-\right)^T\right\} = \Phi(k-1)E\left\{\tilde{x}_{k-1}^+\left(\tilde{x}_{k-1}^+\right)^T\right\}\Phi^T(k-1) + E\left\{w(k)w^T(k-1)\right\}$$
$$= \Phi(k-1)P_{k-1}^+\Phi^T(k-1) + Q(k-1) \quad (7.97)$$

A time-diagram of the various quantities involved in the discrete optimal filter equations is given in the Fig. 7.9.

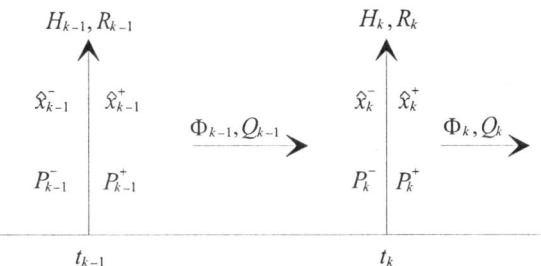

Fig. 7.9. Discrete Kalman filter timing diagram

Fig. 7.10. illustrates the above equations in a block diagram form.

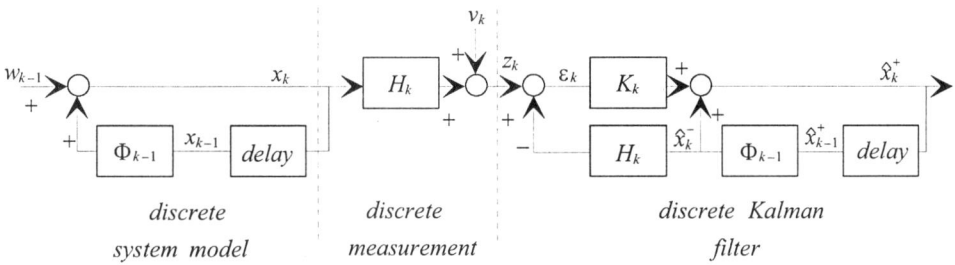

Fig. 7.10. Discrete system and measurement models and discrete-Kalman filter

Similarly as in the continuous-discrete Kalman filter, in the discrete Kalman filter calculations at the covariance level serve to provide K_k, which is then used in the calculation of the conditional mean values, i.e. the estimate $\hat{x}_k$. There is no feedback from the state equations to the covariance equations. This is illustrated in fig. 7.11., which is essentially a simplified computer flow diagram of the discrete Kalman filter.

Summary of the discrete Kalman filter equations is given in the table 7.2, where the noise term Gw, instead of w, is included for completeness. The difference equation for $\tilde{x}_k^+$ can be obtained by subtracting eq. (7.88) from eq. (7.84) and employing eqs. (7.87) and (7.95). Recalling that

$$
\begin{aligned}
\tilde{x}_k^+ &= x_k - \hat{x}_k^+ \\
&= \Phi_{k-1} x_{k-1} + w_{k-1} - \left\{ \Phi_{k-1} \hat{x}_{k-1}^+ + K_k \left[H_k \left(\Phi_{k-1} x_{k-1} + w_{k-1} \right) + v_k - H_k \Phi_{k-1} \hat{x}_{k-1}^+ \right] \right\} \quad (7.98a) \\
&= \left(I - K_k H_k \right) \Phi_{k-1} \tilde{x}_{k-1}^+ + \left(I - K_k H_k \right) w_{k-1} - K_k v_k
\end{aligned}
$$

it can be noted

$$
E\left\{ \tilde{x}_k^+ \right\} = \left(I - K_k H_k \right) \Phi_{k-1} E\left\{ \tilde{x}_{k-1}^+ \right\}
$$

so that $E\left\{ \tilde{x}_k^+ \right\} = 0$ if $E\left\{ \tilde{x}_0^+ \right\} = 0$, i.e. the estimate is unbiased. This is, however, guaranteed by the way we initialize the filter. Hence, the Kalman filter produces an *unbiased estimate*, and so, from eq. (7.98)

$$
P_k^+ = E\left\{ \tilde{x}_k^+ \left(\tilde{x}_k^+ \right)^T \right\} = \left(I - K_k H_k \right) \Phi_{k-1} P_{k-1}^+ \Phi_{k-1}^T \left(I - K_k H_k \right)^T + K_k R_k K_k^T + \left(I - K_k H_k \right) Q_{k-1} \left(I - K_k H_k \right)^T
$$

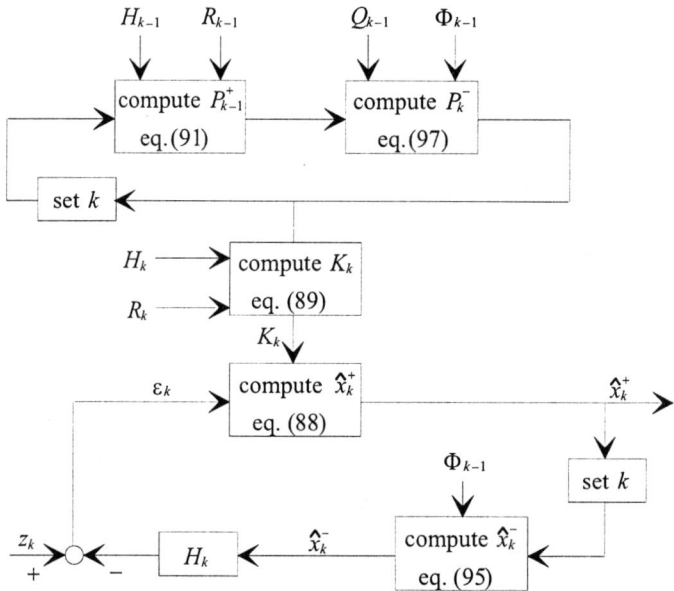

Fig. 7.11. Discrete Kalman filter information flow diagram

By substituting for K_k from eqs. (7.59) or (7.89) and P_{k-1}^+ for eq. (7.97), it is easy to see that we have just found another way to derive the *Ricatti equation* (7.97); that is

$$
\begin{aligned}
P_k^+ &= \left(I - K_k H_k \right) \left(\Phi_{k-1} P_{k-1}^+ \Phi_{k-1}^T + Q_{k-1} \right) \left(I - K_k H_k \right)^T + K_k R_k K_k^T \\
&= \left(I - K_k H_k \right) P_k^- \left(I - K_k H_k \right)^T + K_k R_k K_k^T
\end{aligned}
$$

what is exactly eq. (7.54), from which follows directly eqs. (7.61) or eq. (7.91), after replacing K_k from eq. (7.59) or (7.89).

Analogous to the case of continuos-discrete estimator, it can be shown that

$$E\left\{\hat{x}_k^+\left(\tilde{x}_k^+\right)^T\right\}=0 \tag{7.99}$$

That is, the optimal estimate and its error are *orthogonal* (*orthogonality principle*).

To prove (7.99) lets start from the equations (7.84), (7.87), (7.88), and (7.95) at time $k=1$, yielding

$$x(1)=\Phi(0)x(0)+w(0)$$

$$\hat{x}_1^+=\Phi(0)\hat{x}_0^++K_1\left[-H(1)\Phi(0)\tilde{x}_0^++H(1)w(0)+v(1)\right]$$

where

$$\tilde{x}_0^+=\hat{x}_0^+-x(0)$$

Substituting $x(1)$ from $\hat{x}_1^+$, we have

$$\tilde{x}_1^+=\hat{x}_1^+-x(1)=\left(\Phi(0)-K_1H(1)\Phi(0)\right)\tilde{x}_0^++\left(K_1H(1)-I\right)w(0)+K_1v(1)$$

from which it follows

$$E\left\{\hat{x}_1^+\left(\tilde{x}_1^+\right)^T\right\}=-K_1H(1)\Phi(0)P_0^+\left(\Phi(0)-K_1H(1)\Phi(0)\right)^T$$

$$+K_1H(1)Q(0)\left(K_1H(1)-I\right)^T+K_1R(1)K_1^T$$

where

$$P_0^+=E\left\{\tilde{x}_0^+\left(\tilde{x}_0^+\right)^T\right\}; \ E\left\{w(0)w^T(0)\right\}=Q(0) ; \ E\left\{v(1)v^T(1)\right\}=R(1).$$

Here is used the fact that under the adopted assumptions about the noises and the proper filter initializations

$$\hat{x}_0^+=E\left\{x(0)\right\}$$

we obtain

$$E\left\{w(0)\left(\tilde{x}_0^+\right)^T\right\}=0;$$

$$E\left\{v(1)\left(\tilde{x}_0^+\right)^T\right\}=0;$$

$$E\left\{\hat{x}_0^+w^T(0)\right\}=0 ;$$

$$E\left\{v(1)w^T(0)\right\}=0 ;$$

$$E\left\{\hat{x}_0^+v^T(1)\right\}=0 ;$$

$$E\left\{\hat{x}_0^+\left(\tilde{x}_0^+\right)^T\right\}=0.$$

Having in mind (7.97), one concludes

$$P_1^-=\Phi(0)P_0^+\Phi^T(0)+Q(0).$$

Substituting the last relation into the obtained expression for $E\left\{\hat{x}_1^+\left(\tilde{x}_1^+\right)^T\right\}$ we obtain

$$E\left\{\hat{x}_1^+\left(\tilde{x}_1^+\right)^T\right\}=K_1\left[-H(1)P_1^-+\left(H(1)P_1^-H^T(1)+R(1)\right)K_1^T\right].$$

Since by (7.89)

$$K_1=P_1^-H^T(1)\left[H(1)P_1^-H^T(1)+R(1)\right]^{-1}$$

one concludes that

$$E\left\{\hat{x}_1^+\left(\tilde{x}_1^+\right)^T\right\}=0.$$

Now we may repeat the process for $E\left\{\hat{x}_2^+\left(\tilde{x}_2^+\right)^T\right\}$, since $E\left\{\hat{x}_1^+\left(\tilde{x}_1^+\right)^T\right\}=0$, and, by induction, verify that

$$E\left\{\hat{x}_k^+\left(\tilde{x}_k^+\right)^T\right\}=0, \quad k=1,2,\ldots$$

which completes the proof.

Note that estimate $\hat{x}_k^+$ is a linear function of the measurements up to time k, including the initial information at $k=0$. Thus, in view of geometric property of the minimum mean square error estimator, the estimation error $\tilde{x}_k^+$ is orthogonal into the data space (hyper plane) spanned by the observation set Z^j, $j \le k$.

Tab. 7.2. Summary of discrete Kalman filter

System model	$x_k = \Phi_k x_{k-1} + G_{k-1} w_{k-1}$
	$E\{w_k\}=0 \; ; \; E\left\{w_k w_j^T\right\}=Q_k \delta_{kj}$
Measurement model	$z_k = H_k x_k + v_k$
	$E\{v_k\}=0 \; ; \; E\left\{v_k v_j^T\right\}=R_k \delta_{kj}$
Initial conditions	$E\{x(0)\}=\hat{x}_0 \; ; \; E\left\{\tilde{x}_0 \tilde{x}_0^T\right\}=P_0 \; ; \; \tilde{x}_0 = x_0 - \hat{x}_0$
Other assumptions	$E\left\{w_k v_j^T\right\}=0 \text{ for all } k \text{ and } j$
	$E\left\{w_k \tilde{x}_0^T\right\}=E\left\{v_j \tilde{x}_0^T\right\}$
	$E\left\{w_k \tilde{x}_0^T\right\}=E\left\{v_j \tilde{x}_0^T\right\}=0$
State estimate extrapolation (state update or prediction)	$\hat{x}_k^- = \Phi_{k-1}\hat{x}_{k-1}^+$
Error covariance extrapolation (prediction)	$P_k^- = \Phi_{k-1}P_{k-1}^+\Phi_{k-1}^T + G_{k-1}Q_{k-1}G_{k-1}^T$
State estimate update (measurement update)	$\hat{x}_k^+ = \hat{x}_k^- + K_k\left[z_k - H_k\hat{x}_k^-\right]$
Error covariance update	$P_k^+ = \left[I - K_k H_k\right]P_k^-$
Kalman gain matrix	$K_k = P_k^- H_k^T\left[H_k P_k^- H_k^T + R_k\right]^{-1}$

7.2.1 Optimal discrete-time prediction

Finally, the *optimal filtering* in the absence of measurements can be thought of in terms of *optimal prediction*. Analogous to the continuos-discrete filter, the optimal prediction is equivalent to optimal filtering with arbitrarily large measurement errors, i.e. $R_k^{-1}=0$ and hence $K_k=0$. Therefore, if measurements are unavailable beyond some time, say $t_0=0$, the optimal prediction of $x(1)$ for $t_1 = T$ given $\hat{x}(0)$ must be obtained from eq. (7.88) with $K=0$, or equivalently from eq. (7.95) when the time index $k=1$, i.e. [1, 2, 17, 25, 30, 31, 36, 48]

$$\hat{x}(1)=\Phi(0)\hat{x}(0)$$

The corresponding equation from uncertainty in the optimal prediction, given $P(0)$, is. eq. (7.97) when $k=1$, i.e.

$$P(1) = \Phi(0)P(0)\Phi^T(0) + Q(0)$$

By repeating the procedure, one obtains for two time stage prediction

$$\hat{x}(2) = \Phi(1)\hat{x}(1) = \Phi(1)\Phi(0)\hat{x}(0) = \prod_{i=0}^{1}\Phi(i)\hat{x}(0) = \Phi(2,0)\hat{x}(0)$$

where the prediction error is given by the covariance matrix

$$P(2) = \Phi(1)P(1)\Phi^T(1) + Q(1)$$
$$= \Phi(1)\Phi(0)P(0)\Phi^T(0)\Phi^T(1) + \Phi(1)Q(0)\Phi^T(1) + Q(1)$$
$$= \Phi(2,0)P(0)\Phi^T(2,0) + \sum_{i=1}^{2}\Phi(2,i)Q(i-1)\Phi^T(k,i)$$

Here

$$\Phi(k,0) = \prod_{i=0}^{k-1}\Phi(i) \; ; k = 1,2,...\; ; \; \Phi(k,i) = \prod_{j=i}^{k-1}\Phi(j) \; ; i \le k-1; \; \Phi(k,k) = I \qquad (7.100)$$

Thus, by induction, one concludes that k-stage predictions i given by

$$\hat{x}(k) = \Phi(k,0)\hat{x}(0) \qquad (7.101)$$

where the prediction error covariance matrix is

$$P(k) = \Phi(k,0)P(0)\Phi^T(k,0) + \sum_{i=1}^{k}\Phi(k,i)Q(i-1)\Phi^T(k,i) \qquad (7.102)$$

with $\Phi(k,0)$ and $\Phi(k,i)$ being defined by eq. (7.100).

One special case in prediction, which merits attention, is that of *single-stage prediction* in which we consider the optimal estimate $\hat{x}_{k+1}^-$, $k = 0,1,2,....$As we seen before, this result is extremely useful in development of the equations for *optimal filtering*. Namely, if the optimal estimate $\hat{x}_k^+$ (see fig. 7.9) and the covariance matrix P_k^+ of the corresponding filtering error $\tilde{x}_k^+ = x_k - \hat{x}_k^+$ are known for some $k = 0,1,2,...$ then the single-stage optimal predicted estimate is given by the expression (7.95), i.e. (see Tab. 7.2)

$$\hat{x}_{k+1}^- = \Phi(k)\hat{x}_k^+ \qquad (7.103)$$

while the predicted estimation error covariance matrix $P_{k+1}^- = E\left\{\left(x_{k+1} - \hat{x}_{k+1}^-\right)\left(x_{k+1} - \hat{x}_{k+1}^-\right)^T\right\}$ is given by the expression (7.97); that is

$$P_{k+1}^- = \Phi(k)P_k^+\Phi^T(k) + Q(k) \qquad (7.104)$$

In summary, the operation of the discrete Kalman filter is given in the predictor-corrector form (Tab. 7.2). As an illustration of the design and properties of this filter, let us consider next examples.

Example 7.2 (Discrete-time radar tracking problem): Assume a vehicle being tracked at range $R + \rho(k)$ at time $t_k = kT$, and at range $R + \rho(k+1)$ at time $t_{k+1} = (k+1)T$, T seconds later. We use T to represent the spacing between samples made one scan apart. The average range is denoted by R, and $\rho(k)$, $\rho(k+1)$ represent deviations from the average. We are interested in estimating these deviations, which are assumed to be statistically random with zero-mean value [5, 6, 10, 35]. To a first approximation, if the vehicle is traveling at random velocity $\dot{\rho}(k)$ and T is not too large,

$$\rho(k+1) = \rho(k) + T\dot{\rho}(k) \tag{I}$$

which is the range equation. Similarly, considering acceleration $a_r(k)$ we have

$$Ta_r(k) = \dot{\rho}(k+1) - \dot{\rho}(k) \tag{II}$$

which is the acceleration equation. Assuming that $a_r(k)$ is a zero-mean, stationary white noise process, the acceleration is, on average, zero and uncorrelated between intervals, i.e. $E\{a_r(k+1)a_r(k)\} = 0$, but it has some known variance $E\{a_r^2(k)\}$. Such accelerations might be caused by sudden mind gusts or short-term irregularities in engine thrust. The quantity $u_r(k) = Ta_r(k)$ is also a white noise process, and we have in place of eq. (II)

$$\dot{\rho}(k+1) = \dot{\rho}(k) + u_r(k) \tag{III}$$

We define now a two component signal vector $x_r(k)$ with one component the range, $x_{1r}(k) = \rho(k)$, and the other component the radial velocity, $x_{2r}(k) = \dot{\rho}(k)$. Applying this to equations (I) and (III), we have

$$x_{1r}(k+1) = x_{1r}(k) + Tx_{2r}(k) \tag{IV}$$

$$x_{2r}(k+1) = x_{2r}(k) + u_r(k) \tag{V}$$

or combining them into single vector equation

$$x_r(k+1) = \begin{bmatrix} x_{1r}(k+1) \\ x_{2r}(k+1) \end{bmatrix} = \begin{bmatrix} 1 & T \\ 0 & 1 \end{bmatrix} \begin{bmatrix} x_{1r}(k) \\ x_{2r}(k) \end{bmatrix} + \begin{bmatrix} 0 \\ u_r(k) \end{bmatrix} = A_r x_r(k) + w_r(k) \tag{VI}$$

which is the form of the first-order vector difference equation. We now add two more states concerned with the bearing $\Theta(k)$ and bearing rate, or angular velocity, $\dot{\Theta}(k)$, i.e.

$$\Theta(k+1) = \Theta(k) + T\dot{\Theta}(k) \tag{VII}$$

$$Ta_\Theta(k) = \dot{\Theta}(k+1) - \dot{\Theta}(k) \tag{VIII}$$

where $a_\Theta(k)$ is the angular acceleration. We assume $u_\Theta(k) = Ta_\Theta(k)$ to e random with zero mean and uncorrelated between intervals, i.e. $E\{u_\Theta(k+1)u_\Theta(k)\} = 0$, and also with $u_r(k)$; that is, $E\{u_\Theta(k)u_r(k)\} = 0$.

We define again a two component signal vector $x_\Theta^T(k) = \begin{bmatrix} x_{1\Theta}(k) & x_{2\Theta}(k) \end{bmatrix} = \begin{bmatrix} \Theta(k) & \dot{\Theta}(k) \end{bmatrix}$, and, applying this to eqs. (VII) and (VIII), we have

$$x_{1\Theta}(k+1) = x_{1\Theta}(k) + Tx_{2\Theta}(k) \tag{IX}$$

$$x_{2\Theta}(k+1) = x_{2\Theta}(k) + u_\Theta(k) \tag{X}$$

or in the matrix form

$$x_\Theta(k+1) = \begin{bmatrix} x_{1\Theta}(k+1) \\ x_{2\Theta}(k+1) \end{bmatrix} = \begin{bmatrix} 1 & T \\ 0 & 1 \end{bmatrix} \begin{bmatrix} x_{1\Theta}(k) \\ x_{2\Theta}(k) \end{bmatrix} + \begin{bmatrix} 0 \\ u_\Theta(k) \end{bmatrix} = A_\Theta x_\Theta(k) + w_\Theta(k) \tag{XI}$$

The addition of the states in eq. (XI) augments eq. (VI) into the following state-space equation

$$
x(k+1) = \begin{bmatrix} x_1(k+1) \\ x_2(k+1) \\ x_3(k+1) \\ x_4(k+1) \end{bmatrix} = \begin{bmatrix} x_{1r}(k+1) \\ x_{2r}(k+1) \\ x_{1\Theta}(k+1) \\ x_{2\Theta}(k+1) \end{bmatrix} = \begin{bmatrix} 1 & T & 0 & 0 \\ 0 & 1 & 0 & 0 \\ 0 & 0 & 1 & T \\ 0 & 0 & 0 & 1 \end{bmatrix} \begin{bmatrix} x_1(k) \\ x_2(k) \\ x_3(k) \\ x_4(k) \end{bmatrix} + \begin{bmatrix} 0 \\ u_r(k) \\ 0 \\ u_\Theta(k) \end{bmatrix}
$$

$$
= \begin{bmatrix} A_r & 0 \\ 0 & A_\Theta \end{bmatrix} \begin{bmatrix} x_r(k) \\ x_\Theta(k) \end{bmatrix} + \begin{bmatrix} w_r(k) \\ w_\Theta(k) \end{bmatrix}
\tag{XII}
$$

The radar sensors are assumed to provide noisy estimates of the range $x_{1r}(k) = \rho(k)$ and bearing $x_{1\Theta}(k) = \Theta(k)$ at time interval T. At time $t_k = kT$, the two sensor outputs are then

$$
y_r(k) = x_{1r}(k) + v_r(k)
\tag{XIII}
$$

$$
y_\Theta(k) = x_{1\Theta}(k) + v_\Theta(k)
\tag{XIV}
$$

Therefore, the data or measurements can be written as

$$
y_r(k) = \begin{bmatrix} 1 & 0 \end{bmatrix} \begin{bmatrix} x_{1r}(k) \\ x_{2r}(k) \end{bmatrix} + v_r(k) = h_r^T x_r(k) + v_r(k)
\tag{XV}
$$

$$
y_\Theta(k) = \begin{bmatrix} 1 & 0 \end{bmatrix} \begin{bmatrix} x_{1\Theta}(k) \\ x_{2\Theta}(k) \end{bmatrix} + v_\Theta(k) = h_\Theta^T x_\Theta(k) + v_\Theta(k)
\tag{XVI}
$$

or, combining them into signal vector equation

$$
y(k) = \begin{bmatrix} y_r(k) \\ y_\Theta(k) \end{bmatrix} = \begin{bmatrix} 1 & 0 & 0 & 0 \\ 0 & 0 & 0 & 1 \end{bmatrix} \begin{bmatrix} x_r(k) \\ x_\Theta(k) \end{bmatrix} + \begin{bmatrix} v_r(k) \\ v_\Theta(k) \end{bmatrix}
$$

$$
= \begin{bmatrix} h_r^T & 0 \\ 0 & h_\Theta^T \end{bmatrix} \begin{bmatrix} x_r(k) \\ x_\Theta(k) \end{bmatrix} + \begin{bmatrix} v_r(k) \\ v_\Theta(k) \end{bmatrix} = Hx(k) + v(k)
\tag{XVII}
$$

The additive noise, $v(k)$, is usually assumed to be Gaussian white with zero-mean and variances $E\{v_r^2(k)\} = r_\rho(k)$ and $E\{v_\Theta^2(k)\} = r_\Theta(k)$. So far, we have established vector equations for the process, or system, model given by eq. (XII), and measurement, or data, model given by eq. (XVII). The next step is to formulate noise covariance matrices Q for the system, and R for the measurement model. For the later, using eq. (XVII), we have

$$
R(k) = E\{v(k)v^T(k)\} = \begin{bmatrix} r_\rho(k) & 0 \\ 0 & r_\Theta(k) \end{bmatrix}
\tag{XVIII}
$$

and the system noise covariance, defined in eq. (XII), for this case is given by

$$
Q(k) = E\{w(k)w^T(k)\} = \begin{bmatrix} 0 & 0 & 0 & 0 \\ 0 & \sigma_r^2 & 0 & 0 \\ 0 & 0 & 0 & 0 \\ 0 & 0 & 0 & \sigma_\Theta^2 \end{bmatrix} = \begin{bmatrix} Q_r & 0 \\ 0 & Q_\Theta \end{bmatrix}
\tag{XIX}
$$

where $\sigma_r^2(k) = E\{u_r^2(k)\}$ and $\sigma_\Theta^2(k) = E\{u_\Theta^2(k)\}$ are the variances of $Ta_r(k)$, and $Ta_\Theta(k)$, respectively. Specific values must be substituted for those variances, in order to define the Kalman filter. To do this, we assume that the probability density function (pdf) of the acceleration in either direction (ρ or Θ), is uniform and equal $p(a) = 1/2M$, between limits $\pm M$; therefore the variance $\sigma_a^2 = M^2/3$.

The variances in eq. (XIX) are then

$$\sigma_r^2 = E\{(Ta_r)^2\} = T^2 \sigma_a^2 \ ; \ \sigma_\Theta^2 = \sigma_r^2 / R^2 \tag{XX}$$

The acceleration term σ_a has the units of meters per second per second. Similarly as in example 7.1, the second relation in (XX) follows from the fact that σ_a represents linear acceleration, and we must divide it by the radial distance $R + \rho(k)$ in order to convert i to the desired angular acceleration a_Θ; that is $\sigma_\Theta^2 = E\{(Ta/R)^2\}$, assuming that the range deviation $\rho(k)$ is small compared with the average range R.

Now all we need to do is enter these system and covariance matrices into our computer program which implements the discrete Kalman filter. At each time $t_k = kT$ we give the filter the radar readings $y_r(k)$ and $y_\Theta(k)$ and it will give the best estimates $x_1(k) = x_{1r}(k) = \rho(k)$, $x_2(k) = x_{2r}(k) = \dot{\rho}(k)$, $x_3(k) = x_{1\Theta}(k) = \Theta(k)$ and $x_4(k) = x_{2\Theta}(k) = \dot{\Theta}(k)$. To start Kalman processing we have to initialize the gain matrix K. For this purpose the error covariance matrix P has to be specified in some way. A reasonable ad hoc initialization can be established using two measurements, range and bearing, at times $k=1$ and $k=2$. From these four measurements we can make the following estimates

$$\hat{x}(2) = \begin{bmatrix} \hat{x}_1(2) \\ \hat{x}_2(2) \\ \hat{x}_3(2) \\ \hat{x}_4(2) \end{bmatrix} = \begin{bmatrix} y_r(2) \\ \dfrac{y_r(2) - y_r(1)}{T} \\ y_\Theta(2) \\ \dfrac{y_\Theta(2) - y_\Theta(1)}{T} \end{bmatrix} \tag{XXI}$$

To calculate $P(2)$, we use the general expression

$$P(2) = E\{\tilde{x}(2)\tilde{x}^T(2)\} \ ; \ \tilde{x}(2) = x(2) - \hat{x}(2) \tag{XXII}$$

Values for $\hat{x}(2)$ are given by eq. (XXI), and using eqs. (XII) and (XVII) for $x(2)$, we obtain

$$\tilde{x}(2) = x(2) - \hat{x}(2) = \begin{bmatrix} -v_r(2) \\ w_r(1) - (v_r(2) - v_r(1))/T \\ -v_\Theta(2) \\ w_\Theta(1) - (v_\Theta(2) - v_\Theta(1))/T \end{bmatrix} \tag{XXIII}$$

Taking into account the independence of noise sources w and v, and also the independence between individual noise samples, it can be easily shown that the above matrix is given by

$$P(2) = \left[\begin{array}{cc|cc} p_{11} & p_{12} & 0 & 0 \\ p_{12} & p_{22} & 0 & 0 \\ \hline 0 & 0 & p_{33} & p_{34} \\ 0 & 0 & p_{34} & p_{44} \end{array} \right] = \left[\begin{array}{c|c} P_r(2) & 0 \\ \hline 0 & P_\Theta(2) \end{array} \right] \qquad \text{(XXIV)}$$

where

$$p_{11} = r_\rho \; ; \; p_{12} = r_\rho / T \; ; \; p_{22} = 2r_\rho / T^2 + \sigma_r^2 \qquad \text{(XXV)}$$
$$p_{33} = r_\Theta \; ; \; p_{34} = r_\Theta / T \; ; \; p_{44} = 2r_\Theta / T + \sigma_\Theta^2$$

As a numerical example we take for range $R=160\text{km}$, scan time $T = 1\text{s}$, and a maximum acceleration $M=2.1\,\text{ms}^{-2}$. For the above numerical values, we can calculate noise variances in eq. (XX) as $\sigma_r^2 = 330$ and $\sigma_\Theta^2 = 1.3 \cdot 10^{-8}$. Let the root mean-square (r.m.s.) noise in the range sensor be equivalent to 1km, therefore $r_\rho = 1000m$. Furthermore, let r.m.s. noise r_Θ in the bearing sensor be 1^0 or 0.017rad. In Fig. 7.12 we show a MATLAB program which implements the discrete-Kalman filter for the range subsystem. As in the example 7.1, note that derived discrete equations describe two completely decoupled systems, one for the range dynamics $\{r, \dot{r}\}$ and one for the angle dynamics $\{\Theta, \dot{\Theta}\}$. This means that we can simplify things by running not one Kalman filter on a fourth order system but two parallel Kalman filters on two second-order systems. The number of computations required per iteration by the Kalman filter is of the order of n^3, so one fourth-order filter requires $4^3 = 64$ operations while two second-order filters require about $2 \cdot 2^3 = 16$ operations. Aside from this, using parallel processing a separate microprocessors can be assigned to each filter. Fig. 7.13 shows the actual and estimated values of range and range rate, while Fig. 7.14 represents the time history of covariance matrix error elements.

```
function ex72

% design of aircraft trajectory

tfin=100; %total simulation time

x(1)=0;y(1)=0;r(1)=0;theta(1)=0;
vx(1)=10;vy(1)=5;i=1;
for t=2:1:tfin
  if ((t>20)& (t<25))
    vx(t)=vx(t-1)+0.5;
  else
    if ((t>30) & (t<35))
      vx(t)=vx(t-1)-0.5;
    else
      vx(t)=vx(t-1);
    end
  end
  if ((t>15)& (t<20))
    vy(t)=vy(t-1)+0.05;
  else
    if ((t>20) & (t<25))
      vy(t)=vy(t-1)-0.1;
    else
```

```
      vy(t)=vy(t-1);
    end
  end
  x(t)=x(t-1)+vx(t-1);
  y(t)=y(t-1)+vy(t-1);
  r(t)=sqrt(x(t)^2+y(t)^2);
  theta(t)=atan2(y(t),x(t));
  vr(t)=(x(t)*vx(t)+y(t)*vy(t))/r(t);
end
for i=1:100
  robs(i)=r(i)+sqrt(1000)*randn;
  thetaobs(i)=theta(i)+sqrt(0.017)*randn;
end

% design of Kalman filter for range
rp(1:2,1)=[robs(1);0];pp=[1000 0;0 1000];
A=[1 1;0 1];
h=[1 0];R=1000;Q=.3;G=[0;1];
tmp(1:3,1)=[1000;0;1000];
for i=2:100
  rm(1:2,i)=A*rp(1:2,i-1);
  pm=A*pp*A'+G*Q*G';
  K=pm*h'*inv(h*pm*h'+R);
```

```
  rp(1:2,i)=rm(1:2,i)+K*(robs(i)-
h*rm(1:2,i));
  pp=(eye(2)-K*h)*pm;
  tmp(1:3,i)=[pp(1,1);pp(1,2);pp(2,2)];
end
figure(1);plot(1:100,r,1:100,rp(1,:),'--');
legend('the thrue range','the estimated range');
title('the thrue  and estimated range');
figure(2);plot(1:100,vr(1:100),1:100,rp(2,:),'--
');
legend('the acctual range rate', ...
```

```
'the estimated range rate');
title('the actual and estimated range rate');
figure(3);
plot(1:100,tmp(1,1:100),...
1:100,tmp(2,1:100),'--',1:100,tmp(3,1:100),'-
.');
legend('p(1,1)','p(1,2)','p(2,2)');
title('time  history  of  covariance  matrix
elements');
keyboard;
```

Fig. 7.12: MATLAB program for discrete Kalman filter

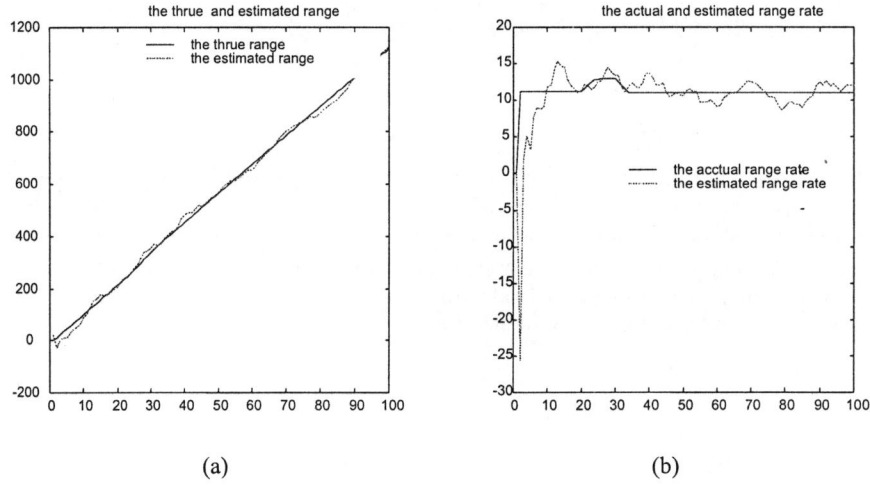

(a)   (b)

Fig. 7.13: (a) Thrue and estimated range; (b) Thrue and estimated range rate

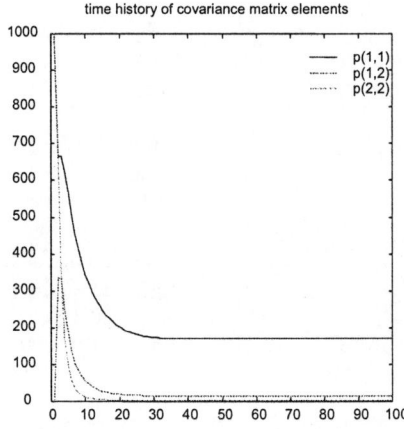

Fig. 7.14: Time history of estimation error covariance matrix elements

Example 7.3 (Kalman filter identifier): The Kalman filter can easily be formulated to solve the well-known system identification problem: Given a set of noisy measurements $\{y(t)\}$ and inputs $\{u(t)\}$, find the minimum variance estimate $\hat{\Theta}(t)$ of a vector $\Theta(t)$ of unknown parameters of a linear time-invariant (LTI) system.

The single input-single output (SISO) LTI system is given by the pulse-transfer function

$$H(z) = \frac{B(z^{-1})}{A(z^{-1})} \tag{I}$$

where A and B are polynomials in z or z^{-1}; that is

$$A(z^{-1}) = 1 + \sum_{i=1}^{n} a_i z^{-i}$$

$$B(z^{-1}) = \sum_{i=0}^{m} b_i z^{-i}$$

If we consider the equivalent time domain representation, then we have a difference equation relating the output sequence $\{y(t)\}$ to the input sequence $\{u(t)\}$

$$A(z^{-1}) y(t) = B(z^{-1}) u(t) \tag{IIa}$$

or

$$y(t) + \sum_{i=1}^{n} a_i z^{-i} y(t) = \sum_{i=0}^{m} b_i z^{-i} u(t) \tag{IIb}$$

If we adopt z^{-1} as the backward-shift operator with the property that $z^{-i} y(t) = y(t-i)$, we obtain from (IIb)

$$y(t) = -\sum_{i=1}^{n} a_i y(t-i) + \sum_{i=0}^{bm} b_i u(t-i) \tag{III}$$

When the system is driven by random disturbance, as it is shown in Fig. 7.15.a, the models are given by the autoregressive model with exogenous inputs, the so-called ARX (see Fig. 7.15.b)

$$A(z^{-1}) y(t) = B(z^{-1}) u(t) + e(t) \tag{IVa}$$

or

$$y(t) = -\sum_{i=1}^{n} a_i y(t-i) + \sum_{i=0}^{m} b_i u(t-i) + e(t) \tag{IVb}$$

where A and B are polynomials, as before, and $\{e(t)\}$ is zero-mean white disturbance, or noise, with variance R.

We must convert the model (IV) to the state-space framework required by the Kalman filter in Tab. 7.2. The measurement model can be expressed from (IV) as

$$y(t) = h^T(t) \Theta(t) + e(t) \tag{V}$$

where

$$h^T(t) = \{-y(t-1), ..., -y(t-n), u(t), ..., u(t-m)\} \tag{VI}$$

and

$$\Theta^T(t) = \{a_1, ..., a_n, b_0, ..., b_m\} \tag{VII}$$

which represents the measurement equation in the Kalman filter formulation (Tab. 7.2).

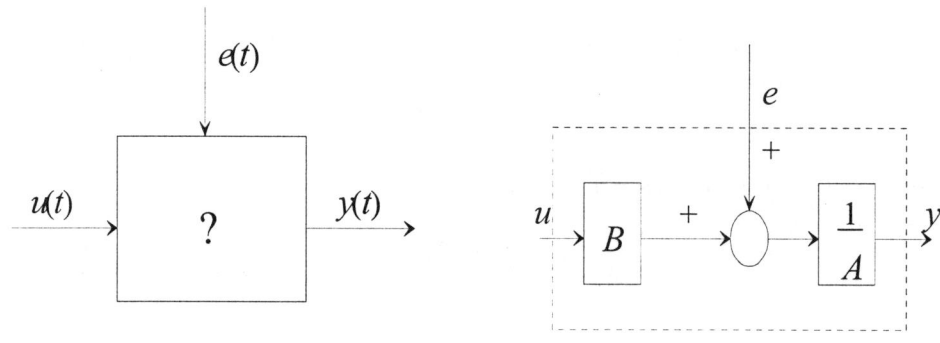

(a)   (b)

Fig. 7.15: (a) Identification problem; (b) ARX system model

The parameters can be modeled as constants with uncertainty $w(t)$; i.e. the state-space model used in this formulation takes the form

$$\Theta(t) = \Theta(t-1) + w(t)$$

where w is zero-mean white noise with covariance (VIII) matrix Q. thus, the Kalman filter identifier is given by Tab. 7.2 where $\Phi = G = I$, with I being an identity matrix.

Thus, the Kalman filter provides the minimum variance estimates of the parameters of the ARX model. Note that the Kalman identifier is identical to the recursive least-squares (LS) algorithm of section 5, with the exception that the parameters can include process noise w. Thus, for the process noise covariance $Q=0$ the Kalman filter identifier reduces to the recursive LS algorithm. However, although the above difference appears minimal, it becomes very significant and necessary in practice, in order to avoid estimator divergence.

As an illustration of the Kalman filter identifier let us apply it to the following example. In this example a first-order over second-order model is used. The system is subjected to a unit step input and noise disturbance. The system pulse-transfer function used is

$$H(z) = \frac{B(z^{-1})}{A(z^{-1})} = \frac{b_1 z^{-1} + b_2 z^{-2}}{1 + a_1 z^{-1} + a_2 z^{-2}}$$

where $b_1 = 1$, $b_2 = 0.5$, $a_1 = -1.8$, $a_2 = 0.82$. The identification model is given by

$$\Theta_k = \Theta_{k-1} + w_k$$
$$y_k = h_k^T \Theta_k + v_k$$

where $h_k^T = \{-y_{k-1} \ -y_{k-2} \ u_{k-1} \ u_{k-2}\}$, $\Theta_k^T = \{a_1 \ a_2 \ b_1 \ b_2\}$. Here v_k is zero-mean process with variance $R = 1/3$ and w_k is also zero-mean noise with variance $Q = 0$. The initial parameter estimate $\hat{\Theta}_0^+ = 0$ and initial covariance matrix $P_0^+ = 10^2 I$. Using given system model and MATLAB we simulated the data with the specified variances. The MATLAB program of the Kalman filter identifier is depicted in Fig. 7.16, while the parameter estimates are shown in Fig. 7.17.

```
function ex73
% system parameters
a(1)=-1.8;a(2)=0.82;b(1)=1;b(2)=0.5;
% the initial conditions of the system
y(1)=0.5;y(2)=-0.5;
% definition of system input signal
N=100; %duration of simulation
u(1:N)=randn(1,N)+2;
% generation of system output
for i=3:N
    y(i)=-a(1)*y(i-1)-a(2)*y(i-2)+b(1)*u(i-
1)+b(2)*u(i-2);
    y(i)=y(i)+sqrt(0.3)*randn;
end
% initialization of the Kalman fitler
R=1/3;p=10*eye(4);
teta=zeros(4,2);
```

```
% Kalman filter
for i=3:N
    h=[-y(i-1) -y(i-2) u(i-1) u(i-2)];
    ni=y(i)-h*teta(:,i-1); z=h*p*h'+R;
    K=p*h'*inv(z); teta(:,i)=teta(:,i-1)+K*ni;
    p=(eye(4)-K*h)*p;
end
figure(1);subplot(221);
plot(1:N,a(1)*ones(1,N),'--',1:N,teta(1,:));
title('a1');
subplot(222);plot(1:N,a(2)*ones(1,N),'--
',1:N,teta(2,:));
title('a2');subplot(223);
plot(1:N,b(1)*ones(1,N),'--',1:N,teta(3,:));
title('b1');subplot(224);
plot(1:N,b(2)*ones(1,N),'--',1:N,teta(4,:));
title('b2');
```

Fig. 7.16: The Kalman filter identifier programm code

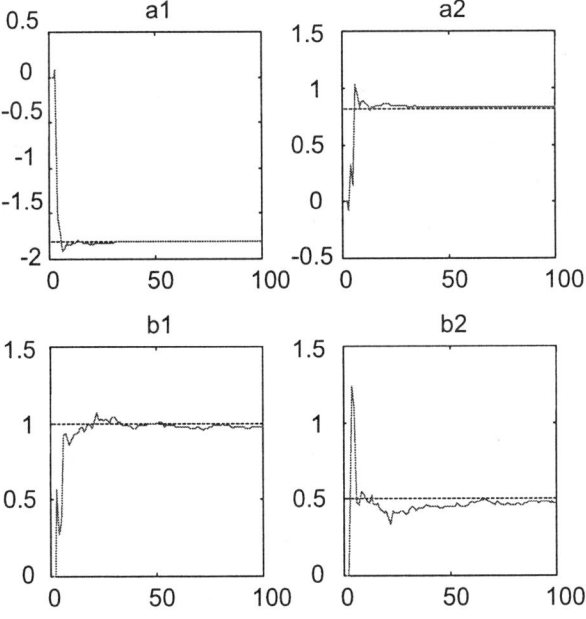

Fig. 7.17: The actual and estimated parameters

7.3. Continuous estimation with continuous observations

The transition from the discrete to the continuous formulation of the Kalman filter is also readily accomplished. Namely, in order to go from the discrete system and measurement model given by the expressions (7.84) and (7.87), respectively, to the continuous system described by eq. (7.83) and continuous measurement model, i.e.

$$\dot{x} = Ax + Gw \tag{7.105}$$

$$z = Hx + v \tag{7.106}$$

where v is zero-mean continuous white noise with spectral density matrix R, it is necessary to observe the equivalencies defined by eq. (7.86), valid in the limit as $t_k - t_{k-1} = T \to 0$; that is

$$\Phi(k) = I + AT \; ; \; Q(k) = GQG^T T \tag{7.107a}$$

What now remains is to establish the equivalence between the discrete white noise sequence $\{v(k)\}$ in eq. (7.87) and the non-physically realizable continuous white noise $v(t)$ in eq. (7.106). Note that, whereas $R(k) = E\{v(k)v^T(k)\}$ is a covariance matrix, $R(t)$ defined $E\{v(t)v^T(\tau)\} = R(t)\delta(t-\tau)$ is a spectral density matrix (the Dirac delta function has units of 1/time). The covariance matrix $R(t)\delta(t-\tau)$ has infinite-valued elements. As mentioned in section 4, the discrete white noise sequence can be made to approximate the continuous white noise process by shrinking the pulse lengths T and increasing their amplitude, such that

$$R(k) = R/T \tag{7.107b}$$

That is, in the limit as $T \to 0$, the discrete noise sequence tends to one of infinite-valued pulses of zero duration, so that its area is equal to the area R under the continuous white noise impulse auto-correlation function.

Using the expressions (7.107), our approach is simply one of writing the appropriate difference equations for discrete Kalman filter (Tab. 7.2) and observing their behavior in the limit as $T \to 0$. Throughout the text, time dependence of all continuous time quantities will often be suppressed for convenient notation (for example, $x(t)$ and $Q(t)$ will be denoted by x and Q, etc.). Furthermore, it shall be assumed that R is nonsingular, i.e. R^{-1} exists. In addition, it is assumed that w and v are uncorrelated.

In discrete form, the state error covariance matrix was shown to propagate according to eqs. (7.97) or (7.104) [1, 2, 25, 35, 36]. The expression (7.104) is now rewritten as

$$P_{k+1}^- = [I + AT] P_k^+ [I + AT]^T + GQG^T T \tag{7.108a}$$

or equivalently

$$P_{k+1}^- = P_k^+ + \left[AP_k^+ + P_k^+ A^T + GQG^T \right] T + O(T^2) \tag{7.108b}$$

where $O(T^2)$ denotes terms of the order T^2. Furthermore, as a consequence of optimal use of a measurement, P_k^+ can be expressed by eq. (7.92), i.e.

$$P_k^+ = [I - K_k H(k)] P_k^- \tag{7.109}$$

Inserting the expression (7.109) into the equation (7.108b) and rearranging terms yields

$$\frac{P_{k+1}^- - P_k^-}{T} = AP_k^- + P_k^- A^T + GQG^T - \frac{1}{T}K_k H(k)P_k^- - AK_k H(k)P_k^- - K_k H(k)P_k^- A^T + O(T)$$

Examing the term $(1/T)K_k$, we note that from eqs. (7.89) and (7.107b)

$$\frac{1}{T}K_k = \frac{1}{T}P_k^- H^T(k)\left[H(k)P_k^- H^T(k) + R(k)\right]^{-1}$$

$$= P_k^- H^T(k)\left[H(k)P_k^- H^T(k)T + R(k)T\right]^{-1} \qquad (7.111)$$

$$= P_k^- H^T(k)\left[H(k)P_k^- H^T(k)T + R\right]^{-1}$$

Thus, in the limit as $T \to 0$ we get

$$\lim_{T \to 0}\frac{1}{T}K_k = PH^T R^{-1} \qquad (7.112)$$

and, simultaneously, from eq. (7.110)

$$\dot{P} = AP + PA^T + GQG^T - PH^T R^{-1}HP \qquad (7.113a)$$

The last equation can be also rewritten as

$$\dot{P} = AP + PA^T - KRK^T \; ; \; K = PH^T R^{-1} \qquad (7.113b)$$

Tracing the development of the terms on the right-hand side of eq. (7.113), it is clear that the term $\left(AP + PA^T\right)$ results from behavior of the autonomous (unforced) system without measurements, the term $\left(GQG^T\right)$ accounts for the increase of uncertainty due to process (state) noise w (this term is positive semidefinite in accordance to eq. (7.8)) and the term $\left(-PH^T R^{-1}HP\right)$ accounts for the decrease of uncertainty as a result of measurements. Equation (7.113) is nonlinear in P; it is referred to as the *matrix continuous Riccati equation*. This equation can only be solved analytically for simple problems. Various numerical techniques are available for more complicated problems. Numerical integration is the direct method of approach for solving this equation. In the absence of measurements we get

$$\dot{P} = AP + PA^T + GQG^T \qquad (7.114)$$

which is the linear variance equation previously defined by eqs. (7.66) or (7.76).

Finally, the discrete form of the Kalman filter is defined by eqs. (7.88) and (7.95), i.e.

$$\hat{x}_k^+ = \Phi(k-1)\hat{x}_{k-1}^+ + K_k\left[z(k) - H(k)\Phi(k-1)\hat{x}_{k-1}^+\right] \qquad (7.115)$$

Replacing $\Phi(k-1)$ by eq. (7.107a), and K_k by $PH^T R^{-1}T$ from eq. (7.112), and rearranging terms yields

$$\frac{\hat{x}_k^+ - \hat{x}_{k-1}^+}{T} = A\hat{x}_{k-1}^+ + PH^T R^{-1}\left[z(k) - H(k)\hat{x}_{k-1}^+\right] + O(T) \qquad (7.116)$$

In the limit, as $T \to 0$, this becomes

$$\dot{\hat{x}} = A\hat{x} + PH^T R^{-1}(z - H\hat{x}) \qquad (7.117)$$

which is the continuous Kalman filter, and in which P is computed according to eq. (7.113). System model and continuous Kalman filter are depicted in Fig. 7.18, while the continuous Kalman filter equations are summarized in table 7.3. The deterministic input term Bu is also included for completeness.

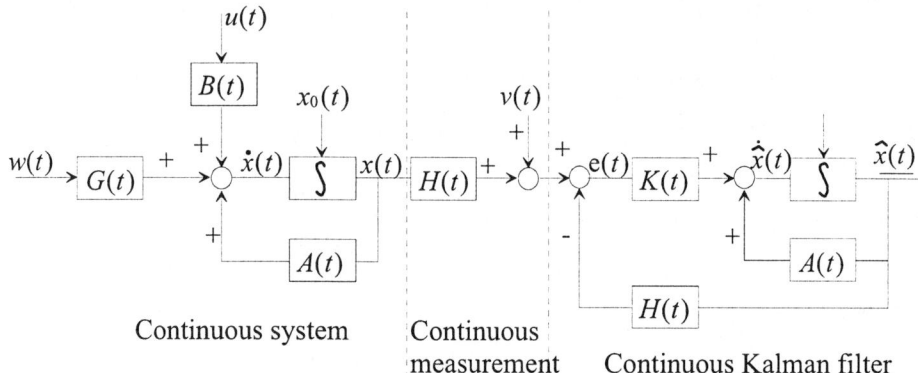

Continuous system   Continuous   Continuous Kalman filter
                         measurement

Fig. 7.18 Continuous system and measurement models, and continuous Kalman filter

Table 7.3. Summary of continuous Kalman filter equations

System model	$\dot{x}(t) = A(t)x(t) + B(t)u(t) + G(t)w(t)$ $E\{w(t)\} = 0;\ E\{w(t)w^T(\tau)\} = Q(t)\delta(t-\tau)$
Measurement model	$z(t) = H(t)x(t) + v(t)$ $E\{v(t)\} = 0;\ E\{v(t)v^T(\tau)\} = R(t)\delta(t-\tau)$
Initial conditions	$E\{x(0)\} = \hat{x}_0;\ E\left\{[x(0)-\hat{x}_0][x(0)-\hat{x}_0]^T\right\} = P_0$
Other assumptions	$E\{x(0)w^T(t)\} = 0;\ E\{x(0)v^T(t)\} = 0;$ $E\{w(t)v^T(\tau)\} = 0$ for all $t, \tau;\ R^{-1}(t)$ exists
State estimate	$\dot{\hat{x}}(t) = A(t)\hat{x}(t) + B(t)u(t) + K(t)\varepsilon(t);\ \hat{x}(0) = \hat{x}_0$
innovation	$\varepsilon(t) = z(t) - H(t)\hat{x}(t)$
error covariance propagation	$\dot{P}(t) = A(t)P(t) + P(t)A^T(t)$ $\qquad + G(t)Q(t)G^T(t) - K(t)R(t)K^T(t);\ P(0) = P_0$
Kalman gain	$K(t) = P(t)H^T(t)R^{-1}(t)$

Error analysis:

The differential equation for the estimation error $\tilde{x} = x - \hat{x}$ can be obtained by subtracting eq. (7.117) from eq. (7.105) and employing eq. (7.106). Recalling that $\hat{x} = x + \tilde{x}$ and $K = PH^T R^{-1}$, this results in

$$\dot{\tilde{x}} = \dot{x} - \dot{\hat{x}} = Ax + Gw - \left\{A\hat{x} + K\left[Hx + v - H(x + \tilde{x})\right]\right\}$$
$$= (A - KH)\tilde{x} + Gw - Kv \tag{7.118}$$

This equation is driven by the process and measurement noise w and v, respectively. If the system matrix $(A-KH)$ tends to an asymptotically stable matrix as $t \to \infty$, then in the limit the estimate $\hat{x}(t)$ converges to the expected true plant state $x(t)$. Let $\Phi(t,\tau)$ be the state-transition matrix for $(A-K(t)H)$, i.e. $\Phi(t,\tau)$ satisfies the autonomous error system

$$\frac{d\Phi(t,\tau)}{dt} = (A-KH)\Phi(t,\tau) \; ; \; \Phi(t,t) = I$$

Then the solution to (7.118) is

$$\tilde{x}(t) = \Phi(t,0)\tilde{x}(0) + \int_0^t \Phi(t,\tau)\left[Gw(\tau) - Kv(\tau)\right]d\tau \tag{7.119}$$

Note that $E\{\tilde{x}(t)\} = 0$ if $E\{\tilde{x}(0)\} = 0$ and noise processes have zero-mean. This is guaranteed by the way we initialize the filter (see table 7.3.). To find a differential equation for error covariance matrix

$$P(t) = E\{\tilde{x}(t)\tilde{x}^T(t)\} \tag{7.120}$$

write

$$\dot{P}(t) = E\{\dot{\tilde{x}}(t)\tilde{x}^T(t)\} + E\{\tilde{x}(t)\dot{\tilde{x}}^T\} \tag{7.121}$$

According to the error dynamics (7.118), the first term is

$$E\{\dot{\tilde{x}}(t)\tilde{x}^T(t)\} = (A-KH)P(t) + GE\{w(t)\tilde{x}^T(t)\} - KE\{v(t)\tilde{x}^T(t)\} \tag{7.122}$$

We used now the cross-correlations

$$R_{v\tilde{x}}(t,t) = E\{v(t)\tilde{x}^T(t)\} \tag{7.123}$$

$$R_{w\tilde{x}}(t,t) = E\{w(t)\tilde{x}^T(t)\} \tag{7.124}$$

According to the error solution (7.119) and the fact that $\tilde{x}(0)$, $v(t)$ and $w(t)$ are uncorrelated random vectors, the last equation can be rewritten as

$$R_{v\tilde{x}}(t,t) = -\int_0^t E\{v(t)v^T(\tau)\}K^T(\tau)\Phi^T(t,\tau)d\tau \tag{7.125}$$

$$R_{w\tilde{x}}(t,t) = \int_0^t E\{w(t)w^T(\tau)\}G^T\Phi^T(t,\tau)d\tau \tag{7.126}$$

Since

$$R_v(t,\tau) = E\{v(t)v^T(\tau)\} = R\delta(t-\tau)$$

$$R_w(t,\tau) = E\{w(t)w^T(\tau)\} = Q\delta(t-\tau)$$

further follows

$$R_{v\tilde{x}}(t,t) = -\int_0^t RK^T(\tau)\Phi^T(t,\tau)\delta(t-\tau)d\tau = -\frac{1}{2}RK^T(t)\Phi^T(t,t) = -\frac{1}{2}RK^T(t) \tag{7.127}$$

$$R_{w\tilde{x}}(t,t) = \int_0^t QG^T\Phi^T(t,\tau)\delta(t-\tau)d\tau = \frac{1}{2}QG^T\Phi^T(t,t) = \frac{1}{2}QG^T \tag{7.128}$$

Here is used the fact that $\Phi(t,t) = I$, as well as that only half of the area of the Dirac impulse $\delta(t)$ should be considered as being to the right of $t = 0$. Therefore, the first term in eq. (7.122) is given by

$$E\{\dot{\tilde{x}}(t)\tilde{x}^T(t)\} = (A - KH)P(t) + \frac{1}{2}GQG^T - \frac{1}{2}KRK^T \qquad (7.129)$$

where we have used (7.123), (7.128). The second term in (7.121) represents the transpose of (7.129). Therefore, add its transpose to get

$$\dot{P} = (A - KH)P + P(A - KH)^T + KRK^T + GQG^T \qquad (7.130)$$

By subtracting for $K = PH^T R^{-1}$, it is easy to see that we have just found another way for deriving the Ricatti equation (7.113).

7.3.1 Orthogonality principle (derivation from Wiener-Hopf equation)

Analogous to the case of discrete version of the estimator, it can be shown that

$$E\{\hat{x}(t)\tilde{x}^T(t)\} = 0 \qquad (7.131)$$

what is well known *orthogonality principle*; that is, the optimal estimate and its error are orthogonal. An intuitive argument for the eq. (7.131) is very easy to give. First, it is claimed that the optimal estimate must be linear function of measurements, and of the form in eq. (7.117), i.e.

$$\dot{\hat{x}} = (A - KH)\hat{x} + Kz$$

whose solution is given by

$$\hat{x}(t) = \Phi(t,0)\hat{x}(0) + \int_0^t K(\tau)z(\tau)d\tau \qquad (7.132)$$

where $\Phi(\cdot,\cdot)$ is the state transition matrix of $(A - KH)$. Then,

$$E\{\tilde{x}(t)\hat{x}^T(t)\} = E\{\tilde{x}(t)\}\hat{x}^T(0)\Phi^T(t,0) + \int_0^t E\{\tilde{x}(t)z^T(\tau)\}K^T(\tau)d\tau$$

$$= \int_0^t E\{\tilde{x}(t)z^T(\tau)\}K^T(\tau)d\tau$$

However, in view of the geometric property of the minimum variance estimate $\hat{x}(t)$, that the estimation error $\tilde{x}(t)$ is orthogonal into the data space (plane) spanned by the observations $\{z(\tau), 0 \le \tau \le t\}$ (in other words, the optimal estimate $\hat{x}(t)$, can be considered geometrically as an orthogonal projector in the sense that the estimate $\hat{x}(t)$ is orthogonal projection of the state vector $x(t)$ onto the data space $\{z(\tau), 0 \le \tau \le t\}$, we have

$$E\{\tilde{x}(t)z^T(\tau)\} = 0 \ ; \ 0 \le \tau \le t \qquad (7.133)$$

from which directly follows the equation (7.131).

The equation (7.133) is referred to as the *Wiener-Hopf equation*, and it provides the connection between the Wiener filter and the Kalman filter. Moreover, this equation provides for an alternative approach to the estimation problem for continuous linear systems, in which the problem is solved entirely in the continuous-time domain without any resort to limiting procedures. Namely, we can solve the Wiener-Hopf equation in order to obtain the algorithm for optimal filtering. As mentioned before (section 6), we consider only linear estimators as candidates for the optimal estimates of the form

$$\hat{x}(t) = \int_{t_0}^{t} h(t,\tau) z(\tau) d\tau \quad \text{.}$$ (7.134)

We have selected the final time for which measurement data $\{z(t), t_0 \leq \tau \leq t\}$ are available as t to focus on the filtering problem. The estimation error is given by

$$\tilde{x}(t) = x(t) - \hat{x}(t)$$ (7.135)

Furthermore, we restrict the performance measure for estimation to be the mean square error

$$J(\tilde{x}(t)) = E\{\tilde{x}^T(t)\tilde{x}(t)\} = \text{trace } E\{\tilde{x}(t)\tilde{x}^T(t)\} = \text{trace } P(t)$$ (7.136)

Our problem statement now follows very simple: given the measurement $\{z(\tau), t_0 \leq \tau \leq t\}$, find an estimate of $x(t)$ of the form (7.134), which minimizes the mean square error as given in eq. (7.136). Additionally, we assume that the unknown $x(t)$ is generated by eq. (7.105) and that data $z(t)$ are related to $x(t)$ by the measurement model (7.106). Clearly, the problem is that of determining the system matrix $h(t,\tau)$ (see Fig. 7.19.).

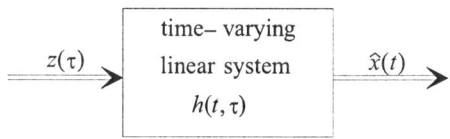

$$z(\tau) \longrightarrow \boxed{\begin{array}{c} \text{time– varying} \\ \text{linear system} \\ h(t,\tau) \end{array}} \xrightarrow{\hat{x}(t)}$$

Fig. 7.19 Presentation of optimal estimator as a time-varying linear system

Very simple, the orthogonality principle provides a necessary and sufficient condition for optimal estimation in the form of the Wiener-Hopf equation (7.133). By utilizing eqs. (7.134) and (7.135), the equation (7.133) can be rewritten as

$$E\left\{\left[\hat{x}(t) - \int_{t_0}^{t} h(t,\tau) z(\tau) d\tau\right] z^T(\sigma)\right\} = 0 \; ; \; t_0 \leq \sigma \leq t$$ (7.137)

where

$$R_{x,z}(t,\sigma) = E\{x(t) z^T(\sigma)\} \; ; \; t_0 \leq \sigma \leq t$$ (7.138)

$$R_z(\tau,\sigma) = E\{z(\tau) z^T(\sigma)\}$$ (7.139)

The required auto-correlations are given by

$$R_{x,z}(t,\sigma) = E\{x(t)[Hx(\sigma) + v(\sigma)]^T\} = E\{x(t) x^T(\sigma)\} H^T + E\{x(t) v^T(\sigma)\}$$
$$= E\{x(t) x^T(\sigma)\} H^T = R_x(t,\sigma) H^T$$ (7.140)

since $E\{xv\} = 0$, and

$$R_z(t,\sigma) = E\{z(t) z^T(\sigma)\} = E\left\{[Hx(t) + v(t)][Hx(\sigma) + v(\sigma)]^T\right\}$$
$$= HE\{x(t) x^T(\sigma)\} H^T + E\{v(t) v^T(\sigma)\} = HR_x(t,\sigma) H^T + R\delta(t-\sigma)$$ (7.141)

By using the expression (7.140) and (7.138), we obtain

$$R_x(t,\sigma)H^T = \int_{t_0} h(t,\tau)R_z(\tau,\sigma)d\tau \; ; \; t_0 < \sigma < t \tag{7.142}$$

Differentiate further the eq. (7.142), using Leibniz's rule

$$\frac{d}{dt}\int_0 f(t,\tau)d\tau = f(t,t) + \int_0 \frac{\partial}{\partial t}f(t,\tau)d\tau \tag{7.143}$$

we get

$$\frac{\partial R_x(t,\sigma)}{\partial t}H^T = h(t,t)R_z(t,\sigma) + \int_{t_0} \frac{\partial h(t,\tau)}{\partial t}R_z(\tau,\sigma)d\tau \tag{7.144}$$

Now use eq. (7.141) to write for $\sigma < t$

$$h(t,t)R_z(t,\sigma) = h(t,t)HR_x(t,\sigma)H^T \tag{7.145}$$

By use of eq. (7.142), this can be written

$$h(t,t)R_z(t,\sigma) = \int_{t_0} h(t,t)Hh(t,\tau)R_z(\tau,\sigma)d\tau \; ; \; t_0 < \sigma < t \tag{7.146}$$

Now, use eq. (7.105) to obtain

$$\frac{\partial R_x(t,\sigma)}{\partial t} = E\left\{\frac{dx(t)}{dt}x^T(\sigma)\right\} = E\left\{[Ax(t)+Gw(t)]x^T(\sigma)\right\}$$
$$= AE\left\{x(t)x^T(\sigma)\right\} = AR_x(t,\sigma) \; ; \; \sigma < t \tag{7.147}$$

since $E\{w(t)x^T(\sigma)\} = 0$ for $\sigma < t$, because $w(t)$ is white noise. Therefore, by eqs. (7.142) and (7.147)

$$\frac{\partial R_x(t,\sigma)}{\partial t}H^T = AR_x(t,\sigma)H^T = \int_{t_0} Ah(t,\tau)R_z(\tau,\sigma)d\tau \; ; \; t_0 < \sigma < t \tag{7.148}$$

Substituting eq. (7.148) into (7.144), using eqs. (7.142) and (7.145), yields

$$\int_{t_0} Ah(t,\tau)R_z(\tau,\sigma)d\tau = h(t,t)H\int_{t_0} h(t,\tau)R_z(\tau,\sigma)d\tau + \int_{t_0} \frac{\partial h(t,\tau)}{\partial t}R_z(\tau,\sigma)d\tau$$

or equivalently

$$0 = \int_{t_0}\left[-Ah(t,\tau)+h(t,t)Hh(t,\tau)+\frac{\partial h(t,\tau)}{\partial t}\right]R_z(\tau,\sigma)d\tau \; ; \; t_0 < \sigma < t \tag{7.149}$$

By eq. (7.141)

$$0 = \int_{t_0}\left[-Ah(t,\tau)+h(t,t)Hh(t,\tau)+\frac{\partial h(t,\tau)}{\partial t}\right]HR_x(\tau,\sigma)H^Td\tau$$
$$+ \left[-Ah(t,\sigma)+h(t,t)Hh(t,\sigma)+\frac{\partial h(t,\sigma)}{\partial t}\right]R \; ; \; t_0 < \sigma < t$$

Since $\det R \neq 0$, this is equivalent to

$$\frac{\partial h(t,\tau)}{\partial t} = Ah(t,\tau) - h(t,t)Hh(t,\tau) \; ; \; t_0 < \tau < t \tag{7.150}$$

This is a differential equation satisfied by the optimal impulse response $h(t,\tau)$. We will not find $h(t,\tau)$ explicitly, but will use eq. (7.150) to find a differential equation for $\hat{x}(t)$. To do this, differentiate eq. (7.134), using Leibniz's rule (7.143) to get

$$\frac{d\hat{x}(t)}{dt} = h(t,t)z(t) + \int_{t_0}^{t} \frac{\partial h(t,\tau)}{\partial t} z(\tau) d\tau \tag{7.151}$$

By using eq. (7.150), this becomes

$$\dot{\hat{x}} = h(t,t)z(t) + \int_{t_0}^{t} \left[Ah(t,\tau) - h(t,t)Hh(t,\tau) \right] z(\tau) d\tau$$

Now substitute from (7.134) to obtain

$$\dot{\hat{x}}(t) = A\hat{x}(t) + h(t,t)\left[z(t) - H\hat{x}(t) \right] \tag{7.152}$$

All that remarks is to find an expression for $h(t,t)$, for then (7.152) provides a realization of the optimal filter $h(t,\tau)$. Unfortunately, we can not do this directly, but we shall show that we can express $h(t,t)$ in terms of the error covariance matrix

$$\begin{aligned}
P(t) &= E\{\tilde{x}(t)\tilde{x}^T(t)\} = E\left\{ \left[x(t) - \int_{t_0}^{t} h(t,\tau)z(\tau)d\tau \right] x^T(t) \right\} \\
&= E\{x(t)x^T(t)\} - \int_{t_0}^{t} h(t,\tau)E\{z(\tau)x^T(t)\}d\tau \\
&= R_x(t,t) - \int_{t_0}^{t} h(t,\tau)R_{zx}(\tau,t)d\tau
\end{aligned} \tag{7.153}$$

Use eq. (7.141) in the eq. (7.142) to obtain

$$R_x(t,\sigma)H^T = \int_{t_0}^{t} h(t,\tau)\left[HR_x(\tau,\sigma)H^T + R\delta(\tau - \sigma) \right] d\tau$$

or with $\sigma = t$

$$R_x(t,t)H^T - \int_{t_0}^{t} h(t,\tau)HR_x(\tau,t)H^T d\tau = h(t,t)R \tag{7.154}$$

According to eqs. (7.140) and (7.153), we have (it should be noted that $R_x(\tau,t) = R_x^T(t,\tau)$)

$$\begin{aligned}
R_x(t,t)H^T - \int_{t_0}^{t} h(t,\tau)HR_x^T(t,\tau)H^T d\tau &= R_x(t,t)H^T - \int_{t_0}^{t} h(t,\tau)R_{xz}^T(t,\tau)H^T d\tau \\
&= \left[R_x(t,t) - \int_{t_0}^{t} h(t,\tau)R_{zx}(\tau,t)d\tau \right] H^T = P(t)H^T
\end{aligned}$$

so that eq. (7.154) reduces to

$$P(t)H^T = h(t,t)R$$

or equivalently

$$h(t,t) = P(t)H^T R^{-1} \tag{7.155}$$

At this point it is clear that $h(t,t)$ is nothing but the Kalman gain

$$K(t) = h(t,t) = P(t)H^T R^{-1} \qquad (7.156)$$

By using $K(t)$ in (7.152), we can manufacture the estimate $\hat{x}(t)$ from the data $z(t)$; that is we derived the eq. (7.117). To compute $K(t)$ we need to find the error covariance $P(t)$, which we do previously in eq. (7.130). Thus, the Kalman filter provides the *optimal estimate* in the general *nonstationary case*. It is time-varying even for a time-invariant system in eq. (7.105), since the Kalman gain $K(t) = h(t,t)$ is dependent on t. This accounts for the fact that the optimal filter cannot be found in the non-steady-state case using frequency domain approach (see chapter 6). Furthermore, the Kalman filter does not explicitly provide an expression for $h(t,\tau)$, however a realization for the optimal filter is given by eq. (7.117) or (7.152).

7.4 Sequential weighting in Kalman filter

In this section we heuristically develop a feel for the operation of Kalman estimator. Namely, the optimality of the Kalman filter is contained in its structure and in the specification of the gain matrix K [13, 17]. There is also an intuitive logic behind the equations for the Kalman filter. Heuristically, the Kalman filter can be viewed simply by the correction equation, i.e.

$$\hat{x}_{new} = \hat{x}_{old} + K\varepsilon_{new} \; ; \; \varepsilon_{new} = z - H\hat{x}_{old} \qquad (7.157a)$$

where $\hat{x}_{old}$ depends on the adopted state space model and ε_{new} is a function of measurement, i.e.

$$\hat{x}_{old} = f(state\ space\ model) \; ; \; \varepsilon_{new} = f(measurement) \qquad (7.157b)$$

Using the model of the Kalman filter, we see that we can view the old (predicted) estimate as a function of the state space model (i.e. system matrix A in the continuous-time case and Φ in the discrete-time case) and the prediction error or innovation ε as a function primarily of the new measurement z, as indicated in Tabs. 7.1 and 7.2. Consider the new estimate under the following cases:

$$K \to small \Rightarrow \hat{x}_{new} = \hat{x}_{old} = f(model)$$

$$K \to large \Rightarrow \hat{x}_{new} = K\varepsilon_{new} = f(measurement)$$

So we can see that the operation of the filter is pivoted about the values of the gain (weighted) matrix K. For small K, the estimate believes the model and for large K the estimator believes the measurement. Let us now investigate the gain matrix and see if its variations are consistent with this heuristic notations. First, recall that the alternative form of the gain equation is given by eqs. (7.63), (7.90) or (7.156); that is

$$K = PH^T R^{-1} \qquad (7.158)$$

assuming that R^{-1} exists. Let us examine the case where K is small. From eq. (7.158) we see that K is small in two cases. First, P is small (R fixed), which is consistent because small P implies that the model is adequate. Second, R is large (P fixed), which is consistent because large R implies that the measurement is noisy, so believe the model.

Next, we examine the case where K is large. Again from eq. (7.158) we see that K is large first when P is large (R fixed), implying that the model is inadequate, so believe the measurement. The second case is when R is small (P fixed), implying that the measurement is good (high signal-to-noise ratio).

We summarize the heuristic Kalman filter operation as follows:

Condition	Gain K	Parameters of Kalman filter
Believe model	Small	☐ P small ; model adequate ☐ R large; measurement noisy
Believe measurement	Large	☐ P small ; measurement good ☐ P large ; model inadequate

Thus, we showed that variation of the Kalman gain K is related to variations in P and R. In summary, if measurement noise is large (R large) and state estimate errors are small (P small), the residual quantity ε is due chiefly to the noise and only small changes in the state estimates should be made (K small). On the other hand, small measurement noise (R small) and large uncertainty in the state estimates (P large) suggest that ε contains considerable information about errors in the estimates (K large). Therefore, ε will be used as a basis for strong corrections to the estimates. Additionally, if we recall the covariance prediction equation (7.65) or (7.104), we see that the uncertainty in the model can be characterized by the process (state) noise covariance Q. For large Q, P is large, indicating high uncertainty, or an inadequate model. For small Q, P is small, indicating an adequate model Therefore, we can heuristically conceive of K as a ratio of process to measurement noise; that is

$$K \sim QR^{-1} \tag{7.159}$$

Using eq. (7.158), we can again see that the variations of Q are consistent with the variations of K. A typical behavior of the root mean square error (rms) in the Kalman filter estimate of a particular state variable can be depicted as in Fig 7.20.

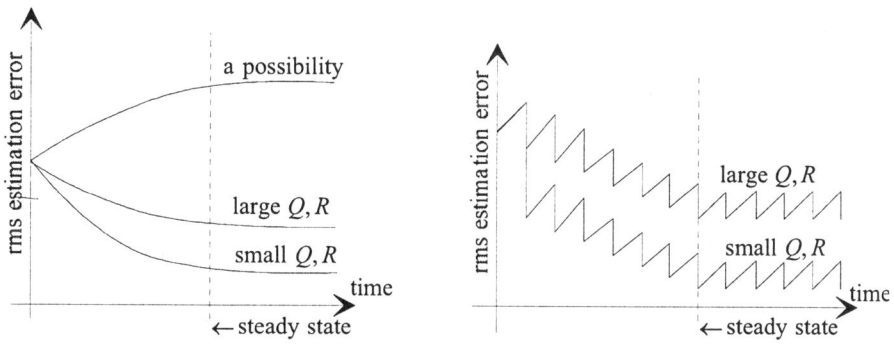

      (a) continuous filter   (b) discrete filter
Fig 7.20 Behavior of the rms error in the Kalman filter estimate of a particular state variable

In the other words, we can perceive the Kalman estimator as a deterministic system with a time-varying bandwidth determined by the filter gain K. Furthermore, if we consider K in terms of the noises ration in (7.159), we see that as Q increases K increases and the filter bandwidth increases. Thus, the filter transient performance is faster, but at the cost of more noise in the state estimates. The same effect is achieved by small R. If Q decreases (R increases) K decreases, decreasing the bandwidth. The filter transient performance is slower but the noise is filtered; that is the state estimates are smoother. The effect of Q and R on the filter bandwidth can be depicted as in Fig 7.21.

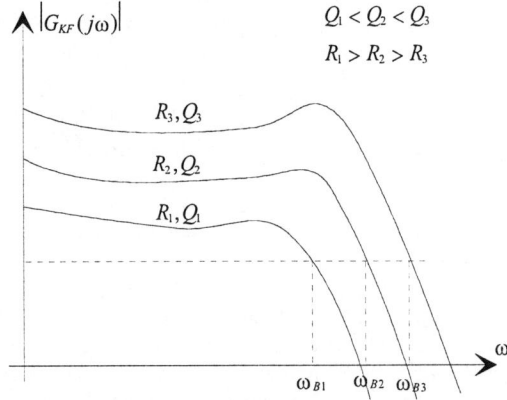

Fig. 7.21. Heuristic Kalman filter bandwidth ω_B ($\left|G_{KF}(j\omega)\right|$ is amplitude-frequency characteristic of the Kalman filter transfer function) and noise covariances

Moreover, it can also be concluded that increasing (decreasing) Q increases (decreases) the steady-state value of the error covariance matrix P and, therefore, increases (decreases) the steady-state Kalman gain K.

7.5 Innovations property of the Kalman filter

As mentioned in the previous chapter, the *innovation*, or *predicted residual*, sequence ε also provides the starting point to check the Kalman filter operation [1, 25]. For a continuous system and measurement given by

$$\dot{x} = Ax + Gw \tag{7.160a}$$

$$z = Hx + v \tag{7.160b}$$

and a continuous Kalman filter defined by

$$\dot{\hat{x}} = A\hat{x} + K\varepsilon \tag{7.161}$$

$$\varepsilon = z - H\hat{x} \tag{7.162}$$

the innovation property states that, if K is the optimal gain, the auto-covariance function of the zero-mean innovation sequence

$$R_\varepsilon(t_1, t_2) = E\left\{\varepsilon(t_2)\varepsilon^T(t_1)\right\} = 0 \; ; \; t_1 \neq t_2 \tag{7.163}$$

In other words, the innovation process, ε, is a zero-mean white noise process. Heuristically, there is no information left in ε, if $\hat{x}$ is an optimal estimate. The equation (7.163) is readily proved. From eqs. (7.160) and (7.162), we see that

$$\varepsilon = H(x - \hat{x}) + v = H\tilde{x} + v \tag{7.164}$$

where $\tilde{x} = x - \hat{x}$ is the estimation error. Furthermore, since $E\{v\} = 0$ and the optimal estimate $\hat{x}$ is unbiased, i.e. $E\{\tilde{x}\} = 0$, one obtains

$$E\{\varepsilon\} = HE\{\tilde{x}\} + E\{v\} = 0 \tag{7.165}$$

Additionally, for $t_2 > t_1$, we get

$$E\{\varepsilon(t_2)\varepsilon^T(t_1)\} = H(t_2)E\{\tilde{x}(t_2)\tilde{x}^T(t_1)\}H^T(t_1)$$
$$+ H(t_2)E\{\tilde{x}(t_2)v^T(t)\} + E\{v(t_2)\tilde{x}^T(t_1)\}H^T(t_1) + E\{v(t_2)v^T(t_1)\}$$

or, taking into account that $v(t_2)$ and $\tilde{x}(t_1)$ are uncorrelated and that $E\{v(t_2)v^T(t_1)\} = R(t_1)\delta(t_2 - t_1)$,

$$E\{\varepsilon(t_2)\varepsilon^T(t_1)\} = H(t_2)E\{\tilde{x}(t_2)\tilde{x}^T(t_1)\}H^T(t_1) + H(t_2)E\{\tilde{x}(t_2)v^T(t_1)\} + R(t_1)\delta(t_2 - t_1) \quad (7.166)$$

From eqs. (7.160)-(7.164) it is seen that $\tilde{x}$ satisfies the equation

$$\dot{\tilde{x}} = \dot{x} - \dot{\hat{x}} = A(x - \hat{x}) + Gw - K[H(x - \hat{x}) + v]$$

or equivalently

$$\dot{\tilde{x}} = (A - KH)\tilde{x} + Gw - Kv \quad (7.167)$$

The solution of this equation is

$$\tilde{x}(t_2) = \Phi(t_2,t_1)\tilde{x}(t_1) + \int_{t_1}^{t_2}\Phi(t_2,\tau)[G(\tau)w(\tau) - K(\tau)v(\tau)]d\tau \quad (7.168)$$

where $\Phi(t_2,t_1)$ is the transition matrix corresponding to the linear continuous system (7.167), i.e. to the system matrix $(A - KH)$. Using (7.168), we directly compute

$$E\{\tilde{x}(t_2)\tilde{x}^T(t_1)\} = \Phi(t_2,t_1)E\{\tilde{x}(t_1)\tilde{x}^T(t_1)\} +$$
$$+ \int_{t_1}^{t_2}\Phi(t_2,\tau)[G(\tau)E\{w(\tau)\tilde{x}^T(t_1)\} - K(\tau)E\{v(\tau)\tilde{x}^T(t_1)\}] =$$
$$= \Phi(t_2,t_1)P(t_1)$$

$$(7.169)$$

Here is used the fact that

$$E\{w(\tau)\tilde{x}^T(t_1)\} = 0; E\{v(\tau)\tilde{x}^T(t_1)\} = 0 ; t_1 < \tau < t_2 ; E\{\tilde{x}(t_1)\tilde{x}^T(t_1)\} = P(t_1)$$

Moreover, we get

$$E\{\tilde{x}(t_2)v^T(t_1)\} = \Phi(t_2,t_1)E\{\tilde{x}(t_1)v^T(t_1)\} +$$
$$+ \int_{t_1}^{t_2}\Phi(t_2,\tau)[G(\tau)E\{w(\tau)v^T(t_1)\} - K(\tau)E\{v(\tau)v^T(t_1)\}]d\tau$$

Taking into account that

$$E\{\tilde{x}(t_1)v^T(t_1)\} = 0; E\{w(\tau)v^T(t_1)\} = 0; t_1 < \tau < t_2 ; E\{v(\tau)v^T(t_1)\} = R(t_1)\delta(\tau - t_1)$$

further follows

$$E\{\tilde{x}(t_2)v^T(t_1)\} = -\Phi(t_2,t_1)K(t_1)R(t_1) \quad (7.170)$$

Therefore, from eqs. (7.166), (7.169) and (7.170), we have

$$E\{\varepsilon(t_2)\varepsilon^T(t_1)\} = H(t_2)\Phi(t_2,t_1)[P(t_1)H^T(t_1) - K(t_1)R(t_1)] + R(t_1)\delta(t_2 - t_1) \quad (7.171)$$

But for optimal filter gain

$$K(t_1) = P(t_1)H^T(t_1)R^{-1}(t_1) \quad (7.172)$$

yielding for the auto-covariance function of the innovation process

$$E\left\{\varepsilon(t_2)\varepsilon^T(t_1)\right\} = R(t_1)\delta(t_2 - t_1) \tag{7.173}$$

In other words, the stochastic process $\varepsilon(\tau)$ is zero-mean white, which is the desired result. Particularly, in the case of stationary or time-invariant systems for which eqs. (7.160) and (7.161) are stable, the auto-covariance function $R_\varepsilon(t_1,t_2)$ is a function of $\tau = t_2 - t_1$, (it should be noted that the autocovariance function reduces to autocorrelation function, since ε is zero-mean process), i.e.

$$R_\varepsilon(\tau) = E\left\{\varepsilon(t_1+\tau)\varepsilon^T(t_1)\right\} = He^{(A-KH)|\tau|}\left(PH^T - KR\right) + R\delta(\tau) = R\delta(\tau)$$

Here is used the fact that (see chapter 4)

$$\Phi(t_2,t_1) = \Phi(t_2 - t_1) = e^{(A-KH)|\tau|} \; ; \; \tau = t_2 - t_1 \; ; \; K(t_1) = P(t_1)H^T R^{-1}.$$

In the stationary, discrete-time case described by the state space model

$$x_{k+1} = \Phi x_k + G w_k \; ; \; w_k \sim (0,Q) \tag{7.174}$$

$$z_k = H x_k + v_k \; ; \; v_k \sim (0,R) \tag{7.175}$$

and a discrete Kalman filter given by

$$\hat{x}_k(+) = \hat{x}_k(-) + K_k \varepsilon_k \tag{7.176}$$

$$\varepsilon_k = z_k - H\hat{x}_k(-) \tag{7.177}$$

$$\hat{x}_k(-) = \Phi \hat{x}_{k-1}(+) \tag{7.178}$$

The innovation property states that

$$R_\varepsilon(j) = E\left\{\varepsilon_k \varepsilon_{k-j}^T\right\} = \begin{cases} HP_k(-)H^T + R & \text{for } j = 0 \\ 0 & \text{for } j > 0 \end{cases} \tag{7.179}$$

This property is readily proved, if we take into account that, using (7.175) and (7.177), the innovation term can be expressed as

$$\varepsilon_k = H\left(x_k - \hat{x}_k(-)\right) + v_k = H\tilde{x}_k(-) + v_k \tag{7.180}$$

where $\tilde{x}_k(-) = x_k - \hat{x}_k(-)$ is the prediction error. Thus, the covariance matrix of the innovation process is given by

$$R_\varepsilon(0) = E\left\{\varepsilon_k \varepsilon_k^T\right\} = HE\left\{\tilde{x}_k(-)\tilde{x}_k^T(-)\right\}H^T + E\left\{v_k v_k^T\right\} = HP_k(-)H^T + R \tag{7.181}$$

which is the desired result for $j = 0$. Here is used the fact that $\tilde{x}_k(-)$ and v_k are uncorrelated, i.e.

$$E\left\{\tilde{x}_k(-)v_k^T\right\} = E\left\{v_k \tilde{x}_k^T(-)\right\} = 0$$

Furthermore, from eqs. (7.174)-(7.178) it is seen that

$$\tilde{x}_{k+1}(-) = \Phi x_k + G w_k - \Phi \hat{x}_k(+)$$

or equivalently

$$\tilde{x}_{k+1}(-) = \Phi x_k + G w_k - \Phi\left\{\hat{x}_k(-) + K_k\left[\left(H x_k + v_k\right) - H\hat{x}_k(-)\right]\right\}$$

from which follows

$$\tilde{x}_k(-) = \Phi(I - K_{k-1}H)\tilde{x}_{k-1}(-) + Gw_{k-1} - \Phi K_{k-1}v_{k-1} \tag{7.182}$$

Using eq. (7.180), we directly compute

$$
\begin{aligned}
E\{\varepsilon_k \varepsilon_{k-1}^T\} &= E\{[H\tilde{x}_k(-) + v_k][H\tilde{x}_{k-1}(-) + v_{k-1}]\} \\
&= HE\{\tilde{x}_k(-)\tilde{x}_{k-1}^T(-)\}H^T + HE\{\tilde{x}_k(-)v_{k-1}^T\}
\end{aligned}
\tag{7.183}
$$

since

$$E\{v_k \tilde{x}_{k-1}^T(-)\} = 0 \; ; \; E\{v_k v_{k-1}^T\} = 0.$$

Therefore, from eq. (7.182) follows

$$E\{\tilde{x}_k(-)\tilde{x}_{k-1}^T(-)\} = \Phi(I - K_{k-1})P_{k-1}(-) \tag{7.184}$$

$$E\{\tilde{x}_k(-)v_{k-1}^T\} = -\Phi K_{k-1}R \tag{7.185}$$

Here is used the fact that $\tilde{x}_{k-1}(-)$ is uncorrelated with w_{k-1} and v_{k-1}, as well as that w_{k-1} and v_{k-1} are also uncorrelated. By substituting (7.184) and (7.185) into (7.183) further follows

$$R_\varepsilon(1) = \{\varepsilon_k \varepsilon_{k-1}^T\} = H\Phi\{P_{k-1}(-)H^T - K_{k-1}[HP_{k-1}(-)H^T + R]\} \tag{7.186}$$

which is the desired result for $j=1$. Note that the optimal choice

$$K_{k-1} = P_{k-1}(-)[HP_{k-1}(-)H^T + R]^{-1} \tag{7.187}$$

makes the expression (7.186) vanish. Additionally, for $j=2$ we have

$$
\begin{aligned}
R_\varepsilon(2) = E\{\varepsilon_k \varepsilon_{k-2}^T\} &= E\{[H\tilde{x}_k(-) + v_k][H\tilde{x}_{k-2}(-) + v_{k-2}]^T\} \\
&= HE\{\tilde{x}_k(-)\tilde{x}_{k-2}^T(-)\}H^T + HE\{\tilde{x}_k(-)v_{k-2}^T\}
\end{aligned}
\tag{7.188}
$$

since

$$E\{v_k \tilde{x}_{k-2}^T(-)\} = E\{v_k v_{k-2}^T\} = 0$$

Moreover, from eq. (7.182)

$$\tilde{x}_{k-1}(-) = \Phi(I - K_{k-2}H)\tilde{x}_{k-2}(-) + Gw_{k-2} - \Phi K_{k-2}v_{k-2} \tag{7.189}$$

and if we substitute this equation into eq. (7.182), we have

$$
\begin{aligned}
\tilde{x}_k(-) &= \Phi(I - K_{k-1}H)\Phi(I - K_{k-2}H)\tilde{x}_{k-2}(-) \\
&\quad + \Phi(I - K_{k-1}H)Gw_{k-2} - \Phi(I - K_{k-1}H)\Phi K_{k-2}v_{k-2} \\
&\quad + Gw_{k-1} - \Phi K_{k-1}v_{k-1}
\end{aligned}
\tag{7.190}
$$

Therefore,

$$E\{\tilde{x}_k(-)\tilde{x}_{k-2}^T(-)\} = \Phi(I - K_{k-1}H)(I - K_{k-2}H)P_{k-2}(-) \tag{7.191a}$$

$$E\{\tilde{x}_k(-)v_{k-2}^T\} = -\Phi(I - K_{k-1}H)\Phi K_{k-2}R \tag{7.191b}$$

and if we replace eqs. (7.191) into (7.187), we obtain

$$R_\varepsilon(2) = H\Phi(I - K_{k-1}H)\Phi\{P_{k-2}(-)H^T - K_{k-2}[HP_{k-2}(-)H^T + R]\} \tag{7.192a}$$

Obviously, $R_\varepsilon(2) = 0$ if we take the optimal gain matrix from eq. (7.187). We may now repeat the process for $j = 3$ and, by induction, verify that

$$R_\varepsilon(j) = H\left[\Phi(I - KH)\right]^{j-1} \Phi\left\{P(-)H^T - K\left[HP(-)H^T + R\right]\right\} \qquad (7.192b)$$

for any $j > 0$, where K and $P(-)$ are the stationary values of the matrices K_k and $P_k(-)$, respectively. Thus, the optimal choice of the gain matrix, in accordance to eq. (7.187); that is

$$K = P(-)H^T\left[HP(-)H^T + R\right]^{-1}$$

makes the expression (7.192b) vanishes for all $j \neq 0$, which is the desired result.

In summary, a necessary and sufficient condition for a Kalman filter to be *optimal* is that the *innovation process* is *zero-mean* and *white*. The innovation quantity is such that it consists of that part of the current measurement containing new information not carried in the preceding measurements. A such definition of the innovation ε suggests that it must be independent of the preceding measurements, and, consequently, of the preceding innovations which are determined on the basis of the preceding measurements. In other words, this suggests that the auto-covariance function of the innovation process $R_\varepsilon(\tau)$ is zero, expect for the zero time-delay ($\tau = 0$), or that this process is white. As mentioned before, this means that there is no information left in the measurements, if $\hat{x}$ is an optimal estimate of the system state.

Generally, for a Kalman filter to yield optimal performance, it is necessary to provide a correct a priori description of the system state-space model matrices A (or Φ), G and H; the noise covariance matrices Q and R; and the initial estimation error covariance $P(0)$. As a practical fact, this

is usually impossible, and guesses of these quantities must be advanced. Hopefully, the filter design will be such that the penalty for misguesses is small. However, a Kalman filter is not functioning properly when the gain K becomes small and the measurements, or innovations, still contain information necessary for the estimates. The filter is said to diverge under these conditions. We may now raise a question if it is possible to deduce non-optimal behavior during the filter operation and thus improve the quality of a priori information. Within certain limits, the answer is yes, and this is a topic of *adaptive Kalman filtering* [18, 22, 52]. In an adaptive Kalman filtering, the innovations property is used as a criterion to test for optimality. Employing tests for whiteness, the experimentally measured steady-state auto-covariance function $R_\varepsilon(\tau)$ is processed to identify unknown noise covariances Q and R for known system model matrices A or Φ, G and H. It can be shown that the value of the gain K which whiteness the innovations process is the optimal gain. However, for the high-order systems of practical interest, the algorithms proposed in the literature may not work as well as theory would predict; other more heuristically motivated approaches may be both computationally simpler and more effective. Such an approach will be discussed in chapter 9.

7.6 Stochastic controllability, observability and stability

Suppose the system under consideration is time-varying and described by the continuous state-space model

$$\dot{x}(t) = A(t)x(t) + B(t)u(t) + G(t)w(t) \qquad (7.193)$$

$$z(t) = H(t)x(t) + v(t) \qquad (7.194)$$

with $w \sim [0, Q(t)]$ and $v \sim [0, R(t)]$. Let $\Phi(t, t_0)$ be the state transition matrix of A, satisfying the autonomous system $\dot{\Phi}(t, t_0) = A\Phi(t, t_0)$; $\Phi(t_0, t_0) = I$. In the absence of measurements $(R = \infty$ or $R^{-1} = 0)$ and with perfect a priori information $(P_0 = 0)$, the continuous Kalman filter estimation error covariance matrix is given by the continuos system matrix Riccati equation (Tab. 7.3)

$$\dot{P} = AP + PA^T + GQG^T \; ; \; P(0) = P_0 = 0 \tag{7.195}$$

for which the solution is

$$P(t) = \int_0^t \Phi(t, \tau) G(\tau) Q(\tau) G^T(\tau) \Phi^T(t, \tau) d\tau \tag{7.196}$$

This is easily verified by substitution into eq. (7.195), using Leibniz's rule

$$\frac{d}{dt} \int_0^t f(t, \tau) d\tau = f(t, t) + \int_0^t \frac{\partial}{\partial t} f(t, \tau) d\tau$$

If the integral (7.196) is positive definite for some $t > 0$, then $P(t) > 0$; that is, the process noise $w(t)$ excites all the states in the system. The system is said to be *uniformly completely controllable* when the integral, the so-called *controllability gramian*, in eq. (7.196) is positive definite and bounded for some $t > 0$. The property of *stochastic controllability* is important for establishing stability of the Kalman filter equations and for obtaining a unique steady-state value P_s of $P(t)$, i.e. $P_s = \lim_{t \to \infty} P(t)$. When the system is stationary (A, B, G, H are constant matrices) and if Q is positive definite ($Q > 0$), the criterion of complete controllability can be expressed algebraically by determining whether the rank of the controllability matrix of (A, G) is equal to n, with n being the dimension of the state vector x (see chapter 4).

In the case of discrete time-varying system described by

$$x_{k+1} = \Phi(k+1, k) x_k + G_k w_k + B_k u_k \; ; \; w_k \sim (0, Q_k) \tag{7.197}$$

$$z_k = H_k x_k + v_k \; ; \; v_k \sim (0, R_k) \tag{7.198}$$

the condition for *complete controllability* is that the *controllability gramian* satisfies

$$\beta_1 I \le \sum_{i=k-N}^{k-1} \Phi(k, i+1) G_i Q_i G_i^T \Phi^T(k, i+1) \le \beta_2 I \tag{7.199}$$

for some value of $N > 0$, where $\beta_1 > 0$ and $\beta_2 > 0$. Again, for the stationary system (Φ, G, B, H are constant matrices) and $Q > 0$, the criterion of complete controllability is that the rank of the controllability matrix of (Φ, G) is equal to the state vector dimension n (see chapter 3).

In the absence of process noise $(Q = 0)$ and a priori information $(P(0) \to \infty)$, the continuous system matrix *Riccati equation* is given by (Tab. 7.3)

$$\dot{P} = AP + PA^T - PH^T R^{-1} HP \; ; \; P(0) \to \infty \tag{7.200}$$

This can be rewritten as

$$\dot{P}^{-1} = -P^{-1}A - A^T P^{-1} + H^T R^{-1} H \tag{7.201}$$

This is easily verified by using the matrix identity $\dot{P}^{-1} = -P^{-1}\dot{P}P^{-1}$. The solution to the linear equation (7.201) in P^{-1} is

$$P^{-1}(t) = \int_0^t \Phi^T(\tau,t) H^T(\tau) R^{-1}(\tau) H(\tau) \Phi(\tau,t) d\tau \tag{7.202}$$

where $\Phi(t,\tau)$ is the transition matrix corresponding to A in (7.193). If the integral (7.202), the so-called *observability gramian*, is positive definite for some $t > 0$, then $P^{-1}(t) > 0$, and it follows that $0 < P(t) < \infty$; that is, through processing measurements it is possible to acquire information, i.e. decrease the estimation error variance, about states that are initially completely unknown. The system is said to be *uniformly completely observable* when the integral (7.202) is positive definite and bounded for some $t > 0$. When the linear system is stationary, or time-invariant, the criterion for complete observability is that the rank of the observability matrix of (A,H) is equal to the state vector dimension.

In the case of discrete time-varying system (7.197), (7.198), the condition for *uniform complete observability* is given by

$$\alpha_1 I \le \sum_{i=k-N}^{k} \Phi^T(i,k) H_i^T R_i^{-1} H_i \Phi(i,k) \le \alpha_2 I \tag{7.203}$$

for some value of $N > 0$, where $\alpha_1 > 0$ and $\alpha_2 > 0$. Particularly, if the discrete system is stationary, the condition for *uniform complete observability* is that the rank of the observability matrix of (Φ,H) is equal to the state vector dimension (see chapter 3). If the system is time-varying, then there is, in general, no constant steady-state solution to the Riccati equation. However, stochastic observability and stochastic controllability, as well as boundedness of A in the continuous-time case, or $\Phi(k+1,k) = \Phi_k$ in the discrete-time case, Q and R guarantee that for large t the behavior of P is unique independent of the initial condition P_0. Additionally, the most important question, from the practical point of view, is when the *estimation error system* (7.167) or (7.182) is *asymptotically stable*; that is, when does the estimate $\hat{x}$ converge to the true state x. Unfortunately, optimality of the Kalman filter does not guarantee its stability. The conditions under which the estimation error system is asymptotically stable are the same as for the existence of unique P, independent of P_0. However, the difference between a stable original system and a stable Kalman filter should be clearly realize. We are not concerned here with regulating the system, but only with estimating its states. With a stable filter the estimate $\hat{x}$ closely tracks the state x of the system, whether it is stable or not. Fig. 7.22 shows the relation between an unstable system and an unstable filter.

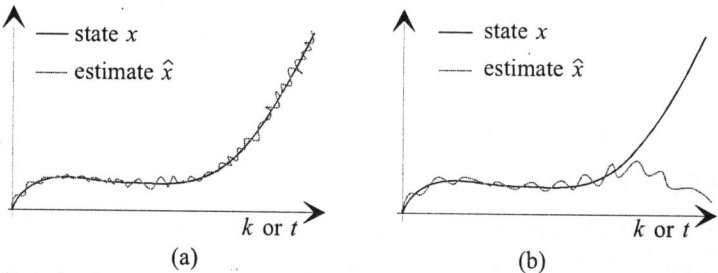

Fig. 7.22: Relation between system and filter stability: (a) unstable system and stable filter; (b) unstable system and unstable filter

In summary, one key result exists which assures both stability of the filter and uniqueness of the behavior of the estimation error covariance P for large t, independently of P_0. It requires stochastic uniform complete observability, stochastic uniform complete controllability, and boundedness of noise covariance matrices Q and R, and system matrix A, or Φ in discrete-time case. It is important to note that complete observability and controllability are quite restrictive requirements and in many cases of practical significance these conditions are not fulfilled, but Kalman filter operates satisfactorily. This is attributed to the fact that the solution to the estimation error system (7.167) or (7.182) frequently tends toward zero over a finite time-interval of interest, even though the error system may not be asymptotically stable in the strict sense of the definition (see chapters 3,4).

7.7 Statistical steady state (the link between Wiener and Kalman filter)

In the case where system and measurement dynamics are linear, constant coefficient equations (A, B, G, H in (7.193) and (7.194) are not functions of time) and the driving noise statistics are stationary (Q, R are not functions of time), the estimation process may reach a *steady-state* wherein the estimation error covariance matrix P is constant. The covariance $P(t) = E\{\tilde{x}(t)\tilde{x}^T(t)\}$, with $\tilde{x}(t)$ being the estimation error defined by eqs. (7.167) and (7.172), is given by the *continuous Riccati equation* (Tab. 7.3)

$$\dot{P} = AP + PA^T + GQG^T - PH^T R^{-1} HP \qquad (7.204)$$

with the initial condition $P(0)$. As mentioned in the preceding section, *completely observability* has been shown to be a sufficient condition for the existence of a *bounded steady-state* solution $P_s = \lim_{t \to \infty} P(t)$. In this case, for large t $(t \to \infty)$, $\dot{P}(t) = 0$ so that (7.204) tends to the *algebraic Riccati equation*

$$0 = AP + PA^T + GQG^T - PHR^{-1}HP \qquad (7.205)$$

Obviously, any limiting solution to (7.204) is a solution to (7.205), although the converse is not true. *Complete controllability* will assure that the *steady state solution* P_s is *unique*, independently of P_0. Thus, in the steady state the rate at which uncertainty builds (GQG^T) is just balanced by the following: (a) the rate at which new information enters the system $(P_s H^T R^{-1} HP_s)$; (b) the system dissipation due to damping expressed in the system matrix A.

Again, if the system is linear, discrete-time and time-invariant ($\Phi(k+1,k) = \Phi, B, G, H$ in (7.197) and (7.198) are matrices with constant coefficients) and the process and measurement noises are stationary (Q and R are not functions of time index k), the estimation process may reach a *steady-state* wherein the estimation-error covariance matrix P is constant. The covariance matrix $P_{k+1}(-) = E\{\tilde{x}_{k+1}(-)\tilde{x}_{k+1}^T(-)\}$ is given by the *discrete Riccati equation*.

$$\begin{aligned}
P_{k+1}(-) &= \Phi_k \left[(I - K_k H_k) P_k(-)(I - K_k H_k)^T + K_k R_k K_k^T \right] \Phi_k^T + G_k Q_k G_k^T \\
&= \Phi_k \left[P_k(-) - P_k(-) H_k^T \left(H_k P_k(-) H_k^T + R_k \right)^{-1} H_k P_k(-) \right] \Phi_k^T + G_k Q_k G_k^T
\end{aligned} \qquad (7.206)$$

with the initial condition P_0. It should be noted that the first equation in (7.206) follows directly from the estimation error equation (7.182) with $\Phi(k+1,k) = \Phi_k = \Phi$, while the second one can be derived by substituting the expression for K_k in (7.187) into the first equation in (7.206). In the stationary situation concerned, the solution to (7.206) tends to a *bounded steady state* value P_s if

$$\lim_{k \to \infty} P_k(-) = P_s \qquad (7.207a)$$

is bounded. In this case, for large k, $P_{k+1}(-) = P_k(-) = P_s$ and (7.206) tends to the *algebraic Riccati equation*

$$P_s = \Phi \left[P_s - P_s H^T \left(H P_s H^T + R \right)^{-1} H P_s \right] \Phi^T + GQG^T \qquad (7.207b)$$

It is clear that any limiting solution to (7.206) is a solution to (7.207). Similarly as in the continuos-time stationary case, *stochastic uniform complete controllability* and *observability*, and *boundedness* of Φ, Q and R, guarantee that the *bounded limiting solutions* to (7.206) or (7.207) exist for all choices of P_0, and this solution is unique and independent of P_0.

The corresponding steady-state optimal continuous Kalman filter is given by

$$\dot{\hat{x}} = A\hat{x} + K_s \left[z - H\hat{x} \right] \qquad (7.208)$$

where K_s is constant steady-state Kalman gain $\left(K_s = P_s H^T R^{-1} \right)$. Here is assumed that a deterministic input $u = 0$.

This equation may be rewritten as

$$\dot{\hat{x}} - \left(A - K_s H \right)\hat{x} = K_s z \qquad (7.209)$$

Laplace transforming both sides and neglecting initial conditions yields

$$\left(sI - A + K_s H \right)\hat{X}(s) = K_s Z(s) \qquad (7.210)$$

where s is the Laplace transform variable, and K_s is the steady-state Kalman gain. In this way, we obtains

$$\hat{X}(s) = \left[\left(sI - A + K_s H \right)^{-1} K_s \right] Z(s) \qquad (7.211)$$

The quantity in brackets, representing the matrix transfer function

$$G_w(s) = \left(sI - A + K_s H \right)^{-1} K_s \qquad (7.212)$$

which operates on the measurement vector $Z(s)$ to produce the state vector estimate $\hat{X}(s)$ is the Wiener optimal filter. Particularly, in the scalar or one-dimensional case, where $\hat{x}$, z, $A=a$ and $H=h$ being the scalar quantities, the equation (7.211) may be written as

$$\frac{\hat{X}(s)}{Z(s)} = G_w(s) = \frac{k_s}{s + \left(k_s h - a \right)} \qquad (7.213)$$

which is the optimum Wiener filter in the conventional scalar transfer function form.

Example 7.5 (comparison of Kalman filter and Wiener filter): Suppose the scalar unknown quantity $x(t)$ has spectral density

$$S_x(\omega) = \frac{2a\sigma^2}{\omega^2 + a^2} \qquad (I)$$

and the data $z(t)$ are generated by

$$z(t) = x(t) + v(t) \qquad (II)$$

where $x(t)$ and $v(t)$ are uncorrelated, and $v(t)$ is a zero-mean white noise with variance r.

I) Wiener filter
Recall that the Wiener filter can be found by first factoring $S_z(\omega)$ to find a whitening filter. Since $x(t)$ and $v(t)$ are uncorrelated, we have

$$
\begin{aligned}
R_{xz}(\tau) &= E\{x(t+\tau)z(t)\} = E\{x(t+\tau)[x(t)+v(t)]\} \\
&= E\{x(t+\tau)x(t)\} = R_x(\tau)
\end{aligned}
\tag{III}
$$

and

$$
\begin{aligned}
R_z(\tau) &= E\{z(t+\tau)z(t)\} = E\{[x(t+\tau)+v(t+\tau)][x(t)+v(t)]\} \\
&= R_x(\tau)+r\delta(\tau)
\end{aligned}
\tag{IV}
$$

with $\delta(\tau)$ being the impulse or delta function. Hence, the spectral density

$$
S_{xz}(\omega) = S_x(\omega)
\tag{V}
$$

and

$$
S_z(\omega) = S_x(\omega)+r
\tag{VI}
$$

Thus, the optimal Wiener filter is given by (see chapter 6)

$$
G_w(j\omega) = \frac{1}{S_z^+(\omega)}\left[\frac{S_{xz}(\omega)}{S_z^-(\omega)}\right]^+
\tag{VII}
$$

where (+) denotes the causal part of the corresponding spectral densities. Since from (I) and (VI) follows

$$
\begin{aligned}
S_z^{-1}(\omega) &= \frac{\omega^2+a^2}{r[\omega^2+a^2(1+\lambda)]} = \frac{j\omega+a}{\sqrt{r}[j\omega+a\sqrt{1+\lambda}]} \cdot \frac{-j\omega+a}{\sqrt{r}[-j\omega+a\sqrt{1+\lambda}]} \\
&= \frac{1}{S_z^+(\omega)} \cdot \frac{1}{S_z^-(\omega)}
\end{aligned}
\tag{VIII}
$$

where signal-to-noise ratio

$$
\lambda = \frac{2\gamma}{a} \; ; \gamma = \frac{\sigma^2}{r}
\tag{IX}
$$

one concludes

$$
\frac{1}{S_z^+(\omega)} = \frac{j\omega+a}{\sqrt{r}[j\omega+a\sqrt{1+\lambda}]}
\tag{X}
$$

and

$$
\frac{1}{S_z^-(\omega)} = \frac{-j\omega+a}{\sqrt{r}[-j\omega+a\sqrt{1+\lambda}]}
\tag{XI}
$$

and

$$\frac{S_{xz}(\omega)}{S_z^-(\omega)} = \frac{2a\sigma^2}{\sqrt{r}} \frac{1}{(j\omega+a)(-j\omega+a\sqrt{1+\lambda})} \qquad \text{(XII)}$$

The expression (XII) can be represented into partial fractions expansion form, yielding

$$\frac{S_{xz}(\omega)}{S_z^-(\omega)} = \frac{2\sigma^2}{\sqrt{r}\left(1+\sqrt{1+\lambda}\right)}\left[\frac{1}{j\omega+a} + \frac{1}{-j\omega+a\sqrt{1+\lambda}}\right] \qquad \text{(XIII)}$$

from which one obtains

$$\left[\frac{S_{xz}(\omega)}{S_z^-(\omega)}\right]^+ = \frac{2\sigma^2}{\sqrt{r}\left(1+\sqrt{1+\lambda}\right)}\frac{1}{j\omega+a} \qquad \text{(XIV)}$$

In this way, the optimal Wiener transfer function (VII) reduces to

$$G_w(j\omega) = \frac{2\gamma}{1+\sqrt{1+\lambda}} \cdot \frac{1}{j\omega+a\sqrt{1+\lambda}} \qquad \text{(XV)}$$

where γ and λ are defined by (IX), and the impulse response of the optimal filter represents the inverse Fourier transform of (XV), i.e.

$$g_w(t) = F^{-1}\{G_w(j\omega)\} = \frac{2\gamma}{1+\sqrt{1+\lambda}}\exp\left(-a\sqrt{1+\lambda}t\right) \; ; \; t \geq 0 \qquad \text{(XVI)}$$

It is clear that the Wiener filter is causal and stable low-pass filter whose bandwidth, i.e. cut-off frequency, depends on a and λ. The cut-off frequency ω_c of the filter is defined by

$$\left|G_w(j\omega_c)\right| = \frac{1}{\sqrt{2}}\left|G_w(j0)\right| \qquad \text{(XVII)}$$

and is equal to

$$\omega_c = a\sqrt{1+\lambda} \qquad \text{(XVIII)}$$

II) Kalman filter

To find a state-space form for the unknown $x(t)$, we can perform a spectral factorization on $S_x(\omega)$ to obtain

$$S_x(\omega) = S_x^+(\omega)S_x^-(\omega)S_w(\omega) = \frac{1}{j\omega+a}\frac{1}{-j\omega+a}(2a\sigma^2) \qquad \text{(XIX)}$$

Introducing $s = j\omega$, the shaping filter has transfer function of

$$H(s) = S_x^+(s) = \frac{1}{s+a} \qquad \text{(XX)}$$

for which a state realization is

$$X(s) = H(s)W(s) = \frac{1}{s+a}W(s) \qquad \text{(XXI)}$$

or equivalently

$$sX(s) = -aX(s) + W(s) \qquad \text{(XXII)}$$

Inverse Laplace transforming both sides of (XXII) and neglecting initial conditions yields

$$\dot{x}(t) = -ax(t) + w(t) \tag{XXIII}$$

with w being a zero-mean white noise with variance $q = 2a\sigma^2$ (see eq. (XIX)). Let $x(0) \sim (m_0, p_0)$. The state-space model on which the Kalman filter should be run is (XXIII), (II). For this system, the continuous Riccati equation (7.204) reduces to

$$\dot{p} = -\frac{1}{r}p^2 - 2ap + 2a\sigma^2 \tag{XXIV}$$

This equation can be rewritten as

$$\frac{dp}{\dfrac{1}{r}p^2 + 2ap - 2a\sigma^2} = -dt \tag{XXV}$$

Since the roots of the quadratic equation

$$p^2 + 2arp - 2ar\sigma^2 = 0 \tag{XXVI}$$

are

$$p_{1,2} = (-a \pm \beta)r \; ; \; \beta = \sqrt{a^2 + 2a\sigma^2/r} \tag{XXVII}$$

further follows $p^2 + 2arp - 2ar\sigma^2 = (p - p_1)(p - p_2)$, from which one obtains

$$\frac{1}{p^2 + 2arp - 2ar\sigma^2} = \frac{1}{p_1 - p_2}\left[\frac{1}{p - p_1} - \frac{1}{p - p_2}\right] \tag{XXVIII}$$

Taking into account (XXV) and (XXVIII), we have

$$\int_{p_0}^{p} \frac{dp}{p - p_1} - \int_{p_0}^{p} \frac{dp}{p - p_2} = -2\beta \int_0^t dt \tag{XXIX}$$

or

$$\ln(p - p_1)\Big|_{p_0}^{p} - \ln(p - p_2)\Big|_{p_0}^{p} = -2\beta t \tag{XXX}$$

Thus,

$$\frac{p - p_1}{p - p_2} \cdot \frac{p_0 - p_2}{p_0 - p_1} = \exp(-2\beta t) \tag{XXXI}$$

yielding

$$p(t)\left[1 - \frac{p_0 - p_1}{p_0 - p_2}e^{-2\beta t}\right] = p_1 - p_2\frac{p_0 - p_1}{p_0 - p_2}e^{-2\beta t} \tag{XXXII}$$

The steady state value p_s of $p(t)$ is given by

$$p_s = \lim_{t \to \infty} p(t) = p_1 = ar\left[\sqrt{1 + \lambda} - 1\right] = \frac{2\sigma^2}{1 + \sqrt{1 + \lambda}} \tag{XXXIII}$$

where λ is defined by (IX). The Kalman gain is

$$K(t) = \frac{1}{r}p(t) \tag{XXXIV}$$

and the Kalman filter is

$$\dot{\hat{x}} = -\left[a + K(t)\right]\hat{x} + K(t)z \tag{XXXV}$$

In the limit as $t \to \infty$ the steady-state filter is reacted. It is

$$K_s = \frac{1}{r}p_s = a\left(\sqrt{1+\lambda} - 1\right) \tag{XXXVI}$$

where p_s is given by (XXXIII), and

$$\dot{\hat{x}} = -a\sqrt{1+\lambda}\,\hat{x} + a\left(\sqrt{1+\lambda} - 1\right)z \tag{XXXVII}$$

The transfer function of the steady-state Kalman filter is

$$G_k(s) = \frac{\hat{X}(s)}{Z(s)} = \frac{a\left(\sqrt{1+\lambda} - 1\right)}{s + a\sqrt{1+\lambda}}$$

which is exactly the Wiener filter (XV). Note that the numerator of the Wiener filter (XV) is the steady-state Kalman gain (XXXVI) (see eq. (7.213)).

As mentioned in the chapter 6, underlying continuous Wiener filter design is the Wiener-Hopf integral equation, and its solution through spectral factorization. The contribution of Kalman was recognition of the fact that the integral equation could be converted into a nonlinear Riccati equation, whose solution contains all the necessary information for designing the optimal filter. The problem of spectral factorization in the Wiener filter is analogous to the requirement for solving nonlinear algebraic Riccati equation in the steady-state Kalman filter.

The corresponding steady-state discrete Kalman filter is given by

$$\hat{x}_{k+1}(-) = \Phi\left(I - K_sH\right)\hat{x}_k(-) + Bu_k + \Phi K_s z_k \tag{7.214}$$

By applying Z-transform on both sides of (7.214), and neglecting initial conditions, one obtains

$$\left[zI - \Phi\left(I - K_sH\right)\right]\hat{X}(z) = \Phi K_s Z(z) \tag{7.215}$$

where z is the Z-transform variable, and K_s is the steady-state Kalman gain in (7.207). In this way, we have

$$\hat{X}(z) = \left[zI - \Phi\left(I - K_sH\right)\right]^{-1} \Phi K_s Z(z) \tag{7.216}$$

The quantity in brackets is the matrix discrete transfer function which represents the Wiener optimal filter in the multi-dimensional case (see chapter 6), i.e.

$$G_w(z) = \left[zI - \Phi\left(I - K_sH\right)\right]^{-1} \Phi K_s \tag{7.217}$$

Particularly, in the scalar case this equation can be rewritten in the conventional scalar discrete transfer function form, i.e.

$$G_w(z) = \frac{\hat{X}(z)}{Z(z)} = \frac{\Phi K_s}{zI + \Phi\left(K_sh - 1\right)} \tag{7.218}$$

EXTENSIONS OF THE OPTIMUM RECURSIVE (KALMAN) FILTER

Chapter 8 discusses the extensions of the Kalman filtering technique to solve problems it was not directly designed to solve. In section 8.1. is shown how to use the linear Kalman filtering approach with minor modifications to solve the coloured-noise-source problems. Nonlinear estimatiors, using the linearized and extended Kalman filters, are developed in section 8.2, as well. Finally, in section 8.3. linear state estimators are improved by simultaneously estimation of the uncertain system parameters and/or noise statistics together with the system states. This results in parameter and noise adaptive filtering. Multiple-model estimation, representing an approach which allows for many possible values of model parameters and noise levels has been also presented in this section.

8.1 Correlated noise and shaping filters

As mentioned before, departure from the ideal case of white uncorrelated noises can occur in three ways: a) process noise can be nonwhite, i.e. correlated with itself; b) the measurement and process noise can be correlated with each other; c) the measurement noise can be nonwhite. In this section, we investigate how the linear Kalman filter can be modified to solve the coloured-noise-source problems [17, 30].

8.1.1 Coloured process noise

Suppose we are prescribed the plant by a linear discrete-time state-space model with discrete measurements

$$x_{k+1} = Ax_k + Bu_k + Gw_k \tag{8.1}$$

$$z_k = Hx_k + v_k \tag{8.2}$$

where v_k is zero-mean white noise sequence with covariance matrix R; x_0 is the initial condition with mean m_0 and covariance matrix P_0; v_k, w_k and x_0 are uncorrelated, but process noise w_k is not white. A noise sequence w_k whose spectral density $S_w(e^{j\omega})$ is not a constant is said to be coloured. From the spectral factorization theorem, if $S_w(z)$ is rational and if the determinant $|S_w(z)| \neq 0$ for almost every z, then there is a rational asymptotically stable spectral factor $H(z)$ with zeros inside or on the unit circle such that (see chapter 6)

$$S_w(z) = H(z)H^T(z^{-1}) \tag{8.3}$$

If $|S_w(z)| \neq 0$ on the unit circle $|z| = 1$, then $H(z)$ is minimum phase, i.e. all its zeros are strictly inside the circle $|z| = 1$. If the linear system $H(z)$ is driven by zero-mean white noise w_k^* with unit spectral density I, i.e. unit covariance matrix $Q^* = I$, then the output of the system has spectral density $S_w(z)$. The system $H(z)$ that manufactures coloured noise w_k with a given spectral density $S_w(z)$ from white noise w_k^* with unit spectral density $S_{w^*}(z) = I$ is called a *spectrum shaping filter*, and is given in Fig 8.1.

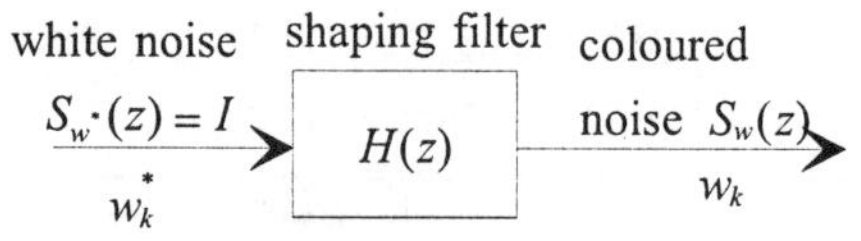

$$\text{white noise} \quad \text{shaping filter} \quad \text{coloured}$$

$$\underset{w_k^*}{\overset{S_{w^*}(z) = I}{\longrightarrow}} \boxed{H(z)} \underset{w_k}{\overset{\text{noise } S_w(z)}{\longrightarrow}}$$

Fig. 8.1. Spectral density shaping filter for discrete system

Given a spectral factorization for the spectral density $S_w(z)$ of w_k, we can find a state-space realization for $H(z)$. Then the noise w_k can be represented by the state-space model (it should be mentioned that the choice of state variables is not unique)

$$x_{k+1}^* = A^* x_k^* + G^* w_k^* \tag{8.4}$$

$$w_k = H^* x_k^* + D^* w_k^* \tag{8.5}$$

where

$$H(z) = H^* \left(zI - A^*\right)^{-1} G^* + D^* \tag{8.6}$$

and w_k^* is zero-mean white noise with the covariance matrix $Q^* = I$. Note that stars denote additional variables. Augmenting the states of (8.1) by the additional states of (8.4), we get

$$\begin{bmatrix} x_{k+1} \\ x_{k+1}^* \end{bmatrix} = \begin{bmatrix} A & GH^* \\ 0 & A^* \end{bmatrix} \begin{bmatrix} x_k \\ x_k^* \end{bmatrix} + \begin{bmatrix} B \\ 0 \end{bmatrix} u_k + \begin{bmatrix} GD^* \\ G^* \end{bmatrix} w_k^* \tag{8.7}$$

$$z_k = \begin{bmatrix} H & 0 \end{bmatrix} \begin{bmatrix} x_k \\ x_k^* \end{bmatrix} + v_k \tag{8.8}$$

with w_k^* and v_k white and uncorrelated. Now the discrete Kalman filter (Tab. 7.2) can be run on this augmented plant. Note that it must estimate the states of the original plant and also of the noise shaping filter.

In the continuous case (8.1) and (8.2) become

$$\dot{x} = Ax + Bu + Gw \tag{8.9}$$

$$z = Hx + w \tag{8.10}$$

with $v(t)$ being a zero-mean white and uncorrelated with $x(0)$. If process noise $w(t)$ is uncorrelated with $x(0)$ and $v(t)$ but not white, then we should proceed as for the discrete situation. Factor the spectral density $S_w(\omega)$ of $w(t)$ as

$$S_w(s) = H(s) H^T(-s) \tag{8.11}$$

If $|S_w(s)| \neq 0$ for $\mathrm{Re}\{s\} = 0$ (real part of s equal to zero), i.e. $|S_w(s)| \neq 0$ on the imaginary axis of the s-plane, then $H(s)$ is minimum phase. Then, find a state-space representation for $H(s)$ (it should be noted again that the choice of state variables is not unique)

$$\dot{x}^* = A^* x^* + G^* w^* \tag{8.12}$$

$$w = H^* x^* + D w^*$$ (8.13)

where

$$H(s) = H^* \left(sI - A^* \right)^{-1} G^* + D^*$$ (8.14)

Here stars denote additional variables, and $w^*(t)$ is zero-mean white noise with the unit spectral density $S_{w^*}(s) = I$. The block diagram of the shaping filter is given in Fig. 8.2.

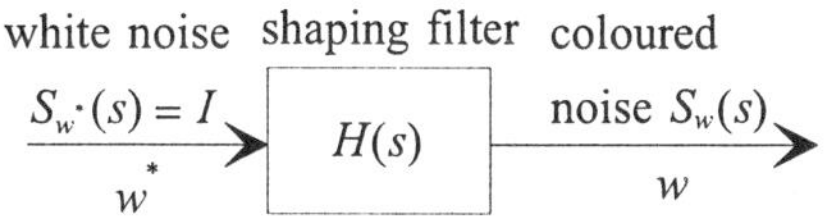

white noise shaping filter coloured

$$\xrightarrow[w^*]{S_{w^*}(s) = I} \boxed{H(s)} \xrightarrow[w]{\text{noise } S_w(s)}$$

Fig. 8.2. Spectral density shaping filter for continuous system

To describe the dynamics of both the unknown $x(t)$ and the process noise $w(t)$ write the augmented system

$$\begin{bmatrix} \dot{x} \\ \dot{x}^* \end{bmatrix} = \begin{bmatrix} A & GH^* \\ 0 & A^* \end{bmatrix} \begin{bmatrix} x \\ x^* \end{bmatrix} + \begin{bmatrix} B \\ 0 \end{bmatrix} u + \begin{bmatrix} GD^* \\ G^* \end{bmatrix} w^*$$ (8.15)

$$z = \begin{bmatrix} H & 0 \end{bmatrix} \begin{bmatrix} x \\ x^* \end{bmatrix} + v$$ (8.16)

The continuous Kalman filter (Tab. 7.3) is now applied to this augmented system.

Example 8.1 (Aircraft longitudinal dynamics with gust noise):

The longitudinal dynamics of an aircraft can be represented in the short period approximation by the harmonic oscillator

$$\dot{x} = \begin{bmatrix} 0 & 1 \\ -\omega_n^2 & -2\delta\omega_n \end{bmatrix} x + \begin{bmatrix} 0 \\ 1 \end{bmatrix} w$$ (I)

where $x = \begin{bmatrix} \Theta & \dot{\Theta} \end{bmatrix}^T$ and Θ is pitch angle. The process noise $w(t)$ represents wind gust which change the angle of attack δ, and so influence pitch rate. The gust noise might have a spectral density which can be approximated by the low-frequency spectrum

$$S_w(\omega) = \frac{2a\sigma^2}{\omega^2 + a^2} \; ; a > 0$$ (II)

Performing a factorization on (II) there results

$$S_w(s)\big|_{s=j\omega} = \frac{\sqrt{2a}\sigma}{s+a} \cdot \frac{\sqrt{2a}\sigma}{-s+a} = H(s)H(-s)$$ (III)

so that the minimum-phase shaping filter is

$$H(s) = \frac{1}{s+a}$$ (IV)

which must be driven by zero-mean white input noise w^* with variance $\sqrt{2a}\sigma$. A state realization of (IV) is

$$W(s) = H(s)W^*(s) = \frac{1}{s+a} W^*(s)$$

or equivalently

$$sW(s) = -aW(s) + W^*(s) \qquad \text{(V)}$$

The last relation can be represented in time-domain as

$$\dot{x}^* = -ax^* + w^* \qquad \text{(VI)}$$

$$w = x^* \qquad \text{(VII)}$$

Augmenting the plant (I) by the shaping filter (VI) and redefining the state as $x = \begin{bmatrix} \Theta & \dot{\Theta} & x^* \end{bmatrix}^T$ there results

$$\dot{x} = \begin{bmatrix} 0 & 1 & 0 \\ -\omega_n^2 & -2\delta\omega_n & 1 \\ 0 & 0 & -a \end{bmatrix} \begin{bmatrix} \Theta \\ \dot{\Theta} \\ x^* \end{bmatrix} + \begin{bmatrix} 0 \\ 0 \\ 1 \end{bmatrix} w^* = Ax + Gw^* \qquad \text{(VIII)}$$

If pitch Θ is measured, then we have the measurement equation

$$z = \begin{bmatrix} 1 & 0 & 0 \end{bmatrix} \begin{bmatrix} \Theta \\ \dot{\Theta} \\ x^* \end{bmatrix} + v = Hx + v \qquad \text{(IX)}$$

where the measurement noise v is zero-mean white with the variance r. Model of the system and measurement is given in Fig. 8.3.

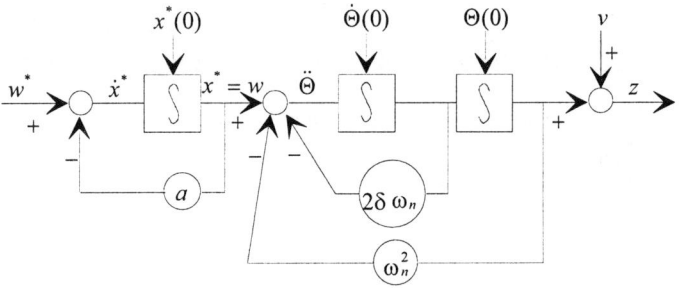

Fig. 8.3. Model of the system and measurement

Now the continuous Kalman filter (Tab. 7.2) can be run on (VIII) and (IX). If pitch Θ is measured every T sec, and the sampling period T is small so that T^2 is negligible, then the discretized plant model is given by (see chapter 4)

$$x_{k+1} = \begin{bmatrix} 1 & T & 0 \\ -\omega_n^2 T & 1 - 2\delta\omega_n T & T \\ 0 & 0 & 1 - aT \end{bmatrix} x_k + \begin{bmatrix} 0 \\ 0 \\ 1 \end{bmatrix} w_k^* = A_d x_k + G_d w_k^* \qquad \text{(X)}$$

$$z_k = \begin{bmatrix} 1 & 0 & 0 \end{bmatrix} x_k + v_k = H_d x_k + v_k \qquad \text{(XI)}$$

where w_k^* is zero-mean white noise with variance $q_d = 2a\sigma^2 T$ and v_k is zero-mean white noise with variance $r_d = r/T$. In practice the discretization would be performed using e^{AT}, not by the Euler's approximation $A_d = I + AT$; $B_d = BT$; $q_d = qT$ and $r_d = r/T$. We use the later method in the example for simplicity only. The actual sampling process with zero-order hold corresponds to discretization by e^{AT}, not $I + AT$. Now the discrete Kalman filer (Tab. 7.2) can be run on (X) and (XI). Model of the discrete system and measurement is given in Fig. 8.4.

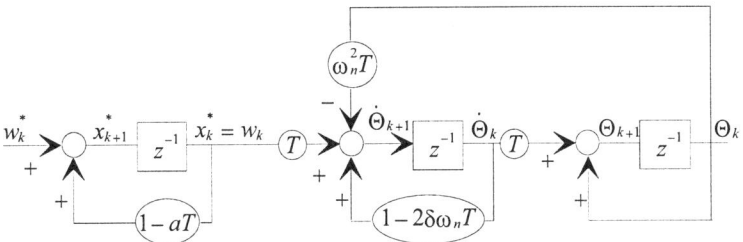

Fig. 8.4. Discrete model of the system and measurement

In practice, to find $S_w(\omega)$ in the first place one could assume $\{w_k\}$ is ergodic and use one sample function w_k to determine the auto-correlation function $R_w(k)$ by time averages. A z-transform yields the spectral density, and then a curve fit yields $S_w(z)$ for $z = e^{j\omega}$.

Some useful continuous and discrete shaping filters are shown in Tabs 8.1. and 8.2. In these tables, $w(t)$ and w_k are white noise processes.

Table 8.1. Some useful discrete spectrum-shaping filters

Process	State Equation	Spectral density
Random bias	$x_{k+1} = x_k$; x_0 random $\quad E\{x_0\} = m_0$; $\mathrm{cov}\{x_0\} = p_0$	
Brownian motion	$x_{k+1} = x_k + w_k$ w_k white, zero$-$mean $\mathrm{var}\{w_k\} = Q$	
First order Markov	$x_{k+1} = ax_k + w_k$ w_k white, zero$-$mean $\mathrm{var}\{w_k\} = (1-a^2)\sigma^2$	$S_x(\omega) = \dfrac{(1-a^2)\sigma^2}{1+a^2 - 2a\cos(\omega)}$
Second order Markov	$x_{k+1} = \begin{bmatrix} 0 & 1 \\ -\omega_n^2 & -2\alpha \end{bmatrix} x_k + \begin{bmatrix} 1 \\ \omega_n - 2\alpha \end{bmatrix} w_k$ $z_k = [1 \; 0] x_k$ w_k zero$-$mean white; $\mathrm{var}\{w_k\} = (1-\alpha^2)\sigma^2$	$S_x(\omega) = \dfrac{(1-\alpha^2)\sigma^2\left[(1+\omega_n^2) + 2\omega_n\cos(\omega)\right]}{(1+4\alpha^2+\omega_n^4) + 4\alpha(1+\omega_n^2)\cos\omega + 2\omega_n^2\cos(2\omega)}$

Table 8.2. Some useful continuous spectrum-shaping filters

Process	State Equation	Spectral density $S_x(\omega)$
Random bias	$\dot{x} = 0$; $x(0)$ random $E\{x(0)\} = m_0$; $\text{cov}\{x(0)\} = p_0$	
Brownian motion (Wiener process)	$\dot{x} = w$; $x(0) = 0$ $w(t)$ white, zero– mean with $\text{cov}\{w(t)\} = Q$	$x(t)$ nonstationary, zero-mean with $\text{cov}\{x(t)\} = Qt$
First-order Markov	$\dot{x} = -ax + w$ $w-$ white, zero– mean with $\text{var}\{w\} = 2a\sigma^2$	 $S_x(\omega) = \dfrac{2a\sigma^2}{\omega^2 + a^2}$
Second-order Markov	$\dot{x} = \begin{bmatrix} 0 & 1 \\ -\omega_n^2 & -2\alpha \end{bmatrix} x + \begin{bmatrix} 1 \\ \omega_n - 2\alpha \end{bmatrix} w$ $z = \begin{bmatrix} 1 & 0 \end{bmatrix} x$ $w-$ zero– mean; $\text{var}\{w\} = 2\alpha\sigma^2$	 $S_z(\omega) = \dfrac{2\alpha\sigma^2\left(\omega^2 + \omega_n^2\right)}{\omega^4 + 2\left(2\alpha^2 - \omega_n^2\right)\omega^2 + \omega_n^4}$

The random bias could be used in a navigation problem. Brownian motion, or the Wiener process, is useful in describing the motion of particles under the influence of diffusion. It should also be used to model biases that are known to vary with time. First-order Markov process is useful to model band-limited noises. The second-order Markov process has a periodic autocorrelation, and is useful to describe oscillatory random processes such as fuel slosh or vibration.

8.1.2 Correlated Measurement and Process Noises

Now we consider the linear continuous system (8.9) and (8.10), i.e.

$$\dot{x} = Ax + Bu + Gw \; ; \; x(0) \sim (m_0, P_0) \; ; \; w \sim (0, Q) \tag{8.17}$$

$$z = Hx + v \; ; \; v \sim (0, R) \tag{8.18}$$

where $v(t)$ is zero-mean white noise with covariance matrix R; $x(0)$ is the initial condition with mean m_0 and covariance matrix P_0, and $w(t)$ is zero-mean white noise with covariance matrix Q. It is assumed that $x(0)$ is uncorrelated with $v(t)$ and $w(t)$, but now the process w, and measurement v, noises are correlated; that is

$$E\left\{w(t)v^T(\tau)\right\} = C(t)\delta(t-\tau) \tag{8.19}$$

where $\delta(\cdot)$ is the Dirac delta function. One approach is to convert this new problem to an equivalent problem with no correlation between process and measurement noises. To do this, add zero to the right-hand-side of eq. (8.17) in the form

$$\dot{x} = Ax + Bu + Gw + D(z - Hx - v) \tag{8.20}$$

where D is some matrix. The filtering problem under consideration now is

$$\dot{x} = (A - DH)x + (Bu + Dz) + (Gw - Dv) \tag{8.21}$$

$$z = Hx + v \tag{8.22}$$

where $(Bu + Dz)$ is treated as a known input and $(Gw - Dv)$ is treated as a process noise. This noise is white, since it is zero-mean and its covariance is given by

$$E\left\{\left[Gw(t) - Dv(t)\right]\left[Gw(\tau) - Dv(\tau)\right]^T\right\} = GE\left\{w(t)w^T(\tau)\right\}G^T + DE\left\{v(t)v^T(\tau)\right\}D^T -$$
$$-DE\left\{v(t)w^T(\tau)\right\}G^T - GE\left\{w(t)v^T(\tau)\right\}D^T = \left(GQG^T + DRD^T - DC^TG^T - GCD^T\right)\delta(t-\tau) \tag{8.23}$$

As mentioned above, the motivation behind these manipulations is that D can be selected to make $(Gw - Dv)$ and v uncorrelated; that is

$$E\left\{\left[Gw(t) - Dv(t)\right]v^T(\tau)\right\} = GE\left\{w(t)v^T(\tau)\right\} - DE\left\{v(t)v^T(\tau)\right\} = (GC - DR)\delta(t-\tau) \tag{8.24}$$

which becomes zero if we select

$$D = GCR^{-1} \tag{8.25}$$

With this choice, the modified system is

$$\dot{x} = (A - GCR^{-1}H)x + GCR^{-1}z + Bu + (Gw - GCR^{-1}v) \tag{8.26}$$

The new process noise auto-correlation function is

$$E\left\{\left[Gw(t) - Dv(t)\right]\left[Gw(\tau) - Dv(\tau)\right]^T\right\} = \left(GQG^T + DRD^T - DC^TG^T - GCD^T\right)\delta(t-\tau)$$
$$= \left[GQG^T + GCR^{-1}R\left(GCR^{-1}\right)^T - GCR^{-1}C^TG^T - GC\left(GCR^{-1}\right)^T\right]\delta(t-\tau) \tag{8.27}$$
$$= G\left(Q - CR^{-1}C^T\right)G^T\delta(t-\tau)$$

which is less than the GQG^T term arising from the original process noise w. Applying the continuous Kalman filter (Tab. 7.3) to (8.26) and (8.22), there results the estimate update

$$\dot{\hat{x}} = (A - GCR^{-1}H)\hat{x} + Bu + GCR^{-1}z + PH^TR^{-1}(z - H\hat{x})$$

or

$$\dot{\hat{x}} = A\hat{x} + Bu + (PH^T + GC)R^{-1}(z - H\hat{x})$$

Defining a modified Kalman gain by

$$K = (PH^T + GC)R^{-1} \tag{8.28}$$

this becomes

$$\dot{\hat{x}} = A\hat{x} + Bu + K(z - H\hat{x})$$ (8.29)

which is identical in form to the estimate update equation in Table 7.3. The error covariance update equation for (8.26) with process noise autocorrelation function (8.27) is (see Tab. 7.3)

$$\dot{P} = (A - GCR^{-1}H)P + P(A - GCR^{-1}H)^T + G(Q - CR^{-1}C^T)G^T - PH^T R^{-1}HP$$

This can be written in terms of the modified gain (8.28) as

$$\dot{P} = AP + PA^T + GQG^T - KRK^T$$ (8.30)

which is identical to the expression in Table 8.3. Nonzero C results in a smaller error covariance (i.e. larger gain K) than the uncorrelated noise case ($C=0$) due to the additional information provided by the cross-correlation term C. Thus, we have found that if the measurement and process noises are correlated then it is only necessary to define a modified Kalman gain and use the formulation (8.30) for the Riccati equation (i.e. error covariance update equation); all other equations are the same as for the uncorrelated case (Tab. 7.3). The results are summarized in Table 8.3.

Table 8.3. Summary of continuos Kalman filter equations (correlated process and measurement noise)

System model Measurement model	$\dot{x}(t) = A(t)x(t) + B(t)u(t) + G(t)w(t)$; $w(t) \sim (0, Q(t))$
	$z(t) = H(t)x(t) + v(t)$; $v(t) \sim (0, R(t))$
Initial conditions	$x(0) \sim (m_0, P_0)$
Other assumptions	$E\{w(t)x^T(0)\} = 0$; $E\{v(t)x^T(0)\} = 0$
	$E\{w(t)v^T(\tau)\} = C(t)\delta(t-\tau)$; R^{-1} exists
State estimate Error Covariance Propagation	$\dot{\hat{x}}(t) = A(t)\hat{x}(t) + B(t)u(t) + K(t)[z(t) - H(t)\hat{x}(t)]$; $\hat{x}(0) = m_0$
	$\dot{P}(t) = A(t)P(t) + P(t)A^T(t) + G(t)Q(t)G^T(t) - K(t)R(t)K^T(t)$; $P(0) = P_0$
Kalman gain matrix	$K(t) = [P(t)H^T(t) + G(t)C(t)]R^{-1}(t)$

This method can also be used to derive results for the corresponding discrete-time problem (8.1), (8.2); that is

$$x_{k+1} = Ax_k + Bu_k + Gw_k$$ (8.31)

$$z_k = Hx_k + v_k$$ (8.32)

with $w_k \sim (0, Q)$ and $v_k \sim (0, R)$, both white, but with

$$E\{w_k v_j^T\} = C\delta_{jk}$$ (8.33)

where δ_{jk} is the Kronecker delta. In this case we could write

$$E\left\{\begin{bmatrix} w_k \\ v_k \end{bmatrix}\begin{bmatrix} w_k^T & v_k^T \end{bmatrix}\right\} = \begin{bmatrix} Q & C \\ C^T & R \end{bmatrix}$$ (8.34)

Similarly as in the preceding case, one approach to rederive the Kalman filter equations is to convert this new problem to an equivalent problem with no correlation between noise process. To do this, add zero to the right hand side of (8.31), i.e.

$$x_{k+1} = Ax_k + Bu_k + Gw_k + D(z_k - Hx_k - v_k) \qquad (8.35)$$

where

$$D = GCR^{-1} \qquad (8.36)$$

The filtering problem under consideration is now

$$x_{k+1} = (A - DH)x_k + (Bu_k + Dz_k) + (Gw_k - Dv_k) \qquad (8.37)$$

$$z_k = Hx_k + v_k \qquad (8.38)$$

where in eq. (8.37) the term $(Bu + Dz)$ is treated as a known input and $(Gw - Dv)$ is treated as a process noise. By the choice of D specified by (8.36), we have

$$E\{(Gw_k - Dv_k)v_k^T\} = GE\{w_k v_k^T\} - DE\{v_k v_k^T\} \qquad (8.39)$$
$$= GC - DR = GC - GCR^{-1}R = 0$$

and, thus, the measurement and process noises in this equivalent problem are indeed uncorrelated. Applying previously derived results for discrete Kalman filtering (Tab. 7.2) to the reformulated problem in eqs. (8.37) and (8.38), we have

$$\hat{x}_{k+1}^- = (A - DH)\hat{x}_k^+ + Bu_k + Dz_k \qquad (8.40)$$

where

$$\hat{x}_k^+ = \hat{x}_k^- + P_k^- H^T \left(HP_k^- H^T + R\right)^{-1}\left(z_k - H\hat{x}_k^-\right) \qquad (8.41)$$

Here is used the fact that x_k and v_k are still uncorrelated (x_k depends on w_{k-1} and hence on v_{k-1}), so that the correlation in (8.33) does not affect the measurement update; that is the relation (8.41) is identical to the corresponding one in Table 7.2.

By using (8.36) and substituting eq. (8.41) into eq. (8.40), further follows

$$\hat{x}_{k+1}^- = \left(A - GCR^{-1}H\right)\hat{x}_k^- + \left(A - GCR^{-1}H\right)P_k^- H^T\left(HP_k^- H^T + R\right)^{-1}\left(z_k - H\hat{x}_k^-\right) + Bu_k + GCR^{-1}z_k$$

$$= A\hat{x}_k^- + Bu_k + GCR^{-1}\left(z_k - H\hat{x}_k^-\right) + \left(A - GCR^{-1}H\right)P_k^- H^T\left(HP_k^- H^T + R\right)^{-1}\left(z_k - H\hat{x}_k^-\right)$$

$$= A\hat{x}_k^- + Bu_k + \left[GCR^{-1} + \left(A - GCR^{-1}H\right)P_k^- H^T\left(HP_k^- H^T + R\right)^{-1}\right]\left(z_k - H\hat{x}_k^-\right)$$

However, taking into account that

$$GCR^{-1} = GCR^{-1}\left(HP_k^- H^T + R\right)\left(HP_k^- H^T + R\right)^{-1}$$

one concludes further

$$GCR^{-1} + \left(A - GCR^{-1}H\right)P_k^- H^T\left(HP_k^- H^T + R\right)^{-1} =$$

$$\left[GCR^{-1}\left(HP_k^- H^T + R\right) + AP_k^- H^T - GCR^{-1}HP_k^- H^T\right]\left(HP_k^- H^T + R\right)^{-1} =$$

$$\left(GC + AP_k^- H^T\right)\left(HP_k^- H^T + R\right)^{-1}$$

so that

$$\hat{x}_{k+1}^- = A\hat{x}_k^- + Bu_k + \left(GC + AP_k^- H^T\right)\left(HP_k^- H^T + R\right)^{-1}\left(z_k - H\hat{x}_k^-\right) \tag{8.42}$$

or equivalently

$$\hat{x}_{k+1}^- = A\hat{x}_k^- + Bu_k + K_k\left(z_k - H\hat{x}_k^-\right) \tag{8.43a}$$

where

$$K_k = \left(GC + AP_k^- H^T\right)\left(HP_k^- H^T + R\right)^{-1} \tag{8.43b}$$

The equation (8.43a) is identical to the corresponding equation from Table 7.2. except for the gain term (4.83b). Therefore, the only effect of noises correlated as in (8.33) is to modify the Kalman gain.

To find the effect of the correlation (8.33) on the error covariance, write the new a priori error system as

$$\begin{aligned}\tilde{x}_{k+1}^- &= x_{k+1} - \hat{x}_{k+1}^- \\ &= Ax_k + Bu_k + Gw_k - A\hat{x}_k^- - Bu_k - K_k\left(z_k - H\hat{x}_k^-\right) \\ &= \left(A - K_k H\right)\tilde{x}_k^- + Gw_k - K_k v_k \end{aligned} \tag{8.44}$$

with K_k as in (8.43). Now, since $E\{\tilde{x}_{k+1}^-\} = 0$, we have

$$\begin{aligned}P_{k+1}^- = E\left\{\tilde{x}_{k+1}^-\left(\tilde{x}_{k+1}^-\right)^T\right\} &= \left(A - K_k H\right)E\left\{\tilde{x}_k^-\left(\tilde{x}_k^-\right)^T\right\}\left(A - K_k H\right)^T \\ &+ GE\left\{w_k w_k^T\right\}G^T + K_k E\left\{v_k v_k^T\right\}K_k^T + \left(A - K_k H\right)E\left\{\tilde{x}_k^- w_k^T\right\}G^T \\ &- \left(A - K_k H\right)E\left\{\tilde{x}_k^- v_k^T\right\}K_k^T - GE\left\{w_k v_k^T\right\}K_k^T - K_k E\left\{v_k w_k^T\right\}G^T \\ &+ GE\left\{w_k\left(\tilde{x}_k^-\right)^T\right\}\left(A - K_k H\right)^T - K_k E\left\{v_k\left(\tilde{x}_k^-\right)^T\right\}\left(A - K_k H\right)^T \end{aligned}$$

However, as mentioned before, x_k and v_k are uncorrelated (x_k depends on w_{k-1} and hence v_{k-1}) and this is also the case with x_k and w_k (x_k depends on w_{k-1}); that is

$$E\left\{\tilde{x}_k^- v_k^T\right\} = 0 \; ; \; E\left\{\tilde{x}_k^- w_k^T\right\} = 0,$$

Thus, the last relation reduces to

$$P_{k+1}^- = \left(A - K_k H\right)P_k^-\left(A - K_k H\right)^T + GQG^T + KRK^T - GCK_k^T - K_k C^T G^T$$

or equivalently

$$P_{k+1}^- = AP_k^- A^T - K_k\left(HP_k^- A^T + C^T G^T\right) - \left(AP_k^- H^T + GC\right)K_k^T + K_k\left(HP_k^- H^T + R\right)K_k^T + GQG^T$$

This can be rewritten as

$$\begin{aligned}P_{k+1}^- = AP_k^- A^T &- K_k\left(HP_k^- H^T + R\right)\left(HP_k^- H^T + R\right)^{-1}\left(HP_k^- A^T + C^T G^T\right) \\ &- \left(AP_k^- H^T + GC\right)\left(HP_k^- H^T + R\right)^{-1}\left(HP_k^- H^T + R\right)K_k^T + K_k\left(HP_k^- H^T + R\right)K_k^T + GQG^T \end{aligned}$$

or, after replacing the expression (8.43b) for K_k,

$$P_{k+1}^- = AP_k^- A^T - K_k \left(HP_k^- H^T + R \right) K_k^T - K_k \left(HP_k^- H^T + R \right) K_k^T + K_k \left(HP_k^- H^T + R \right) K_k^T + GQG^T$$
$$= AP_k^- A^T - K_k \left(HP_k^- H^T + R \right) K_k^T + GQG^T$$

Finally, since the correlation in (8.33) does not affect the measurement update, the relation for P_k^+ remains the same as in Table 7.2; that is

$$P_k^+ = P_k^- - P_k^- H^T \left(HP_k^- H^T + R \right)^{-1} HP_k^- \tag{8.46}$$

In eqs. (8.43) and (8.45) the time and measurement updates are combined, and this result in the alternative Kalman filter formulation, which manufactures the a priori estimate $\hat{x}_k^-$ and error covariance P_k^-. This formulation, known as *apriori Kalman filter*, is summarized in Table 8.4.

Table 8.4.: Discrete Kalman filter: A Priori Formulation

State Model	$x_{k+1} = Ax_k + Bu_k + Gw_k \ ; \ w_k \sim (0, Q)$
Measurement Model	$z_k = Hx_k + v_k \ ; \ v_k \sim (0, R)$
Initial conditions	$x_0 \sim (\hat{x}_0, P_0)$
Other Assumptions	$E\{w_k x_0^T\} = 0; \ E\{v_k x_0^T\} = 0 \ ; \ E\{w_k v_j^T\} = C\delta_{kj}$
State Estimate	$\hat{x}_{k+1}^- = A\hat{x}_k^- + Bu_k + K_k \left(z_k - H\hat{x}_k^- \right)$
Error Covariance Propagation	$P_{k+1}^- = AP_k^- A^T - K_k \left(HP_k^- H^T + R \right) K_k^T + GQG^T$
Kalman gain	$K_k = \left(AP_k^- H^T + GC \right) \left(HP_k^- H^T + R \right)^{-1}$

We have suppressed the time index on the system matrices A, B, G, H as we shall usually do hence forth for convenience. The *apriori filter formulation* is identical in structure to the deterministic *observer*, as it is shown in Fig. 8.5 (the only difference is that the constant gain matrix of the deterministic observer is replaced by the time-varying Kalman gain K_k). Here, even if system is time invariant, the gain K_k depends on time.

The error covariance propagation equation in Table 8.4 is a Ricccati equation, which is a matrix quadratic equation. It is a time-varying version of the corresponding equation for designing the deterministic state observer. From the error propagation equation it is clear that the error covariance is decreased due to the new information provided by the cross correlation term C in K_k. Even if $C \neq 0$, we could still implement the discrete Kalman filter of Table 7.2, but it would be suboptimal in this case.

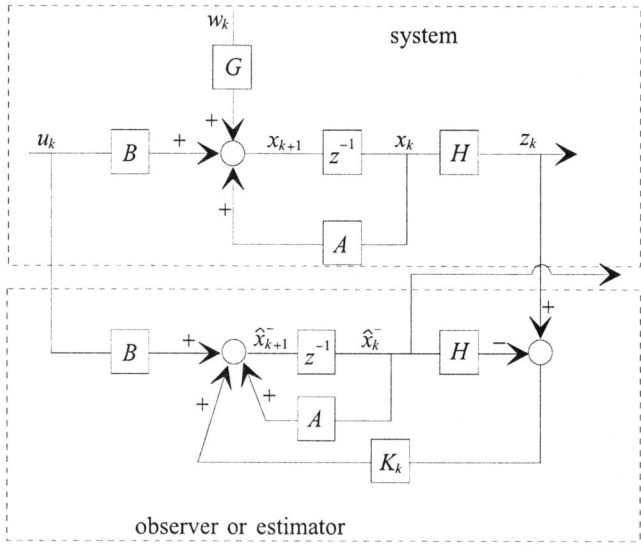

Fig. 8.5. Apriori Discrete Kalman Filter as State Estimator

8.1.3 Correlated Measurement Noise

Measurements may contain errors whose correlation times are significant. Let us assume that, through the use of a shaping filter, these measurements errors are described as the output of a first-order vector differential equation forced by white noise. The technique of state vector augmentation could then be used to recast this problem into a form where the solution has already been obtained (chapter 7). Suppose now that we are prescribed the system (8.1) and (8.2) with $w_k \sim (0, Q)$ white and w_k and v_k uncorrelated, but that the measurement noise is

$$v_k = n_k + v_k^*$$

where n_k is coloured and v_k^* is white; that is

$$x_{k+1} = Ax_k + Bu_k + Gw_k \;\; ; \;\; w_k \sim (0, Q) \tag{8.47}$$

$$z_k = Hx_k + v_k \tag{8.48}$$

$$v_k = n_k + v_k^*; \quad n_k \text{ coloured}; \quad v_k^* \sim (0, R) \text{ white} \tag{8.49}$$

If $S_n(z)$ is the spectral density of n_k, then a spectral factorization (chapter 3) yields

$$S_n(z) = H(z)H^T(z^{-1})$$

and we can find a state realization (A^*, G^*, H^*, D^*) for the shaping filter $H(z)$; that is

$$x_{k+1}^* = A^* x_k^* + G^* w_k^* \tag{8.50}$$

$$n_k = H^* x_k^* + D^* w_k^* \tag{8.51}$$

$$v_k = n_k + v_k^* = H^* x_k^* + D^* w_k^* + v_k^* \tag{8.52}$$

where $w_k^* \sim (0, I)$ is zero-mean white noise with unit covariance matrix. The new measurement noise in (8.52) is $(D^* w^* + v^*)$, and since w^* and v^* are uncorrelated, i.e. $E\{w_k^* v_k^{*T}\} = 0$, its covariance matrix

$$R = E\left\{(D^* w_k^* + v_k^*)(D^* w_k^* + v_k^*)^T\right\}$$
$$= D^* E\{w_k^* w_k^{*T}\} D^{*T} + E\{v_k^* v_k^{*T}\} = D^* D^{*T} + R^* \tag{8.53}$$

Augmenting the state and measurement equations of (8.47)-(8.49) by (8.50)-(8.52) there results the new equations

$$\begin{bmatrix} x_{k+1} \\ x_{k+1}^* \end{bmatrix} = \begin{bmatrix} A & 0 \\ 0 & A^* \end{bmatrix} \begin{bmatrix} x_k \\ x_k^* \end{bmatrix} + \begin{bmatrix} B \\ 0 \end{bmatrix} u_k + \begin{bmatrix} G & 0 \\ 0 & G^* \end{bmatrix} \begin{bmatrix} w_k \\ w_k^* \end{bmatrix} \tag{8.54}$$

$$z_k = \begin{bmatrix} H & H^* \end{bmatrix} \begin{bmatrix} x_k \\ x_k^* \end{bmatrix} + (D^* w_k^* + v_k^*) \tag{8.55}$$

The new process (state) noise $\begin{bmatrix} w_k^T & w_k^{*T} \end{bmatrix}^T$ is zero-mean white with the covariance matrix

$$E\left\{\begin{bmatrix} w_k \\ w_k^* \end{bmatrix} \begin{bmatrix} w_k^T & w_k^{*T} \end{bmatrix}\right\} = \begin{bmatrix} E\{w_k w_k^T\} & E\{w_k w_k^{*T}\} \\ E\{w_k^* w_k^T\} & E\{w_k^* w_k^{*T}\} \end{bmatrix} = \begin{bmatrix} Q & 0 \\ 0 & I \end{bmatrix} \tag{8.56}$$

and the new measurement noise $(D^* w_k^* + v_k^*)$ is zero-mean white with the covariance matrix R given by (8.53). Now, however, the measurement and process noises are correlated with

$$E\left\{\begin{bmatrix} w_k \\ w_k^* \end{bmatrix} [D^* w_k^* + v_k^*]^T\right\} = \begin{bmatrix} E\{w_k w_k^{*T}\} D^{*T} + E\{w_k v_k^{*T}\} \\ E\{w_k^* w_k^{*T}\} D^{*T} + E\{w_k^* v_k^{*T}\} \end{bmatrix} = \begin{bmatrix} 0 \\ D^{*T} \end{bmatrix} \tag{8.57}$$

so that the modified gain of the previous subsection should be used for optimality. If the relative degree of $S_n(z)$ (the difference of the degrees of the polynomials in the denumerator and numerator of the real rational function $S_n(z)$) is greater than zero, then $D^* = 0$ and the new measurement and process noises are uncorrelated. If $v_k^* = 0$ in (8.52) then $R^* = 0$. However, this approach can often still be used, since the discrete Kalman filter requires that $\left\{[H \; H^*] P [H \; H^*]^T + R\right\}^-$ exists; that is $\det\left\{[H \; H^*] P [H \; H^*]^T + R\right\} \neq 0$ where R is given by (8.53).

Unfortunately, an augmented state approach can not be applied in the case of continuous-time systems. Consider the continuous system and measurement described by

$$\dot{x} = Ax + Bu + Gw \; ; \; w \sim (0, Q) \tag{8.58}$$

$$z = Hx + v \tag{8.59}$$

and suppose that, by spectral factorization or some other means, a shaping filter for v has been found to be

$$\dot{v} = A^* v + G^* w^* \; ; \; w^* \sim (0, I) \tag{8.60}$$

where w^* is white and uncorrelated with the process noise w, i.e. $E\{ww^{*T}\}=0$. The eq. (8.60) can be rewritten as

$$\dot{x}^* = A^*x^* + G^*w^* \tag{8.61a}$$

$$v = x^* \tag{8.61b}$$

The augmented state vector $x_a = \begin{bmatrix} x^T & x^{*T} \end{bmatrix}^T$ satisfies the differential equation

$$\dot{x}_a = \begin{bmatrix} \dot{x} \\ \dot{x}^* \end{bmatrix} = \begin{bmatrix} A & 0 \\ 0 & A^* \end{bmatrix}\begin{bmatrix} x \\ x^* \end{bmatrix} + \begin{bmatrix} B \\ 0 \end{bmatrix}u + \begin{bmatrix} G & 0 \\ 0 & G^* \end{bmatrix}\begin{bmatrix} w \\ w^* \end{bmatrix} \tag{8.62}$$

and the measurement equation (8.59) becomes

$$z = \begin{bmatrix} H & I \end{bmatrix}x_a \tag{8.63}$$

In this reformulated problem, the equivalent measurement noise is zero. Correspondingly, the equivalent noise covariance matrix R is singular $(R \equiv 0)$ and, thus, the Kalman gain matrix $K = PH^T R^{-1}$ in Table 7.3 does not exist. However, there is another approach to this problem which avoids both the difficulty of singular R and the undesirability of working with a high order augmented system. From eq. (8.60), where we see that $\dot{v} - A^*v = \dot{x}^* - A^*x^* = w^*$ is a white zero-mean noise process with the unit covariance matrix I, we are led to define the derived measurement z_d by

$$z_d = \dot{z} - A^*z - HBu \tag{8.64}$$

Then, if H is time invariant

$$z_d = \frac{d}{dt}(Hx+v) - A^*(Hx+v) - HBu$$

$$= H\dot{x} + \dot{v} - A^*Hx - A^*v - HBu$$

$$= H(Ax + Bu + Gw) + A^*v + G^*w^* - A^*Hx - A^*v - HBu$$

$$= (HA - A^*H)x + G^*w^* + HGw$$

Defining a new measurement matrix by

$$H_d = HA - A^*H \tag{8.65}$$

we can write

$$z_d = H_dx + G^*w^* + HGw \tag{8.66}$$

This is a new output equation in terms of the derived measurement. The reason for this manipulation is that the new measurement noise $(G^*w^* + HGw)$ is zero-mean white with the covariance matrix

$$R_d = E\{(G^*w^* + HGw)(G^*w^* + HGw)^T\}$$

$$= G^*E\{w^*w^{*T}\}G^{*T} + HGE\{ww^T\}G^TH^T \tag{8.67}$$

$$= G^*G^{*T} + HGQG^TH^T$$

so that (8.58) and (8.66) define a new system driven by white noise. The order of this system is unchanged (there is no augmentation of the state vector). In this new system, the process and measurement noises are correlated with the cross-covariance matrix

$$E\left\{w\left[G^*w^* + HGw\right]^T\right\} = E\left\{ww^{*T}\right\}G^{*T} + E\left\{ww^T\right\}G^TH^{TT} = (HGQ)^T \qquad (8.68)$$

so that for optimality the modified Kalman gain in Table 8.3 should be used. The optimal filter is therefore given by Table 8.3, i.e.

$$\dot{\hat{x}} = A\hat{x} + Bu + K\left(z_d - H_d\hat{x}\right)$$
$$= A\hat{x} + Bu + K\left(\dot{z} - A^*z - HBu - H_d\hat{x}\right) \qquad (8.69)$$

$$\dot{P} = AP + PA^T + GQG^T - KR_dK^T$$
$$= AP + PA^T + GQG^T - K\left(G^*G^{*T} + HGQG^TH^T\right)K^T \qquad (8.70)$$

$$K = \left[PH_d^T + G(HGQ)^T\right]\left(G^*G^{*T} + HGQG^TH^T\right)^{-1} \qquad (8.71)$$

A disadvantage of (8.69) is that the data must be differentiated. To avoid this, note that

$$\frac{d}{dt}(Kz) = \dot{K}z + K\dot{z} \qquad (8.72)$$

so that we can write (8.69) in the alternative form

$$\frac{d}{dt}(\hat{x} - Kz) = \dot{\hat{x}} - \dot{K}z - K\dot{z}$$
$$= A\hat{x} + Bu - \dot{K}z - K\left(A^*z + HBu + H_d\hat{x}\right) \qquad (8.73)$$

The block diagram of this version of the optimal filter is given in Fig. 8.6.

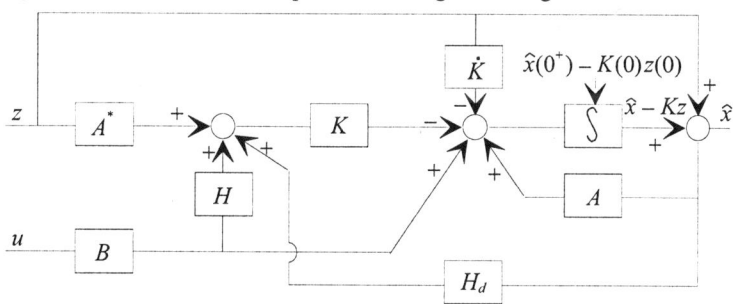

Fig. 8.6. Continuous Kalman filter with derived measurements

Regarding the initialization of this filter, not that the instant after measurement data are available is $t = 0^+$, so that to initialize the filter we must use

$$\hat{x}(0^+) = \hat{x}(0) + P_0H^T\left(HP_0H^T + R\right)^{-1}\left[z(0) - H\hat{x}(0)\right] \qquad (8.74)$$

$$P(0^+) = P_0 - P_0H^T\left(HP_0H^T + R\right)^{-1}HP_0 \qquad (8.75)$$

where $x(0) \sim (\hat{x}_0, P_0)$ and $E\left\{v(0)v^T(0)\right\} = R$. Thus, the initial condition of the filter is dependent upon the initial measurement $z(0)$ and cannot be determined a priori; that is

$$\text{initial   condition} = \hat{x}(0^+) - K(0)z(0) \tag{8.76}$$

Example 8.2 (Integrated gyroscope drift rate test data processor):

For a certain class of integrating gyroscopes, all units have a constant but a priori unknown drift rate, ε, once thermally stabilized. The gyroscopes are instrumented to stabilize a single-axis test table. Continuous indications of table angle, Θ, which is a direct measure of the integrated gyrosocope drift rate, are available. The table angle readout has an error, e, which is described by an exponential autocorrelation function model of standard deviation σ and short correlation time T. Design a real-time drift rate test data processor [17, 35].

The equation of motion of the test table, neglecting servo errors, is

$$\Theta = \varepsilon t + \Theta_0 \tag{I}$$

or equivalently

$$\dot{\Theta} = \varepsilon \; ; \; \dot{\varepsilon} = 0 \tag{II}$$

while the measurement is described by

$$z = \Theta + e \tag{III}$$

with

$$\dot{e} = -\frac{1}{T}e + w \; ; \; w \sim (0, q) \tag{IV}$$

where w is a white noise of zero-mean and spectral density $q = 2\sigma^2 / T$. As it is shown in section 4, the random system disturbance e has the autocorrelation function which is a decreasing exponential; that is

$$R_e(\tau) = E\{e(t+\tau)e(t)\} = \sigma^2 e^{-\frac{|\tau|}{T}} \tag{V}$$

Since the measurement noise e in (III) is correlated with accordance to (V), one possibility is to use the augmented system state approach, i.e. to augment the two differential equations in (II) with eq. (IV), resulting in a third-order system (i.e. augmented state vector is now three-state row vector)

$$\dot{x} = \begin{bmatrix} \dot{\Theta} \\ \dot{\varepsilon} \\ \dot{e} \end{bmatrix} = \begin{bmatrix} 0 & 1 & 0 \\ 0 & 0 & 0 \\ 0 & 0 & -\frac{1}{T} \end{bmatrix} \begin{bmatrix} \Theta \\ \varepsilon \\ e \end{bmatrix} + \begin{bmatrix} 0 \\ 0 \\ 1 \end{bmatrix} w = Ax + Gw \tag{VI}$$

$$z = \begin{bmatrix} 1 & 0 & 1 \end{bmatrix} \begin{bmatrix} \Theta \\ \varepsilon \\ e \end{bmatrix} = Hx \tag{VII}$$

A model of the system and measurement is given in Fig 8.7. Thus, in the reformulated problem, the matrix B in (8.58) is equal to zero and the equivalent measurement noise v in (8.59) is also zero. Correspondingly, in the measurement noise shaping filter (8.60) or (8.61a) we have $A^* = B^* = 0$ and the derived measurement is given by (8.64),

$$z_d = \dot{z} \tag{VIII}$$

According to (8.69)-(8.71) the optimal filter is now

$$\dot{\hat{x}} = A\hat{x} + K(\dot{z} - HA\hat{x}) = (I - KH)A\hat{x} + K\dot{z} \tag{IX}$$

$$\dot{P} = AP + PA^T + qGG^T - K\left(qHGG^T H^T\right)K^T$$
$$= AP + PA^T + qGG^T - qKK^T \tag{X}$$

$$K = \left(PA^T + qGG^T\right)H^T\left(qHGG^T H^T\right)^{-1} = \left(PA^T + qGG^T\right)H^T q^{-1} \tag{XI}$$

Here is used the fact that $HGG^T H^T = 1$. Generally, if $\left(HGG^T H^T\right)^{-1}$ does not exist, i.e. $\det\left(HGG^T H^T\right) = 0$, then we must differentiate the data once more to define a new derived measurement, and obtain the resulting filter.

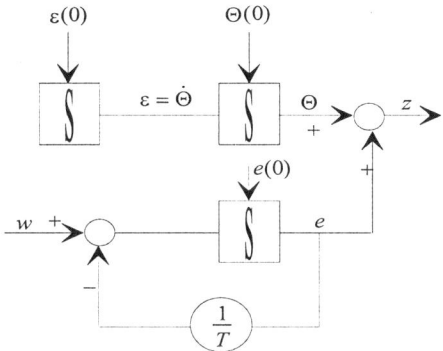

Fig. 8.7. Mathematical model of system and measurement

The same approach can be applied in the case of discrete-time systems, in order to avoid both the difficulty of singular measurement covariance matrix R and the undesirability of working with a high order, or augmented, system.

Suppose we are given system (8.47), (8.48) with $w_k \sim (0, Q)$ white, but with colored v_k generated by the shaping filter

$$x_{k+1}^* = A^* x_k^* + G^* w_k^* \tag{8.77}$$

$$v_k = x_k^* \tag{8.78}$$

with $w_k^* \sim (0, I)$ white and uncorrelated with w_k. This shaping filter can be found by factoring the spectrum of v_k. In this case, the derived measurement can be defined as (see eq. (8.64))

$$z_{k+1}^d = z_{k+1} - A^* z_k - HBu_k \tag{8.79}$$

Then

$$z_{k+1}^d = Hx_{k+1} + v_{k+1} - A^* Hx_k - A^* v_k - HBu_k$$
$$= H\left(Ax_k + Bu_k + Gw_k\right) + \left(A^* x_k^* + G^* w_k^*\right) - A^* Hx_k - A^* x_k^* - HBu_k \tag{8.80}$$
$$= \left(HA - A^* H\right)x_k + G^* w_k^* + HGw_k$$

Defining a new measurement matrix as

$$H^d = HA - A^* H \tag{8.81}$$

we can write eq. (8.80) as

$$z_{k+1}^d = H^d x_k + \left(G^* w_k^* + HGw_k \right) \tag{8.82}$$

In this equation the measurement noise $G^* w_k^* + HGw_k$ is zero-mean white with the covariance matrix

$$
\begin{aligned}
E\left\{ \left(G^* w_k^* + HGw_k \right) \left(G^* w_k^* + HGw_k \right)^T \right\} &= G^* E\left\{ w_k^* w_k^{*T} \right\} G^{*T} + HGE\left\{ w_k w_k^T \right\} G^T H^T \\
&= G^* G^{*T} + HGQG^T H^T
\end{aligned} \tag{8.83}
$$

Here is used the fact that w_k^* and w_k are uncorrelated. Therefore, (8.47) and (8.82) define a new system driven by zero-mean white noise. However, in this system, the measurement and process noises are correlated with

$$E\left\{ \left(G^* w_k^* + HGw_k \right) w_k^T \right\} = G^* E\left\{ w_k^* w_k^T \right\} + HGE\left\{ w_k w_k^T \right\} = HGQ \tag{8.84}$$

so the modified Kalman gain of the previous section should be used for optimality. It should be noted again that the advantage of the *derived measurement approach* is that the dimension of the state is not increased by adding a shaping filter.

To see how to implement the discrete Kalman filter with derived measurements, by Table 8.4 and (8.79), we have

$$
\begin{aligned}
\hat{x}_{k+1}(-) &= A\hat{x}_k(-) + Bu_k + K_k \left(z_k^d - H^d \hat{x}_k(-) \right) \\
&= A\hat{x}_k(-) + Bu_k + K_k \left[\left(z_k - A^* z_{k-1} - HBu_{k-1} \right) - H^d \hat{x}_k(-) \right] \\
&= \left(A - K_k H^d \right) \hat{x}_k(-) + Bu_k - K_k A^* z_{k-1} + K_k \left(z_k - HBu_{k-1} \right)
\end{aligned} \tag{8.85}
$$

An implementation of the recursion (8.85) is given in Fig. 8.8.

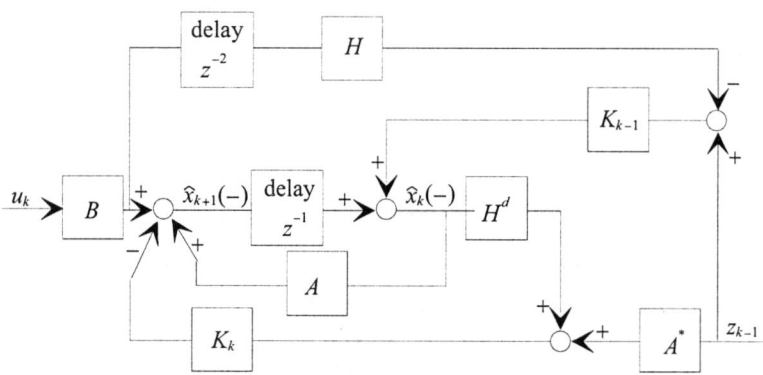

Fig. 8.8. An implementation of the discrete Kalman filter with derived measurements

To see this, note that the signal at the input to the one unit delay z^{-1} is

$$\left(A - K_k H^d \right) \hat{x}_k(-) + Bu_k - K_k A^* z_{k-1}$$

so the output of the delay is the signal delayed by 1, i.e.

$$\left(A - K_{k-1} H^d \right) \hat{x}_{k-1}(-) + Bu_{k-1} - K_{k-1} A^* z_{k-2}$$

Now it is easy to see from the figure that

$$\hat{x}_k(-) = \left(A - K_{k-1}H^d\right)\hat{x}_{k-1}(-) + Bu_{k-1} - K_{k-1}A^*z_{k-2} + K_{k-1}\left(z_{k-1} - HBu_{k-2}\right)$$

which is delayed version of (8.85).

8.2. Nonlinear minimum variance estimation (the extended Kalman filter)

In this section we develop an approximate solution to the nonlinear filtering problems that we define below. The solutions involve the linearization of the continuous and discrete nonlinear processes about a reference trajectory, and the modification or extension of the linear Kalman filtering algorithms using the linearized models. In this way, we develop the linearized and extended Kalman filters for a continuous-discrete, discrete and continuous nonlinear estimation problems.

8.2.1 Continuous-discrete linearized and extended Kalman filters

This subchapter extends the discussion of optimal estimation for linear systems to the more general case described by the nonlinear stochastic differential equation

$$\dot{x}(t) = a\left(x(t),t\right) + G(t)w(t) \tag{8.86}$$

The vector a is a nonlinear function of the state $x(t)$, and $\{w(t)\}$ is zero-mean white noise process having spectral density matrix $Q(t)$, i.e.

$$E\{w(t)\} = 0 \; ; \; E\{w(t)w^T(\tau)\} = Q(t)\delta(t-\tau) \tag{8.87}$$

with δ being the Dirac delta function. Given the continuous nature of a dynamical system and the requirement of microprocessors for discrete data, the continuous-discrete model is usually the most useful formulation for modern signal processing and control purposes. Therefore, we shall investigate the problem of estimating $x(t)$ from sampled nonlinear measurements of the form

$$z(k) = h_k\left(x(t_k)\right) + v(k) \; ; \; k = 1,2,... \tag{8.88}$$

where h_k depends upon both the time index k and the state at each sampling time t_k, and $\{v(k)\}$ is a white random sequence of zero-mean random variables with associated covariance matrix $R(k)$, i.e.

$$E\{v(k)\} = 0 \; ; \; E\{v(k)v^T(j)\} = R(k)\delta(k,j) \tag{8.89}$$

with $\delta(k,j)$ being the Kronecker's delta symbol. Moreover, it is assumed that the white noise processes $\{w(t)\}$ and $\{v(k)\}$ are uncorrelated with each others, and with the initial condition $x(0)$, representing a random vector with mean value $\bar{x}_0$ and the associated covariance matrix $P_0 = E\left\{\left[x(0) - \bar{x}_0\right]\left[x(0) - \bar{x}_0\right]^T\right\}$.

For simplicity, we assume the process noise matrix $G(t)$ is independent of $x(t)$. Theoretical treatments of this subject often deal with a more general version of eq. (8.86), where $G(\cdot)$ is a nonlinear matrix function of $x(t)$; that is $G = g\left(x(t),t\right)$, and $w(t)$ is again a vector white noise process. In this case a theory for estimating $x(t)$ cannot be developed within the traditional framework of mean-square stochastic calculus, which is used so far in this text, because the right hand side of eq. (8.86) is not integrable in the mean square sense, owing to the statistical properties of the term $g(x(t),t)w(t)$. This difficulty is overcome by formulating the nonlinear filtering problem within the context of Ito stochastic calculus, which provides consistent mathematical rules for integrating eq (8.86) in this case. However, a theoretical discussion of the later topic is beyond the scope of this text. Moreover, when the white noise term in eq. (8.86) is independent of $x(t)$, i.e. is equal to $G(t)w(t)$, the manipulations associated with mean square stochastic calculus can be applied, and we shall use this approach to obtain a computationally feasible algorithms for nonlinear estimation. Therefore, the main goal of this chapter is to provide insight into principles of nonlinear estimation theory, which will be useful for most practical problems. In this spirit, we circumvent the mathematical issues, raised by eq. (8.86), and take some liberties with notation, in order to make clear the essential ideas [1, 17, 23].

For several reasons, the problem of filtering for nonlinear systems is considerably more difficult and admits a wider variety of solutions than does the linear estimation problem. First of all, in the optimal filtering for linear stochastic systems with Gaussian statistics the optimal estimate of $x(t)$ for most reasonable Bayesian optimization criteria is the conditional mean of $x(t)$ given the measurement data (chapter 5). Furthermore, the Gaussian property implies that the conditional mean can be computed from a unique linear operation on the measurement data, e.g. the Kalman filter algorithm (chapter 7). Consequently, there is little theoretical justification for using a different data processing technique. By contrast, in the nonlinear problem $x(t)$ is generally not Gaussian; hence many Bayesian criteria lead to estimates that are different from the conditional mean. In addition, optimal estimation algorithms for nonlinear systems often cannot be expressed in closed form, requiring methods for approximating optimal nonlinear filters. In summary, the optimal estimation problem for nonlinear systems is in general very complicated, and only in a few special cases the algorithms exist which are easy to implement or understand.

Within the framework of the model in eqs. (8.86)-(8.89), this chapter considers the minimum variance estimation criteria, i.e. those which calculate the conditional mean of $x(t)$, that result in the extended Kalman filter. Fortunately, the extended Kalman filter has been found to yield accurate estimates in a number of important practical applications. Because of this experience and its similarity to the conventional Kalman filter, it is usually one of the first methods to be tried for any nonlinear filtering problem.

Thus, given equation of motion (8.86) and measurement data (8.88), we seek algorithms for calculating the minimum variance estimate of $x(t)$ as a function of time and the accumulated measurement data. Since the *minimum variance estimate* is always the *conditional mean* of the state vector (see chapter 5), regardless of its probability density function, suppose that the measurement at time t_{k-1} has just been processed and the corresponding value $\hat{x}(t_{k-1})$ of the conditional mean is known. Between times t_{k-1} and t_k, no measurements are taken and the state propagates according to eq. (8.86). By formally integrating eq. (8.86), we obtain

$$x(t) = x(t_{k-1}) + \int_{t_{k-1}}^{t} a[x(\tau),\tau]d\tau + \int_{t_{k-1}}^{t_k} G(\tau)w(\tau)d\tau \qquad (8.90)$$

Taking the expectation of both sides of eq. (8.90) conditioned on all the measurements $\{z(i), t_i < t_{k-1}\}$ taken up until time t_{k-1}, i.e.

$$\varepsilon(\cdot) = E\{\cdot|z(i); t_i \le t_{k-1}\} \qquad (8.91)$$

interchanging the order of expectation and integration, produces

$$\varepsilon\big[x(t)\big]=\varepsilon\big[x(t_{k-1})\big]+\int_{t_{k-1}}^{t}\varepsilon\big[a\big(x(\tau),\tau\big)\big]d\tau+\int_{t_{k-1}}^{t}G(\tau)\varepsilon\big(w(\tau)\big)d\tau$$

$$=\hat{x}(t_{k-1})+\int_{t_{k-1}}^{t}\varepsilon\big[a\big(x(\tau),\tau\big)\big]d\tau$$

since $\varepsilon\big[x(t_{k-1})\big]$ is the minimum variance estimate $\hat{x}(t_{k-1})$ and $\varepsilon\big[w(\tau)\big]=0$, or after differentiation

$$\frac{d}{dt}\varepsilon\big[x(t)\big]=\varepsilon\big(a\big(x(t),t\big)\big)\ ;\ t_{k-1}\le t<t_{k} \tag{8.92}$$

with the initial conditions for $t=t_{k-1}$

$$\varepsilon\big[x(t_{k-1})\big]=\hat{x}(t_{k-1}) \tag{8.93}$$

Therefore, on the interval $t_{k-1}\le t<t_{k}$, the conditional mean (8.91) of $x(t)$ is the solution to eq. (8.92), with the initial condition (8.93), which can be written more compactly as

$$\dot{\hat{x}}(t)=\hat{a}\big[x(t),t\big] \tag{8.94}$$

where the caret $(\hat{\ })$ denotes the conditional expectation operation (8.91) in eq. (8.92). Similarly, a differential equation for the conditional estimation error covariance matrix

$$P(t)=\varepsilon\big[\tilde{x}(t)\tilde{x}^{T}(t)\big]=\varepsilon\big[\big(x(t)-\hat{x}(t)\big)\big(x(t)-\hat{x}(t)\big)^{T}\big]$$

$$=\varepsilon\big[x(t)x^{T}(t)\big]-\hat{x}(t)\hat{x}^{T}(t) \tag{8.95}$$

is derived by differentiating both sides of eq. (8.95), interchanging the order of conditional expectation ε and differentiation, and substituting for $\dot{x}(t)$ and $\dot{\hat{x}}(t)$ from eqs. (8.86) and (8.94), respectively. The result is

$$\dot{P}=\frac{d}{dt}\varepsilon\big[xx^{T}\big]$$

or since (see chapter 4)

$$d\big(xx^{T}\big)=\big(x+dx\big)\big(x+dx\big)^{T}-xx^{T}=xdx^{T}+dxx^{T}+dxdx^{T}$$

$$=x\big(adt+Gwdt\big)^{T}+\big(adt+Gwdt\big)x^{T}+\big(adt+Gwdt\big)\big(adt+Gwdt\big)^{T}$$

one concludes

$$d\varepsilon\big[xx^{T}\big]=\varepsilon\big[xa^{T}\big]dt+\varepsilon\big[ax^{T}\big]dt+G\varepsilon\big[dw^{*}\big(dw^{*}\big)^{T}\big]G^{T}+o\big(dt^{2}\big)$$

where $wdt=dw^{*}$ with $\operatorname{cov}\{dw^{*}\}=Qdt$ (see chapter 4). Dividing by dt and taking the limits as dt goes to zero, gives the differential equation

$$\dot{P}=\varepsilon\big[ax^{T}\big]+\varepsilon\big[xa^{T}\big]+GQG^{T}-\hat{a}\hat{x}^{T}-\hat{x}\hat{a}^{T}\ ;\ t_{k-1}\le t<t_{k} \tag{8.96}$$

where the dependence of $x(\cdot)$ upon t, and $a(\cdot)$ upon x and t, is suppressed for convenient notation. The equation (8.96) can be rewritten as

$$\dot{P}(t)=\big(\widehat{xa^{T}}\big)-\hat{x}\hat{a}^{T}+\big(\widehat{ax^{T}}\big)-\hat{a}\hat{x}^{T}+G(t)Q(t)G^{T}(t)\ ;\ t_{k-1}\le t<t_{k} \tag{8.97}$$

where the caret $(\hat{\cdot})$ denotes the conditional expectation $\varepsilon(\cdot)$ in (8.91). Equations (8.94) and (8.97) are generalizations of the propagation equations for the linear estimation problem (see chapter 7). If $\hat{x}(t)$ and $P(t)$ can be calculated, they will provide both estimate of the state vector between measurements times and a measure of the estimation accuracy. Equations (8.94) and (8.97) are, in general, unsolvable, since they are not ordinary differential equations. To see this, note that (8.94) is really

$$\dot{\hat{x}} = \hat{a}(x,t) = \int_{-\infty}^{\infty} a(x,t) f\left(x(t)\big|z(i)\ ;\ t_i \le t_{k-1} < t\right)dx \tag{8.98}$$

which depends on the entire conditional probability density function f; that is on all the conditional moments of $x(t)$. Note the important point that in general

$$\hat{a}(x,t) \ne a(\hat{x},t) \tag{8.99}$$

In the linear case

$$\dot{x} = a(x,t) + G(t)w = Ax + Gw \tag{8.100}$$

we do have (note that $\varepsilon\{w\} = 0$)

$$\hat{a}(x,t) = \varepsilon\big[a(x,t)\big] = \varepsilon[Ax] = A\varepsilon[x] = A\hat{x} = a(\hat{x},t) \tag{8.101}$$

so the mean update is the familiar

$$\dot{\hat{x}} = A\hat{x} \tag{8.102}$$

That is, $\dot{\hat{x}}$ depends on upon $A(t)$ and $\hat{x}(t)$. Furthermore, in the linear case (8.97) reduces to the familiar Lyapunov equation,

$$\dot{P} = PA^T + AP + GQG^T \tag{8.103}$$

It follows that $\dot{P}(t)$ depends only upon $A(t), P(t)$ and $Q(t)$. Therefore, the differential equations (8.102) and (8.103) for the estimate and its covariance matrix, respectively, can be readily integrated in the linear case. However, in the more general nonlinear case (8.98) and (8.99) the entire conditional probability density function $f(\cdot|\cdot)$ for $x(t)$ must be known, in order to compute $\hat{a}(x,t)$. Therefore, to obtain practical estimation algorithms, methods of computing the mean $\hat{x}$ and covariance matrix P which do not depend upon knowing probability density function $f(\cdot|\cdot)$ for $x(t)$ are needed. A method often used to achieve this goal is to expand nonlinear system characteristics $a(\cdot)$ in a Taylor series about a known vector $x^*(t)$ that is close to $x(t)$. In particular, if $a(\cdot)$ is expanded about the current estimate, i.e., conditional mean, of the state vector $x(t)$, then $x^*(t) = \hat{x}(t)$ and

$$a(x,t) = a(\hat{x},t) + \frac{\partial a}{\partial x}\bigg|_{x=\hat{x}} (x-\hat{x}) + \cdots \tag{8.104}$$

where it is assumed that the required partial derivatives exist. Taking conditional expectation ε in (8.91) yields

$$\hat{a}(x,t) = a(\hat{x},t) + O(\cdot) + \dots$$

As a first order approximation to $\hat{a}(x,t)$ we can therefore take $a(\hat{x},t)$. Update (8.94) is thus replaced by

$$\dot{\hat{x}} = a(\hat{x},t) \; ; \; t_{k-1} \le t < t_k \qquad (8.105)$$

Up to this point $\hat{x}$ has denoted the exact conditional mean $\varepsilon[x(t)]$; henceforth, $\hat{x}$ denotes an estimate of the state that is an approximation to the conditional mean.

To determine a first-order approximation to (8.97), substitute for $a(x,t)$ using the first two terms of (8.104) and then take the indicated conditional expected values. Representing the Jacobian as

$$A(x,t) = \frac{\partial\, a(x,t)}{\partial\, x} \qquad (8.106)$$

this results in

$$a(x,t) = a(\hat{x},t) + A(\hat{x},t)(x-\hat{x}) + \cdots \qquad (8.107)$$

yielding

$$\dot{P} = \overline{x[a(\hat{x},t)+A(\hat{x},t)(x-\hat{x})]^T} - \hat{x}a^T(\hat{x},t) + \overline{[a(\hat{x},t)+A(\hat{x},t)(x-\hat{x})]x^T} - a(\hat{x},t)\hat{x}^T + GQG^T$$

$$= \overline{x(x-\hat{x})}A^T + \overline{A(x-\hat{x})x^T} + GQG^T + \left\{-\overline{\hat{x}(x-\hat{x})}A^T - \overline{A(x-\hat{x})\hat{x}^T}\right\}$$

$$= \overline{(x-\hat{x})(x-\hat{x})^T}A(\hat{x},t) + A(\hat{x},t)\overline{(x-\hat{x})(x-\hat{x})^T} + GQG^T$$

$$= PA^T(\hat{x},t) + A(\hat{x},t)P + GQG^T$$

where the sign ($\overline{\;\cdot\;}$) denotes the conditional expectation $\varepsilon(\cdot)$. Here we added the null terms in $\{\cdot\}$, i.e. $\overline{\hat{x}(x-\hat{x})} = \hat{x}\overline{(x-\hat{x})} = \hat{x}(\hat{x}-\hat{x}) = 0$ and $\overline{(x-\hat{x})\hat{x}^T} = (\hat{x}-\hat{x})\hat{x}^T = 0$.

Equations (8.105) and (8.108) represent an approximation, computationally feasible time update for the estimate $\hat{x}$ and error covariance P. The estimate simple propagates according to the nonlinear dynamics, and the error covariance propagates like that of a linear system with plant matrix $A(\hat{x},t)$. Thus, these equations have a structure similar to the Kalman filter propagation equations for linear systems. Consequently, they are referred to as the *extended Kalman filter* propagation equations.

Note that Jacobian $A(\cdot,t)$ is evaluated for each t at the current estimate, which is provided by (8.105), so that there is coupling between (8.105) and (8.108). To eliminate this coupling it is possible to introduce a further approximation and solve not (8.108) but instead

$$\dot{P}(t) = A\left[\hat{x}(t_{k-1}),t\right]P(t) + P(t)A^T\left[\hat{x}(t_{k-1}),t\right] + G(t)QG^T(t) \; ; \; t_{k-1} \le t < t_k \qquad (8.109)$$

In this equation, the Jacobian $A(\cdot,\cdot)$ is evaluated once using the estimate $\hat{x}(t_{k-1})$ after updating at t_{k-1} to include $z(k-1)$. This is used as the plant matrix for the time propagation over the entire interval $t_{k-1} \le t < t_k$ until the next measurement time t_k. Note that the time update (8.105), (8.108) or (8.109) can be accomplished very easily on a digital computer using a Runge-Kutta numerical integration method (see chapter 7).

To obtain a complete filtering algorithm, update equations which account for measurement data are needed. To develop update equations, assume that the estimate of $x(t)$ and its associated covariance matrix have been propagated using eqs. (8.105) and (8.108) or (8.109), and denote the solutions ate time t_k by $\hat{x}_k(-)$ and $P_k(-)$; that is $\hat{x}_k(-) = \hat{x}(t_k)$ and $P_k(-) = P(t_k)$. When the measurement $z(k)$ is taken, an improved estimate of the state is sought. Motivated by the linear estimation problem discussed in the chapter 7, we require that the updated estimate be a linear function of the measurement, i.e.

$$\hat{x}_k(+) = \hat{x}_k(-) + K_k\left[z(k) - \hat{z}_k(-)\right] \tag{8.110}$$

where $\hat{z}_k(-)$ is an anticipated value, on the basis of measurement equation (8.88), for the data vector $z(k)$

$$\begin{aligned}\hat{z}_k(-) &= \varepsilon^-\left[z(k)\right] = \varepsilon^-\left[h_k\left(x(t_k)\right) + v(k)\right] \\ &= \varepsilon^-\left[h_k\left(x(t_k)\right)\right]\end{aligned} \tag{8.111}$$

where

$$\varepsilon^-\left[h_k\left(x(t_k)\right)\right] = E\left\{h_k\left(x(t_k)\right)\middle|\left(z_i\,;\,t_i \le t_{k-1}\right)\right\} \tag{8.112}$$

and

$$\varepsilon^-\left[v(k)\right] = E\left\{v(k)\middle|\left(z(i)\,;\,t_i \le t_{k-1}\right)\right\} = E\left\{v(k)\right\} = 0$$

For notational ease define

$$x_k = x(t_k) \tag{8.113}$$

and suppress the k index of $h(\cdot)$. The eqs. (8.111) and (8.88) are then written as

$$\hat{z}_k(-) = \varepsilon^-\left[h_k(x_k)\right] = \hat{h}^-(x_k) \tag{8.114}$$

$$z(k) = h(x_k) + v(k) \tag{8.115}$$

so that the eq. (8.110) reduces to

$$\hat{x}_k(+) = \hat{x}_k(-) + K_k\left[z(k) - \hat{h}^-(x_k)\right] \tag{8.116}$$

Proceeding with arguments similar to those used in the section 7, we define the estimation error just before and just after update $\hat{x}_k(-)$ and $\hat{x}_k(+)$, respectively, by

$$\tilde{x}_k(+) = \hat{x}_k(+) - x_k \tag{8.117}$$

$$\tilde{x}_k(-) = \hat{x}_k(-) - x_k \tag{8.118}$$

Then eqs. (8.117) and (8.118) are combined with eqs. (8.115) and (8.116) to produce the following expression for the estimation error

$$\tilde{x}_k(+) = \tilde{x}_k(-) + K_k\left[h(x_k) - \hat{h}^-(x_k)\right] + K_k v(k) \tag{8.119}$$

It is desired that the a posterior estimate $\hat{x}_k(+)$ be *unbiased* if the a priori estimate $\hat{x}_k(-)$ is ; i.e. that $E\{\tilde{x}_k(+)\} = 0$ if $E\{\tilde{x}_k(-)\} = 0$. By taking expected values, and recognizing that

$$E\{\tilde{x}_k(-)\} = E\{v(k)\} = 0 \tag{8.120}$$

and

$$E\{h(x_k)\} = \int h(x_k) f(x_k, z_1, ..., z_{k-1}, z_k) dx_k dz_1 \cdots dz_k$$

$$= \int h(x_k) f(x_k z_k | z_1, ..., z_{k-1}) f(z_1, ..., z_{k-1}) dx_k dz_k dz_1 \cdots dz_{k-1}$$

$$= \int h(x_k) \int f(x_k, z_k | z_1, ..., z_{k-1}) dz_k f(z_1, ..., z_{k-1}) dx_k dz_1 \cdots dz_{k-1}$$

$$= \int h(x_k) f(x_k | z_1, ..., z_{k-1}) f(z_1, ..., z_{k-1}) dx_k dz_1 \cdots dz_{k-1} = E\{\varepsilon^-[h(x_k)]\} = E\{\hat{h}^-(x_k)\}$$

one obtains

$$E\{\tilde{x}_k(+)\} = 0 \tag{8.121}$$

This is consistent with the fact that the desired estimate is an approximation of the conditional mean. To determine the optimal gain matrix K_k, the same procedure used for the linear estimation problem in section 7 is employed. First an expression is obtained for the estimation error covariance matrix

$$P_k(t) = E\{\tilde{x}_k(+)\tilde{x}_k^T(+)\} \tag{8.122}$$

in terms of K_k; then K_k is chosen to minimize an appropriate function of $P_k(t)$. Applying the definition (8.122) to eq. (8.119), recognizing that $v(k)$ is uncorrelated with $\tilde{x}_k(-)$ and x_k, and using the relations

$$P_k(-) = E\{\tilde{x}_k(-)\tilde{x}_k^T(-)\} \tag{8.123}$$

$$R(k) = E\{v(k)v^T(k)\} \tag{8.124}$$

we obtain

$$P_k(+) = P_k(-) + K_k E\{[h(x_k) - \hat{h}^-(x_k)][h(x_k) - \hat{h}^-(x_k)]^T\} K_k^T$$

$$+ E\{\tilde{x}_k(-)[h(x_k) - \hat{h}^-(x_k)]^T\} K_k^T + K_k E\{[h(x_k) - \hat{h}^-(x_k)]\tilde{x}_k^T(-)\} + K_k R(k) K_k^T$$

The estimate being sought, the approximate conditional mean of $x(t)$, is a *minimum variance estimate*; that is, it minimizes the criterion

$$J_k = E\{\tilde{x}_k^T(+)\tilde{x}_k(+)\} = trace\{P_k(+)\} \tag{8.126}$$

Taking the trace of both sides of eq. (8.125), substituting the result into partial derivative of the trace of the product of matrices, one obtains (see chapter 7)

$$\frac{\partial J_k}{\partial K_k} = 2E\{[h(x_k) - \hat{h}^-(x_k)][h(x_k) - \hat{h}^-(x_k)]^T\} K_k$$

$$+ 2E\{\tilde{x}_k(-)[h(x_k) - \hat{h}^-(x_k)]^T\} + 2R(k) K_k = 0$$

from which follows the desired optimal gain matrix

$$K_k = -E\{\tilde{x}_k(-)[h(x_k) - \hat{h}(x_k)]^T\}\left[E\{[h(x_k) - \hat{h}^-(x_k)][h(x_k) - \hat{h}^-(x_k)]^T\} + R(k)\right]^{-1}$$

Substituting eq. (8.127) into eq. (8.125) produces, after some manipulation,

$$P_k(+) = P_k(-) + K_k \left[E\left\{ \left[h(x_k) - \hat{h}^-(x_k) \right] \left[h(x_k) - \hat{h}^-(x_k) \right]^T \right\} + R(k) \right] K_k^T$$
$$+ E\left\{ \tilde{x}_k(-) \left[h(x_k) - \hat{h}^-(x_k) \right]^T \right\} K_k^T + K_k E\left\{ \left[h(x_k) - \hat{h}^-(x_k) \right] \tilde{x}_k^T(-) \right\} \tag{8.128}$$

$$= P_k(-) - E\left\{ \tilde{x}_k(-) \left[h(x_k) - \hat{h}^-(x_k) \right]^T \right\} \left[E\left\{ \left[h(x_k) - \hat{h}^-(x_k) \right] \left[h(x_k) - \hat{h}^-(x_k) \right]^T \right\} + R(k) \right]^{-1} \times$$
$$\times \left[E\left\{ \left[h(x_k) - \hat{h}^-(x_k) \right] \left[h(x_k) - \hat{h}^-(x_k) \right]^T \right\} + R(k) \right] K_k^T$$
$$+ E\left\{ \tilde{x}_k(-) \left[h(x_k) - \hat{h}^-(x_k) \right]^T \right\} K_k^T + K_k E\left\{ \left[h(x_k) - \hat{h}^-(x_k) \right] \tilde{x}_k^T(-) \right\}$$

$$= P_k(-) + K_k E\left\{ \left[h(x_k) - \hat{h}^-(x_k) \right] \tilde{x}_k^T(-) \right\}$$

Equations (8.116), (8.127) and (8.128) together provide updating algorithms for the estimate when a measurement is taken. However, they are impractical to implement in this form because they depend upon the probability density function for $x(t)$ to calculate $h(x_k)$.

It can quickly be demonstrated that in the case of linear measurements
$$z(k) = Hx_k + v(k) \tag{8.129}$$
our equations reduces to the Kalman filter measurement update. In general, $\hat{h}^-(x_k) \neq h(\hat{x}_k(-))$; however, in the linear case
$$\hat{h}^-(x_k) = \varepsilon^- \left[h(x_k) \right] = \varepsilon^- (Hx_k) = H\varepsilon^-(x_k) = H\hat{x}_k(-) = h[\hat{x}_k(-)] \tag{8.130}$$
Substituting this into (8.127), (8.128) and (8.116) yields the measurement updates for the linear Kalman filter, i.e.
$$K_k = E\left\{ \tilde{x}_k(-) \tilde{x}_k^T(-) \right\} H^T \left[HE\left\{ \tilde{x}_k(-) \tilde{x}_k^T(-) \right\} H^T + R \right]^{-1}$$
$$= P_k(-) H^T \left[HP_k(-) H^T + R \right]^{-1}$$
$$P_k(+) = P_k(-) - K_k HE\left\{ \tilde{x}_k(-) \tilde{x}_k^T(-) \right\} = (I - K_k H) P_k(-)$$
$$\hat{x}_k(+) = \hat{x}_k(-) + K_k \left[z(k) - H\hat{x}_k(-) \right]$$
In order to find a measurement update that can be conveniently programmed, expand $h(x_k)$ in a Taylor series about $\hat{x}_k(-)$, the prior estimate at time t_k :
$$h(x_k) = h[\hat{x}_k(-)] + \left. \frac{\partial h}{\partial x} \right|_{x=\hat{x}_k(-)} (x_k - \hat{x}_k(-)) + \cdots \tag{8.131}$$
Represent the Jacobian as
$$H_k(x) = \frac{\partial h_k(x)}{\partial x} \tag{8.132}$$
To find a first-order approximation update, substitute for $h(x_k)$ in (8.116), (8.127) and (8.128), using the first two terms of (8.131) and carrying out the indicated expectation operations. Since
$$\hat{h}^-(x_k) = \varepsilon^- \left[h(x_k) \right] = \varepsilon^- \left\{ h(\hat{x}_k(-)) + H(\hat{x}_k(-)) \left[x_k - \hat{x}_k(-) \right] \right\}$$
$$= h[\hat{x}_k(-)] + H(\hat{x}_k(-)) \left[\varepsilon^-(x_k) - \hat{x}_k(-) \right] = h[\hat{x}_k(-)]$$
and
$$h(x_k) - \hat{h}^-(x_k) = H(\hat{x}_k(-)) \tilde{x}_k(-) \tag{8.133}$$

the result is

$$\hat{x}_k(t) = \hat{x}_k(-) + K_k\left[z(k) - h_k\left(\hat{x}_k(-)\right)\right]$$

$$K_k = P_k(-)H_k^T\left(\hat{x}_k(-)\right)\left[H_k\left(\hat{x}_k(-)\right)P_k(-)H_k^T\left(\hat{x}_k(-)\right) + R(k)\right]^{-1} \qquad (8.134)$$

$$P_k(+) = \left[I - K_kH_k\left(\hat{x}_k(-)\right)\right]P_k(-) \qquad (8.135)$$

Equations (8.133)-(8.135) constitute the *extended Kalman filtering* algorithm for nonlinear continuous-time system with discrete measurements. A summary of the mathematical model and the filter equations is given in the table 8.5. A deterministic input $u(t)$ is included in the equations for completeness.

Table 8.5. Continuous-discrete extended Kalman filter

system mode	$\dot{x}(t) = a\left(x(t), u(t), t\right) + G(t)w(t)$; $E\{w(t)\} = 0$; $E\{w(t)w^T(\tau)\} = Q(t)\delta(t-\tau)$
measurement model	$z(k) = h_k\left(x(t_k)\right) + v(k)$; $k = 1,2,\dots$ $E\{v(k)\} = 0$; $E\{v(k)v^T(j)\} = R(k)\delta(k,j)$
initial conditions	$E\{x(0)\} = \hat{x}_0$; $\tilde{x}(0) = x(0) - \hat{x}(0)$ $E\left\{\left[x(0) - \hat{x}_0\right]\left[x(0) - \hat{x}_0\right]^T\right\} = P_0$;
other assumptions	$E\{w(t)v^T(k)\} = 0$ for all k and t; $E\{w(t)\tilde{x}^T(0)\} = E\{v(k)\tilde{x}^T(0)\} = 0$
state estimate propagate	$\dot{\hat{x}}(t) = a\left(\hat{x}(t), u(t), t\right)$; $\hat{x}_k(-) = \hat{x}(t_k)$
error covariance matrix	$\dot{P}(t) = A\left(\hat{x}(t), t\right)P(t) + P(t)A^T\left(\hat{x}(t), t\right) + G(t)Q(t)G^T(t)$; $P_k(-) = P(t_k)$
state estimate update	$\hat{x}_k(+) = \hat{x}_k(-) + K_k\left[z(k) - h_k\left(\hat{x}_k(-)\right)\right]$
error covariance update	$P_k(+) = \left[I - K_kH_k\left(\hat{x}_k(-)\right)\right]P_k(-)$
gain matrix	$K_k = P_k(-)H_k^T\left(\hat{x}_k(-)\right)\left[H_k\left(\hat{x}_k(-)\right)P_k(-)H_k^T\left(\hat{x}_k(-)\right) + R(k)\right]^{-1}$
definitions	$A\left(\hat{x}(t), t\right) = \left.\dfrac{\partial\, a\left(x(t), u(t), t\right)}{\partial x(t)}\right|_{x(t) = \hat{x}(t)}$ $H_k\left(\hat{x}_k(-)\right) = \left.\dfrac{\partial\, h_k\left(x(t_k)\right)}{\partial x(t_k)}\right|_{x(t_k) = \hat{x}_k(-)}$

An important advantage of the *continuous-discrete extended Kalman filter* (EKF) is that the optimal estimate is available continuously at all times, including times between the measurements times t_k. Note that if $a(x, u, t)$ and $h_k(x)$ are linear, then the EKF reverts to the continuous-discrete Kalman filter in section 7.

Since the time and measurement updates depend on Jacobians evaluated at the current estimate, the error covariance P and Kalman gain K cannot be computed off-line for the EKF. They must be computed in real time as the data become available.

It should be realized that the measurement times t_k need not be equally spaced. The time update (state estimate propagate) is performed over any interval during which no data are available; when data become available, a measurement update (state estimate update) is performed. This means that the cases of intermittent or missing measurements, and pure prediction in the absence of data, can easily be dealt with using the EKF.

Higher-order approximations to the optimal nonlinear updates can also be derived by retaining higher-order terms in the Taylor series expansions.

Table 8.6: Summary of continuous-discrete linearized Kalman filter

System model	$\dot{x}(t) = a\big(x(t), u(t), t\big) + G(t)w(t);$ $E\{w(t)\} = 0 \,;\, E\{w(t)w^T(\tau)\} = Q(t)\delta(t - \tau)$
Measurement model	$z(k) = h_k\big(x(t_k)\big) + v(k) \,;\, k = 1, 2, \ldots$ $E\{v(k)\} = 0 \,;\, E\{v(k)v^T(j)\} = R(k)\delta(k, j)$
Initial conditions	$E\{x(0)\} = \hat{x}_0 \,;\, E\{\tilde{x}(0)\tilde{x}^T(0)\} = P_0 \,;\, \tilde{x}(0) = x(0) - \hat{x}_0$
Other assumptions	$E\{w(t)v^T(k)\} = 0 \;\text{ for all } k \text{ and } t$ $E\{w(t)\tilde{x}^T(0)\} = E\{v(k)\tilde{x}^T(0)\} = 0 \;\text{ for all } t \text{ and } k$
State estimate propagate (time update)	$\dot{\hat{x}}(t) = a\big(x_n(t), u(t), t\big) + A\big(x_n(t), t\big)\big[\hat{x}(t) - x_n(t)\big]$
Error covariance matrix	$\dot{P}(t) = A\big(x_n(t), t\big)P(t) + P(t)A^T\big(x_n(t), t\big) + G(t)Q(t)G^T(t)$
State estimate update (measurement update)	$\hat{x}_k(+) = \hat{x}_k(-) + K_k\big[z(k) - h_k\big(x_n(t_k)\big) - H_k\big(x_n(t_k)\big)\big[\hat{x}_k(-) - x_n(t_k)\big]\big]$
Error covariance update	$P_k(+) = \big[I - K_k H_k\big(x_n(t_k)\big)\big]P_k(-)$
Gain matrix	$K_k = P_k(-)H_k^T\big(x_n(t_k)\big)\big[H_k\big(x_n(t_k)\big)P_k(-)H_k^T\big(x_n(t_k)\big) + R(k)\big]^{-1}$
Definitions	$A\big(x_n(t), t\big) = \dfrac{\partial a\big(x(t), u(t), t\big)}{\partial x(t)}\bigg|_{x(t) = x_n(t)}$ $H_k\big(x_n(t_k)\big) = \dfrac{\partial h_k\big(x(t_k)\big)}{\partial x(t_k)}\bigg|_{x(t_k) = x_n(t_k)}$

In some cases the nominal trajectory $x_n(t)$ of the state $x(t)$ is known a priori. An example is a robot arm for welding a car door which always moves along the same path. In this case the EKF equations can be used with $\hat{x}(t)$ replaced by $x_n(t)$. Since $x_n(t)$ is known beforehand, the error covariance and K_k can now be computed off-line before the measurements are taken. This procedure of linearizing about a known nominal trajectory results in what is called the *linearized Kalman filter* (LKF). If a controller is employed to keep the state $x(t)$ on the nominal trajectory $x_n(t)$, the LKF performs quite well. Summary of continuous-discrete linearized Kalman filter (LKF) is given in the table below.

Generally speaking, the LKF procedure yields less filtering accuracy than the EKF because $x_n(t)$ is usually not as close to the actual trajectory as is $\hat{x}(t)$.

A block diagram of a software implementation of the continuous-discrete EKF is shown in Fig. 8.9. A driver program, implementing the EKF along the lines of Fig. 8.9., can be written, which is the same for every problem. Subroutines containing the continuous dynamics and the measurement details depend on the particular problem. The next example illustrates how to write the MATLAB program in order to use the EKF.

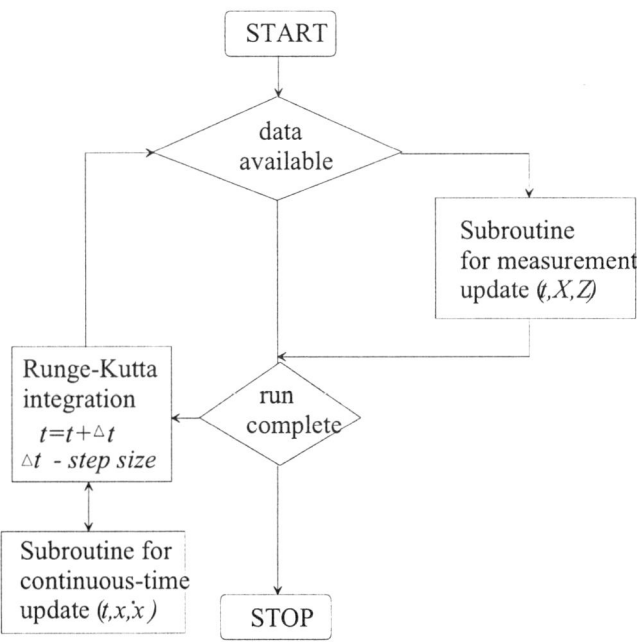

Fig. 8.9. Software implementation of the continuous-discrete extended Kalman filter (EKF)

Example 8.3 (Two dimensional tracking problem): In this example, we consider the problem of tracking a body falling freely through the atmosphere. The motion is modeled in two dimensions by assuming the body falls in a straight line, and the range measurement is made along the line-of-sight of radar, as illustrated in Fig. 8.10.

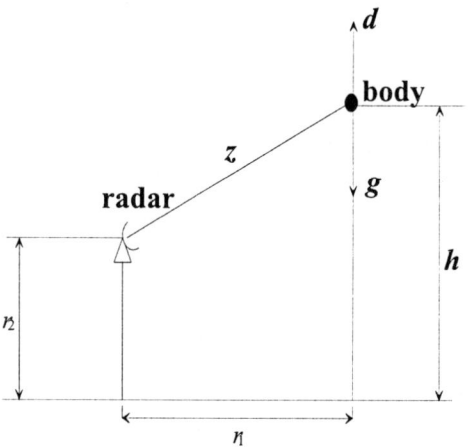

Fig. 8.10: Tracking a free-falling body

The state variables for this problem are designed as

$$x_1 = h \ , \ x_2 = \dot{h} \ , \ x_3 = \beta \tag{I}$$

where β is the so-called ballistic coefficient of the falling body and h is its height above the earth. The ballistic coefficient is included as a state variable cause it is not well known and so it must be estimated. The state equations are

$$\dot{x}_1 = \dot{h} = x_2 \tag{IIa}$$

$$\dot{x}_2 = \ddot{h} = d - g \tag{IIb}$$

$$\dot{x}_3 = 0 \tag{IIc}$$

where g is acceleration of gravity, and d is drag deceleration

$$d = \frac{\rho \dot{h}^2}{2\beta} = \frac{\rho x_2^2}{2x_3} \ ; \ \rho = \rho_0 \exp\left(-\frac{h}{k_\rho}\right) \tag{III}$$

with ρ being atmospheric density (ρ_0 is the atmospheric density at the sea level) and k_ρ is a decay constant. The differential equation for velocity, x_2, is nonlinear through the dependence of drag on velocity, air density and ballistic coefficient. The equations of motion for body (II) can be rewritten in the matrix form

$$\dot{x} = \begin{bmatrix} \dot{x}_1 \\ \dot{x}_2 \\ \dot{x}_3 \end{bmatrix} = \begin{bmatrix} x_2 \\ d-g \\ 0 \end{bmatrix} + \begin{bmatrix} 0 \\ 0 \\ w_3 \end{bmatrix} = a(x) + w \tag{IV}$$

where we introduced the white noise $w_3 \sim (0, q)$ to account for the small changes in β. Initial values of the state variables are assumed to have mean value $\hat{x}_0$ and the covariance matrix P_0, i.e. $x(0) \sim (\hat{x}_0, P_0)$ with $\hat{x}_0^T = [\hat{x}_{10} \ \hat{x}_{20} \ \hat{x}_{30}]$; $P_0 = diag\{P_{110} \ P_{220} \ P_{330}\}$. Suppose range measurements are taken every second (Fig. 8.11), is

$$z_k = \sqrt{r_1^2 + \left(h_k - r_k\right)^2} + v_k = \sqrt{r_1^2 + \left(x_1\left(t_k\right) - r_k\right)^2} + v_k = h\left(t_k\right) + v_k \qquad \text{(V)}$$

with $t_k = kT$, $T=1$s and $v_k \sim \left(0, r\right)$.

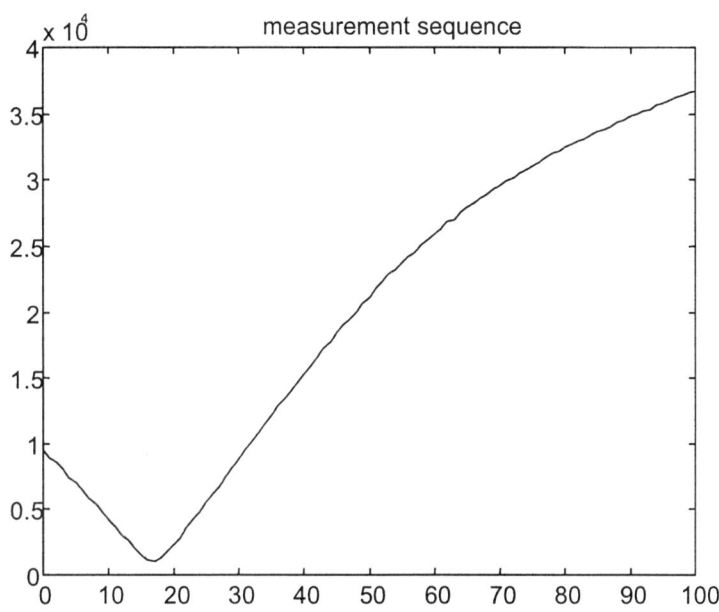

Fig. 8.11: Range measurements

The problem of estimating all the states may be solved using both the extended and linearized Kalman filters (Tabs. 8.5 and 8.6). Recall that the extended Kalman filter is linearized about the current estimate of the state, whereas the linearized filter utilities a precomputed nominal trajectory $x_n\left(t\right)$. The latter can be derived by solving eq. (IV) with $w = 0$, i.e.

$$\dot{x}_n = a\left(x_n\right); \; x_n\left(0\right) = \hat{x}_0 \qquad \text{(VI)}$$

Actual and estimated states are shown in Fig. 8.12, while MATLAB programs for realizing the extended filter are given in Fig 8.13. The simulation parameter values are the following:

$$\rho_0 = 1220\, g/m^3; K_\rho = 10263m; g = 9.81m/s^2; q = 1000g^2/m^2s^6$$
$$x_{10} \sim \left(10000m, 50m^2\right); \; x_{20} \sim \left(-500m/s, 200\, m^2/s^2\right);$$
$$x_{30} \sim \left(6\cdot10^7\, g/ms^2, 2\cdot10^{12}\, g^2/m^2s^4\right); r = 5000m^2/Hz; r_1 = 1000m; r_2 = 500m$$

We note that neither filter tracks β accurately early in the trajectory. This is due to the fact that the thin atmosphere at high altitude produces a small drag force on the body, so that the measurements contain little information about β. Evidently, both the filters yield better estimates as the body falls into the earth's denser atmosphere. This occurs because both the measurement and dynamic nonlinearities become stronger as the altitude decreases. Moreover, the extended Kalman filter gives better estimates than the linearized one. This is consistent with the behaviour of the associated covariance matrices computed from Tabs. 8.5 and 8.6. This is demonstrated by the fact that the estimated states tend to remain between the lines, which are the square roots $\pm\sqrt{P_{ii}}$ of the corresponding diagonal elements in the computed covariance error matrix P. In contrast, the estimates for the linearized filter tend to have larger errors than predicted by the covariance error matrix P. In summary, the extended Kalman filter generally performs better than a filter linearized about a precomputed nominal trajectory. However, the latter is more easily mechanized because filters gains can be precomputed and stored. Thus, there is a trade-off between filter complexity and estimation accuracy.

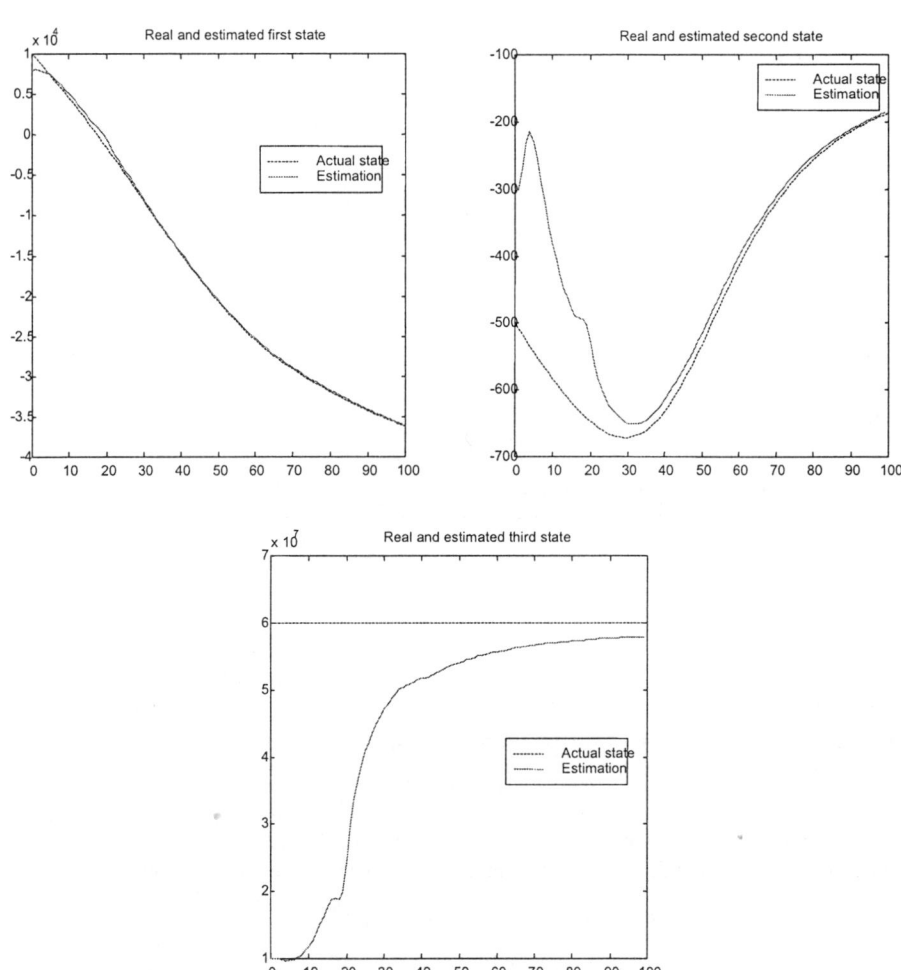

Fig. 8.12: The actual and estimated system states

```
function ex83                                    [t,x]=ode45('model',[t0:T:tf],x0);
t0=0; %starting simulation time
tf=100; %final simulation time                   figure(1);plot(t0:T:tf,x(:,1));
x0=[10000;-500;6*10^7]; %initial state           title('time history of the first state');
T=1; %sampling period                            figure(2);plot(t0:T:tf,x(:,2));
```

```
title('time history of the second state');
figure(3);plot(t0:T:tf,x(:,3));
title('time history of the third state');
% forming of measurements
r1=1000;r2=500;r=5000;
for t=t0:T:tf
    i=round(t/T)+1;
    z(i)=sqrt(r1^2+(x(i,1)-
r2)^2)+sqrt(r)*randn;
end
figure(4);plot(t0:T:tf,z);
title('measurement sequence');

% initialization of Kalman fitler
xp(1:3,1)=[8000; -300; 10000000];
pp=diag([500 200 2*10^12]);
xm(1:3,1)=xp(1:3,1);pm=pp;
q=1000;r=5000;

%Kalman filter
N=round((tf-t0)/T);
for i=2:N
    % state estimation and error covariance
update
    h=sqrt(r1^2+(xm(1,i-1)-r2)^2);
    H=[(xm(1,i-1)-r2)/h  0 0];
    K=pm*H'*inv(H*pm*H'+r);
    xp(1:3,i)=xm(1:3,i-1)+K*(z(i)-h);
    pp=(eye(3)-K*H)*pm;
    % state estimate propagate and error
covariance matrix
    tmp0=[xp(1:3,i)'    pp(1,1:3)    pp(2,2:3)
pp(3,3)]';
    [t,tmp]=ode45('model1',[i*T
(i+1)*T],tmp0);
    nn=length(t);
    xm(1:3,i)=tmp(nn,1:3)';
    pm=[tmp(nn,4:6);tmp(nn,5)    tmp(nn,7:8);
tmp(nn,6) tmp(nn,8:9)];
```

```
end
figure(5);
plot(t0:T:tf,x(:,1),'--',t0:T:tf-T,xp(1,:));
title('Real and estimated first state');
legend('Actual state', 'Estimation');
figure(6);
plot(t0:T:tf,x(:,2),'--',t0:T:tf-T,xp(2,:));
title('Real and estimated second state');
legend('Actual state', 'Estimation');
figure(7);
plot(t0:T:tf,x(:,3),'--',t0:T:tf-T,xp(3,:));
title('Real and estimated third state');
axis([0,100,10000000,70000000]);
legend('Actual state', 'Estimation');

function xdot=model(t,x);
% definition of parameters
r0=1220; kr=10263; g=9.81;
xdot(1)=x(2);
xdot(2)=r0*exp(-x(1)/kr)*x(2)^2/(2*x(3))-g;
xdot(3)=0;
xdot=xdot';

function xdot=model1(t,x);
ro=1220;kr=10263; g=9.81; q=1000; r=5;
a1=-ro*x(2)^2*exp(-x(1)/kr)/(2*kr*x(3));
a2=ro*x(2)*exp(-x(1)/kr)/x(3);
a3=-ro*exp(-x(1)/kr)*x(2)^2/(2*x(3)^2);
xdot(1)=x(2);
xdot(2)=ro*exp(-x(1)/kr)*x(2)^2/(2*x(3))-g;
xdot(3)=0;
xdot(4)=2*x(5);
xdot(5)=x(7)+a1*x(4)+a2*x(5)+a3*x(6);
xdot(6)=x(8);
xdot(7)=2*(a1*x(5)+a2*x(7)+a3*x(8));
xdot(8)=a1*x(6)+a2*x(8)+a3*x(9);
xdot(9)=q;
xdot=xdot';
```

Fig. 8.13: MATLAB programe code for extended Kalman filter

It is interesting to comment the way of calculating the quantities $\hat{x}_k^-$ and P_k^- for the problem concerned. Namely, these quantities are calculated as the solutions of a vector nonlinear differential equation, using Runge-Kutta method, that is realized by MATLAB function *ode45.m*. For our example, the first three differential equations correspond to the system states and are given by:

$$\dot{x}_1 = x_2$$

$$\dot{x}_2 = \frac{\rho_0 \exp\left(-x_1 / k_\rho\right) x_2^2}{2 x_3} - g$$

$$\dot{x}_3 = 0$$

If we assume the form of the covariance matrix:

$$P = \begin{bmatrix} p_1 & p_2 & p_3 \\ p_2 & p_4 & p_5 \\ p_3 & p_5 & p_6 \end{bmatrix}$$

the matrix differential equation:

$$\dot{P} = AP + PA^T + GQG^T$$

reduces to the set of six scalar differential equations:

$$\dot{p}_1 = 2 p_2$$
$$\dot{p}_2 = p_4 + a_1 p_1 + a_2 p_2 + a_3 p_3$$
$$\dot{p}_3 = p_5$$
$$\dot{p}_4 = 2\left(a_1 p_2 + a_2 p_4 + a_3 p_5\right)$$
$$\dot{p}_5 = a_1 p_3 + a_2 p_5 + a_3 p_6$$
$$\dot{p}_6 = q$$

where

$$A = \frac{\partial a(x)}{\partial x} = \begin{bmatrix} 0 & 1 & 0 \\ a_1 & a_2 & a_3 \\ 0 & 0 & 0 \end{bmatrix}; \ a_i = \frac{\partial(\dot{x}_2)}{\partial x_i}; \ i = 1, 2, 3.$$

The expressions for a_i; $i = 1, 2, 3$ are given in the MATLAB code.

8.2.2 Discrete linearized and extended Kalman filters

In this chapter we develop the linearized and extended Kalman filters for a discrete nonlinear systems described by a stochastic nonlinear difference equation in state-space form

$$x(k) = a\big(x(k-1)\big) + b\big(u(k-1)\big) + w(k-1) \tag{8.136}$$

with the corresponding measurement model

$$z(k) = h\big(x(k)\big) + v(k) \tag{8.137}$$

where $a(\cdot)$, $b(\cdot)$ and $h(\cdot)$ are nonlinear vector functions of vector arguments x, u and x, and $w(k)$ and $v(k)$ are zero-mean white with covariance matrices $Q(k)$ and $R(k)$, respectively.

First, we linearize the process and measurement models (8.136) and (8.137) using a deterministic reference trajectory $\left(x_n(k), u_n(k)\right)$. In practice, the reference trajectory is obtained either by developing a mathematical model of the process or by simulating about some reasonable operating conditions to generate the trajectory using

$$x_n(k) = a\big(x_n(k-1)\big) + b\big(u_n(k-1)\big)$$ (8.138)

Consider deviations from this trajectory given by

$$\delta x(k) = x(k) - x_n(k)$$ (8.139)

$$\delta u(k) = u(k) - u_n(k)$$ (8.140)

Substituting eqs. (8.136) and (8.138) into eq. (8.139), we obtain the deviation or perturbation trajectory as

$$\delta x(k) = a\big(x(k-1)\big) - a\big(x_n(k-1)\big) + b\big(u(k-1)\big) - b\big(u_n(k-1)\big) + w(k-1)$$ (8.141)

Expanding $a(\cdot)$ and $b(\cdot)$ in a Taylor series about the reference (x_n, u_n) and using only the first two terms of the series expansion, i.e. first-order Taylor series expansion, we have

$$a\big(x(k-1)\big) = a\big(x_n(k-1)\big) + A\big(x_n(k-1)\big)\delta x(k-1)$$ (8.142)

$$b\big(u(k-1)\big) = b\big(u_n(k-1)\big) + B\big(x_n(k-1)\big)\delta u(k-1)$$ (8.143)

where A and B are the Jacobian matrices defined by

$$A\big(x_n(k-1)\big) = \frac{da\big(x_n(k-1)\big)}{dx(k-1)} \; ; \; B\big(u_n(k-1)\big) = \frac{db\big(u_n(k-1)\big)}{du(k-1)}$$ (8.144)

Substituting eqs. (8.142) and (8.143) into eq. (8.141), we obtain the linearized system

$$\delta x(k) = A\big(x_n(k-1)\big)\delta x(k-1) + B\big(u_n(k-1)\big)\delta u(k-1) + w(k-1)$$ (8.145)

Similarly, the measurement system (8.137) can be linearized, by applying the first-order Taylor series expansion

$$h\big(x(k)\big) = h\big(x_n(k)\big) + H\big(x_n(k)\big)\delta x(k)$$ (8.146)

so that the measurement deviation

$$\delta z(k) = z(k) - z_n(k) = z(k) - h\big(x_n(k)\big)$$ (8.147)

is given by

$$\delta z(k) = H\big(x_n(k)\big)\delta x(k) + v(k)$$ (8.148)

where H is the measurement Jacobian

$$H\big(x_n(k)\big) = \frac{dh\big(x_n(k)\big)}{dx(k)}$$ (8.149)

We can also use linearization technique to approximate the statistics of eqs. (8.136) and (8.137). If we use the first-order Taylor series expansions of (8.142), (8.143) and (8.146) about the expectation $m_x(k) = E\{a\big(x(k-1)\big)\}$ rather than $x_n(k)$, then

$$m_x(k) = E\{a\big(x(k-1)\big)\} + E\{b\big(u(k-1)\big)\} + E\{w(k-1)\}$$ (8.150)

becomes

$$m_x(k) = a\big(m_x(k-1)\big) + b\big(u(k-1)\big)$$ (8.151)

since $E\{w(k)\} = 0$ and

$$
\begin{aligned}
E\{a(x(k-1))\} &= E\{a(m_x(k-1)) + A(m_x(k-1))(x(k-1) - m_x(k-1))\} \\
&= a(m_x(k-1)) + A(m_x(k-1))(m_x(k-1) - m_x(k-1)) \\
&= a(m_x(k-1))
\end{aligned}
$$

Similarly, a difference equation for the estimation error covariance matrix

$$
P_x(k) = E\{\tilde{x}(k)\tilde{x}^T(k)\} \tag{8.152}
$$

where

$$
\tilde{x}(k) = x(k) - m_x(k) \tag{8.153}
$$

can also be derived in a similar fashion. Substituting eqs. (8.136) and (8.153), we have

$$
\tilde{x}(k) = a(x(k-1)) - a(m_x(k-1)) + w(k-1) \tag{8.154}
$$

The use of the first-order Taylor series expansion of $a(x(k-1))$ about $m_x(k-1)$ yields

$$
a(x(k-1)) = a(m_x(k-1)) + A(m_x(k-1))\tilde{x}(k-1) \tag{8.155}
$$

where A is the state Jacobian in (8.144), so that

$$
\tilde{x}(k) = A(m_x(k-1))\tilde{x}(k-1) + w(k-1) \tag{8.156}
$$

Substituting (8.156) into (8.152), and taking into account that $\tilde{x}(k-1)$ and $w(k-1)$ are uncorrelated, one obtains

$$
P_x(k) = A(m_x(k-1))P_x(k-1)A^T(m_x(k-1)) + Q(k-1) \tag{8.157}
$$

Using the same approach, we can derive the measurement statistics $m_z(k) = E\{z(k)\}$ and $P_z(k) = E\{\tilde{z}(k)\tilde{z}^T(k)\}$ where $\tilde{z}(k) = z(k) - m_z(k)$. If we use the first order Taylor series expansion of $h(x(k))$ around $m_x(k)$, we have

$$
h(x(k)) = h(m_x(k)) + H(m_x(k))\tilde{x}(k) \tag{8.158}
$$

where H is the measurement Jacobian in (8.149) and $\tilde{x}$ is defined by (8.153), so that

$$
\begin{aligned}
m_z(k) &= E\{h(x(k))\} + E\{v(k)\} \\
&= h(m_x(k)) + H(m_x(k))E\{\tilde{x}(k)\} + E\{v(k)\} \\
&= h(m_x(k))
\end{aligned} \tag{8.159}
$$

Here is used the fact that $E\{v(k)\} = 0$ and $E\{\tilde{x}(k)\} = 0$. Furthermore,

$$
\tilde{z}(k) = z(k) - m_z(k) = h(x(k)) + v(k) - h(m_x(k)) \tag{8.160}
$$

Substituting (8.158) into (8.160), we have

$$
\tilde{z}(k) = H(m_x(k))\tilde{x}(k) + v(k) \tag{8.161}
$$

from which one obtains

$$
P_z(k) = H(m_x(k))P_x(k)H^T(m_x(k)) + R(k) \tag{8.162}
$$

since $\tilde{x}(k)$ and $v(k)$ are uncorrelated.

We summarize these results into Table 8.7. Note that one can use the MATLAB software package and these relations to simulate nonlinear stochastic discrete system (see example 8.4).

Suppose we utilize the perturbation model of eqs. (8.145) and (8.148) to construct a *linearized discrete Kalman* estimator using (A, B, C) Jacobians linearized about the nominal trajectory (x_n, u_n). Substituting the Jacobians and $\delta\hat{x}$ for the matrices and $\hat{x}$ in Table 7.2 we obtain a form of linearized discrete Kalman filter for $\delta\hat{x}$; that is, the estimated perturbation. This filter gives the *minimum variance estimate* $\delta\hat{x}$ of δx. From eq. (8.139), we obtain the linearized estimate of x; that is

$$\hat{x}_k(-) = \delta\hat{x}_k(-) + x_n(k) \tag{8.163}$$

The linearized filtering algorithm is not typically implemented in this manner because we are usually interested in $\hat{x}$ not $\delta\hat{x}$. We can rewrite the algorithm in terms of $\hat{x}$ if we substitute the prediction equation of the linearized filter for $\delta\hat{x}$ (Tab. 7.2) and eq. (8.138) in eq. (8.163).

Table 8.7. Approximate nonlinear stochastic discrete-time system model

State propagation	$x(k) = a(x(k-1)) + b(u(k-1)) + w(k-1)$; $w(k) \sim (0, Q(k))$
State mean propagation	$m_x(k) = a(m_x(k-1)) + b(u(k-1))$
State covariance propagation	$P_x(k) = A(m_x(k-1)) P_x(k-1) A^T(m_x(k-1)) + Q(k-1)$
Measurement propagation	$z(k) = h(x(k)) + v(k); v(k) \sim (0, R(k))$
Measurement mean propagation	$m_z(k) = h(m_x(k))$
Measurement covariance propagataion	$P_z(k) = H(m_x(k)) P_x(k) H^T(m_x(k)) + R(k)$
Initial conditions	$x(0), P(0)$
Other assumptions	$E\{w(k)v^T(j)\} = E\{w(k)x^T(0)\} = E\{v(k)x^T(0)\} = 0$ for all k, j
Definitions	$A(m_x(k)) = \dfrac{da(m_x(k))}{dm_x(k)}$; $H(m_x(k)) = \dfrac{dh(m_x(k))}{m_x(k)}$

That is, substitute (see Tab. 7.2)

$$\delta\hat{x}_k(-) = A(x_n(k-1))\delta\hat{x}_{k-1}(+) + B(u_n(k-1))\delta u(k-1)$$

and

$$x_n(k) = a(x_n(k-1)) + b(u(k-1))$$

into eq. (8.163) to obtain

$$\begin{aligned}
\hat{x}_k(-) &= a(x_n(k-1)) + A(x_n(k-1))[\hat{x}_{k-1}(+) - x_n(k-1)] \\
&+ b(u_n(k-1)) + B(u_n(k-1))[u(k-1) - u_n(k-1)]
\end{aligned} \tag{8.164}$$

The corresponding innovation can be found by substituting eqs. (8.147) and (8.163) into

$$\delta\varepsilon(k)=\delta z(k)-H\left(x_n(k)\right)\delta\hat{x}_k(-)=z(k)-h\left(x_n(k)\right)-H\left(x_n(k)\right)\left[\hat{x}_k(-)-x_n(k)\right] \quad (8.165)$$

Since $\delta\hat{x}_k(+)$ and $\delta\hat{x}_k(-)$ are both functions of $x_n(k)$, the standard corrected-state estimate equation (Tab. 7.2) evolves. Note also that $\delta P_k(-)=P_k(-)$ and $\delta P_k(+)=P_k(+)$, since

$$\delta\tilde{x}_k(-)=\delta x(k)-\delta\hat{x}_k(-)=\left[x(k)-x_n(k)\right]-\left[\hat{x}_k(-)-x_n(k)\right]$$
$$=x(k)-\hat{x}_k(-)=\tilde{x}_k(-) \quad (8.166a)$$

and

$$\delta\tilde{x}_k(+)=\delta x(k)-\delta\hat{x}_k(+)=\left[x(k)-x_n(k)\right]-\left[\hat{x}_k(+)-x_n(k)\right]$$
$$=x(k)-\hat{x}_k(+)=\tilde{x}_k(+) \quad (8.166b)$$

We summarize this *discrete linearized Kalman filter* algorithm in Table 8.8.

If, instead of using the reference trajectory $\left(x_n(k),u_n(k)\right)$, we choose to linearize about each new estimate, as soon as it becomes available, then *discrete extended Kalman filter* algorithm results, instead of the discrete linearized one. As mentioned before, the reason for choosing the estimate to linearize about each new estimate is to use a better reference trajectory, as soon as one is available. As a consequence, large initial estimation errors do not propagate and, therefore, linearity assumption are less likely to be violated.

Table 8.8 : Discrete linearized Kalman filter

State prediction	$\hat{x}_k(-)=a\left(x_n(k-1)\right)+A\left(x_n(k-1)\right)\left(\hat{x}_{k-1}(+)-x_n(k-1)\right)$ $+b\left(u_n(k-1)\right)+B\left(u_n(k-1)\right)\left(u(k-1)-u_n(k-1)\right)$
Covariance prediction	$P_k(-)=A\left(x_n(k-1)\right)P_{k-1}(+)A^T\left(x_n(k-1)\right)+Q(k-1)$
Innovation	$\varepsilon(k)=z(k)-h\left(x_n(k)\right)-H\left(x_n(k)\right)\left(\hat{x}_k(-)-x_n(k)\right)$
Innovation covariance	$R_\varepsilon(k)=H\left(x_n(k)\right)P_k(-)H^T\left(x_n(k)\right)+R(k)$
Kalman gain	$K_k=P_k(-)H^T\left(x_n(k)\right)R_\varepsilon^{-1}(k)$
State-correction	$\hat{x}_k(+)=\hat{x}_k(-)+K_k\varepsilon(k)$
Covariance correction	$P_k(+)=\left[I-K_kH\left(x_n(k)\right)\right]P_k(-)$
Initial conditions	$\hat{x}_0(+)=\hat{x}_0 \; ; P_0(+)=P_0$

Thus, if we choose in eq. (8.164)

$$x_n(k-1)=\hat{x}_{k-1}(+) \; ; u_n(k-1)=u(k-1)$$

we have

$$\hat{x}_k(-)=a\left(\hat{x}_{k-1}(+)\right)+b\left(u(k-1)\right) \quad (8.167)$$

Similarly, if we take in eq. (8.165)

$$x_n(k) = \hat{x}_k(-)$$

we obtain the following

$$\delta\varepsilon(k) = z(k) - h(\hat{x}_k(-)) \tag{8.168}$$

Additionally, the covariance and gain equations are identical to those in Table 8.8, but with Jacobians A, B, C linearized about $\hat{x}_k(-)$. We summarize the discrete extended Kalman filter algorithm in Table 8.9. Note that the error covariance P and the Kalman gain K are now functions of the current state estimate, which is under Gaussian assumptions on the noises $w(k)$, $v(k)$ and the   initial   condition

$x(0)$ a conditional mean and, therefore, a single realization of a stochastic process. Thus, ensemble, or Monte Carlo, techniques should be used to evaluate estimator performance. Additionally, this algorithm is usually implemented in practice using sequential processing and U-D factorization techniques. Before we close this section, consider the following simple examples.

Table 8.9 : Discrete extended Kalman filter

State prediction	$\hat{x}_k(-) = a(\hat{x}_{k-1}(+)) + b(u(k-1))$
Covariance prediction	$P_k(-) = A(\hat{x}_k(-))P_{k-1}(+)A^T(\hat{x}_k(-)) + Q(k-1)$
Innovation	$\varepsilon(k) = z(k) - h(\hat{x}_k(-))$
Innovation covariance	$R_\varepsilon(k) = H(\hat{x}_k(-))P_k(-)H^T(\hat{x}_k(-)) + R(k)$
Kalman gain	$K_k = P_k(-)H^T(\hat{x}_k(-))R_\varepsilon^{-1}(k)$
State-correction	$\hat{x}_k(+) = \hat{x}_k(-) + K_k\varepsilon(k)$
Covariance correction	$P_k(+) = \left[I - K_kH(\hat{x}_k(-))\right]P_k(-)$
Initial conditions	$\hat{x}_0(+) = \hat{x}_0 \; ; \; P_0(+) = P_0$

Example 8.4 (passive localization and tracking problem): Consider the application of the extended Kalman filter to the passive localization and tracking problem that arises frequently in sonar and navigation applications. Therefore, consider a maneuvering observer O monitoring noisy bearing measurements Θ from a target t assumed to be traveling at a constant velocity (Fig. 8.14). These measurements are to be used to estimate the target position r and velocity v. The velocity and position of the target relative to the observer are defined by

$$v_x(t) = v_{tx}(t) - v_{Ox}(t) \; ; \; v_y(t) = v_{ty}(t) - v_{Oy}(t)$$
$$r_x(t) = r_{tx}(t) - r_{Ox}(t) \; ; \; r_y(t) = r_{ty}(t) - r_{Oy}(t) \tag{I}$$

We derive the discrete equations of motion by using the following numerical approximation (first-order Euler approximation)

$$v(t) = \frac{dr(t)}{dt} \approx \frac{r(t) - r(t-1)}{T} \tag{IIa}$$

for T being the sampling interval, or equivalently

$$r(t) = r(t-1) + Tv(t-1) \tag{IIb}$$

or equivalently

$$r_x(t) = r_x(t-1) + Tv_x(t-1) \; ; \; r_y(t) = r_y(t-1) + Tv_y(t-1) \tag{IIc}$$

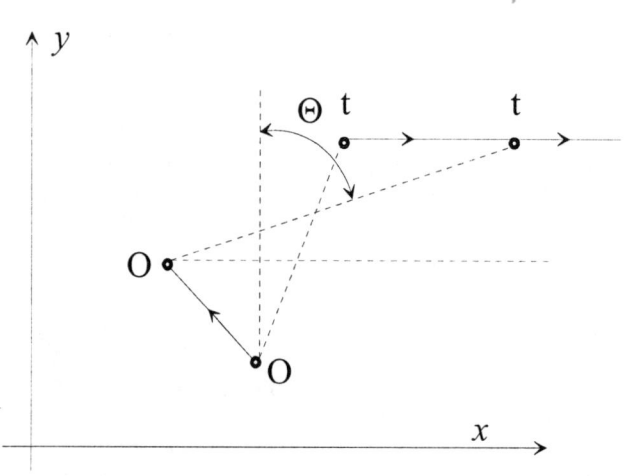

Fig. 8.14: Observer/target ground track geometry

Furthermore, since $v_t(t) = v_t(t-1) = v_t$ (constant target velocity), we also have

$$
\begin{aligned}
v(t) &= v_t(t) - v_O(t) = v_t(t-1) - v_O(t) + v_O(t-1) - v_O(t-1) \\
&= \left[v_t(t-1) - v_O(t-1) \right] - \left[v_O(t) - v_O(t-1) \right] \\
&= v(t-1) - \Delta v_O(t-1)
\end{aligned} \tag{III}
$$

where Δv_O is the observer incremental change in velocity. Thus, from (III) follows

$$v_x(t) = v_x(t-1) - \Delta v_{Ox}(t-1) \tag{IVa}$$

$$v_y(t) = v_y(t-1) - \Delta v_{Oy}(t-1) \tag{IVb}$$

Relations (IIc) and (IV) define the state-space model of the equations of motions in two dimensions (x,y) where the state vector $X^T = \left[r_x \, r_y \, v_x \, v_y \right] = \left[x_1 \, x_2 \, x_3 \, x_4 \right]$.

The bearing-only measurement consists of passive bearings Θ (Fig. 8.14) estimated from a single sensor. For the problem concerned we have the bearing by

$$\Theta(x,t) = \arctan \frac{r_x(t)}{r_y(t)} \tag{V}$$

Using the eqs. (IIb), (IV) and (V) the entire system can be represented by the linear state-space form

$$X(t) = AX(t-1) + Bu(t-1) + w(t-1) \tag{VIa}$$

where

$$A = \begin{bmatrix} 1 & 0 & T & 0 \\ 0 & 1 & 0 & T \\ 0 & 0 & 1 & 0 \\ 0 & 0 & 0 & 1 \end{bmatrix}; B = \begin{bmatrix} 0 & 0 \\ 0 & 0 \\ 1 & 0 \\ 0 & 1 \end{bmatrix}; u = \begin{bmatrix} -\Delta v_{Ox} \\ = \Delta v_{Oy} \end{bmatrix} \qquad \text{(VIb)}$$

with the nonlinear measurement model

$$y(t) = \arctan\left(\frac{x_1(t)}{x_2(t)}\right) + v(t) = h(X(t)) + v(t) \qquad \text{(VII)}$$

Here $w \sim (0, Q)$ and $v \sim (0, R)$ are white noises representing uncertainties in the states and measurements.

The extended Kalman filter of Tab. 8.9 is implemented using the derived model and the following Jacobian matrices:

$$A(X) = A;$$

$$H(X) = \frac{dh(X)}{dX} = \left[\frac{\partial h}{\partial x_1} \frac{\partial h}{\partial x_2} \frac{\partial h}{\partial x_3} \frac{\partial h}{\partial x_4}\right] = \left[\frac{x_2}{R} \; -\frac{x_1}{R} \; 0 \; 0\right] \qquad \text{(VIII)}$$

where $R = \sqrt{x_1^2 + x_2^2}$

The MATLAB software in Fig. 8.15 was used to simulate the system model and the Kalman filter for the following tracking scenario (Fig. 8.16). The sampling interval $T = 0.0333h$ was used. The simulated bearing measurements are shown in Fig. 8.17. The initial conditions for the run where $X^T(0) = [13 \; 13 \; 30 \; -10]$ and the noise parameters $Q = diag\{10^{-6}\}$, $R = 3.05 \cdot 10^{-2}$. The filter is initialized with $\hat{X}(0) = [13 \; 13 \; 30 \; -10]^T$ and $P_0 = diag\{10^{-6}\}$. The results of this run are shown in Fig. 8.18, where the respective position estimates and the corresponding actual values are depicted. The innovations sequence appears statistically white and zero-mean (Figs. 8.19 and 8.20) indicating satisfactory performance.

```
function ex84
T=0.0333; %the sampling period
% scenario design
N=100; % duration of simulation in sampling
...
        periods
vtx=5; %constant x coordinate of the target
velocity
vty=0; %constant y coordinate of the target
velocity
vox(1)=-25;voy(1)=10;
% initial velocity of the object
rtx(1)=15;rty(1)=15;rox(1)=2;roy(1)=2;
%initial positions of the target and observer
for i=2:N
   vox(i)=vox(i-1)+.3;
   voy(i)=voy(i-1)-.15;
   rox(i)=rox(i-1)+T*vox(i-1);
   roy(i)=roy(i-1)+T*voy(i-1);
   rtx(i)=rtx(i-1)+T*vtx(i-1);
   vtx(i)=vtx(i-1);
   rty(i)=rty(i-1);
end
figure(1);plot(rox,roy,rtx,rty,'--');
title('target and observer moving');
legend('trajectory of observer','trajectory of
target');
% forming of measurements
R=3.05*10^(-2);
for i=1:N
   y(i)=atan((rtx(i)-rox(i))/(rty(i)-...
   roy(i)))+sqrt(R)*randn;
end
figure(2);plot(1:N,y);title('measurement
sequence');

% Kalman filter design
xp(1:4,1)=[rtx(1)-rox(1) rty(1)-roy(1) ...
vtx(1)-vox(1) vty(1)-voy(1)]';
xm(1:4,1)=xp;
```

```
pp=(10^(-6))*eye(4);pm=pp;Q=10^(-
6)*eye(4);
A=[1 0 T 0;0 1 0 T;0 0 1 0;0 0 0 1];
B=[0 0;0 0;1 0;0 1];
eps(1)=0;
for i=2:N
  eps(i)=y(i)-atan(xm(1,i-1)/xm(2,i-1));
  rad=sqrt(xm(1,i-1)^2+xm(2,i-1)^2);
  H=[xm(2,i-1)/rad -xm(1,i-1)/rad 0 0];
  K=pm*H'*inv(H*pm*H'+R);
  xp(1:4,i)=xm(1:4,i-1)+K*eps(i);
  pp=(eye(4)-K*H)*pm;
    xm(1:4,i)=A*xp(1:4,i)+B*...
    [vox(i-1)-vox(i); voy(i-1)-voy(i)];
  pm=A*pp*A'+Q;
end

figure(3);plot(1:N,rtx-rox,1:N,xp(1,:));
title('the actual an estimated first coordinate');

figure(4);plot(1:N,rty-roy,1:N,xp(2,:));
title('the actual an estimated second
coordinate');
```

```
figure(5);plot(1:N,vtx-vox,1:N,xp(3,:));
title('the actual an estimated third
coordinate');

figure(6);plot(1:N,vty-voy,1:N,xp(4,:));
title('the actual an estimated fourth
coordinate');

figure(7);plot(1:N,eps);
title('the innovation sequence');

for i=1:10
  Re(i)=0;
  for j=1:N-i+1
    Re(i)=Re(i)+eps(j)*eps(j+i-1);
  end
  Re(i)=Re(i)/(N-i);
end

figure(8);plot(0:9,Re);grid;
title('the estimated auto-correlation function ...
of the residual sequence');
keyboard;
```

Fig. 8.15: MATLAB program code for the example 8.4

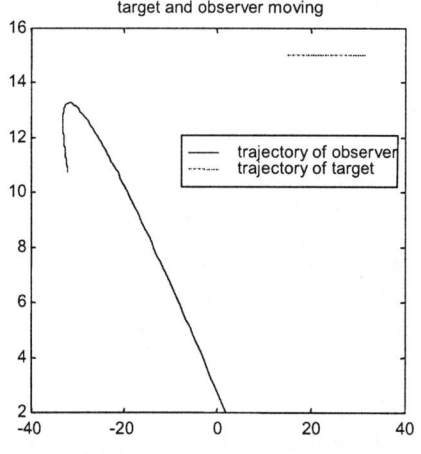
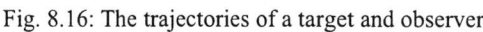

Fig. 8.16: The trajectories of a target and observer

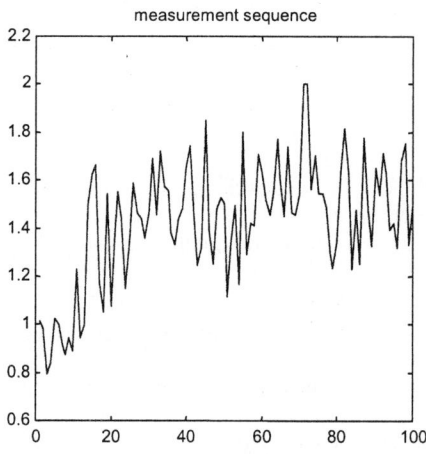

Fig. 8.17: The bearing measurements sequence

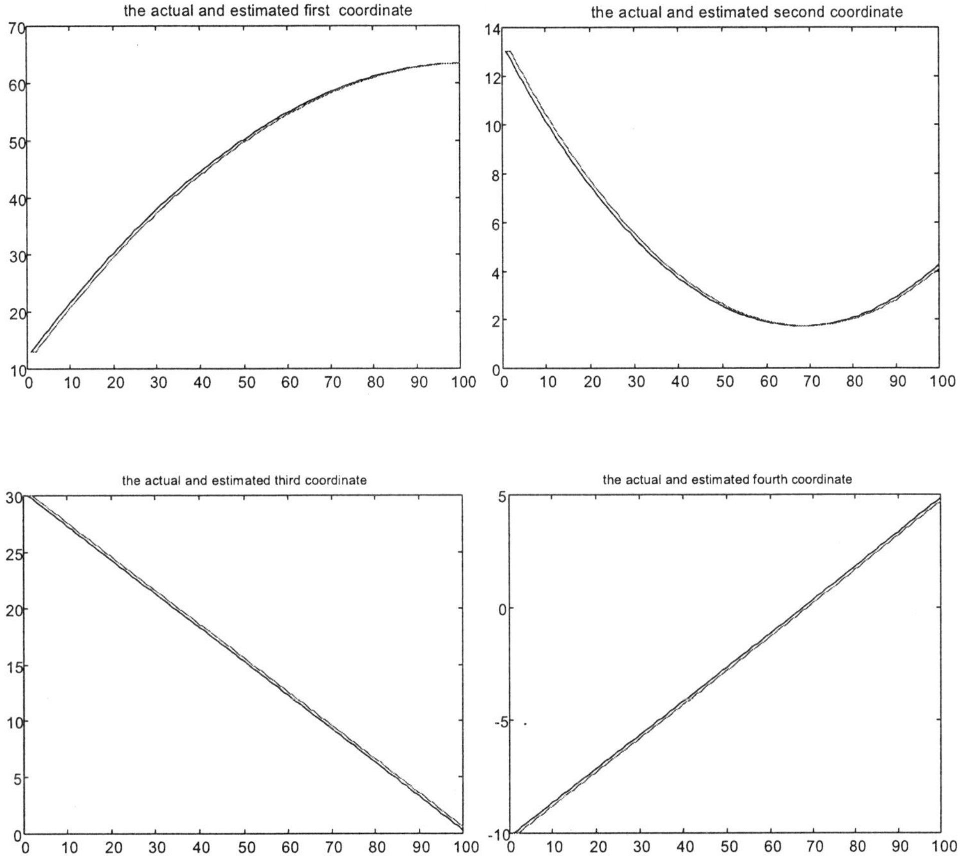

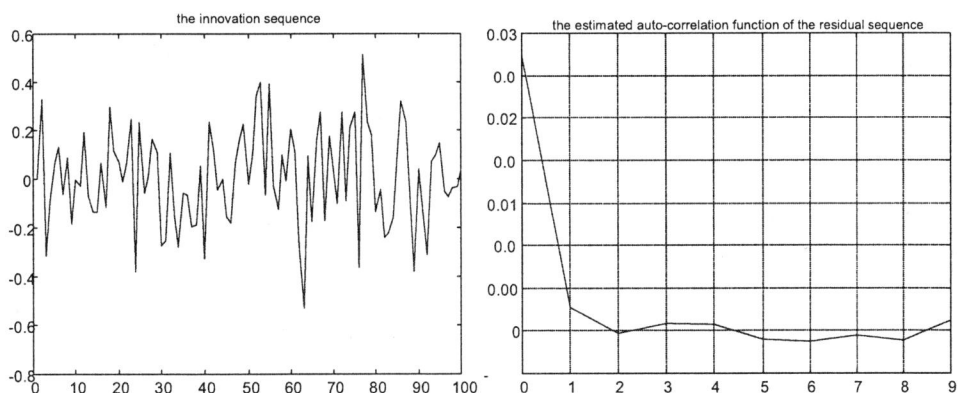

Fig. 8.18: Actual and estimated states

Fig. 8.19: Innovation sequence

Fig. 8.20: Autocorrelation function of the
innovations

Example 8.5 (extended Kalman filter as real-time parameter identifier): Consider the discrete nonlinear process given by

$$x(t) = (1 + \Theta_1 T) x(t-1) + \Theta_2 T x^2(t-1) + w(t-1) \tag{I}$$

with corresponding measurement model

$$y(t) = x^2(t) + x^3(t) + v(t) \tag{II}$$

where $w \sim (0, Q) = (0, 0)$, $v \sim (0, R) = (0, 0.09)$, $x(0) \sim (\hat{x}_0, P_0) = (1.2, 0.01)$. The true parameters are

$$\Theta_{true} = [\Theta_1 \quad \Theta_2]^T = [-0.05 \quad 0.04]^T$$

The extended Kalman filter can be applied to solve the problem of estimating the unknown parameters by augmenting these parameters into the state-space form using the parameter model

$$\Theta(t) = \Theta(t-1) \; ; \; \text{(a constant parameter vector)}$$

The new state equations become

$$X(t) = \begin{bmatrix} x(t) \\ \Theta_1(t) \\ \Theta_2(t) \end{bmatrix} = \begin{bmatrix} (1 + \Theta_1(t-1)T) x(t-1) + \Theta_2(t-1) T x^2(t-1) + w(t-1) \\ \Theta_1(t-1) \\ \Theta_2(t-1) \end{bmatrix} = a(X(t-1)) \tag{III}$$

and the nonlinear measurement model is given by

$$y(t) = x^2(t) + x^3(t) + v(t) = h(x(t)) + v(t) \tag{IV}$$

The extended Kalman filter equations follow directly from Tab. 8.9, with the Jacobian matrices

$$A(X(t-1)) = \frac{\partial a(X(t-1))}{\partial X} = \begin{bmatrix} 1 + \Theta_1(t-1) + 2T\Theta_2(t-1)x(t-1) & Tx(t-1) & Tx^2(t-1) \\ 0 & 1 & 0 \\ 0 & 0 & 1 \end{bmatrix} \tag{V}$$

$$H(X(t-1)) = \frac{\partial h(X(t-1))}{\partial X} = [2x(t-1) + 3x^2(t-1) \quad 0 \quad 0] $$

Using MATLAB software in Fig. 8.21 the extended Kalman filter is applied to solve this problem for 15000 measurements with the sampling period $T = 0.01s$. Initially, we used in the filter algorithm

$$\hat{x}_0^T = [1.2 \quad 0 \quad 0] \; ; \; P_0 = 0.001 * diag\{1,1,1\} \tag{VI}$$

```
function ex85
% generation of the measurements

t1=-0.05; t2=0.04; N=15000;
x(1)=1;T=0.01;R=0.09;
for i=2:N
  x(i)=(1+t1*T)*x(i-1)+t2*T*x(i-1)^2;
end
for i=1:N
  y(i)=x(i)^2+x(i)^3+sqrt(R)*randn;
end

% design of Kalman filter
```

```
xp=[1. -.05 0.04]';xm=xp;pp=.00001*eye(3);
pm=pp;eps(1)=0;Q=[.00001 0 0;0 0. 0;0 0 0.];
for i=2:N
  h=[2*xm(1,i-1)+3*xm(1,i-1)^2 0 0];
  eps(i)=y(i)-xm(1,i-1)-xm(1,i-1)^3;
  K=pm*h'*inv(h*pm*h'+R);
  xp(1:3,i)=xm(1:3,i-1)+K*eps(i);
  pp=(eye(3)-K*h)*pm;

  xm(1,i)=(1+xp(2,i)*T)*xp(1,i)+xp(3,i)*T*xp(
1,i)^2;
  xm(2,i)=xp(2,i);
  xm(3,i)=xp(3,i);
```

```
A(1,1)=1+xm(2,i)*T+2*T*xm(3,i)*xm(1,i);
  A(1,2)=T*xm(1,i);
  A (1,3)=T*xm(1,i)^2; A(2:3,1:3)=[0 1 0; 0
0 1];
  pm=A*pp*A'+Q;
end
```

```
figure(1);plot(1:N,x,1:N,xp(1,:));

title('the first state');
figure(2);plot(1:N,t1*ones(1,N),1:N,xp(2,:));
title('the first parameter');
figure(3);plot(1:N,t2*ones(1,N),1:N,xp(3,:));
title('the second parameterr');
keyboard;
```

Fig. 8.21: The MATLAB program for example 8.5

In Fig. 8.22 we depict the state and parameter estimates. Obviously, after a short transient of approximately 25 samples, the state estimates begins tracking the true state (Fig. 8.22.a). However, the parameter estimates slowly converge to their true values (Fig. 8.22. b and c). The final parameter estimates are $\hat{\Theta} = -0.047$ and $\hat{\Theta}_2 = 0.033$. Although not converged, it is clear that the estimator will track the values eventually. A reason for the slow convergence may result from the lack of sensitivity of the measurements, or equivalently innovations, to parameter variations. The innovations sequence appears statistically white, but a slight bias is detected.

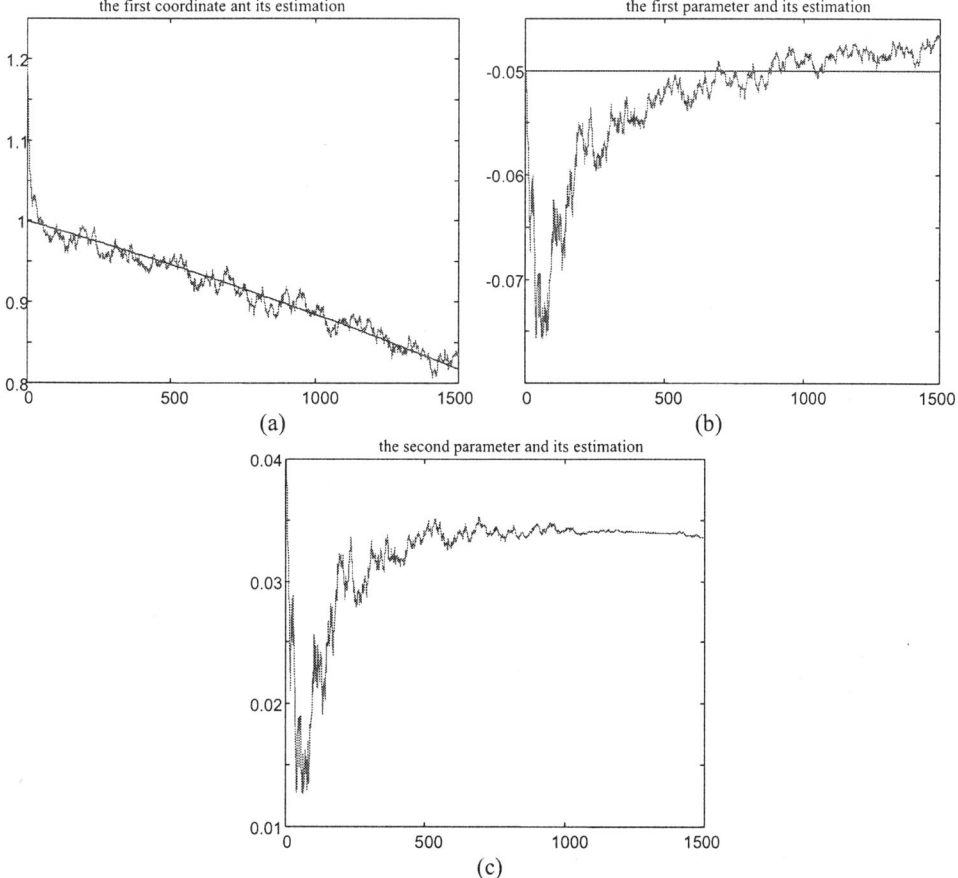

Fig. 8.22: The actual states and their estimations related to example 8.5

8.2.3 Continuous linearized and extended Kalman filters

In this subchapter we develop the linearized and extended Kalman filters for a continuous nonlinear systems described by a stochastic nonlinear differential equation in a state-space form

$$\dot{x}(t) = a(x(t),t) + G(t)w(t) \; ; \; w(t) \sim (0, Q(t)) \tag{8.169}$$

with the corresponding measurement model

$$z(t) = h(x(t),t) + v(t) \; ; \; v(t) \sim (0, R(t)) \tag{8.170}$$

The *continuous linearized Kalman filter*, which represents a heuristic extension of the continuous Kalman filter presented in section 7, is a *minimum variance filter* based on linearization about a nominal trajectory. We assume that a differential equation for the nominal trajectory, $x_n(t)$, is specified by eq. (8.169) with $w(t) = 0$. If the deviation from the nominal observation

$$\delta z(t) = z(t) - z_n(t) \tag{8.171}$$

where

$$z_n(t) = h(x_n(t),t) \tag{8.172}$$

and deviation from the nominal trajectory

$$\delta x(t) = x(t) - x_n(t) \tag{8.173}$$

is small enough, the estimate $\delta\hat{x}(t)$ of $\delta x(t)$ is given by the continuous Kalman filter algorithm (Tab. 7.1). For the linearized state space model described by

$$\begin{aligned}
\delta\dot{x}(t) &= \dot{x}(t) - \dot{x}_n(t) \\
&= a(x(t),t) + G(t)w(t) - a(x_n(t),t) \\
&= a(x_n(t),t) + \frac{da(x_n(t),t)}{dx_n(t)}(x(t) - x_n(t)) + G(t)w(t) - a(x_n(t),t)
\end{aligned}$$

or

$$\delta\dot{x}(t) = A(t)\delta x(t) + G(t)w(t) \tag{8.174}$$

and the measurement model

$$\begin{aligned}
\delta z(t) &= z(t) - z_n(t) = h(x(t),t) + v(t) - h(x_n(t),t) \\
&= h(x_n(t),t) + \frac{dh(x_n(t),t)}{dx_n(t)}\delta x(t) + v(t) - h(x_n(t),t)
\end{aligned}$$

or

$$\delta z(t) = H(t)\delta x(t) + v(t) \tag{8.175}$$

the continuous Kalman filter is described by Tab. 7.1, i.e.

$$\delta\dot{\hat{x}}(t) = A(t)\delta\hat{x}(t) + K(t)\left[\delta z(t) - H(t)\delta\hat{x}(t)\right] \tag{8.176}$$

$$K(t) = P(t)H^T(t)R^{-1}(t) \tag{8.177}$$

$$\dot{P}(t) = A(t)P(t) + P(t)A^T(t) - P(t)H^T(t)R^{-1}(t)H(t)P(t) + G(t)Q(t)G^T(t) \quad (8.178)$$

with the initial conditions

$$\delta\hat{x}(0) = 0 \; ; \; P(0) = P_0 \quad (8.179)$$

The Jacobian matrices A and H are defined by (8.174) and (8.175); that is

$$A(t) = \frac{\partial a(x_n(t),t)}{\partial x_n(t)} \; ; \; H(t) = \frac{\partial h(x_n(t),t)}{\partial x_n(t)} \quad (8.180)$$

The complete state estimation is given by

$$\hat{x}(t) = x_n(t) + \delta\hat{x} \quad (8.181)$$

where $x_n(0) = E\{x(0)\}$, the a priori mean of the state $x(t)$. Note also that $\delta P(t) = E\{\delta\tilde{x}(t)\delta\tilde{x}^T(t)\}$ is equal to $P(t) = E\{\tilde{x}(t)\tilde{x}^T(t)\}$, since

$$\delta\tilde{x}(t) = \delta x(t) - \delta\hat{x}(t) = [x(t) - x_n(t)] - [\hat{x}(t) - x_n(t)]$$
$$= x(t) - x_n(t) = \tilde{x}(t)$$

For convenience, we summarize the *linearized continuous Kalman filter* algorithm in Table 8.10.

Table 8.10. Linearized continuous Kalman filter algorithm

Nominal trajectory	$\dot{x}_n(t) = a(x_n(t),t) \; ; \; x_n(0) = E\{x(0)\} = \hat{x}_0$
Nominal observation	$z_n(t) = h(x_n(t),t)$
Linearized state-space model	$\delta\dot{x}(t) = A(t)\delta x(t) + G(t)w(t)$
Linearized observation model	$\delta z(t) = H(t)\delta x(t) + v(t)$
Linearized Kalman filter algorithm	$\delta\dot{\hat{x}} = A(t)\delta\hat{x}(t) + K(t)[\delta z(t) - H(t)\delta\hat{x}(t)]$
Kalman gain	$K(t) = P(t)H^T(t)R_v^{-1}(t)$
Error-covariance algorithm	$\dot{P} = AP + PA^T - PH^TR^{-1}HP + GQG^T$
Initial conditions	$\delta\hat{x}(0) = 0 \; ; \; P(0) = P_0$
Complete linearized state space	$\hat{x}(t) = x_n(t) + \delta\hat{x}(t)$
Definitions	$A(t) = \frac{\partial a(x_n(t),t)}{\partial x_n(t)} \; ; \; H(t) = \frac{\partial h(x_n(t),t)}{\partial x_n(t)}$

A different set of continuous-filter algorithms may be obtained by assuming that the minimum variance estimate $\hat{x}(t)$ is known and is used to expand the nonlinear state and observation models in the first-order Taylor series about $x_n(t) = \hat{x}(t)$, such that we obtain a set of equations linear in $x(t)$; that is

$$\dot{x}(t) = a(\hat{x}(t),t) + \frac{da(\hat{x}(t),t)}{d\hat{x}(t)}(x(t) - \hat{x}(t)) + G(t)w(t) \quad (8.182)$$

$$z(t) = h(\hat{x}(t),t) + \frac{dh(\hat{x}(t),t)}{d\hat{x}(t)}(x(t) - \hat{x}(t)) + v(t) \tag{8.183}$$

By comparing (8.182) and (8.183) with the basic linear state-space model (chapter 7)

$$\dot{x}(t) = A(t)x(t) + G(t)w(t) + u(t) \tag{8.184}$$

$$z(t) = H(t)x(t) + v(t) + y(t) \tag{8.185}$$

we have the known input and known disturbance

$$u(t) = a(\hat{x}(t),t) - \frac{da(\hat{x}(t),t)}{d\hat{x}(t)}\hat{x}(t) \tag{8.186}$$

$$y(t) = h(\hat{x}(t),t) - \frac{dh(\hat{x}(t),t)}{d\hat{x}(t)}\hat{x}(t) \tag{8.187}$$

and the linearized Jacobian matrices

$$A(t) = \frac{\partial a(\hat{x}(t),t)}{\partial \hat{x}(t)} \; ; \; H(t) = \frac{\partial h(\hat{x}(t),t)}{\partial \hat{x}(t)} \tag{8.188}$$

Determination of the resultant continuous Kalman filter is now a trivial task. We simply use the relations from Table 7.1

$$\dot{\hat{x}}(t) = A(t)\hat{x}(t) + u(t) + K(t)\left[z(t) - y(t) - H(t)\hat{x}(t)\right] \tag{8.189}$$

$$\dot{P}(t) = A(t)P(t) + P(t)A^T(t) - P(t)H^T(t)R_v^{-1}(t)H(t)P(t) + G(t)Q(t)G^T(t) \bullet \tag{8.190}$$

with the definitions of A and H in eqs. (8.188) and u and y in eqs. (8.186) and (8.187), respectively. The resulting *continuous extended Kalman filter* is presented in Table 8.11.

Table 8.11 Continuous extended Kalman filter algorithm

State-space model	$\dot{x}(t) = a(x(t),t) + G(t)w(t) \; ; w(t) \sim (0,Q(t))$
Observation model	$z(t) = h(x(t),t) + v(t) \; ; \; v(t) \sim (0,R(t))$
Filter algorithm	$\dot{\hat{x}} = a(\hat{x}(t),t)\hat{x}(t) + K(t)\left[z(t) - h(\hat{x}(t),t)\delta\hat{x}(t)\right]$
Kalman gain	$K(t) = P(t)H^T(t)R^{-1}(t)$
Error-covariance algorithm	$\dot{P} = AP + PA^T - PH^TR^{-1}HP + GQG^T$
Initial conditions	$\hat{x}(0) = E\{x(0)\} = \hat{x}(0) \; ;$ $P(0) = P_0 = E\left\{\left[x(0) - \hat{x}(0)\right]\left[x(0) - \hat{x}(0)\right]^T\right\}$
Other assumptions	$E\{w(t)v^T(\tau)\} = E\{w(t)x^T(0)\} = E\{v(t)x^T(0)\} = 0$
Definitions	$A(t) = \frac{\partial a(\hat{x}(t),t)}{\partial \hat{x}(t)} \; ; \; H(t) = \frac{\partial h(\hat{x}(t),t)}{\partial \hat{x}(t)}$

8.3. Adaptive Filtering

State estimators have been formulated under the assumption that dynamic system parameters and process/measurement noise statistics are known but there is usually some degree of uncertainty of this knowledge . If the actual values of system parameters and noise covariances are different from those used in estimation , then the filter is suboptimal . In these cases state estimates may contain more error than is necessary and, in some instances , diverge from the neighborhood of true values. Therefore , state estimates could be improved by simultaneously estimating the uncertain system parameters and noise statistics and using this additional information to adapt the filter gains and system model parameters to the measurements . Adaptive filters may perform as well as optimal filters in the limit . However , the less that is known about a system prior to estimation , the greater the error in the resulting suboptimal estimates . Our attention is restricted here to adaptive state estimators for linear systems, although the general approach could be applied to nonlinear systems with minor modifications. Furthermore , it is assumed that the unknown system parameters and noise statistics are constant, although their estimates will vary in time . In any case , adaptive filters prove to be nonlinear, as it will be shown in the following [18, 23, 35].

8.3.1. Parameter-Adaptive Filtering

Let us consider a linear continuos-time system model

$$\frac{dx(t)}{dt} = A(p,t)x(t) + B(p,t)u(t) + G(p,t)w(t)$$

$$y(t) = H(t)x(t) + v(t)$$

(8.191)

where

$$E\{w(t)\} = 0 \; ; \; E\{v(t)\} = 0; E\left\{\begin{bmatrix} w(t) \\ v(t) \end{bmatrix} \begin{bmatrix} w'(\tau) & v'(\tau) \end{bmatrix}\right\} = \begin{bmatrix} Q & 0 \\ 0 & R \end{bmatrix} \delta(t-\tau) \qquad (8.192)$$

A parameter-adaptive filter estimates the unknown system parameter vector p as well as the system state vector $x(t)$ by augmenting the state vector , as it is described earlier . Thus , even if the systems matrices A,B and G are linear functions of p , nonlinearity is introduced by product op the elements of p and x or u . The extended Kalman filter provides a basis for adaptive filtering that is straightforward and approximate optimal in the minimum variance sense . This is not to say , however , that satisfactory results are guaranteed for every realization of the filter . The success of parameter estimation depends on many characteristics of the system in question : the number of the uncertain parameters (dimension of vector p) , the magnitude of uncertainty , the functional dependence of system outputs on the uncertain parameters , the quality of output measurements , and the knowledge of system inputs . An augmented state vector $x_a(t)$ is formed from the original n-dimensional state vector $x(t)$ and the l-dimensional vector $p(t)$ of parameters to be estimated as

$$x_a(t) = \begin{bmatrix} x(t) \\ p(t) \end{bmatrix} \qquad (8.193)$$

and the nonlinear dynamic equation has dimension $(l+n)$

$$\frac{dx_a(t)}{dt} = f_a(x_a(t), u(t), w_a(t), t); x_a(0) \, given \qquad (8.194)$$

with m-dimensional input vector $u(t)$ and s-dimensional augmented disturbance input (state or process noise) $w_a(t) = \{w'(t) w'_p(t)\}'$ with $w_p(t)$ being the parameter noise as described below. Alternatively,

$$\begin{bmatrix} \dfrac{dx(t)}{dt} \\ \dfrac{dp(t)}{dt} \end{bmatrix} = \begin{bmatrix} f(x(t),p(t),u(t),w_a(t),t)_1 \\ f_2(p(t),w_a(t),t) \end{bmatrix}; x(0), p(0) \, given \qquad (8.195)$$

where

$$f_1(.) = A(p(t),t)x(t) + B(p(t),t)u(t) + G_1(p(t),t)w_a(t) \qquad (8.196)$$

Here $G_1(.) = \begin{bmatrix} G(.) & 0 \end{bmatrix}$ with 0 being the zero matrix of corresponding order and $f_2(.)$ depends of the model parameter vector , described below . The observation equation is

$$y(t) = \begin{bmatrix} H(t) & 0 \end{bmatrix} \begin{bmatrix} x(t) \\ p(t) \end{bmatrix} + v(t) = H_a(t)x_a(t) + v(t) \qquad (8.197)$$

assuming that no additional measurements are made . The noise statistics are defined as before, bearing in mind that $w(.)$, $G(.)$ and the disturbance input covariance matrix Q are redefined in order to account for the disturbance $w_p(.)$ that drive the parameter dynamics. Parametric variations in $H(.)$ may not be distinguished from variations in $A(.)$ without prior information because the measurement time history could result from any one of an infinite number of state-space systems models obtained by similarity (linear) transformation . As long as the principal objective is an improved state vector estimate , neglecting parameter variations in $H(.)$ may be of little significance , as the parameters of $A(.)$ can adapt to those of the best similar system models . To the extent that p represents an input-output sensitivity , which in turn provides the basis for parameter estimation , specific knowledge of the input signal $u(.)$ as well as the output $y(.)$ is needed. Deterministic inputs can provide better parameter estimates than stochastic inputs, as long as the effects of parameter variations are observable in the output . If the input is only inferred by its statistics , and if the inference itself is uncertain , then the parameter estimates will suffer.

The extended Kalman filter follows section 8.2 directly. The state estimates is found by integrating the nonlinear differential equation that models system and parameter dynamics with optimal forcing by the measurement residual

$$\frac{d\hat{x}_a(t)}{dt} = f_a(\hat{x}_a(t),u(t),0,t) + K(t)\big[y(t) - H_a(t)\hat{x}_a(t)\big] \qquad (8.198)$$

The zero in the third element of $f_a(.)$ represents the assumed zero-mean value of the disturbance input. The $(l \times n) \times r$ filter gain matrix is

$$K(t) = P(t)H'_a(t)R^{-1} \qquad (8.199)$$

where $P(t)$ is the integral of the $(l \times n) \times (l \times n)$ matrix Riccati equation

$$\frac{dP(t)}{dt} = A_a(t)P(t) + P(t)A'_a(t) + G_a(t)Q_a G'_a(t) - P(t)H'_a(t)R^{-1}H_a(t)P(t); P(0) = P_0 \qquad (8.200)$$

Here the sensitivity to process noise is described by

$$G_a(t) = \begin{bmatrix} G(p(t),t) & 0 \\ 0 & \partial f_2(.)/\partial w_p \end{bmatrix}\Bigg|_{x=x(t),p=p(t),u=u(t)} \tag{8.201}$$

and the covariance matrix of zero-mean augmented disturbance input $w_a(t)$, say Q_a, is a block-diagonal matrix containing the process noise covariance Q and parameter noise covariance Q_p, i.e.

$$Q_a = \begin{bmatrix} Q & 0 \\ 0 & Q_p \end{bmatrix} \tag{8.202}$$

The system Jacobian matrix $A_a(t)$ is formed in the usual way

$$A_a(t) = \begin{bmatrix} \partial f_1/\partial x & \partial f_1/\partial p \\ \partial f_2/\partial x & \partial f_2/\partial p \end{bmatrix}\Bigg|_{x=x(t),p=p(t),u=u(t)} \tag{8.203}$$

From eqs. (8.194)-(8.196) follows

$$\frac{\partial f_1}{\partial x} = A(p(t),t)$$

$$\frac{\partial f_1}{\partial x} = \left\{ \left[\frac{\partial A(.)}{\partial p}\right]x(t) + \left[\frac{\partial B(.)}{\partial p}\right]u(t) \right\}\Bigg|_{p=p(t)} \tag{8.204}$$

while $\partial f_2/\partial x, \partial f_2/\partial p$ depend upon the dynamic model for the parameter vector p. It should be noted that the dimension of p is problem-dependent and has no general relationship to the dimensions of $A(.)$, $B(.)$ and $G(.)$. The number of uncertain parameters must be small in some sense , i.e. on the order of the state vector dimension n or less , because model uncertainty diminishes the filter ability to distinguish between variations resulting from disturbance input (process noise) and those due to measurements errors (observation noise) . The magnitude of each parameters uncertainty should be small for similar reasons . Assume that the parameters are a set of random variables p , specified only by prior estimates of its mean-value p_0 and covariance matrix $P_p(0)$, and satisfying the differential equation

$$\frac{dp(t)}{dt} = f_2(p(t), w_a(t), t) = 0; p(0) = p_0 \tag{8.205}$$

The augmented state covariance matrix is calculated by integrating eq. (8.200) with the corresponding elements of $P(0)$ initialized by $P_p(0)$. While eq. (8.195) may appear to assure that $p(t)$ will always be $p(0)$, recall that the estimate $\hat{p}(t)$ of $p(t)$ is forced by the measurement residual in eq. (8.198) . With no parameter noise in eq. (8.205) $Q_p = 0$ in eq. (8.202) and the elements of the Kalman gain matrix $K(t)$ in eq.(8) that update $\hat{p}(t)$ in the augmented state vector estimate $\hat{x}_a' = \begin{bmatrix} \hat{x}' & \hat{p}' \end{bmatrix}$ should go to zero as time increases . An alternative method useful in accounting for time-varying parameters is to assume that the time variation in the parameter vector is partially random in nature . In particular , we replace the constant parameter model (8.205) with the expression

$$\frac{dp(t)}{dt} = f_2(p(t), w_a(t), t) = \begin{bmatrix} 0 & I \end{bmatrix} w_a(t) = w_p(t) \tag{8.206}$$

where $w_p(t)$ is zero mean with the covariance matrix Q_p. The strength of the noise should correspond roughly to the possible range of parameter variation. For example , if it is known that the i-th element p_i of p is likely to change by an amount Δp_i over the interval of interest Δt , then the *i-th* diagonal element of Q_p in eq. (8.202) is given by

$$Q_p(i,i) = \frac{(\Delta p_i)^2}{\Delta t}; i = 1, \cdots, s \qquad (8.207)$$

In practice , it is frequently observed that eq. (8.206) is a good model for the purpose of filter design, even though the parameter changes may actually be deterministic in nature. However, the reader should be aware that there are many different methods in the literature that can be used for parameter identification , such as maximum likelihood, least-squares, equation error , stochastic approximation and correlation. Some of the alternative identification methods mentioned above are related to situations where the unknown parameters are assumed to be constant with unknown statistics , and where an algorithm is desired that yields perfect (unbiased , consistent) parameter estimates in the limit , as an infinite number of the measurements is taken. The equation errors , stochastic approximation and correlation methods are all in this category . Other methods are based upon various optimization criteria, e.g. maximum likelihood and least-squares methods. The maximum likelihood estimate is one that maximizes the joint probability density function (pdf) for the set of unknown parameters . However , it is typically calculated by a non-real time (nonrecursive) algorithm that is not well suited for control system applications. In addition, the pdf can have several peaks , in which case the optimal estimate may require extensive searching. The least-squares method seeks an estimate that best fits the measurement data , in the sense that it minimizes a quadratic cost function . In general , the estimate is different for each quadratic cost function . In contrast with the minimum variance estimate , the least-squares estimate does not require knowledge of noise statistics . However , the extended Kalman filtering algorithm , representing an approximate solution for the minimum variance estimate , provides a logical choice of parameter identification techniques when the random process statistics are known and requirement exists for real-time estimates. The block diagram of the parameter-adaptive extended Kalman filter is given in the next fiqure.

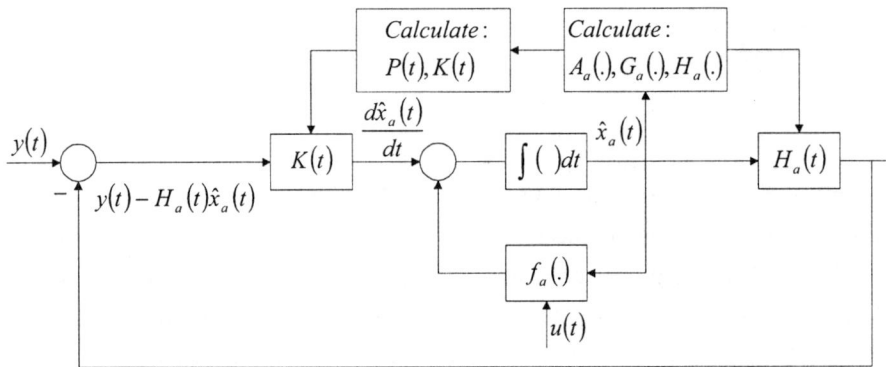

Fig. 8.23 Block diagram of the parameter-adaptive extended Kalman filter

The performance of the parameter-adaptive filter are illustrated in the following example .

Example 8.6 (damped harmonic oscillator): Many systems can be modeled , at least to a first approximation , by the equation

$$\frac{d^2y}{dt^2} + 2\xi\omega_n \frac{dy}{dt} + \omega_n^2 y = \omega_n^2 u + \omega_n^2 w \tag{I}$$

For instance , if $y(t)$ is aircraft pitch and $u(t)$ is elevator deflection , then (I) corresponds to the short-period approximation to the longitudinal dynamics, where the disturbance input $w(t)$ compensates for unmodeled dynamics . On the other hand , the angular motion of a weathervane is also modeled by (I) , where $y(t)$ represents angular position , dy/dt represents angular rate , $w(t)$ is the random angular direction of the wind , while the deterministic control input $u(t)=0$. By selecting the state components as $x_1 = y, x_2 = dy/dt$ the plant (I) can be put into controllable canonical form

$$\frac{dx}{dt} = \begin{bmatrix} 0 & 1 \\ -\omega_n^2 & -2\xi\omega_n \end{bmatrix} x + \begin{bmatrix} 0 \\ \omega_n^2 \end{bmatrix} w = Ax + Gw; x = \begin{bmatrix} x_1 \\ x_2 \end{bmatrix} \tag{II}$$

Since $u=0$, this represents a difficult estimation problem , in that the disturbance input w is known only by its mean and covariance . Furthermore , let us assume that the damping ratio   and that the natural frequency $\omega_n = 2\,rad/s$, as well as that the both state components are measured each $T=0.1s$, giving rise to the discrete measurement model

$$\tag{III}$$

The initial conditions for the state mean and covariance are zero , while the zero-mean white continuous state (process) noise w and the zero-mean white discrete measurement noise v are represented by the spectral density Q and covariance matrix R as

$$Q = 1; \; R = \begin{bmatrix} r_1 & 0 \\ 0 & r_2 \end{bmatrix} = I \tag{IV}$$

Assuming that ω_n^2 is an uncertain constant system parameter , i.e. $p = -\omega_n^2$, the system equations (8.194) and (8.195) for the extended Kalman filter become

$$\tag{V}$$

where

$$A(p) = \begin{bmatrix} 0 & 1 \\ p & a \end{bmatrix}; G(p) = \begin{bmatrix} 0 \\ -p \end{bmatrix} \tag{VI}$$

or more explicitly

$$\begin{bmatrix} \frac{dx_1}{dt} \\ \frac{dx_2}{dt} \\ \frac{dp}{dt} \end{bmatrix} = \begin{bmatrix} 0 & 1 & 0 \\ p & a & 0 \\ 0 & 0 & 0 \end{bmatrix} \begin{bmatrix} x_1 \\ x_2 \\ p \end{bmatrix} + \begin{bmatrix} 0 \\ -p \\ 0 \end{bmatrix} w \tag{VII}$$

with $a = -2\xi\omega_n$ being the known system parameter. The sensitivity matrices (8.197), (8.201)-(8.204) for the Riccati equation (8.200) are as follows

$$
A_a = \begin{bmatrix} \partial f_1 / \partial x & \partial f_1 / \partial p \\ \partial f_2 / \partial x & \partial f_2 / \partial p \end{bmatrix} = \begin{bmatrix} A(\hat{p}) & \dfrac{\partial A(\hat{p})}{\partial p}\hat{x} \\ 0 & 0 \end{bmatrix} = \begin{bmatrix} 0 & 1 & 0 \\ \hat{p} & a & \hat{x}_1 \\ 0 & 0 & 0 \end{bmatrix}
$$

$$
G_a = \begin{bmatrix} \partial f_1 / \partial w \\ \partial f_2 / \partial w \end{bmatrix} = \begin{bmatrix} G(\hat{p}) \\ 0 \end{bmatrix} = \begin{bmatrix} 0 \\ -\hat{p} \\ 0 \end{bmatrix} ; H_a = \begin{bmatrix} H & 0 \end{bmatrix} = \begin{bmatrix} 1 & 0 & 0 \\ 0 & 1 & 0 \end{bmatrix}
$$

(VIII)

where the sign $\wedge$ denotes an estimated quantity . The Riccati equation can be expressed as six scalar equations

$$
\frac{dp_{11}}{dt} = p_{12}\left(2 - \frac{p_{12}}{r_2}\right) - \frac{p_{11}^2}{r_1}
$$

$$
\frac{dp_{12}}{dt} = \hat{p}p_{11} + p_{12}\left(a - \frac{p_{11}}{r_1} - \frac{p_{22}}{r_2}\right) + \hat{x}_1 p_{13} + p_{22}
$$

$$
\frac{dp_{13}}{dt} = p_{23}\left(1 - \frac{p_{12}}{r_1}\right) - \frac{p_{11}p_{13}}{r_1}
$$

(IX)

$$
\frac{dp_{22}}{dt} = p_{12}\left(2\hat{p} - \frac{p_{12}}{r_1}\right) + p_{22}\left(2a - \frac{p_{22}}{r_2}\right) + 2\hat{x}_1 p_{23} + \frac{\hat{p}^2}{q}
$$

$$
\frac{dp_{23}}{dt} = p_{13}\left(\hat{p} - \frac{p_{12}}{r_1}\right) + p_{23}\left(a - \frac{p_{22}}{r_2}\right) + \hat{x}_1 p_{33}
$$

$$
\frac{dp_{33}}{dt} = -\frac{p_{13}^2}{r_1} - \frac{p_{23}^2}{r_2}
$$

where

$$
P = \begin{bmatrix} p_{11} & p_{12} & p_{13} \\ p_{12} & p_{22} & p_{23} \\ p_{13} & p_{23} & p_{33} \end{bmatrix}
$$

(X)

The optimal gain matrix (8.199) is

$$
K = \begin{bmatrix} k_{11} & k_{12} \\ k_{21} & k_{22} \\ k_{31} & k_{32} \end{bmatrix} = \begin{bmatrix} \dfrac{p_{11}}{r_1} & \dfrac{p_{12}}{r_2} \\ \dfrac{p_{12}}{r_1} & \dfrac{p_{22}}{r_2} \\ \dfrac{p_{13}}{r_1} & \dfrac{p_{23}}{r_2} \end{bmatrix}
$$

(XI)

and the state estimator (8.198) is

$$
\begin{bmatrix} \dfrac{d\hat{x}_1}{dt} \\[2ex] \dfrac{d\hat{x}_2}{dt} \\[2ex] \dfrac{d\hat{p}}{dt} \end{bmatrix} = \begin{bmatrix} \hat{x}_2 \\[1ex] \hat{p}\hat{x}_1 + a\hat{x}_2 \\[1ex] 0 \end{bmatrix} + \begin{bmatrix} k_{11} & k_{12} \\ k_{21} & k_{22} \\ k_{31} & k_{32} \end{bmatrix} \begin{bmatrix} y_1 - \hat{x}_1 \\[1ex] y_2 - \hat{x}_2 \end{bmatrix} \qquad \text{(XII)}
$$

The initial expected value of $\hat{p}$ is $p(0) = -4.4$ and the assumed covariance covariance of the uncertainty is $P_p(0) = 20$. Thus , the initial conditions for the differential equations (IX) and (XII) are as follows

$$
P(0) = \begin{bmatrix} 0 & 0 & 0 \\ 0 & 0 & 0 \\ 0 & 0 & 20 \end{bmatrix} ; \ \hat{x}(0) = \begin{bmatrix} 1.1 \\ 0.9 \\ -4.4 \end{bmatrix} \qquad \text{(XIII)}
$$

The MATLAB program code under which the simulation results are generated is given in Fig. 8.24, while the obtained experimental results are depicted in Fig. 8.25. The results have shown that the average value of the estimated parameter during the $2s$ simulation is closer to the true parameter value (4) than is the initial estimate (4.4), and the corresponding estimation error variance p_{33} is reduced by 35% ($p_{33}(0) = 20$) .However, convergence to the actual value is irregular and by no means guaranteed. Initially , the parameter estimate appears to be converging to the true value of 4. Inspection of the Riccati equation (IX) reveals that the parameter estimate is strongly dependent on the estimate $\hat{x}_1$. The rate of changes of p_{23} contains the product of $\hat{x}_1$ and p_{33} ; hence , there is substantial variation in not only the magnitude but the sign of the corresponding Kalman gain matrix element in (XI)

Due to the oscillations of the lightly damped system ($\xi = 0.1$) ; k_{31} and k_{32} ultimately will go to zero as p_{33} goes to zero , but if $\hat{p}$ is not in the neighborhood of its true value as these gains vanishes , a bias in $\hat{p}$ will remain .

```
function ex86
Y0=[1 1 1.1 .9 -4.2 0 0 0 0 20]';
[T,Y]=ode45('ex86fun',[0 2],Y0);
keyboard;
function xdot=ex86fun(t,x)
wn=2;w=randn;v1=randn;v2=randn;ceta=0.1;
r1=1;r2=1;a=-2*ceta*wn;
xdot(1)=x(2);
xdot(2)=-wn*wn*x(1)-
2*ceta*wn*x(2)+wn*wn*w;
y1=x(1)+v1;y2=x(2)+v2;
xdot(3)=x(4)+x(6)*(y1-x(3))/r1+x(7)*(y2-
x(4))/r2;
xdot(4)=x(3)*x(5)+a*x(4);+x(7)*(y1-
x(3))/r1+x(9)*(y2-x(4))/r2;

xdot(5)=x(8)*(y1-x(3))/r1+x(10)*(y2-
x(4))/r2;
xdot(6)=x(8)*(2-x(8)/r2)-x(6)*x(6)/r1;
xdot(7)=x(5)*x(6)+x(7)*(a-x(6)/r1-
x(9)/r2)+x(3)*x(8)+x(9);
xdot(8)=x(10)*(1-x(7)/r1)-x(6)*x(8)/r1;
xdot(9)=x(7)*(2*x(5)-x(7)/r1)+x(9)*(2*a-
x(9)/r2)+2*x(3)*x(10)+x(5)*x(5)/1;
xdot(10)=x(8)*(x(5)-x(7)/r1)+x(10)*(a-
x(9)/r2)+x(3)*x(11);
xdot(11)=-x(8)*x(8)/r1-x(10)*x(10)/r2;

xdot=xdot';
```

Fig. 8.24: MATLAB program code for Example 8.6

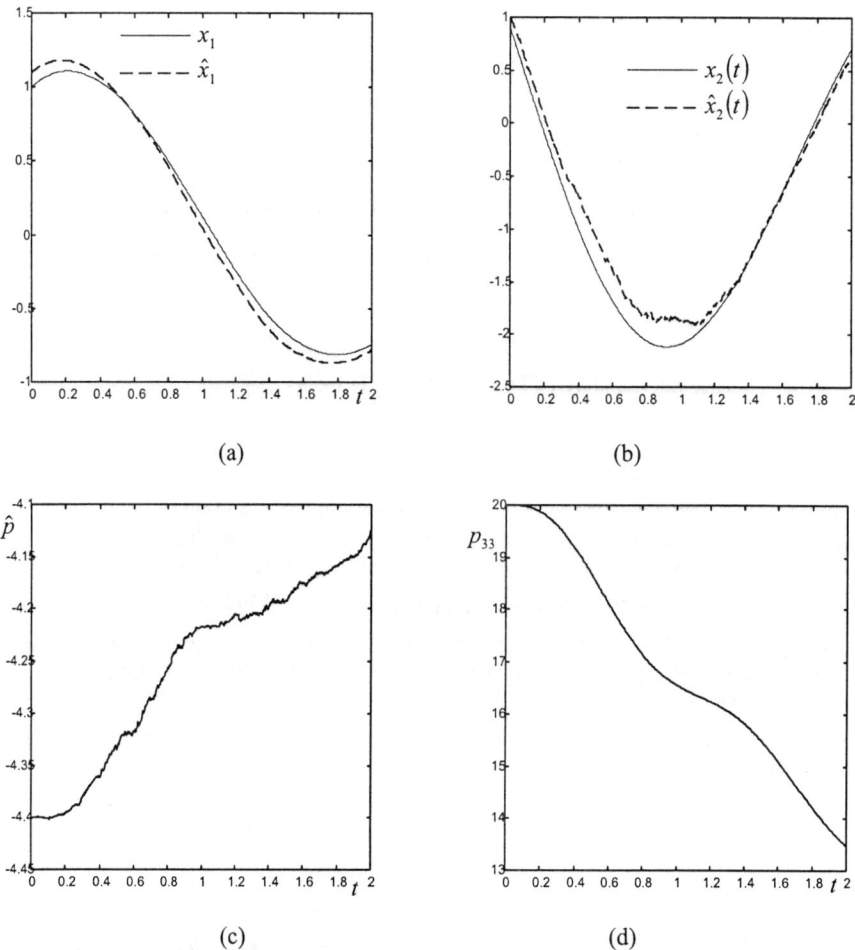

(a)   (b)

(c)   (d)

Fig. 8.25 Simulation results for the dumped harmonic oscillator: a) true and estimated state x_1; b) true and estimated parameter x_2; c) estimated parameter $p = -\omega_n^2$; d) estimation error variance p_{33}.

8.3.2 Noise-adaptive Filtering

The linear optimal Kalman filter is said to have innovations property ; that is , the filter extracts all the available information from the measurement , leaving only zero-mean white noise in the measurement residual. Considering a discrete-time model with constant system and noise matrices

$$x_k = Fx_{k-1} + w_{k-1}$$
$$y_k = Hx_k + v_k \qquad (8.208)$$

where x is the n-dimensional state vector , F is the state transition matrix from t_{k-1} to t_k , y_k is the p-dimensional observation vector and H is the observation mapping matrix . The stochastic disturbance vectors w_{k-1} and v_k are treated as independent , stationary , Gaussian, white noise sequences with properties

$$E\{w_k\} = q \ ; \ \ E\{(w_i - q)(w_j - q)^t\} = Q\delta_{ij} \tag{8.209}$$

$$E\{v_k\} = r \ ; \ \ E\{(v_i - r)(v_j - r)^t\} = R\delta_{ij} \tag{8.210}$$

where q and r are the true means and Q and R are the true moments about the mean (covariance matrices) of the process (state) and observation noises sequences , respectively . Given an a priori estimate of the state $\hat{x}_0$ and state covariance P_0 and given a priori information represented by (8.209) , (8.210) , the best linear minimum variance unbiased estimate of the state for the system defined by (18) is obtained sequentially for $k=1,2 \ldots$ with the discrete-Kalman filter

state propagation (time update) :

$$\hat{x}_k(-) = F\hat{x}_{k-1}(+) + q$$
$$P_k(-) = FP_{k-1}(+)F^t + Q \tag{8.211a}$$

innovations (measurement residual) :

$$\varepsilon_k = y_k - H\hat{x}_k(-) - r \tag{8.211b}$$

Kalman gain :

$$K_k = P_k(-)H^t\left[HP_k(-)H^t + R\right]^{-1} \tag{8.211c}$$

state estimation (measurement update) :

$$\hat{x}_k(+) = \hat{x}_k(-) + K_k\varepsilon_k$$
$$P_k(+) = (I - K_kH)P_k(-) \tag{8.211d}$$

The optimal residual covariance is

$$S_k = E\{\varepsilon_k\varepsilon_k^t\} = HP_k(-)H^t + R \tag{8.212}$$

because $E\{\hat{x}_k v_k^t\} = E\{x_k v_k^t\} = 0$. If the residual is not white or if its covariance does not equal (8.212) , the filter is suboptimal . This could be result of system parameters uncertainty , or it may be the result of actual noise moments that are different from the quantities (q, r, Q, R) used in the filter implementation (8.211) . Tests of whiteness , mean and covariance can indicate unsatisfactory behaviour of the filter and the results of these tests can be used to improve filter performances.

A *test of whiteness* can be based upon the *autocorrelation function* matrix $C(\tau)$ of the *measurement residual* or the *observation noise sample*

$$r_k = y_k - H\hat{x}_k(-) \tag{8.213}$$

Namely , for the linear observation state relationship at a given observation time t_k , given by (8.208), where y_k is a p-vector and v_k is the measurement white noise with mean r and covariance matrix R. The true state x_k is unknown, so v_k cannot be determined; but, an intuitive approximation for v_k is given by (24). Assuming that the random process represented r_k is not only stationary but ergodic, an estimate of $C(\tau)$ based on N samples can be calculated as

$$C(\tau) = \frac{1}{N}\sum_{n=\tau}^{N} r_n r_{n-\tau}^t \tag{8.214}$$

where N is large compared to τ. Elements of the autocorrelation function $c_{ij}(\tau)$ can be normalized by their zero-lag values, producing correlation coefficients, say $\rho_{ij}(\tau)$, that theoretically range between plus and minus one. Assuming negligible measurement cross-correlation (by hypothesis, the $v_j, j=1,\cdots,N$ are independent and the quantities r and R are constant; if the noise samples r_j are assumed to be representative of v_j, they may be considered independent and identically distributed), the test for whiteness can be constructed on the diagonal elements alone

$$\rho_{ii}(\tau) = \frac{c_{ii}(\tau)}{c_{ii}(0)}; i=1,\cdots,p \tag{8.215}$$

The normalized diagonal elements should be one for $\tau=0$ and zero otherwise. However, variations occur as a result of finite sample length. It can be shown that the 95% confidence limits on $\rho_{ij}(\tau)$ for $\tau \neq 0$ are

$$|\rho_{ii}(\tau)| \leq \frac{1.96}{\sqrt{N}}; i=1,\cdots,p \tag{8.216}$$

Consequently, if less than 5% of the $\rho_{ij}(\tau)$ exceed this threshold, then the i-th residual can be considered white, and the filter is processing the i-th measurement in optimal fashion. The test is clearly more specific to errors in R than in Q; single-element errors in the latter presumably could cause nonwhiteness in all elements of r_k. The observation noise statistics (measurement bias or mean value and covariance) can be estimated approximately with relatively simple batch-processing techniques. Since the noise samples r_k in (8.213) are assumed to be representative of the measurement noise v_k in (8.208), a simple estimation problem can be constructed. Define a random variable R on the sample space Ω_R from which the data $r_k; k=1,\cdots,N$ are obtained. Based on these empirical measurements, the unknown distribution of R, characterized by a mean r and covariance C_R, is to be estimated. An unbiased estimator for r is taken as a sample mean (see section V)

$$\hat{r} = \frac{1}{N}\sum_{k=1}^{N} r_k \tag{8.217}$$

An unbiased estimator for R is obtained by first constructing an estimator for C_R, the covariance of R, as follows (see chapter V)

$$\hat{C}_R = \frac{1}{N-1}\sum_{k=1}^{N}(r_k-\hat{r})(r_k-\hat{r})' \tag{8.218}$$

or, after substituting (8.217),

$$\hat{C}_R = \frac{1}{N-1}\sum_{k=1}^{N} r_k r_k' - \frac{N}{N-1}\hat{r}\hat{r}' \tag{8.219}$$

from which it follows

$$E\{\hat{C}_R\} = \frac{1}{N-1}\sum_{k=1}^{N} E\{r_k r_k'\} - \frac{N}{N-1}\frac{1}{N^2}\sum_{j=1}^{N}\sum_{k=1}^{N} E\{r_j r_k'\}$$

Taking into account (8.212) and the fact that r_k and r_j are independent, i.e. $E\{r_k r_k'\} = E\{r_k r_k'\}\delta_{kj}$, further follows

$$E\left\{\hat{C}_R\right\} = \frac{1}{N}\sum_{k=1}^{N}\left[HP_k\left(-\right)H'\right] + R \tag{8.220}$$

Thus , an unbiased estimate of R , after substitution of (8.219) , is given by

$$\hat{R} = \hat{C}_R - \frac{1}{N}\sum_{k=1}^{N}\left[HP_k\left(-\right)H'\right] + R$$

$$= \frac{1}{N-1}\sum_{k=1}^{N}\left\{\left(r_k - \hat{r}\right)\left(r_k - \hat{r}\right)' - \frac{N-1}{N}HP_k\left(-\right)H'\right\} \tag{8.221}$$

where $P_k\left(-\right)$ is computed in the Kalman filtering algorithm (8.211). If the Riccati eq. (8.211d) has reached steady-state , i.e. $P_k\left(-\right) = P_{ss}\left(-\right)$, from (8.221) follows

$$\hat{R}_s = \hat{C}_{Rs} - HP_{ss}\left(-\right)H' \tag{8.222}$$

Similar approximation estimates can be formed for process (state) noise mean (input disturbance bias) and covariance . For the state noise statistics consider the linear dynamical state relation at a given time t_{k+1}, where x_{k+1} is an n-dimensional state vector and w_k is disturbance input, i.e. process or state white noise , with mean q and covariance Q . The true states x_{k+1}, x_k are unknown , so w_k cannot be determined , but an intuitive approximation for w_k is (see eq. (8.208))

$$q_k = x_{k+1} - F\hat{x}_k\left(+\right) \tag{8.223}$$

where q_k is defined as *forcing residual* or *state noise sample* at time t_{k+1} . Unlike the measurement residual , the forcing residual must be approximated using the estimate $\hat{x}_{k+1}\left(+\right)$ in eq. (8.223) instead of the true state x_{k+1} . By hypothesis , the w_k for $k = 1, \cdots, N$ are independent , and the parameters q and Q are constant . If the q_k are assumed to be representative of the w_k, they may be considered independent and identically distributed . Again , define a parameter estimation problem : let Q be a random variable on the sample space Ω_Q from which is obtained the data set q_k , $k = 1, \cdots, N$. Based on the measurements , the unknown distribution of Q, characterized by q and Q , has to be estimated . An unbiased estimator for q is the sample mean

$$\hat{q} = \frac{1}{N}\sum_{k=1}^{N}q_k \tag{8.224}$$

An unbiased estimator for Q is obtained by first constructing the estimator for C_Q , the covariance of random variable Q ,

$$\hat{C}_Q = \frac{1}{N-1}\sum_{k=1}^{N}\left(q_k - \hat{q}\right)\left(q_k - \hat{q}\right)' \tag{8.225}$$

Following the same steps as in (8.221) , we have

$$E\left\{\hat{C}_Q\right\} = \frac{1}{N-1}\sum_{k=1}^{N}E\left\{q_kq_k'\right\} - \frac{N}{N-1}E\left\{\hat{q}\hat{q}'\right\}$$

or , after substituting $\hat{q}$ from (8.225) ,

$$E\left\{\hat{C}_Q\right\} = \frac{1}{N-1}\sum_{k=1}^{N}E\left\{q_kq_k'\right\} - \frac{1}{N(N-1)}\sum_{j=1}^{N}\sum_{k=1}^{N}q_jq_k'$$

Since from eq. (8.223)

$$q_k = F\left[x_k - \hat{x}_k\left(+\right)\right] + w_k$$

as well as q_k and q_j are independent, i.e. $E\left\{q_k q_j'\right\} = E\left\{q_k q_k'\right\}\delta_{kj}$, further follows

$$E\left\{\hat{C}_Q\right\} = \frac{1}{N}\sum_{k=1}^{N}\left[FP_k\left(+\right)F'\right] + Q \tag{8.226}$$

leading to the following unbiased estimate for Q

$$\hat{Q} = \hat{C}_Q - \frac{1}{N}\sum_{k=1}^{N}\left[FP_k\left(+\right)F'\right]$$

$$= \frac{1}{N-1}\sum_{k=1}^{N}\left\{\left(q_k - \hat{q}\right)\left(q_k - \hat{q}\right)' - \frac{N-1}{N}FP_k\left(+\right)F'\right\} \tag{8.227}$$

In steady-state , this is equivalent to

$$\hat{Q}_s = \hat{C}_{Qs} - FP_{ss}\left(+\right)F' \tag{8.228}$$

In summary , unbiased estimator for the first and second order noise statistics $\left(r, q, R, Q\right)$ are presented in (8.217), (8.222), (8.225) and (8.227), respectively. These estimates are based on N observation noise samples (measurement residuals), $r_k, k = 1, \cdots, N$ and state noise samples (forcing residuals) $q_k, k = 1, \cdots, N$, which are assumed to be statistically independent and identically distributed. However, it is assumed that the systems matrices F and H are exactly known. If the systems matrices are also unknown, the equations for noise statistics identification are much more complicated. Care must be exercised in the identification of unknown systems matrices from the residual sequence, since it is known that whiteness of the residual sequence is not a sufficient condition to estimate an unknown system matrix F (nonunique solution for F can be obtained from an identification scheme) . Moreover , it is possible to obtain the noise-adaptive filter gain directly, without explicit estimation of the noise covariance matrices Q and R , but all these discussions are beyond the present scope , and further information can be found in the literature. The organization of noise-adaptive filter is given in the next figure .

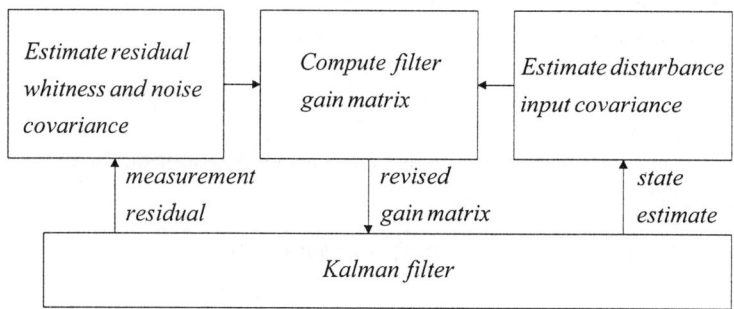

Fig. 8.26. Block scheme of noise-adaptive filter

The preceding results for $\hat{r}, \hat{q}, \hat{R}, \hat{Q}$ can be recast in *recursive limited memory* form for continual estimates of slowly varying noise statistics . These estimates are based upon the last l_r and l_q noise samples $r_j, j = k - l_r + 1, \cdots, k$ and $q_j, j = k - l_q + 1, \cdots, k$ at time   , respectively (the so-called sliding window or frame of the corresponding length l). Starting from eq. (8.217) , the observation noise mean at time   is given by

$$\hat{r}_k = \frac{1}{l_r}\sum_{j=k-l_r+1}^{k} r_j$$

from which it follows

$$\hat{r}_k = \frac{1}{l_r}\sum_{j=k-l_r+1}^{k-1} r_j + \frac{1}{l_r}r_k + \frac{1}{l_r}r_{k-l_r} - \frac{1}{l_r}r_{k-l_r}$$

$$= \frac{1}{l_r}\sum_{j=k-l_r}^{k-1} r_j + \frac{1}{l_r}\left(r_k - r_{k-l_r}\right)$$

or equivalently

$$\hat{r}_k = \hat{r}_{k-1} + \frac{1}{l_r}\left(r_k - r_{k-l_r}\right) \tag{8.229}$$

Taking into account (8.221) , the observation noise covariance at time is given by

$$\hat{R}_k = \frac{1}{l_r-1}\sum_{j=k-l_r+1}^{k}\left\{\left(r_j - \hat{r}_k\right)^2 - \frac{l_r-1}{l_r}\Gamma_j\right\};\Gamma_j = HP_j\left(-\right)H' \tag{8.230}$$

where

$$\left(r_j - \hat{r}_k\right)^2 = \left(r_j - \hat{r}_k\right)\left(r_j - \hat{r}_k\right)' \tag{8.231}$$

The eq. (8.230) can be rewritten as

$$\hat{R}_k = \frac{1}{l_r-1}\sum_{j=k-l_r+1}^{k-1}\left[\left(r_j - \hat{r}_k\right)^2 - \frac{l_r-1}{l_r}\Gamma_j\right] + \frac{1}{l_r-1}\left(r_k - \hat{r}_k\right)^2 - \frac{1}{l_r}\Gamma_k$$

$$+ \frac{1}{l_r-1}\left(r_{k-l_r} - \hat{r}_k\right)^2 - \frac{1}{l_r-1}\left(r_{k-l_r} - \hat{r}_k\right)^2 + \frac{1}{l_r}\Gamma_{k-l_r} - \frac{1}{l_r}\Gamma_{k-l_r}$$

$$= \frac{1}{l_r-1}\sum_{j=k-l_r}^{k-1}\left[\left(r_j - \hat{r}_k\right)^2 - \frac{l_r-1}{l_r}\Gamma_j\right]$$

$$+ \frac{1}{l_r-1}\left[\left(r_k - \hat{r}_k\right)^2 - \left(r_{k-l_r} - \hat{r}_k\right)^2 + \frac{l_r-1}{l_r}\left(\Gamma_{k-l_r} - \Gamma_k\right)\right] \tag{8.232}$$

However , since from (8.229) follows

$$\left(r_j - \hat{r}_k\right)^2 = \left[\left(r_j - \hat{r}_{k-1}\right) - \frac{1}{l_r}\left(r_k - r_{k-l_r}\right)\right]^2$$

$$= \left(r_j - \hat{r}_{k-1}\right)^2 - \frac{1}{l_r}\left(r_k - r_{k-l_r}\right)\left(r_j - \hat{r}_{k-1}\right)'$$

$$- \frac{1}{l_r}\left(r_j - \hat{r}_{k-1}\right)\left(r_k - r_{k-l_r}\right)' + \frac{1}{l_r^2}\left(r_k - r_{k-l_r}\right)^2$$

one concludes

$$\sum_{j=k-l_r}^{k-1}\left(r_j - \hat{r}_k\right)^2 = \sum_{j=k-l_r}^{k-1}\left(r_j - \hat{r}_{k-1}\right)^2 - \frac{1}{l_r}\left(r_k - r_{k-l_r}\right)\sum_{j=k-l_r}^{k-1}\left(r_j - \hat{r}_{k-1}\right)$$

$$- \frac{1}{l_r}\sum_{j=k-l_r}^{k-1}\left(r_j - \hat{r}_{k-1}\right)\left(r_k - r_{k-l_r}\right)' + \frac{1}{l_r}\left(r_k - r_{k-l_r}\right)^2 \tag{8.233}$$

Bearing in mind that

$$\sum_{j=k-l_r}^{k-1} \left(r_j - \hat{r}_{k-1} \right) = \sum_{j=k-l_r}^{k-1} r_j - \left(l_r - 1 \right)\hat{r}_{k-1} = 0 \qquad (8.234)$$

further follows

$$\sum_{j=k-l_r}^{k-1} \left(r_j - \hat{r}_k \right)^2 = \sum_{j=k-l_r}^{k-1} \left(r_j - \hat{r}_{k-1} \right)^2 + \frac{1}{l_r}\left(r_k - r_{k-l_r} \right)^2 \qquad (8.235)$$

Taking into account (8.230) , (8.232) and (8.235), one obtains

$$\hat{R}_k = \hat{R}_{k-1} + \frac{1}{l_r - 1}\left[\left(r_k - \hat{r}_k \right)^2 - \left(r_{k-l_r} - \hat{r}_k \right)^2 + \frac{1}{l_r}\left(r_k - r_{k-l_r} \right)^2 + \frac{l_r - 1}{l_r}\left(\Gamma_{k-l_r} - \Gamma_k \right) \right] \qquad (8.236)$$

Following the same steps as (8.229) and (8.236), the recursive estimators for mean and covariance of the process (state) noise are given by

$$\hat{q}_k = \hat{q}_{k-1} + \frac{1}{l_q}\left(q_k - q_{k-l_q} \right)$$

$$\hat{Q}_k = \hat{Q}_{k-1} + \frac{1}{l_q - 1}\left[\left(q_k - \hat{q}_k \right)^2 - \left(q_{k-l_q} - \hat{q}_k \right)^2 + \frac{1}{l_q}\left(q_k - q_{k-l_q} \right)^2 + \frac{l_q - 1}{l_q}\left(\Delta_{k-l_q} - \Delta_k \right) \right] \qquad (8.237)$$

$$\Delta_j = FP_j\left(+ \right)F'$$

Noise-adaptive limited memory filtering algorithm is presented in (8.229), (8.230), (8.236) and (8.237), as it is shown in Tab. 8.27. This procedure requires some extra storage and a shifting operation on the noise samples, such that r_{k-l_r}, q_{k-l_q} are discarded as r_k, q_k are incorporated into the processor. It is obvious that the noise covariance estimators $\hat{R}$ and $\hat{Q}$ in (8.236) and (8.237), respectively, may become negative in numerical applications, especially when a small amount of data has been processed. For this reason , the diagonal elements of $\hat{R}$ and $\hat{Q}$ are always reset to the absolute value of their estimates. During filter initialization , the noise samples r_k, q_k are poor indicators of the local ,noise environment. This requires a *fading memory approach* in which successive noise samples are multiplied by a growing weight factor given by

$$g_k = \left(k-1 \right)\left(k-2 \right)\cdots\left(k-\beta \right)/k^\beta \qquad (8.238)$$

which has the property

$$\lim_{k\to\infty} g_k = 1 \qquad (8.239)$$

and the use of the invalid noise samples is delayed for the first β stages. In fact, heuristically motivated and computationally simple approaches may have a great deal to offer in practice . The feasibility of the proposed approach is demonstrated by simulation results presented in the next example .

Example 8.7 (three state track model with position only measurement): Consider the one dimensional radar tracking problem , where for simplicity a target is moving in the x-direction with the velocity v and constant acceleration a . An estimate of the target position s, velocity v and acceleration a is required every Ts $(T=4s)$. Thus , if s_k, v_k, a_k indicate target position , velocity and acceleration at time $t_k = kT, k = 0,1,\cdots$, respectively , we are required to find estimates $\hat{s}_k, \hat{v}_k, \hat{a}_k$.

The first task is to model the systems dynamics . Since by hypothesis a is constant , then by integrating the equation $d^2s/dt^2 = a$ over the interval $[t_k,t], t_k \leq t \leq t_{k+1}$, one obtains

$$s_t = s_k + v_k(t - t_k) + \frac{a_k(t - t_k)^2}{2}$$
$$v_t = v_k + a_k(t - t_k) \tag{I}$$
$$a_t = a_k$$

from which it follows , for $t = t_{k+1}$,

$$x_{k+1} = \begin{bmatrix} 1 & T & T^2/2 \\ 0 & 1 & T \\ 0 & 0 & 1 \end{bmatrix} x_k + \begin{bmatrix} 0 \\ 0 \\ 1 \end{bmatrix} w_k = F x_k + G w_k \tag{II}$$

where the state vector $x_k = [s_k, v_k, a_k]^t$. Here w_k is the disturbance input (state or process noise), which compensates for unmodeled system dynamics . Namely , for the problem in question there is no state noise , but in the presence of target maneuver the state model (I) is inadequate and we compensate for a such model error by the zero-mean white noise term w_k , as it is done in eq. (II). We need also to model the measurement process . Since by assumption only the target position is observed , the output equation is given by

$$y_k = \begin{bmatrix} 1 & 0 & 0 \end{bmatrix} x_k + n_k = H x_k + n_k \tag{III}$$

where n_k is zero-mean white measurement noise sequence , uncorrelated with the process noise sequence w_k . The sequence w_k is adopted to be Gaussian with unit variance , while the random variable n_k is generated from the heavy-tailed Gaussian probability density function (pdf)

$$f(n) = (1 - \varepsilon) N(n/0,1) + \varepsilon N(n/0, \sigma_0^2); 0 \leq \varepsilon \leq 1, \sigma_0^2 \gg 1 \tag{IV}$$

where $N(./a,b)$ denotes the Gaussian pdf with mean a and variance b .In monopulse radars this heavy-tailed behaviour is presented because of the target glint , i.e. the heavy-tailed aspects of the underlying pdf is associated with the large glint spikes .

Observations are generated in the computer program from a set of true kinematic equations of motion , while the filter-world equations of motion are given by (II) .

The performance of the noise-adaptive Kalman filter are compared with the standard Kalman filter . Simulation results are compared in terms of the estimated noise statistics and cumulative estimation error (CEE criterion)

$$CEE(k) = \frac{1}{k} \sum_{i=1}^{k} \frac{\|\hat{x}_i(+) - x_i\|}{\|x_i\|} \tag{V}$$

where $\|.\|$ is the Euclidean norm . Fig. 8.28 depicts the true and estimated states , while the CEE results are plotted in Fig. 8.29 , and the estimated noise statistics are presented in Fig. 8.30. The simulation results have shown that the adaptive Kalman filter , based on an estimation of the noise statistics simultaneously with the systems states , can improve the state estimation performance when a priori statistics of the measurement noise are erroneous and there exist significant dynamical model error (disturbance input or state noise) , owing to the target maneuver .

Tab. 8.27. Summary of noise-adaptive Kalman filter

system model measurement model	$x_k = F\,x_{k-1} + G\,w_{k-1}$; $Gw \approx (q, Q)$ $y_k = H\,x_k + v_k$; $v \approx (r, R)$
state propagation (time update)	$\hat{x}_k(-) = F\,\hat{x}_{k-1}(+) + \hat{q}_{k-1}$ $P_k(-) = F\,P_{k-1}(+)\,F^t + \hat{Q}_{k-1}$
observation noise samples (l_r - sliding window length)	$r_k = y_k - H\,\hat{x}_k(-)$; $k \geq l_r$
observation noise statistic estimates (mean value and covariance matrix)	$\hat{r}_k = \hat{r}_{k-1} + \dfrac{1}{l_r}\left(r_k - r_{k-l_r}\right)$; $\Gamma_k = H P_k(-) H^t$ $\hat{R}_k = \hat{R}_{k-1} + \dfrac{1}{l_r - 1}[\left(r_k - \hat{r}_k\right)^2 - \left(r_{k-l_r} - \hat{r}_k\right)^2$ $+\dfrac{1}{l_r}\left(r_k - r_{k-l_r}\right)^2 + \dfrac{l_r - 1}{l_r}\left(\Gamma_{k-l_r} - \Gamma_k\right)^2]$
Kalman gain matrix	$K_k = P_k(-) H^t \left(\Gamma_k + \hat{R}_k\right)^{-1}$
state estimation measurement update	$\hat{x}_k(+) = \hat{x}_k(-) + K_k\left(r_k - \hat{r}_k\right)$ $P_k(+) = (I - K_k H) P_k(-)$
state noise samples (l_q - sliding window length)	$q_k = \left[\hat{x}_{k+1}(+) - F\,\hat{x}_k(+)\right]$; $k \geq l_q$
state noise statistics estimates (mean value and covariance matrix)	$\hat{q}_k = \hat{q}_{k-1} + \dfrac{1}{l_q}\left(q_k - q_{k-l_q}\right)$; $\Delta_k = F P_k(+) F^t$ $\hat{Q}_k = \hat{Q}_{k-1} + \dfrac{1}{l_q - 1}[\left(q_k - \hat{q}_k\right)^2 - \left(q_{k-l_q} - \hat{q}_k\right)^2$ $+\dfrac{1}{l_q}\left(q_k - q_{k-l_q}\right)^2 + \dfrac{l_q - 1}{l_q}\left(\Delta_{k-l_q} - \Delta_k\right)^2]$
initial assumptions	$\hat{x}_0(+), P_0(+), \hat{r}_0, \hat{w}_0, \hat{R}_0, \hat{Q}_{w0} \Rightarrow$ $\hat{q}_0 = G\,\hat{w}_0$; $\hat{Q}_0 = G\,\hat{Q}_{w0}\,G^t$
other assumptions	$\left(r - \hat{r}\right)^2 = \left(r - \hat{r}\right)\left(r - \hat{r}\right)^t$

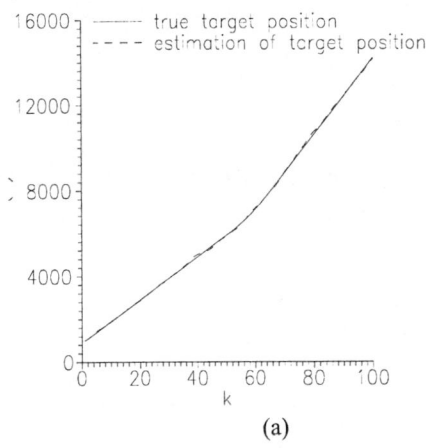

(a)

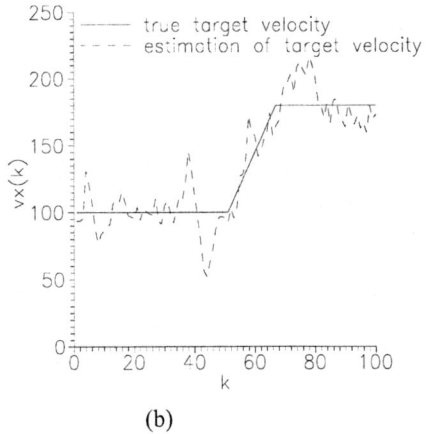

(b)

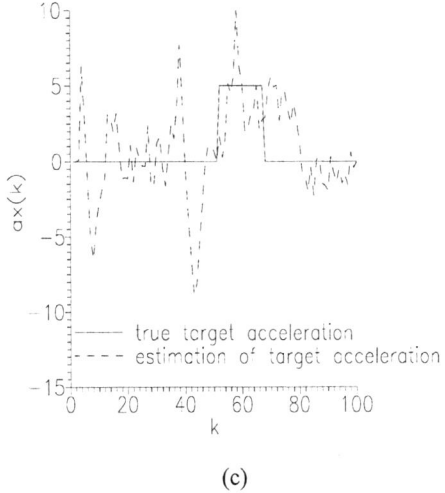

(c)

Fig. 8.28 The true and estimated target states : a) position ; b) velocity ; c) acceleration

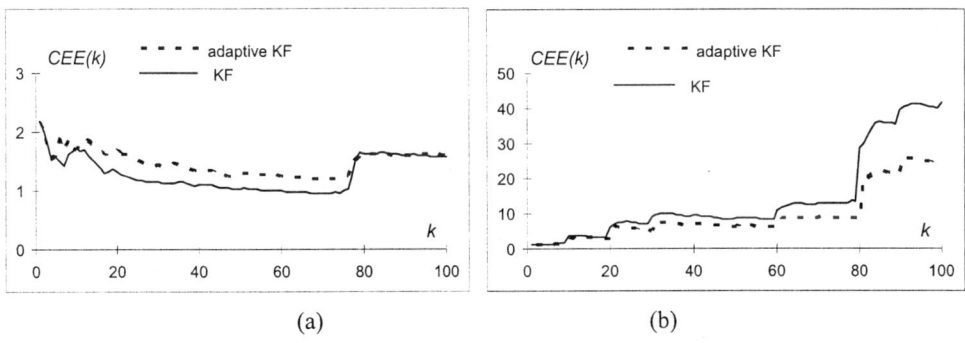

(a)   (b)

Fig. 8.29 Estimation error comparison of adaptive and standard Kalman filter : a) in the case of pure
Gaussian noise : b) in the case of heavy-tailed Gaussian mixture

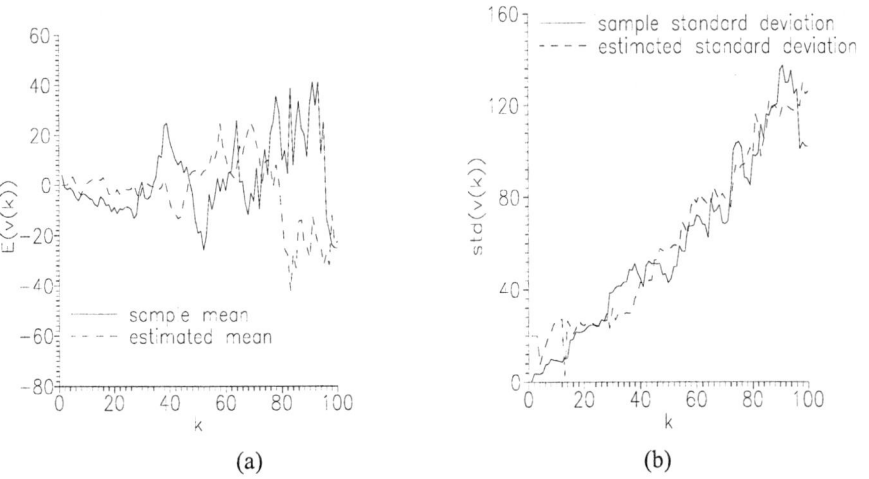

(a)   (b)

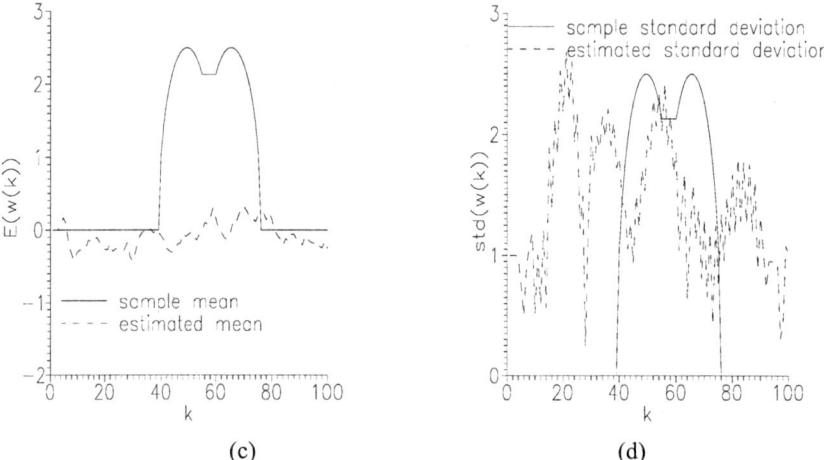

(c)   (d)

Fig. 8.30. Estimated the first and second ordered moments of state and observation noises:
a) observation noise expectation (mean value); b) observation noise standard deviation; c) state
noise expectation ; d) state noise standard deviation

8.3.3. Multiple Model Estimation

Multiple model estimation is an approach which allows for many possible values of system
parameters and noise levels. In this approach, a hypothesis is defined by specific values of the
system matrices, noise statistics or initial conditions, and a separate filter is assigned to each
hypothesis. However, the computational burden is great: given l independent parameters that each
can have k possible values, k^l *parallel filters* would be required. On the other hand, the
discrimination of multiple-model algorithms is somewhat better than that of the adaptive filtering
techniques discussed before. The success of multiple-model estimation is dependent not only on the
initial choice of likely models, but on the criterion chosen for *hypothesis testing*. Although the
statistics of the measurement residual should be involved, neither the whiteness test nor a
comparison of residual covariances can discriminate between models with high confidence.
Therefore, a better approach is to use the measurement residual and its covariance for each filter to
estimate the *likelihood* that the associate hypothesis is the correct one. The *optimal estimate* is then
computed as the weighted sum of the individual estimates , where each estimated is weighted by its
hypothesis probability. Recursive estimation of these probabilities is based on *sequential parallel*
application of *Bayes rule*: given I hypothetical parameter vector p of dimension l and the r-
component measurement vector y_k, where the subscript k denotes the $k\text{-}th$ sample of sequence , the
conditional probability mass function for the $j\text{-}th$ parameter set is (see chapter II)

$$P\left(p_j / y_k\right) = \frac{f\left(y_k / p_j\right)P\left(p_j\right)_{k-1}}{\sum_{i=1}^{I} f\left(y_k / p_j\right)P\left(p_j\right)_{k-1}} \qquad (8.240)$$

where $P(.)$ and $f(.)$ denotes the probability and probability density function (pdf) , respectively .
Since

$$P\left(p_j\right)_{k-1} = P\left(p_j / y_{k-1}\right)P\left(y_{k-1}\right)$$

and the probability that the measurement has been obtained at instant $k\text{-}1$ is one , further follows

$$P\big(p_j\big)_{k-1} = P\big(p_j / y_{k-1}\big) \qquad (8.241)$$

The eqs. (8.240) and (8.241) form the basis of a recursion, beginning with a given value of $P\big(p_j / y_0\big) = P\big(p_j / 0\big); 1 \le j \le I$. The conditional pdf of y_k given p_j, i.e. $f\big(y_k / p_j\big)$, remains to be found . Recall that the equations for a stationary discrete time linear system with true parameter set are (see section IV)

$$\begin{aligned} x_k &= F\,x_{k-1} + B\,u_{k-1} + G\,w_{k-1}\,; x_0 \\ y_k &= H\,x_k + v_k \end{aligned} \qquad (8.242)$$

where w_k is zero mean white Gaussian with covariance matrix Q_k , while v_k is the zero-mean white Gaussian with covariance matrix R_k , i.e. $w_k \approx N\big(w_k / 0, Q_k\big)$. and $v_k \approx N\big(v_k / 0, R_k\big)$. Thus , if x_k is known , the underlying conditional pdf would be Gaussian , that is

$$\begin{aligned} f\big(y_k / p\big) = f\big(y_k / x_k\big(p\big)\big) &= \frac{1}{(2\pi)^{n/2}\det^{1/2} R_k} \exp\left\{-\frac{1}{2}\big(y_k - Hx_k\big)' R_k^{-1}\big(y_k - Hx_k\big)\right\} \\ &= \frac{1}{(2\pi)^{n/2}\det^{1/2} R_k} \exp\left\{-\frac{1}{2} v_k' R_k^{-1} v_k\right\} \end{aligned} \qquad (8.243)$$

where n is the dimension of the state vector x_k. However , the true state is unknown , so it is necessary to use the pdf of y_k conditioned by the optimal estimate $\hat{x}_k$, generated by the Kalman filter (see chapter VII)

$$\begin{aligned} \hat{x}_k(-) &= F\hat{x}_{k-1}(+) + Bu_{k-1} \\ P_k(-) &= FP_{k-1}(+)F' + GQ_{k-1}G' \\ K_k &= P_k(-)H_t\big[HP_k(-)H' + R_k\big]^{-1} \\ \hat{x}_k(+) &= \hat{x}_k(-) + K_k\big[y_k - H\hat{x}_k(-)\big] \\ P_k(+) &= \big[P_k^{-1}(-) + H'R_k^{-1}H\big]^{-1} \end{aligned} \qquad (8.244)$$

This allows the required pdf of y_k given $\hat{x}_k(p)$ to be estimated by

$$f\big(y_k / \hat{x}_k(p)\big) = \frac{1}{(2\pi)^{n/2}\det^{1/2} S_k} \exp\left\{-\frac{1}{2} r_k' S_k^{-1} r_k\right\} \qquad (8.245)$$

where the measurement residual or innovation

$$r_k = y_k - H\hat{x}_k(-) \qquad (8.246)$$

and the corresponding residual covariance matrix

$$S_k = H P_k(-)H' + R_k \qquad (8.247)$$

S_k accounts for the estimation error in $\hat{x}_k$ and $f\big(y_k / \hat{x}_k(p)\big)$ has a greater spread than does $f\big(y_k / x_k\big)$. Starting from the above facts , the construction of the multiple-model algorithm is based on the following steps :
1) the Kalman filter (8.244) is formed for each hypothetical parameter set p_i providing state $\hat{x}_k(\pm; p_i)$ residual $r_k(p_i)$ and covariance estimates $P_k(\pm; p_i)$ and $S_k(p_i)$, as it is denoted in eqs. (8.244), (8.246) and (8.247), respectively .

2) the conditional pdf of the observations based on each hypothesis are estimated using (8.245)-(8.247) as

$$f\left(y_k / \hat{x}_k\left(p_i\right)\right) = \frac{1}{\left(2\pi\right)^{n/2} \det^{1/2} S_k\left(p_i\right)} \exp\left\{-\frac{1}{2} r_k'\left(p_i\right) S_k^{-1}\left(p_i\right) r_k\left(p_i\right)\right\} \tag{8.248}$$

where

$$r_k\left(p_i\right) = y_k - H\,\hat{x}_k\left(-;p_i\right) \tag{8.249}$$

and

$$S_k\left(p_i\right) = H\,P_k\left(-;p_i\right)H' + R_k \tag{8.250}$$

for $i = 1, \cdots, I$.

3) the conditional pdfs of $p_j; 1 \le j \le I$ given y_k are estimated by (8.240), (8.241), (8.244)-(8.250) as

$$P\left(p_j / y_k\right) = \frac{f\left(y_k / \hat{x}_k\left(p_j\right)\right) P\left(p_j / y_{k-1}\right)}{\sum\limits_{i=1}^{I} f\left(y_k / \hat{x}_k\left(p_i\right)\right) P\left(p_i / y_{k-1}\right)} \tag{8.251}$$

4) the *conditional mean estimate* (see chapter V) is the weighted sum of *parallel estimates*

$$\hat{x}_k\left(+\right) = \sum_{i=1}^{I} P\left(p_i / y_k\right) \hat{x}_k\left(+;p_i\right) \tag{8.252}$$

5) the estimate of *p* is formed as the weighted sum of the hypotethical parameter sets

$$\hat{p}_k = \sum_{i=1}^{I} P\left(p_i / y_k\right) \hat{p}_i \tag{8.253}$$

The block diagram of multiple – model estimator is given in the following figure .

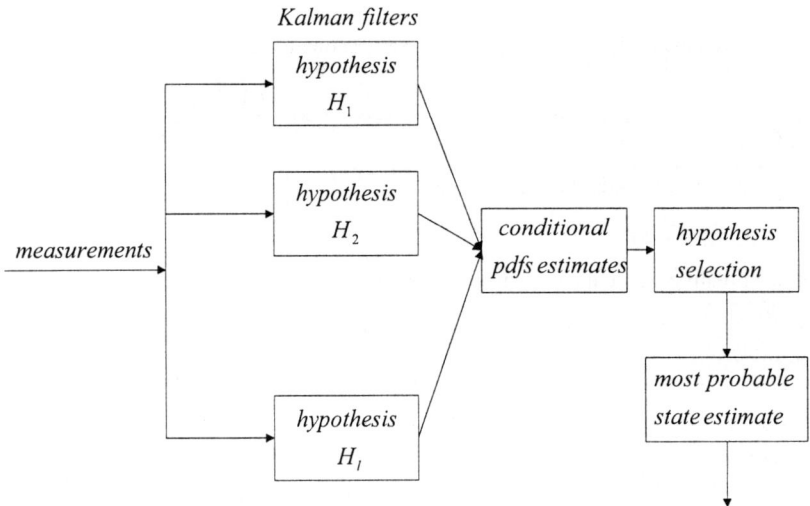

Fig. 8.31 Organization of multiple – model estimation

The multiple model algorithm can be simplified in a number of ways. It is always desirable to minimize the number of hypothesis. Of course , the choice of the number is a matter for engineering judgment. For example, if $Q = \alpha Q_0$ satisfactory adaptation may be afforded by varying the single parameter α rather than the individual elements of Q_0. Furthermore, once probability estimates have pinpointed a small number of most likely hypothesis, Kalman filters for the less likely hypothesis can be dropped. Moreover, the process of refining filter adaptation could be facilitated by introducing a new filter based on the most probable parameters . Finally , parallel – processing computers obviously would be very suited to multiple – model estimation. The feasibility of the approach has been presented in the next example .

Example 8.8. (revisited example 8.6.): The weathervane dynamics is described by

$$\frac{dx}{dt} = \begin{bmatrix} 0 & 1 \\ -\omega_n^2 & -2\xi\omega_n \end{bmatrix} x + \begin{bmatrix} 0 \\ \omega_n^2 \end{bmatrix} w = Ax + Gw; x = \begin{bmatrix} x_1 \\ x_2 \end{bmatrix} \tag{I}$$

where x_1 is the weathervane angle , x_2 represents the angular rate and w is the random wind input. Both the angle and rate are measured , yielding the following measurement model

$$y = \begin{bmatrix} 1 & 0 \\ 0 & 1 \end{bmatrix} x + v = Hx + v; y = \begin{bmatrix} y_1 \\ y_2 \end{bmatrix}; v = \begin{bmatrix} v_1 \\ v_2 \end{bmatrix} \tag{II}$$

The true natural frequency is $\omega_n = 2$, while the damping ratio is in reality $\xi = 0.1$. Whereas a continuous-discrete algorithm was used in the example 8.6., the current example is based on discrete-time estimation, and three discrete Kalman filters are implemented with the hypothesis $II_i, i = 1, 2, 3$ that the unknown parameter $p = \omega_n^2$ belongs to the parameter set $(p_1, p_2, p_3) = (4, 4.4, 4.8)$, while the predicted probabilities are $P(p_1 / 0) = 0.2, P(p_2 / 0) = 0.6, P(p_3 / 0) = 0.2$ $P(p_1 / 0) = 0.2, P(p_2 / 0) = 0.6, P(p_3 / 0) = 0.2$, respectively. With a sampling interval $T=0.1s$, the terms of order T^2 can be ignored , so the sampled plant is approximately given by (see chapter IV)

$$x_{k+1} = \begin{bmatrix} 1 & T \\ -\omega_n^2 T & 1 - 2\xi\omega_n T \end{bmatrix} x_k + w_k = F x_k + w_k ; F = I + AT \tag{III}$$

where the zero-mean discrete process noise covariance matrix

$$Q = E\{w_k w_k^t\} = TGQ_c G^t ; Q_c = 1 \deg^2 / s \tag{IV}$$

with Q_c being the spectral density of the continuous process noise w in (I) . The discrete plant (III) is observable through the measurement model

$$y_k = H x_k + v_k \tag{V}$$

where the covariance matrix of the zero-mean white discrete measurement noise v_k is given by

$$R = E\{v_k v_k^t\} = \frac{1}{T} R_c ; R_c = \begin{bmatrix} 1 & 0 \\ 0 & 1 \end{bmatrix} \tag{VI}$$

with R_c being the spectral density matrix of the continuous measurement noise v in (II). Thus, the measurement error covariance is effectively 10 times higher than the continuous measurement error due to sampling effects . The MATLAB program for simulating the system and multiple-model estimator is given in Fig. 8.32., while Fig. 8.33. depicts the simulation results, i.e. estimated hypothesis probabilities and estimated parameter values. The problem as posed is almost too easy for the multiple-model estimator concerned, because the parameter appears in the disturbance input matrix G, and therefore has a large effect on the sampled disturbance covariance Q. This, in turn, has a large effect on $P_k(-)$ and S_k, and the accompanying simulation demonstrates fast convergence to the true parameter value. If the parameter does not appear in the disturbance input matrix G, the convergence would have been much less certain.

```
function ex88


N=100; %the duration of simulation
T=0.1; %sampling peirod


% system's parameters
ceta=0.1;wn=2;
F=[1 T;-wn*wn*T 1-2*ceta*wn*T];
G=[0;wn*wn];H=eye(2);R=0.001/T;
p1=4;p2=4.4;p3=4.8;
wn1=sqrt(p1);wn2=sqrt(p2);wn3=sqrt(p3);


% filter's parameters
F1=[1 T;-wn1*wn1*T 1-2*ceta*wn1*T];
G1=[0;wn1*wn1];H1=eye(2);
F2=[1 T;-wn2*wn2*T 1-2*ceta*wn2*T];
G2=[0;wn2*wn2];H2=eye(2);
F3=[1 T;-wn3*wn3*T 1-2*ceta*wn3*T];
G3=[0;wn3*wn3];H3=eye(2);


% filter's initialization
x(1:2,1)=[1;-1];
y(1:2,1)=x(1:2,1)+randn(2,1)*sqrt(1/T);
xp1(1:2,1)=y(1:2,1);xp2=xp1;xp3=xp1;
xm1(1:2,1)=y(1:2,1);xm2=xm1;xm3=xm1;
Pp1=100*eye(2);Pm1=Pp1;
Pp2=100*eye(2);Pm2=Pp2;
Pp3=100*eye(2);Pm3=Pp3;
prob1(1)=0.2;prob2(1)=0.6;prob3(1)=0.2;
p(1)=[prob1(1) prob2(1) prob3(1)]'*[p1 p2
p3]';


for i=2:N

  % definition of the  measurements
  x(1:2,i)=F*x(1:2,i-1)+[0;randn]*...
  sqrt(T*G'*G*pi*pi/(180*180));
  y(1:2,i)=x(1:2,i)+randn(2,1)*sqrt(R);

  % definition of the first filter
  K=Pm1*H1'*inv(H1*Pm1*H1+R*eye(2));
```

```
  xp1(1:2,i)=xm1(1:2,i-1)+...
  K*(y(1:2,i)-H1*xm1(1:2,i-1));
  Pp1=(eye(2)-K*H1)*Pm1;
  xm1(1:2,i)=F1*xp1(1:2,i);

  Pm1=F1*Pp1*F1'+G1*(T*pi*pi/(180*180))*
  G1';

  % definition of the second filter
  K=Pm2*H2'*inv(H2*Pm2*H2+R*eye(2));
  xp2(1:2,i)=xm2(1:2,i-1)+...
  K*(y(1:2,i)-H2*xm2(1:2,i-1));
  Pp2=(eye(2)-K*H2)*Pm2;
  xm2(1:2,i)=F2*xp2(1:2,i);

  Pm2=F2*Pp2*F2'+G2*(T*pi*pi/(180*180))*
  G2';


  % definition of the third filter
  K=Pm3*H3'*inv(H3*Pm3*H3+R*eye(2));
  xp3(1:2,i)=xm3(1:2,i-1)+...
  K*(y(1:2,i)-H3*xm3(1:2,i-1));
  Pp3=(eye(2)-K*H3)*Pm3;
  xm3(1:2,i)=F3*xp3(1:2,i);

  Pm3=F3*Pp3*F3'+G3*(T*pi*pi/(180*180))*
  G3';


  r1=y(1:2,i)-H1*xm1(1:2,i-1);
  S1=H1*Pm1*H1'+R*eye(2);
  f1=exp(-
  0.5*r1'*inv(S1)*r1)/(2*pi*sqrt(det(S1)));

  r2=y(1:2,i)-H2*xm2(1:2,i-1);
  S2=H2*Pm2*H2'+R*eye(2);
  f2=exp(-
  0.5*r2'*inv(S2)*r2)/(2*pi*sqrt(det(S2)));

  r3=y(1:2,i)-H3*xm3(1:2,i-1);
  S3=H3*Pm3*H3'+R*eye(2);
  f3=exp(-
  0.5*r3'*inv(S3)*r3)/(2*pi*sqrt(det(S3)));
```

```
psum=[prob1(i-1) prob2(i-1) prob3(i-1)]*...
[f1 f2 f3]';

prob1(i)=prob1(i-1)*f1/psum;
prob2(i)=prob2(i-1)*f2/psum;
```

```
prob3(i)=prob3(i-1)*f3/psum;

p(i)=[prob1(i)  prob2(i)  prob3(i)]*[p1  p2
p3]';
end
```

Fig. 8.32 MATLAB program code for the harmonic oscillator and multiple model algorithm for estimating the natural frequency .

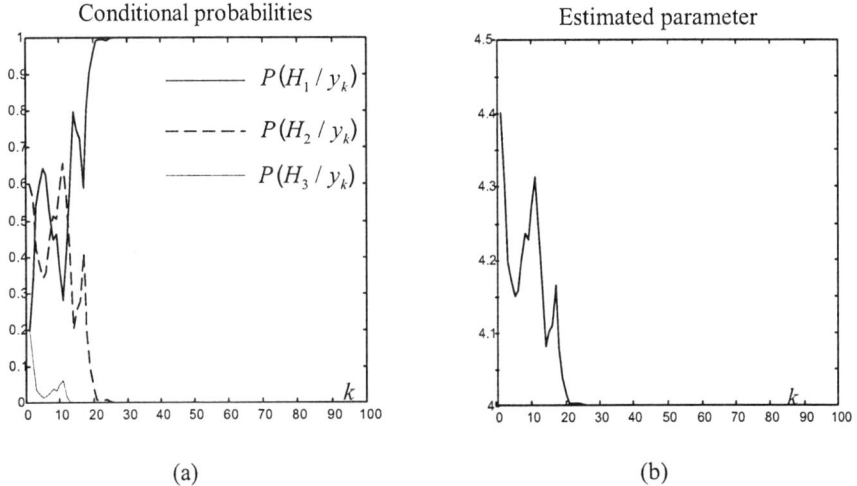

(a)   (b)

Fig. 8.33 Simulation results: a) estimated hypothesis probabilities; b) estimated parameter

APPENDICES

APPENDIX A: PROPERTIES OF LAPLACE AND Z TRANSFORMS

A brief introduction to the Laplace and Z transforms is presented in this appendix. For those readers wanting to delve more clearly into these transforms, a more detailed text is suggested for supplemental reading [22, 37].

By definition, *the (unilateral) Laplace transform* of a function of time $f(t)$ is

$$F(s) = L\{f(t)\} = \int_0^\infty f(t)e^{-st}dt \tag{A1}$$

Note that the variable time has been integrated out of the equation and that the Laplace transform is a function of the complex variable s. Table A.1 lists several useful theorems of the Laplace transform

Table A.1: Laplace transform theorems

Name	Theorem
Derivative	$L\left\{\dfrac{df(t)}{dt}\right\} = sF(s) - f(0^+)$
n-th order derivative	$L\left(\dfrac{d^n f(t)}{dt^n}\right) = s^n F(s) - s^{n-1}\dfrac{d^{n-1}f(0^+)}{dt^{n-1}} - \cdots - s\dfrac{df(0^+)}{dt} - f(0^+)$
Integral	$L\left\{\int_0^t f(\tau)d\tau\right\} = \dfrac{F(s)}{s}$
Shifting	$L\{f(t-t_0)h(t-t_0)\} = e^{-t_0 s}F(s)\ ;\ h(t-t_0) = \begin{cases} 1\ ; t \ge t_0 \\ 0\ ; t < t_0 \end{cases}$
Initial value	$\lim\limits_{t\to 0} f(t) = \lim\limits_{s\to\infty} sF(s)$
Final value	$\lim\limits_{t\to\infty} f(t) = \lim\limits_{s\to 0} sF(s)$
Frequency shift	$L\{e^{-at}f(t)\} = F(s+a)$
Convolution	$L\left\{\int_0^t f_1(t-\tau)f_2(\tau)d\tau\right\} = L\left\{\int_0^t f_1(\tau)f_2(t-\tau)d\tau\right\} = F_1(s)F_2(s)$

The proofs of these theorems are not given here, and interested readers should see the literature.

The inverse Laplace transform is given by

$$f(t) = L^{-1}\{F(s)\} = \frac{1}{2\pi j} \int_{\sigma-j\infty}^{\sigma+j\infty} F(s)e^{st}ds \qquad (A2)$$

where $L^{-1}\{\cdot\}$ indicates the inverse transform and $j = \sqrt{-1}$. Equations (A1) and (A2) form the Laplace transform pair. Given a function $f(t)$ we integrate (A1) to find its Laplace transform $F(s)$. Then if this function $F(s)$ is used to evaluate (A2), the result will be the original value of $f(t)$. The value of σ in (A2) is determined by the singularities, i.e. poles of $F(s)$, provided that $F(s)$ is a rational function (poles are the roots of denominator of $F(s)$). We seldom use (A2) to evaluate an inverse Laplace transform; instead we use (A1) to construct a table of transforms for common time functions. Then, when possible, we use this table to find the inverse transform rather than integrating (A2). A short table of commonly required transforms is given in Table A2.

Table A2: Laplace transform pairs

Time function $f(t)$	Laplace transformation $F(s)$
Dirac impulse $\delta(t)$	1
Step function $h(t)$	$\dfrac{1}{s}$
$th(t)$	$\dfrac{1}{s^2}$
$\dfrac{t^2 h(t)}{2}$	$\dfrac{1}{s^3}$
$t^{k-1}h(t)$	$\dfrac{(k-1)!}{s^k}$
$\exp(-at)h(t)$	$\dfrac{1}{s+a}$
$t\exp(-at)h(t)$	$\dfrac{1}{(s+a)^2}$
$t^k \exp(-at)h(t)$	$\dfrac{(k-1)!}{(s+a)^k}$
$(1-\exp(-at))h(t)$	$\dfrac{a}{s(s+a)}$
$\left(t - \dfrac{1-\exp(-at)}{a}\right)h(t)$	$\dfrac{a}{s^2(s+a)}$
$(1-(1+at)\exp(-at))h(t)$	$\dfrac{a^2}{s(s+a)^2}$

$\left(\exp(-at)-\exp(-bt)\right)h(t)$	$\dfrac{(b-a)}{(s+a)(s+b)}$
$\sin(bt)h(t)$	$\dfrac{b}{s^2+b^2}$
$\cos(bt)h(t)$	$\dfrac{s}{s^2+b^2}$
$t\sin(bt)h(t)$	$\dfrac{2bs}{\left(s^2+b^2\right)^2}$
$t\cos(bt)h(t)$	$\dfrac{s^2-b^2}{\left(s^2+b^2\right)^2}$
$\exp(-at)\sin(bt)h(t)$	$\dfrac{b}{\left(s+a\right)^2+b^2}$
$\exp(-at)\cos(bt)h(t)$	$\dfrac{s+a}{\left(s+a\right)^2+b^2}$
$\left(1-\exp(-at)\left(\cos(bt)+\dfrac{a}{b}\sin(bt)\right)\right)h(t)$	$\dfrac{a^2+b^2}{s\left((s+a)^2+b^2\right)}$

As it can be seen from Tab. A2, we usually work with the Laplace transform expressed as a ratio of polynomials in the variable s (we call this ratio of polynomials as a *rational function*). However, the table used to find the inverse transform contain only low-order rational functions. Hence, a method is required for converting from a general rational function to the forms that appears in the Tab. A2. This method is called the *partial-fraction expansion* method. To perform this expansion, first the roots of the denominator of $F(s)$ must be found. Then, $F(s)$ can be expressed as

$$F(s)=\frac{B(s)}{A(s)}=\frac{B(s)}{\displaystyle\prod_{j=1}^{p}(s-s_j)^{k_j}}=\sum_{j=1}^{p}\sum_{i=1}^{k_j}\frac{K_{ji}}{(s-s_j)^i} \tag{A3}$$

where Π indicates the product of terms, and s_j is the repeated-root of order k_j. The coefficient of the repeated roots terms are calculated from the equation

$$K_{ji}=\frac{1}{(k_j-1)!}\lim_{s\to s_j}\frac{d^{k_j-1}}{ds^{k_j-1}}\left[(s-s_j)^{k_j}F(s)\right] \tag{A4}$$

From Tab. A2, the terms in (A3) has the inverse transform of the form

$$L^{-1}\left\{\frac{K_{ji}}{(s-s_j)^i}\right\}=\frac{K_{ji}}{(i-1)!}t^{i-1}\exp\{s_jt\} \ ; \ t\ge 0 \tag{A5}$$

yielding

$$f(t) = L^{-1}\{F(s)\} = \sum_{j=1}^{p}\sum_{i=1}^{k_j}\frac{K_{ji}}{(i-1)!}t^{i-1}\exp(s_jt) \; ; \; t \geq 0 \tag{A6}$$

An alternative way of calculating inverse transform can be based on Cauchy's theorem

$$f(t) = \sum_{j=1}^{p}\operatorname*{Res}_{s_j}\{F(s)\exp(st)\} \; ; \; t \geq 0 \tag{A7}$$

where the residue for the repeated-root s_j of order k_j is given by

$$\operatorname*{Res}_{s_j}\{F(s)\exp(st)\} = \frac{1}{(k_j-1)!}\lim_{s\to s_j}\frac{d^{k_j-1}}{ds^{k_j-1}}\left[(s-s_j)^{k_j}F(s)\exp(st)\right] \tag{A8}$$

By definition, *the (unilateral) Z-transform* of a number sequence $\{f(k)\}$ is defined as a power series in z^{-k} with coefficients equal to the values of $f(k)$, i.e. [15, 38]

$$F(z) = Z\{f(k)\} = \sum_{k=0}^{\infty}f(k)z^{-k} \tag{A9}$$

If the sequence $\{f(k)\}$ is generated from a continuous-time function $f(t)$ by sampling every T seconds, $f(k)$ is understood to be $f(kT)$; that is, the T is dropped for convenience (T=1). Let $F(s)$ be the Laplace transform of $f(t)$ and $F(z)$ be the Z-transform of $f(k)$. Then, the properties of Z-transform are given in the table A3.

Table A3: Properties of Z-transform

Laplace transform	Samples	Z-transform	Comment
$F(s)$	$f(kT) = f(k)$	$F(z)$	
$k_1F_1(s) + k_2F_2(s)$	$k_1f_1(k) + k_2f_2(k)$	$k_1F_1(z) + k_2F_2(z)$	Z-transform is linear
$F_1(e^{Ts})F_2(e^{Ts})$	$\sum_{l=-\infty}^{\infty}f_1(l)f_2(k-l)$	$F_1(z)F_2(z)$	Discrete convolution corresponds to the product of Z-transforms, $z = e^{sT}$ is a link between z and s complex variables, i.e. $L\{f(k)\} = F(z)\big|_{z=e^{sT}}$
$e^{-nTs}F(s)$	$f(k-n)$	$z^{-n}F(z)$	shift in time
$e^{nTs}F(s)$	$f(k+n)$	$z^{n}\left[F(z) - \sum_{k=0}^{n-1}f(k)z^{-k}\right]$	
$F(s+a)$	$e^{-akT}f(k)$	$F(e^{aT}z)$	shift in frequency

	$f(\infty)=\lim_{k\to\infty} f(k)$	$\lim_{z\to 1}(z-1)F(z)$	In all poles of $(z-1)F(z)$ are inside the unit circle $|z|=1$, a $F(z)$ converges for $|z|\geq 1$
-	$f(0)=\lim_{k\to 0} f(k)$	$\lim_{z\to\infty} F(z)$	
$F\left(\dfrac{s}{\omega_n}\right)$	$f(\omega_n k)$	$F\left(\dfrac{z}{\omega_n T}\right)$	Time and frequency scaling
	$f_1(k)f_2(k)$	$\dfrac{1}{2\pi j}\oint_C F_1(\xi)F_2(z/\xi)\dfrac{d\xi}{\xi}$	Time product
$F_3(s)=F_1(s)F_2(s)$	$f_3(k)=\displaystyle\int_{-\infty}^{\infty} f_1(\tau)f_2(k-\tau)d\tau$	$F_3(z)$	Continuous convolution does not correspond to the product of Z-transform

A short table of Z-transform of discrete sequences is given in Table A4.

Table A4:

$f(t)$	$F(z)$
$h(t)$	$\dfrac{z}{z-1}$
$\dfrac{t^2}{2}$	$\dfrac{T^2 z(z+1)}{2(z-1)^3}$
t^{n-1}	$\lim_{a\to 0}(-1)^{n-1}\dfrac{\partial^{n-1}}{\partial a^{n-1}}\left[\dfrac{z}{z-\exp(-aT)}\right]$
$\exp(-at)$	$\dfrac{z}{z-\exp(-aT)}$
$t\exp(-aT)$	$\dfrac{Tz\exp(-aT)}{(z-\exp(-aT))^2}$
$t^n\exp(-at)$	$(-1)^n\dfrac{\partial^n}{\partial a^n}\left[\dfrac{z}{z-\exp(-aT)}\right]$
$1-\exp(-at)$	$\dfrac{z(1-\exp(-aT))}{(z-a)(z-\exp(-aT))}$

$t - \dfrac{1 - \exp(-aT)}{a}$	$\dfrac{z\left[\left(aT - 1 + \exp(-aT)\right)z + \left(1 - \exp(-aT) - aT\exp(-aT)\right)\right]}{a(z-1)^2(z - \exp(-aT))}$
$1 - (1 + at)\exp(-at)$	$\dfrac{z\left[\left(1 - \exp(-aT) - aT\exp(-aT)\right)z + \left(\exp(-2aT) + (aT - 1)\exp(-aT)\right)\right]}{(z-1)(z - \exp(-aT))^2}$
$\exp(-at) - \exp(-bt)$	$\dfrac{z\left(\exp(-aT) - \exp(-bT)\right)}{(z - \exp(-aT))(z - \exp(-bT))}$
$\sin(bt)$	$\dfrac{z\sin(bT)}{z^2 - 2z\cos(bT) + 1}$
$\cos(bt)$	$\dfrac{z(z - \cos(bT))}{z^2 - 2z\cos(bT) + 1}$
$\exp(-at)\sin(bt)$	$\dfrac{z\exp(-aT)\sin(bT)}{z^2 - 2z\exp(-aT)\cos(bT) + \exp(-2aT)}$
$\exp(-at)\cos(bt)$	$\dfrac{z^2 - z\exp(-aT)\cos(bT)}{z^2 - 2z\exp(-aT)\cos(bT) + \exp(-2aT)}$

The inverse Z-transform is the closed complex integral

$$f(k) = \frac{1}{2\pi j}\oint_C F(z) z^{k-1} dz \tag{A10}$$

where the contour is a circle in the region of convergence of $F(z)$. If $f(k)$ is causal, i.e. $f(k) = 0$ for $k < 0$, then the region of convergence is outside the smallest circle which contains all the poles of rational function $F(z)$ (the poles are the roots of the denominator of $F(z)$). It is this property which permits inversion by partial-fraction expansion and long division. Similarly, as in the case of Laplace transform, the *partial-fraction procedure* is to expand $F(z)$ into the forms of the terms that appear in the Z-transform tables, such as Tab. A4. The procedure is then the same as far the Laplace transform. However, note that we generally expand $F(z)/z$ into partial fractions, and then multiply by z to obtain the expression in the proper form. Another possibility for performing of inversion is to apply the *Cauchy's procedure* (A7) and (A8) where the function $F(s)\exp(st)$ has to be replaced by $F(z)z^{k-1}$. The procedure is then the same as for the Laplace transform. Finally, the *power series method* for finding the inverse Z-transform involves dividing the denominator of $F(z)$ into the numerator such that a power series of the form $\sum_{i=0}^{\infty} a_i z^{-i}$ is obtained. The values of the discrete sequence $\{f(k)\}$ are seen to be the coefficients in this series, i.e. $f(k) = a_k, k = 0, 1, \ldots$.

Z transform relation to the Laplace transform: In communications and control practice, the discrete-time sequence $f(k)$ are samples of a continuous-time waveform $f(t)$. These can be interpreted as a form of modulation of the signal $f(t)$ by a sequence of impulses $\sum_{k=0}^{\infty} \delta(t-kT)$, i.e.

$$f_s(t) = \sum_{k=0}^{\infty} f(kT)\delta(t-kT)$$

where T is a sampling interval. Since the Laplace transform of $\delta(t-kT)$ is $L\{\delta(t-kT)\} = \exp\{-skT\}$, we have $F_s(s) = L\{f_s(kT)\} = \sum_{k=0}^{\infty} f(kT)\exp\{-skT\}$. On the other hand, the Z-transform of the sequence $f(kT)$ is defined by $F(z) = \sum_{k=0}^{\infty} f(kT)z^{-k}$. Comparing the last two equations, we have

$$F(z) = F_s(s)\big|_{z=\exp\{sT\}}$$

which establishes the relationship between z and s complex frequencies, i.e. $z = \exp(sT)$. Therefore, a point $s_1 = \sigma_1 + j\omega_1$ in the s-plane transforms to a point $z_1 = \exp\{s_1 T\} = \exp\{\sigma_1 T\}\exp\{j\omega_1 T\}$ in the z-plane. This means $|z_1| = \exp\{\sigma_1 T\}$ and $\arg\{z_1\} = \omega_1 T$. To obtain more information on s-z plane relationship, we consider now the case $\sigma_1 = 0$ and find ω_1 for which $\arg\{z_1\}$ changes from $-\pi$ to π. Thus we have $-\pi < \omega_1 T < \pi$, yielding $-\pi/T < \omega_1 < \pi/T$ or $-\omega_s/2 < \omega_1 < \omega_s/2$ with $\omega_s = 2\pi/T$. Therefore, the path of imaginary axis between $\pm(\omega_s/2)$ transforms into the unit circle $|z| = 1$. It follows that a left-hand-side strip $(\sigma < 0)$ ω_s wide in the s-plane transforms inside the unit circle, while the corresponding right-hand-side strip $(\sigma > 0)$ transforms outside the unit circle. Successive ω_s wide strips in the s-plane transform into the same place.

APPENDIX B: A REVIEW OF MATRIX ANALYSIS

This appendix presents a very brief review of matrices. Those readers requiring more details are referred to the literature [7, 43].

The general $m \times n$ matrix A (m rows, n columns) is written as

$$A = \begin{bmatrix} a_{11} & a_{12} & \cdots & a_{1n} \\ a_{21} & a_{22} & \cdots & a_{2n} \\ \vdots & \vdots & \ddots & \vdots \\ a_{m1} & a_{m2} & \cdots & a_{mn} \end{bmatrix} = \left[a_{ij} \right]_{m \times n} \tag{B1}$$

where a_{ij} denotes the element that is common to both the i-th row and the j-th column. If either m or n is equal to 1, the matrix is often called a vector; that is, a vector is a matrix of either one row or one column. If both m and n are equal to 1, the quantity A is a scalar. Generally, for a vector, only one subscript is used to indicate the row of a column vector or the column of a row vector.

Some useful definitions are now given.

Diagonal matrix is a square matrix $n \times n$ in which all elements $a_{ij} = 0$ for $i \neq j$; that is, all elements of the matrix not on the main diagonal are zero. An example of a diagonal matrix is

$$D = diag\{d_1, d_2, ..., d_n\} = \begin{bmatrix} d_1 & 0 & \cdots & 0 \\ 0 & d_2 & \cdots & 0 \\ \vdots & \vdots & \ddots & \vdots \\ 0 & 0 & \cdots & d_n \end{bmatrix} = \left[d_i \right]_{n \times n} \tag{B2}$$

Zero matrix The matrix $[0]$, herein simply denoted 0, is defined as the matrix in which each element is the number 0.

Identity matrix is a diagonal matrix in which all diagonal elements are equal to 1. This matrix is usually denoted by I. The identity matrix has the property that, for A a square matrix

$$AI = IA = A \tag{B3}$$

If A is $m \times n$, then

$$AI_n = A \; ; \; I_m A = A \tag{B4}$$

where I_r denotes the $r \times r$ identity matrix. Note that the identity matrix is to matrix algebra what the value of unity is to scalar algebra.

Transpose of a matrix is formed by interchanging the rows and columns; that is, $A = \left[a_{ij} \right]_{m \times n}$ and $A^T = \left[a_{ji} \right]_{n \times m}$, where A^T denotes the transpose of A. For square matrices, if $A = A^T$, then A is said to be *symmetric*. If $A^T = -A$, A is said to be *skew-symmetric*. A property of a skew-symmetric matrix is that all elements on the main diagonal are zero. The transpose has the properties

$$\left(A+B\right)^{T} = A^{T} + B^{T} \;\; ; \;\; \left(AB\right)^{T} = B^{T} A^{T} \tag{B5}$$

Trace of a square matrix A, denoted by $tr\{A\}$ is the sum of the diagonal elements of A; that is

$$tr\{A\} = \sum_{i=1}^{n} a_{ii} \tag{B6}$$

For square matrices A and B, we have

$$tr\{AB\} = tr\{BA\} \tag{B7}$$

Hence

$$tr\{ab^{T}\} = b^{T} a = a^{T} b \tag{B8}$$

$$a^{T} Aa = tr\{Aaa^{T}\} \tag{B9}$$

where a, b are column vectors, while a^{T} and b^{T} are row vectors.

Determinant of a square matrix A, denoted as $|A|$, is a scalar defined by

$$|A| = \sum_{\substack{i=1}}^{n} \sum_{\substack{j=1 \\ j\neq i}}^{n} \cdots \sum_{\substack{l=1 \\ l\neq i,j,...,k}}^{n} a_{1i} a_{2j} \cdots a_{nl} \tag{B10}$$

In each terms the second subscripts $i, j,...,l$ are permutations of the numbers $1,2,...,n$. Terms whose subscripts are even permutations are given plus signs, and those with odd permutations are given minus signs. The determinant has the property

$$|AB| = |A||B| \; ; \; |I| = 1 \tag{B11}$$

where I is an identity matrix. If A and B are $m \times n$ and $n \times m$ matrices, respectively, then

$$|I_{m} + AB| = |I_{n} + BA| \tag{B12}$$

where I_{r} is $r \times r$ identity matrix.

Minor m_{ij} is defined for the element a_{ij} of a square matrix A as the determinant of the matrix that results from deleting the t-th row and j-th column of A.

Cofactor c_{ij} of an element a_{ij} of the square matrix A is defined as

$$c_{ij} = \left(-1\right)^{i+j} m_{ij} \tag{B13}$$

Adjoint $adj(A)$ of a square matrix A is defined as the transpose matrix of cofactors

Inverse A^{-1} of a square matrix A is given by

$$A^{-1} = \frac{adj(A)}{|A|} \tag{B14}$$

The properties of the inverse matrix A are

$$AA^{-1} = A^{-1}A = I \tag{B15}$$

$$\left(AB\right)^{-1} = B^{-1}A^{-1} \tag{B16}$$

$$\left(A^{-1}\right)^{T} = \left(A^{T}\right)^{-1} \tag{B17}$$

$$\left|A^{-1}\right| = 1/\left|A\right| \tag{B18}$$

Note that the matrix inverse is defined only for a square matrix and exists only if the determinant $\left|A\right|$ of the matrix A is nonzero. In other words, all square matrices do not have inverses; only those that are nonsingular have an inverse. For purposes of clarification, a nonsingular matrix may be defined as follows: the determinant of the matrix is not zero, i.e. $\left|A\right| \neq 0$, or no row (column) is a linear combination of other rows (columns).

Rank of a matrix A is the dimension of the largest square matrix contained in A (formed by deleting rows and columns) which has a nonzero determinant. In other words, the rank of a matrix is the maximum number of linearly independent columns or rows in the matrix (it makes no difference whether we take rows or columns). A nonzero vector has rank 1. A matrix which rank equals the number of rows or columns (whichever is less) is said to be of *full rank*. A square matrix of full rank is nonsingular, and vice versa. Moreover, we have

$$rank\left(A+B\right) \leq rank\left(A\right) + rank\left(B\right) \tag{B19}$$

$$rank\left(AB\right) \leq \min\left(rank\left(A\right), rank\left(B\right)\right) \tag{B20}$$

$$rank\left(ab^{T}\right) = \begin{cases} 1 \; ; \; a \neq 0 \neq b \\ 0 \; ; \; a = 0 \text{ or } b = 0 \end{cases} \tag{B21}$$

$$rank\left(\sum_{i=1}^{n} a_i b_i^{T}\right) \leq n \tag{B22}$$

Matrix pseudoinverse: The matrix inverse is defined for square matrices only. For nonsquare matrices the equivalent operator is the pseudoinverse. If a matrix A has more rows than columns, the pseudoivnerse is defined as

$$A^{\#} = \left(A^{T} A\right)^{-1} A^{T} \tag{B23}$$

for nonsingular $A^{T} A$. If A has more columns than rows, the pseudoinverse is defined as

$$A^{\#} = A^{T}\left(AA^{T}\right)^{-1} \tag{B24}$$

Several useful relations concerning the pseudoinverse are the following:

$$AA^{\#} A = A \tag{B25}$$

$$A^{\#} AA^{\#} = A^{\#} \tag{B26}$$

$$\left(A^{\#} A\right)^{T} = A^{\#} A \tag{B27}$$

$$\left(AA^{\#}\right)^{T} = AA^{\#} \tag{B28}$$

If A is square nonsingular, then $A^{\#} = A^{-1}$. If A is $m \times n$, then $A^{\#}$ is $n \times m$. If the linear equation $Ax = y$ have a solution, then $x = A^{\#} y$ is the solution of minimum length $\left\|x\right\| = \sqrt{x^{T}x} = \sqrt{\sum_{i=1}^{n} x_i^{2}}$, where n is the dimension of a column vector x (it should be noted that $A^{\#}$ is $n \times m$ matrix and y is $m \times 1$ column vector). If $Ax = y$ has no solution, then $x = A^{\#}y$ minimizes the sum of squares of the deviations $y - Ax$; that is, it minimizes the criterion $(y - Ax)^T (y - Ax)$; and of all vectors having this property, $x = A^{\#}y$ has minimum length $\|x\|$.

Eigenvalues (characteristic values) of a square matrix A are the roots of the polynomial equation, the so-called characteristic equation

$$f(\lambda) = |\lambda I - A| = 0 \tag{B29}$$

where $|\cdot|$ denotes the determinant, as does $\det(\cdot)$. By expanding determinant, it can be seen that $f(\lambda)$ is a polynomial in λ. The Cayley-Hamilton Theorem states that, for the same polynomial expression,

$$f(A) = 0 \tag{B30}$$

That is, every square matrix satisfies its own characteristic equation. Two properties of an $n \times n$ square matrix A are

$$|A| = \prod_{i=1}^{n} \lambda_i \tag{B31}$$

$$tr(A) = \sum_{i=1}^{n} \lambda_i \tag{B32}$$

Let A be a square matrix. Suppose $\lambda_{\min}$ and $\lambda_{\max}$ are the eigenvalues of A with smallest and largest absolute values, respectively. Then for any (column) vector b

$$|\lambda_{\min}| \|b\| \leq \|Ab\| \leq |\lambda_{\max}| \|b\| \tag{B33}$$

$$|\lambda_{\min}| \|b\|^2 \leq |b^T Ab| \leq |\lambda_{\max}| \|b\|^2 \tag{B34}$$

where $\|b\| = \sqrt{b^T b}$ is the length of the vector b. Finally, the rank of a square matrix A equals the number of nonzero eigenvalues.

Eigenvectors (characteristic vectors) of the square matrix A are the vectors x_i that satisfy the equation

$$\lambda_i x_i = A x_i \tag{B35}$$

where the λ_i are the eigenvalues of A. If A is symmetric $n \times n$, then one can find from (B35) n mutually orthogonal eigenvectors $x_1, ..., x_n$ of A, satisfying $x_i^T x_j = 0$ for $i \neq j$; $i,j=1,...,n$. Usually, we normalize the vectors so that

$$x_i^T x_j = \delta_{ij} = \begin{cases} 1 \; ; \; i = j \\ 0 \; ; \; i \neq j \end{cases} \tag{B36}$$

where δ_{ij} is the Kronecker delta symbol. The x_i then form a set of orthonormal eigenvectors of A.

Let X be the $n \times n$ matrix whose t-th column is x_i. In view of the last equation, we have $X^T X = XX^T = I$, i.e. $X^T = X^{-1}$. The matrix X is said to be *unitary*, or *orthogonal*. If $Ax_i = 0$,

$(x_i \neq 0)$, then x_i is called a *null vector* of A. If A is square, then it can possess null vectors only if it is singular. A singular matrix has at least one zero eigenvalue λ_i.

Functions of square matrices: It is possible to define special polynomial functions of a square matrix A, two of which are

$$e^A = \sum_{k=0}^{\infty} \frac{A^k}{k!} = I + A + \frac{1}{2!}A^2 + \frac{1}{3!}A^3 + \cdots \tag{B37}$$

$$\sin(A) = A - \frac{1}{3!}A^3 + \frac{1}{5!}A^5 - \cdots \tag{B38}$$

where $A^2 = A \cdot A$; $A^3 = A \cdot A \cdot A$, etc. The matrix exponential e^A occurs in the study of constant coefficient matrix differential equations, and is utilized in Chapter 3. Some useful relations for the matrix exponential are the following

$$e^{A+B} = e^A e^B \text{ if } AB = BA \tag{B39}$$

$$e^{TFT^{-1}} = Te^F F^{-1} \text{ for } |T| \neq 0 \tag{B40}$$

$$\left| e^F \right| = e^{tr(F)} \tag{B41}$$

VECTOR-MATRIX OPERATIONS

Vector and matrices can be combined in mathematical expressions in various ways. Since a vector of dimension n can be thought of as $n \times 1$ matrix, the rules developed above for matrix operations can be readily applied. Several of the more common operations are briefly considered in the following.

Addition: To form the sum of two matrices $A = \begin{bmatrix} a_{ij} \end{bmatrix}$ and $B = \begin{bmatrix} b_{ij} \end{bmatrix}$, the matrices must be of the same order, and we add the corresponding elements a_{ij} and b_{ij}; that is, $C = \begin{bmatrix} c_{ij} \end{bmatrix} = A + B$ where $c_{ij} = a_{ij} + b_{ij}$.

Multiplication by a scalar: The product of a scalar k and a matrix $A = \begin{bmatrix} a_{ij} \end{bmatrix}$ is a matrix B formed by multiplying each element a_{ij} of the matrix A by the scalar k, i.e. $B = \begin{bmatrix} b_{ij} \end{bmatrix}$ where $b_{ij} = ka_{ij}$. Thus, for matrix subtraction, $A - B = A + (-1)B$; that is, one simply subtracts corresponding elements.

Vector multiplications: The multiplication of $1 \times n$ vector (row) x^T with a $n \times 1$ vector (column) y is defined as

$$x^T y = \begin{bmatrix} x_1 \cdots x_n \end{bmatrix} \begin{bmatrix} y_1 \\ \vdots \\ y_n \end{bmatrix} = \sum_{i=1}^{n} x_i y_i \tag{B42}$$

This multiplication is referred to as the inner product, or dot product, and yields the scalar. If $x^T y = 0$, x and y are said to be orthogonal. In addition, $x^T x$, the squared length of the vector x, is

$$x^T x = \sum_{i=1}^{n} x_i^2 \tag{B43}$$

The length of the vector x is denoted by

$$\|x\| = \sqrt{x^T x} \tag{B44}$$

The quantity xy^T is referred to as the outer product and yields the matrix

$$xy^T = \begin{bmatrix} x_1 y_1 & x_1 y_2 & \cdots & x_1 y_n \\ x_2 y_1 & x_2 y_2 & \cdots & x_2 y_n \\ \vdots & \vdots & \ddots & \vdots \\ x_n y_1 & x_n y_2 & \cdots & x_n y_n \end{bmatrix} \tag{B45}$$

Similarly, we can form the matrix xx^T, which is called the scatter matrix of the vector x.

Matrix multiplication: A $m \times p$ matrix A may be multiplied only by a $p \times n$ matrix B; that is, the number of columns of A must equal the number of rows of B (we say that the matrices A and B are conformable). The product matrix $C = AB$ has m rows and n columns, and its ij-th element c_{ij} is given by

$$c_{ij} = \sum_{k=1}^{p} a_{ik} b_{kj} \tag{B46}$$

It is noted that two matrices are equal if, and only if, all of their corresponding elements are equal. Thus, $A = B$ implies $a_{ij} = b_{ij}$ for all $i = 1, \ldots, m$ and $j = 1, \ldots, n$. For square matrices A and B of equal dimension, the products AB and BA are both defined, but in general $AB \neq BA$; that is, the matrix multiplication is noncommutative.

Vector-matrix product: If a vector x and matrix A are conformable, the product

$$y = Ax \tag{B47}$$

is defined such that

$$y_i = \sum_{j=1}^{n} a_{ij} x_j \tag{B48}$$

The partitioning of a matrix: We write the matrix A in partitioned form as

$$A = \begin{bmatrix} B & C \\ D & E \end{bmatrix} \tag{B49}$$

Matrices in partitioned form may be multiplied as though the submatrices were elements, provided the resulting expressions make sense. For instance, let x be a n-dimensional vector partitioned as follows

$$x = \begin{bmatrix} a \\ b \end{bmatrix}; \ a = \begin{bmatrix} x_1 \\ \vdots \\ x_l \end{bmatrix}; \ b = \begin{bmatrix} x_{l+1} \\ \vdots \\ x_n \end{bmatrix} \tag{B50}$$

Then one may easily verify that

$$Ax = \begin{bmatrix} B & C \\ D & E \end{bmatrix} \begin{bmatrix} a \\ b \end{bmatrix} = \begin{bmatrix} Ba + Cb \\ Da + Eb \end{bmatrix} \tag{B51}$$

Note that this makes sense only if x is partitioned so that the dimension of a equals the number of columns in B and D. The partitioning of a matrix into more then four submatrices proceeds analogously.

Matrix inversion lemma using partitioned matrices: Given the partitioned matrix

$$A = \begin{bmatrix} A_{11} & A_{12} \\ A_{21} & A_{22} \end{bmatrix} \tag{B52}$$

where A_{11} and A_{22} are invertible, then

$$A^{-1} = \begin{bmatrix} A_{11} + A_{11}A_{12}\left(A_{22} - A_{21}A_{11}^{-1}A_{12}\right)^{-1} & -A_{11}A_{12}\left(A_{22} - A_{21}A_{11}^{-1}A_{12}\right)^{-1} \\ -\left(A_{22} - A_{21}A_{11}^{-1}A_{12}\right)^{-1}A_{21}A_{11}^{-1} & \left(A_{22} - A_{21}A_{11}^{-1}A_{12}\right)^{-1} \end{bmatrix} \tag{B53}$$

Other useful properties can also be shown. This is the *matrix inversion lemma*

$$\left[A_{11} - A_{12}A_{22}^{-1}A_{21}\right]^{-1} = A_{11}^{-1} + A_{11}^{-1}A_{12}\left[A_{22} - A_{21}A_{11}^{-1}A_{12}\right]^{-1}A_{21}A_{11}^{-1} \tag{B54}$$

and, in addition,

$$\left[I + A\right]^{-1} = I - \left[I + A^{-1}\right]^{-1} \tag{B55}$$

Determinants can also be calculated as

$$|A| = |A_{11}||A_{22} - A_{21}A_{11}^{-1}A_{12}| = |A_{22}||A_{11} - A_{12}A_{22}^{-1}A_{21}| \tag{B56}$$

Quadratic forms: The scalar quantity

$$J = x^T A x = \sum_{i=1}^{n} a_{ii}x_i^2 + 2\sum_{i=1}^{n}\sum_{j=i+1}^{n} a_{ij}x_i x_j \tag{B57}$$

is referred to as a *quadratic form*, where A is a $n \times n$ symmetric matrix and x is a n-dimensional (column) vector. An orthogonal matrix Q can always be found such that

$$A' = Q^T A Q \tag{B58}$$

is diagonal with $a_{ii}' = \lambda_i$, where the λ_i are the eigenvalues of A (see, the spectral decomposition in this chapter). It can be shown that the quadratic form reduces to

$$J = \sum_{i=1}^{n} \lambda_i x_i'^2 \tag{3.59}$$

where

$$x' = Q^T x \tag{3.60}$$

The quadratic form is further used to define properties of the matrix A:

    if $x^T A x > 0$ for all real $x \neq 0$, A is said to be *positive definite*;
    if $x^T A x \geq 0$ for all real $x \neq 0$, A is said to be *positive semidefinite*;
    if $x^T A x \leq 0$ for all real $x \neq 0$, A is said to be *negative semidefinite*;
    if $x^T A x < 0$ for all real $x \neq 0$, A is said to be *negative definite*

All eigenvalues of a positive definite or positive semidefinite matrix are positive or nonnegative, respectively. A matrix of the form $A = aa^T$, where a is a column vector, is positive semidefinite, because for every vector x

$$x^T A x = x^T aa^T x = \left(x^T a\right)^2 \geq 0 \tag{B61}$$

The sum of positive semidefinite matrices is positive semidefinite. Hence, $\sum_{i=1}^{n} a_i a_i^T$ is positive semidefinite. If A is positive semidefinite, then so is $B^T A B$ where B is any matrix or vector. The matrix-associated quantity, analogous to the length of a vector, is called the norm and is defined by

$$\|A\| = \max_{\text{all } x} \frac{\|Ax\|}{\|x\|} \tag{B62}$$

where $\|x\|$ is the vector length or norm. With vector length defined by $\|x\| = \sqrt{x^T x}$, the matrix norm is readily computed as

$$\|A\| = \sqrt{\lambda_i} \tag{B63}$$

where λ_i is the maximum eigenvalue of the matrix product $A^T A$.

Gradient operations: The *gradient* or derivative of a scalar function f, with respect to a vector $x^T = \{x_1, ..., x_n\}^T$, is the vector

$$\frac{\partial f}{\partial x} = a \tag{B64a}$$

with the components

$$a_i = \frac{\partial f}{\partial x_i} \; ; \; i = 1, ..., n \tag{B64b}$$

A case of special importance is the vector gradient of the inner product, and the quadratic form; that is

$$\frac{\partial (y^T x)}{\partial x} = y \; ; \; \frac{\partial (x^T y)}{\partial x} = y \; ; \; \frac{\partial x^T A x}{\partial x} = Ax + A^T x \tag{B65}$$

The second partial derivative of a scalar f, with respect to a vector $x = [x_1, ..., x_n]^T$ is a matrix denoted by

$$\frac{\partial^2 f}{\partial x^2} = A \tag{B66a}$$

with the elements

$$a_{ij} = \frac{\partial^2 f}{\partial^2 x_i \partial x_j} \; ; \; i, j = 1, ..., n \tag{B66b}$$

The determinant of A, $|A|$ is called the Hessian of f. In general, the vector gradient of a vector is the matrix defined as

$$\frac{\partial^2 f^T}{\partial x} = A = \left[a_{ij} \right] \tag{B67a}$$

with $f = [f_1, ..., f_m]^T, x = [x_1, ..., x_n]^T$ and the element of A

$$a_{ij} = \frac{\partial f_j}{\partial x_i} ; \; i = 1, ..., n; j = 1, ..., m \tag{B67b}$$

If f and x are of equal dimension ($n=m$), the determinant of A, $|A|$ can be found and is called the *Jacobian* of f.

The matrix gradient of a scalar f is defined by

$$\frac{\partial f}{\partial A} = B = \left[b_{ij} \right]$$

(B68a)

with the component

$$b_{ij} = \frac{\partial f}{\partial a_{ij}}.$$

(B68b)

Two scalar functions of special note are the matrix trace and determinant. Some of gradient functions for the trace and determinant, for square matrices A, B and C, are the following:

$$\frac{\partial}{\partial A} tr(A) = I$$

(B69)

$$\frac{\partial}{\partial A} tr(BAC) = B^T C^T \; ; \; \frac{\partial}{\partial A} tr(BA^T C) = CB$$

(B70)

$$\frac{\partial}{\partial A} tr(ABA^T) = AB + A^T B$$

(B71)

$$\frac{\partial}{\partial A} tr(e^A) = e^A$$

(B72)

$$\frac{\partial}{\partial A} |BAC| = |BAC| \left(A^{-1} \right)^T$$

(B73)

$$\frac{\partial |A|}{\partial A} = \left(A^{-1} \right)^T |A|$$

(B74)

$$\frac{\partial \log|A|}{\partial A} = \frac{1}{|A|} \frac{\partial |A|}{\partial A} = \left(A^{-1} \right)^T$$

(B75)

Another two scalar functions of particular interest are the following:

$$\frac{\partial x^T A y}{\partial A} = xy^T$$

(B76)

$$\frac{\partial x^T A^{-1} x}{\partial A} = -\left(A^{-1} \right)^T xx^T \left(A^{-1} \right)^T$$

(B77)

Spectral decomposition: Let A be a symmetric $l \times l$ matrix. Suppose D and E are, respectively, diagonal and nonsingular $l \times l$ matrices satisfying

$$A = EDE^T$$

(B78a)

Then, EDE^T is referred to as a *spectral decomposition* of A. In component form, a spectral decomposition is given by

$$a_{ij} = \sum_{k=1}^{l} d_{kk} e_{ik} e_{jk} \qquad \text{(B78b)}$$

If none of the d_{ii} are zero, then A is nonsingular, and we can form

$$A^{-1} = \left(E^T\right)^{-1} D^{-1} E^{-1} \qquad \text{(B79)}$$

since E was assumed nonsingular and $D^{-1} = diag\left\{d_{ii}^{-1}\right\}$.

Any symmetric matrix possesses infinitely many spectral decompositions. Of these, the following play important roles:

a) *The eigen value decomposition*: Suppose E is unitary (orthogonal) matrix X satisfying

$$X^T = X^{-1} \qquad \text{(B80)}$$

In this case, we denote D by Λ and d_{ii} by λ_i. Then, we have

$$AX = X\Lambda X^T X = X\Lambda X^{-1} X = X\Lambda \qquad \text{(B81)}$$

Let x_i denote the i-th column of X. Then the last equations equivalent to

$$Ax_i = \lambda_i x_i \qquad \text{(B82)}$$

which states that λ_i and x_i are, respectively, the eigenvalues and eigen vectors of A. The equation

$$A = X\Lambda X^T = \sum_{i=1}^{l} \lambda_i x_i x_i^T \qquad \text{(B83)}$$

represents the eigenvalue decomposition of A. Inverting this equation, we have

$$A^{-1} = \left(X^T\right)^{-1} \Lambda^{-1} X^{-1} = X\Lambda^{-1} X^T = \sum_{i=1}^{l} \lambda_i^{-1} x_i x_i^T \qquad \text{(B84)}$$

provided all $\lambda_i \neq 0$. If we omit from the summation all the terms for which $\lambda_i = 0$, we obtain a matrix $A^{\#}$ named the pseudoinverse of A. This definition of pseudoinverse applies only to symmetric matrices. Here we show how both a matrix and its inverse can be reconstructed when the eigenvalues and eigenvectors are known. We now consider the quadratic form

$$J = x^T A x = x^T X\Lambda X^T x = y^T \Lambda y = \sum_{i=1}^{l} \lambda_i y_i^2 \qquad \text{(B85)}$$

where $y = X^T x = [y_1,...,y_l]^T$. Since X is unitary (orthogonal) the transformation of coordinates $y = X^T x$ does not affect the shape of the contours of the function $J = J(x)$, i.e., the shape of the surfaces on which $J(x) = const$.

b) *The scaled and inverse scaled decompositions:* Given a matrix A, we define a diagonal matrix B with

$$b_{ii} = \begin{cases} \sqrt{|a_{ii}|} & \text{for } a_{ii} \neq 0 \\ 1 & \text{for } a_{ii} = 0 \end{cases} \qquad \text{(B86)}$$

Then, the matrix

$$C = B^{-1} A B^{-1} \; ; \; B^{-1} = diag\left\{b_{ii}^{-1}\right\} \qquad \text{(B87)}$$

has elements

$$c_{ij} = \frac{a_{ij}}{\sqrt{\left|a_{ii}a_{jj}\right|}} \; ; \; c_{ii} = 1 \tag{B88}$$

We refer to C as the *scaled version* of A. If A is a covariance matrix, then C is the matrix of correlation coefficients. Let the eigenvalue decomposition of C be given by

$$C = V\Pi V^T \tag{B89}$$

where $\Pi = diag\{\pi_i\}$; with π_i being the i-th eigenvalue of C, and V is the matrix whose i-th column v_i is the i-th eigenvector of C $\left(V^T = V^{-1}\right)$. We now have

$$A = BCB = BV\Pi V^T B = F\Pi F^T \tag{B90}$$

where $F = BV$. We call the relation

$$A = F\Pi F^T \tag{B91}$$

the *scaled decomposition of A*. Inverting, we obtain

$$A^{-1} = B^{-1}\left(V^T\right)^{-1}\Pi^{-1}V^{-1}B^{-1} = B^{-1}V\Pi^{-1}V^T B^{-1} = G\Pi^{-1}G^T \tag{B92}$$

where $G = B^{-1}V^T$. We call the relation

$$A^{-1} = G\Pi^{-1}G^T \tag{B93}$$

the *inverse scaled decomposition of A*.

c) *The square root decomposition:* If A is positive definite, i.e. $J(x) = x^T A x > 0$ for any column vector x, it is possible to obtain spectral decomposition in which $D = I$, the identity matrix, i.e. $A = E^T$. Of particular interest is the decomposition in which E is a symmetric matrix S, whence $A = SS^T = S^2$. The matrix S is named the *square root* of A. If $A = X\Lambda X^T$ is the eigenvalue decomposition of A, then we have, because $X^T X = X^{-1}X = I$,

$$A = \left(X\Lambda^{1/2}X^T\right)\left(X\Lambda^{1/2}X^T\right) = \left(X\Lambda^{1/2}X^T\right)^2 \tag{B94}$$

so that

$$A^{1/2} = S = X\Lambda^{1/2}X^T \tag{B95}$$

Here $\Lambda^{1/2} = diag\left\{\sqrt{\lambda_i}\right\}$.

d) *Cholesky decomposition:* Again, we assume that A is positive definite, i.e. $x^T A x > 0$, and choose $D = I$. Now, however, we specify that E should be a lower diagonal matrix L; that is, a matrix whose elements above the main diagonal are all zero

$$l_{ij} = 0 \text{ , for } j > i \tag{B96}$$

Since $A = LL^T$, we have

$$a_{ij} = \sum_{k=1}^{l} l_{ik}l_{jk} \tag{B97}$$

which in view of the preceding equation becomes

$$a_{ij} = \sum_{k=1}^{j} l_{ik} l_{jk} \;\; ; \;\; j < i \qquad \text{(B98)}$$

$$a_{ii} = \sum_{k=1}^{j} l_{ik}^2 \qquad \text{(B99)}$$

These equations may be solved recursively for the components l_{ij}. From the last equation, we have

$$\sqrt{a_{11}} = l_{11} \qquad \text{(B100)}$$

yielding, from eq. (B98)

$$l_{i1} = a_{i1} / l_{11} \;\; ; \;\; i = 2,3,...,l \qquad \text{(B101)}$$

Then, using eq. (B98) and (B99) alternately for $i = 2,3,...,l$

$$l_{ij} = \left(a_{ij} - \sum_{k=1}^{j-1} l_{ik} l_{jk} \right) / l_{jj} \;\; ; \;\; j = 2,3,...,i-1; \; \text{skip } i = 2 \qquad \text{(B102)}$$

$$l_{ii} = \left(a_{ii} - \sum_{k=1}^{i-1} l_{ik}^2 \right)^{1/2} \qquad \text{(B103)}$$

This procedure can be carried through provided all of the square roots arguments are positive. This occurs if and only if A is positive definite. Of all the decompositions discussed, the last is the only one that can be accomplished in a finite procedure. All other decompositions depend on the evaluation of eigenvalues, which require an iterative procedure.

Differentiation: The derivative of a matrix is obtained by differentiating the matrix element by element. For example, let $x = [x_1 \; x_2]^T$, then

$$\frac{dx}{dt} = \begin{bmatrix} \dfrac{dx_1}{dt} \\ \dfrac{dx_2}{dt} \end{bmatrix} \qquad \text{(B104)}$$

Integration: The integral of a matrix is obtained by integrating the matrix element by element. For example, for the vector $x = [x_1 \; x_2]^T$ we have

$$\int x dt = \begin{bmatrix} \int x_1 dt \\ \int x_2 dt \end{bmatrix} \qquad \text{(B105)}$$

APPENDIX C: TABLE OF COMMON PROBABILITY DENSITY FUNCTIONS

This appendix describes some of the most common probability density functions (pdf's).For comletness, the table includes the pdf $f(x)$, the mean value $m_x = E\{X\}$, the variance $\sigma_x^2 = E\{(X - m_x)^2\}$, or the covariance matrix $P_X = E\{(X - m_x)(X - m_x)'\}$ in the case of multidimensional X, as well as the characteristic function $\Phi_x(\omega) = E\{j\omega(X - m_x)\}$, that is $\Phi_x(\omega) = E\{j\omega'(X - m_x)\}$ for multidimensional X [26, 41, 52].

Uniform pdf

$$f(x) = \begin{cases} \dfrac{1}{b-a} & ; a \le x \le b \\ 0 & ; \text{elsewhere} \end{cases}$$

$$\text{mean value: } \frac{a+b}{2}$$

$$\text{variance: } \frac{(b-a)^2}{12}$$

$$\text{characteristic function: } \Phi_x(\omega) = \frac{2}{\omega(b-a)} \sin\left(\frac{\omega(b-a)}{2}\right) \exp\left(\frac{j\omega(b+a)}{2}\right)$$

Scalar Gaussuian pdf

$$f(x) = \frac{1}{\sqrt{2\pi}\sigma} \exp\left(-\frac{(x-m)^2}{2\sigma^2}\right)$$

$$\text{mean value: } m$$
$$\text{variance: } \sigma^2$$

$$\text{characteristic function: } \Phi_x(\omega) = \exp\left(\frac{j\omega m - \omega^2\sigma^2}{2}\right)$$

Vector Gaussuian pdf

$$f(x) = \frac{1}{\sqrt{(2\pi)^n \det P}} \exp\left(-\frac{1}{2}(x-m)' P^{-1}(x-m)\right)$$

$$\text{mean value vector: } m$$
$$\text{covariance matrix: } P$$

$$\text{characteristic function: } \Phi_x(\omega) = \exp\left(j\omega'm - \frac{1}{2}\omega'P\omega\right)$$

Exponential pdf

$$f(x) = \begin{cases} \lambda \exp(-\lambda x) & ; x \geq 0 \\ 0 & ; x \leq 0 \end{cases}$$

mean value: $\dfrac{1}{\lambda}$

variance: $\dfrac{1}{\lambda^2}$

characteristic function: $\Phi_x(\omega) = \dfrac{\lambda}{(\lambda - j\omega)}$

Laplacian pdf

$$f(x) = \frac{\lambda}{2} \exp(-\lambda|x|)$$

mean value: 0

variance: $\dfrac{2}{\lambda^2}$

characteristic function: $\Phi_x(\omega) = \dfrac{\lambda^2}{\lambda^2 + \omega^2}$

Rayleigh pdf

$$f(x) = \begin{aligned} & \frac{x}{\sigma^2} \exp\left(-\frac{x^2}{2\sigma^2}\right) ; x \geq 0 \\ & 0 \qquad\qquad ; x \leq 0 \end{aligned}$$

mean value: $\sigma\sqrt{\dfrac{\pi}{2}}$

variance: $\left(2 - \dfrac{\pi}{2}\right)\sigma^2$

characteristic function: does not exist

Cauchy pdf

$$f(x) = \frac{\alpha}{\pi} \frac{1}{\alpha^2 + x^2}$$

mean value: 0

variance: ∞

characteristic function: $\Phi_x(\omega) = \exp(-\alpha|x|)$

Bernoulli: The random variable with Bernoulli distribution is the simplest of the discrete random variables whose values is 1 with probability p and 0 with probability $(1-p)$. The corresponding pdf is

$$f(x) = (1-p)\delta(x) + p\delta(x); \delta(x) = \begin{cases} 1; x = 0 \\ 0; x \neq 0 \end{cases}$$

mean value: p

variance: $p(1-p)$

characteristic function: $1 - p + p\exp(j\omega)$

Binomial: If we have a set of n independent Bernuolli variables then the number of such variables that are equal to 1 is binomial random variable with corresponding pdf:

$$f(x) = \sum_{i=0}^{n} \binom{n}{i} p^i (1-p)^{n-i} \delta(x-i); \delta(x) = \begin{cases} 1; x = 0 \\ 0; x \neq 0 \end{cases}$$

mean value: pn

variance: $p(1-p)n$

characteristic function: $[1 - p + p \exp(j\omega)]^n$

Poisson: If we consider a large number of independent events that can occur uniformly over a long interval, then the number of such events that occur in a fixed interval T is a Poisson variable. The time between pairs of occurrences is an exponential variable with parameter λ. Assuming $\alpha = \lambda T$, Poisson probability density function is given by the relation:

$$f(x) = \sum_{i=0}^{\infty} \frac{\alpha^i e^{-\alpha}}{i!} \delta(x-i); \delta(x) = \begin{cases} 1; x = 0 \\ 0; x \neq 0 \end{cases}$$

mean value: α

variance: α

characteristic function: $\exp\{\alpha[\exp(j\omega) - 1]\}$

References

[1] Anderson,B.D.,and J.B.Moore, Optimal Filtering, Prentice Hall, Englewood Cliffs,N.J.,1979.

[2] Astrom,K.J., Introduction to Stohastic Control Theory, Academic Press, N.Y.,1970.

[3] Astrom,K.J.,and B.Wittenmark, Computer Controlled Systems, Prentice Hall, Englewood Cliffs,N.J.,1984.

[4] Bard,Y., Nonlinear Parameter Estimation, Academic Press, N.Y.,1974.

[5] Bar-Shalom,Y.,and X.R.Li, Estimation and Tracking:Principles,Techniques and Software, Artech House, Boston, MA, 1993.

[6] Bar-Shalom,Y.,and X.R.Li, Multitarget-Multisensor Tracking:Principles and Techniques, Clearance Center, Danvers, MA,1995.

[7] Bellman,R.E., Introduction to Matrix Analysis, McGraw Hill, N.Y.,1970.

[8] Bendat,J.S., and A.G.Piersol, Random Data Analysis and Measurement Procedures, J.Wiley, N.Y.,1986.

[9] Bierman,G.J., Factorization Methods of Discrete Sequential Estimation, Academic Press, N.Y.,1977.

[10] Blackman,S.S., and R. Popoli, Design and analysis of modern tracking systems, Artech House, Boston, MA, 1999.

[11] Bozic,S.M., Digital and Kalman Filtering, Edward Arnold, London,1979.

[12] Bryson,A.E.,and Y.C.Ho, Applied Optimal Control, J.Wiley, N.Y.,1975.

[13] Candy,J.V., Model-Based Signal Processing, Joh Wiley, N.Y., 2006.

[14] Chipperfield,A.J., and P.J.Fleming, MATLAB Toolboxes and Applications for Control, Peter Peregrinus Ltd., London,1993.

[15] Franklin G.F., and J.D.Powel, Digital Control of Dynamic Systems, Addison Wesley, Reading,MA,1980.

[16] Gauss,K.F., Theory of Motion of the Heavenly BodiesMoving about the Sun, Dover Pub., N.Y.,1963.

[17] Gelb,A., Applied Optimal Estimation, MIT Press,Boston,MA,1975.

[18] Goodwin,G., and K.Sin, Adaptive Filtering,Prediction and Control, Prentice Hall, Englewood Cliffs,N.J.,1984.

[19] Grewal, M.S., and A.P. Andrews, Kalman Filtering: Theory and Practice using MATLAB, John Wiley, N.Y., 2001.

[20] Haight,F.A., Applied Probability, Plenum Press,N.Y.,1981.

[21] Hamming,R., Numerical Methods for Scientists and Engineers, McGraw Hill, N.Y.,1962.

[22] Haykin, S., Adaptive Filter Theory, Prentice Hall, N.J., 2002.

[23] Jazwinski,A.H., Stochastic Processes and Filtering Theory, Academic Press, N.Y.,1970.

[24] Kailath,T., Linear Systems, Prentice Hall,Englewood Cliffs,N.J.,1980.

[25] Kailath,T., Lectures on Wiener and Kalman Filtering, Springer Verlag, N.Y.,1981.

[26] Kalman,R.E., A New Approach to Linear Filtering and Prediction Problems, J.Basic Eng.,Trans. ASME,Series D,Vol.82,No.1,pp.35-45,1960.

[27] Kalman,R.E., and R.S.Bucy, New Results in Linear Filtering and Prediction Theory, J.Basic Eng.,Trans. ASME,Series D,Vol.83,No.3,pp.95-108,1961.

[28] Kendall,M.G., and A.Stuart, The Advanced Theory of Statistics, Charles Griffin, London,1962.

[29] Lee,R.C.K., Optimal Estimation,Identification and Control, MIT Press,Boston,MA,1964.

[30] Lewis,F.L., Optimal Estimation with an Introduction to Stochastic Control Theory, J.Wiley, N.Y.,1986.

[31] Lewis,F.L., Applied Optimal Control and Estimation:Digital Design and Implementation, Prentice Hall,Englewood Cliffs,N.J.,1992.

[32] Ljung,L.,and T.Soderstrom, Theory and Practice of Recursive Identification, MIT Press,Boston,MA,1983.

[33] Ljung,L., System Identification:Theory for the User, Prentice Hall,Englewood Cliffs,N.J.,1987.

[34] Magrab, E.B., S. Azarm, B. Balachandran, J.H. Duncan, K.E. Herold and G.C. Walsh, An Engineers Guide to MATLAB with Applications from Mechanical, Aerospace, Electrical and Civil Engineering, Prentice Hall, N.J., 2005.

[35] Maybeck,P., Stochastic Models,Estimation and Control, Academic Press,N.Y.,1979.

[36] Meditch,J.S., Stochastic Optimal Linear Estimation and Control, McGraw Hill, N.Y.,1969.

[37] Melsa,J.L., and D.L.Cohn, Decision and Estimation Theory, McGraw Hill,N.Y.,1976.

[38] Mendel,J., Discrete Techniques of Parameter Estimation, Marcel-Dekker,N.Y.,1973.

[39] Odoni, A. and W.B.Davenport, Probability and Random Processes, Addison Wesley,Reading,MA,1970.

[40] Ogata,K., Solving Control Engineering Problems with MATLAB, Prentice Hall,Englewood Cliffs,N.J.,1994.

[41] Oppenheim,A.V., and R.W.Schafer, Digital Signal Processing, Prentice Hall, Englewood Cliffs, N.J., 1999.

[42] Papoulis, A., Probability,Random Variables and Stochastic Processes, McGraw Hill, N.Y.,2002.

[43] Papoulis, A., Signal Analysis, McGraw Hill, N.Y.,1977.

[44] Pugachev,V.S., Probability Theory and Mathematical Statistics for Engineers, Pergamon Press, N.Y.,1984.

[45] Rabiner,L.R., and B.Gold, Theory and Application of Digital Signal Processing, Prentice Hall, Englewood Cliffs, N.J.,1975.

[46] Rice,J.R., Matrix Computations and Mathematical Software, McGraw Hill, N.Y.,1981.

[47] Ross,S.M., Introduction to Probability and Statistics for Engineers and Scientists, J.Wiley, N.Y.,1987.

[48] Sage,A., and J.Melsa, Estimation Theory with Applications to Communications and Control, McGraw Hill, N.Y.,1971.

[49] Schwartz,M., and L.Shaw, Signal Processing: Discrete Analysis, Detection and Estimation, McGraw Hill, N.Y.,1975.

[50] Smith, S.W., Digital Signal Processing: A practical guide for engineers and scientists, Elsevier, N.Y. 2003.

[51] Soderstrom,T., Discrete-Time Stochastic Systems:Estimation and Control, Springer, N.Y., 2002.

[52] Stengal,R.F., Stochastic Optimal Control, J.Wiley, N.Y.,1986.

[53] Therrien, C.W. and M. Tummala, Probability for electrical and computer engineers, CRC Press, N.Y., 2004.

[54] Tsypkin,Ya.Z., Foundations of the Theory of Learning Systems, Academic Press, N.Y.,1973.

[55] Tsypkin,Ya.Z., Foundations of the Informational Theory of Identification, Nauka, Moscow,1984.

[56] White, R. E., Elements of matrix modelling and computing with MATLAB, Taylor and Francis, London, 2007.

[57] Wiener, N., Extrapolation,Interpolation and Smoothing of Stationary Time Series, MIT Press, Boston,MA,1949.

[58] Yakowitz,S.J., Computational Probability and Simulation, McGraw Hill, N.Y.,1977.

[59] Zayezdny,A.,D.Tabak and D.Wulich, Engineering Applications of Stochastic Processes:Theory,Problems and Solutions, J.Wiley, N.Y.,1989.

Index

CIP - Каталогизација у публикацији
Народна библиотека Србије, Београд

519.21/.24
681.51
621.391:004

KOVAČEVIĆ, Branko
     Fundamentals of Stochastic Signals,
Systems and Estimation Theory with Worked
Examples / Branko Kovačević, Željko Đurović.
- 2nd ed. - Beograd : Academic mind ; Berlin
Heidelberg : Springer Verlag, 2008 (Beograd :
Planeta print). - 414 str. : graf. prikazi ; 24 cm

Tiraž 700. - Bibliografija: str. 407-409. -
Registar.

ISBN 978-86-7466-323-3 (AM)
ISBN 978-3-540-70990-9 (SV)

1. Đurović, Željko [autor]
a) Стохастички процеси b) Системи
аутоматског управљања c) Дигитална обрада
сигнала
COBISS.SR-ID 149670668